OXFORD

endorsed by edexcel

GCSE Maths

Higher

Teacher's Guide

Christopher Green
Marling Boys Grammar School, Stroud

OXFORD
UNIVERSITY PRESS

Great Clarendon Street, Oxford OX2 6DP

Oxford University Press is a department of the University of Oxford.
It furthers the University's objective of excellence in research, scholarship, and education by publishing worldwide in

Oxford New York

Auckland Cape Town Dar es Salaam Hong Kong
Karachi Kuala Lumpur Madrid Melbourne Mexico City
Nairobi New Delhi Shanghai Taipei Toronto

With offices in

Argentina Austria Brazil Chile Czech Republic France Greece
Guatemala Hungary Italy Japan South Korea Poland Portugal
Singapore Switzerland Thailand Turkey Ukraine Vietnam

Oxford is a registered trade mark of Oxford University Press in the UK and in certain other countries

First published 2006

British Library Cataloguing in Publication Data

Data available

ISBN-10: 0-19-915091-5

ISBN-13: 978-0-19-915091-5

10 9 8 7 6 5 4 3 2 1

Typeset by MCS Publishing Service Ltd., Salisbury Wiltshire

Printed and bound by Ashford Colour Press

Acknowledgements
This high quality material is endorsed by Edexcel and has been through a rigorous quality assurance programme to ensure that it is a suitable companion to the specification for both learners and teachers.
This does not mean that its contents will be used verbatim when setting examinations nor is it to be read as being the official specification – a copy of which is available at www.edexcel.org.uk

The Publisher would like to thank the following for permission to reproduce photographs:

p11 Oxford University Press; **p17** Paul Doyle/Alamy; **p81** Science Photo Library; **p86** Brendan Regan/Corbis; **p127** TopFoto UK Ltd; **p135** Oxford University Press; **p149** Hemera/Oxford University Press; **p150** Photodisc/Oxford University Press; **p155** Oxford University Press; **p176** NASA/Oxford University Press; **p182** Oxford University Press; **p195** Oxford University Press; **p197** The Photolibrary Wales/Alamy; **p199** Araldo de Luca/Corbis; **p243** John La Gette/Alamy; **p285** India Images/Dinodia Images/Alamy; **p287** Stockfolio/Alamy; **p304** World Pictures Ltd/Alamy **p308** Oxford University Press; **p315** Andrew Syred/Science Photo Library; **p316l** Bob Croxford/Atmosphere Picture Library/Alamy; **p316r** Ordnance Survey.

Figurative artwork is by Dylan Gibson.
Technical artwork is by MCS Publishing Services Ltd.

About this book

This Teacher's Guide has been specifically written for the Higher tier of the two-tier linear Edexcel GCSE Mathematics specification. It is designed to accompany the Higher Students' Book in the same series, which is written for students who have achieved level 6 or a strong level 5 at Key Stage 3 and are looking to progress to a grade B or C at GCSE.

The series authors are experienced teachers and examiners who have an excellent understanding of the Edexcel two-tier specification and so are well qualified to help you successfully implement the objectives in your classroom.

The book is made up of units which are based on the Edexcel specification, and provide full lesson plans for each of the lessons in the corresponding Students' Book.

The units are:

How to use this guide

This guide is made up of units of work which are grouped in four main strands: Algebra (A), Data (D), Number (N) and Shape, Space and Measures (S). Each unit starts with an introduction to help you in planning your teaching, and contains:

- **Edexcel objectives**, so you can ensure full and appropriate coverage
- **Unit overview**, to give you a convenient summary of the unit
- **Prior knowledge**, so you know what students should already have learned
- **Differentiation**, so you know what is covered in each of the parallel student books

The main content of the book is in the form of double page spreads, which mirror the corresponding Students' Book. The left-hand page provides a suggested lesson plan, containing:

- **Objectives** for the lesson, as contained in the Students' Book
- **Useful resources**, ranging from materials that are very simple and effective to electronic materials that can be found on the accompanying interactive CD-ROM
- **Mental starter**, designed to be inclusive and providing a lead-in to the main concepts of the lesson
- **Introductory activity**, designed to help you bring the topic in the accompanying Students' Book to life, including diagnostic questions to stimulate engagement and provoke discussion
- **Exercise commentary and misconceptions**, providing hints and tips for the exercise, with a particular emphasis on helping students to overcome common difficulties
- **Plenary**, suggesting a way of summarising or extending the learning

The right-hand page of each spread contains a miniature version of the corresponding Students' Book page for reference.

At the end of the unit is an exam review page which identifies the key Edexcel objectives and provides worked solutions to the summary questions in the Students' Book, including commentary and misconceptions.

Numerical answers to the Higher Students' Book and also the Higher Homework Book are contained at the end of this guide.

Coursework guidance

The specification requires each student to undertake two pieces of coursework, amounting to 20% of the overall assessment:

- A project covering AO4 (handling data)
- A task covering AO1 (using and applying mathematics) in the context of AO2 (number and algebra) or AO3 (shape, space and measures).

Handling data project

Students may only submit a single data handling project. Centres taking option A may either use Edexcel's suggested starting points and sample data, or they may make up their own projects, or they may choose to do a bit of both. Centres taking option B are required to submit projects based on material set by Edexcel.

The assessment criteria for data handling projects are broadly categorised as follows:

- Specify the problem and plan
- Collect, process and represent data
- Interpret and discuss results

The handling data project accounts for 10% of the overall assessment weighting.

Using and applying mathematics task

Students may submit more than one task to provide evidence of attainment in AO1. The best performance of each strand will be counted in whichever task it occurs. The strands are:

- Making and monitoring decisions to solve problems
- Communicating mathematically
- Developing skills of mathematical reasoning

The tasks may be practical or investigative, and may involve the use of ICT. Centres taking option A may either use Edexcel's suggested tasks, or they may make up their own tasks, or they may choose to do a bit of both. Centres taking option B are required to submit tasks set by Edexcel.

The using and applying mathematics task accounts for 10% of the overall assessment weighting.

For further information on coursework, including grade descriptions, please visit the Edexcel website at www.edexcel.org.uk

Contents

N1 Integers and decimals

Objectives

F/H Use their previous understanding of integers and place value to round and order integers and decimals

F Multiply or divide any number by powers of 10

F Add, subtract, multiply and divide integers and then any number

F/H Understand highest common factor, least common multiple, prime number and prime factor decomposition

Unit overview

This unit consolidates Key Stage 3 knowledge of place value and decimals, before extending to calculations with negative numbers. Factors and primes, including HCF and LCM, are also covered, laying the basis for later work on adding and subtracting fractions in N3.

Prior knowledge

Before your students start this unit they should be able to:

- Understand and use decimal notation and place value
- Order positive and negative numbers
- Recognise and use multiples, factors and primes (less than 100)

Differentiation

- **Foundation** focuses on place value with integers, reading scales, negative numbers, and factors
- **Foundation Plus** extends to decimal place value, and includes multiples
- **Higher Plus** extends the Higher book to include standard form

N1.1 Place value and ordering numbers

Objectives
- Use place value to round and order whole numbers and decimals
- Multiply and divide any number by powers of 10

Useful resources
- Place value table

Mental starter
Dylan is on his trampoline. He can bounce to a height of 3.67 m, measured from the trampoline to the top of his head. He is 124 cm tall. His sister Alice can bounce to a height of 4.21 m, measured in the same way, and is 178 cm tall.
Who gets more height on their bounces?
(Answer: the same height – 243 cm from trampoline to feet.)

Introductory activity
Discuss how and why you would round large numbers – use populations of countries as an example. Extend to decimals, and use a place value table.
Remind students of the terms descending and ascending. Put in ascending order
14.8, 14.98, 14.899, 14.989, 13.999, 14.88, 14.9888, 14.898, 14.0999
(Answer 13.999, 14.0999, 14.8, 14.88, 14.888, 14.898, 14.98, 14.9888, 14.989, 14.999.)
Remind students to examine the decimals a place value at a time (tens, units, tenths ...)
Use a place value table to help.

When you multiply or divide by 10, 100, 1000, is it the numbers that move or the decimal point?
Draw up columns for place value up to thousandths and discuss the answer for 0.43×1000 and $7.2 \div 100$. Discuss who in the class moves the point, and who moves the numbers. Reinforce the key point that the digits move, not the decimal point.

Exercise commentary and misconceptions
In questions 1 and 2, some students will count the number of digits instead of comparing each place value one at a time.
In questions 3 and 4, encourage students to think 'is my answer getting bigger or smaller?'
Questions 5, 6, and 7 focus on ordering decimals.
Questions 8eii and 9c give answers of zero – discuss when this would be a misleading answer. (For example, the number of people in class today is zero to the nearest hundred.)
In question 10, a few students may confuse significant figures with decimal places.

Plenary
When a prize of £430 is shared evenly between seven people my calculator tells me that each receives £61.42857143. What is a sensible amount to give each person? How many decimal places did you use? How much money is left over?
(Answer: £61.43 to 2 dp. But this means you give out £430.01.) Allow other possible solutions.

N1.1 Place value and ordering numbers

This spread will show you how to:

- Use place value to round and order whole numbers and decimals
- Multiply and divide any number by powers of 10

Keywords
Ascending
Descending
Digit
Integer
Order
Round

- You can **round** large numbers to the nearest hundred, thousand or any other power of ten and round decimal numbers to any number of decimal places.
  - Identify the final **digit** required
  - Round it up if the following digit is a 5 or more
  - Write the rounded number, including any zeros needed to make the place value correct.

Example

Round 72 456 to the nearest **a** 10 **b** 100 **c** 1000.

a 72 456 = 72 460 to the nearest 10
b 72 456 = 72 500 to the nearest 100
c 72 456 = 72 000 to the nearest 1000

Example

Round 6.0374 to the nearest **a** tenth **b** hundredth **c** thousandth.

a 6.0374 = 6.0 to the nearest tenth
b 6.0374 = 6.04 to the nearest hundredth
c 6.0374 = 6.037 to the nearest thousandth

To **order** decimals, look at the tenths digit first, then the hundredths digit, then the thousandths and so on.

Example

Write these numbers in **ascending** order.

0.3 0.275 0.28 0.3005 0.269 997

0.275, 0.28 and 0.269 997 are smaller than 0.3 and 0.3005.
0.269 997 is smallest.
0.275 is smaller than 0.28.
0.3 is smaller than 0.3005.
In ascending order, the numbers are 0.269 997, 0.275, 0.28. 0.3, 0.3005.

Do not be misled by the number of digits. 0.28 is equal to 0.280, and is larger than 0.275.

- Multiplying a number by 10 moves the digits one place to the left. Multiplying by 100 moves the digits two places to the left.
- Dividing a number by 10 moves the digits one place to the right. Dividing by 100 moves the digits two places to the right.

Example

Work out **a** 3.72 ÷ 100 **b** 0.0349 × 10 000 **c** 17.3 ÷ 1000

a 3.72 ÷ 100 = 0.0372 — Move the digits 2 places right.
b 0.0349 × 10 000 = 349 — Move the digits 4 places left.
c 17.3 ÷ 1000 = 0.0173 — Move the digits 3 places right.

Exercise N1.1

1 Write these sets of numbers in ascending order.
a 0.3, 3.1, 1.3, 2, 1, 0.1 **b** 607, 77.2, 27.6, 7.06, 6.07

2 Write these sets of numbers in descending order.
a 6008, 682.8, 862.6, 6000.8, 8000.6 **b** 47.9, 94.7, 49.7, 79.4, 74.9, 97.4

3 Multiply these numbers by 10.
a 16.7 **b** 24.8 **c** 0.716 **d** 1.095 **e** 243 **f** 281.3

4 Divide these numbers by 10.
a 214 **b** 67.3 **c** 4106 **d** 200.7 **e** 6.025 **f** 86

5 Decide which number in each pair is bigger.
Explain your answers.
a 4.52 and 4.05 **b** 5.5 and 5.05 **c** 16.8 and 16.75 **d** 16.8 and 16.15

6 Write these sets of numbers in ascending order.
a 7.83, 7.3, 7.8, 7.08, 7.03, 7.38 **b** 4.2, 8.24, 8.4, 4.18, 2.18, 2.4

7 Write these sets of numbers in descending order.
a 16.7, 18.16, 16.18, 17.16, 18.7, 17.6 **b** 1.06, 13.145, 1.1, 2.38, 13.2, 2.5

8 Round these numbers to the nearest **i** 10 **ii** 100.
a 3048 **b** 1763 **c** 294 **d** 51 **e** 43 **f** 743

9 Round these numbers to the nearest 1000.
a 2964 **b** 1453 **c** 17 **d** 24 598 **e** 16 344 **f** 167 733

10 Round these numbers to **i** 1 decimal place **ii** 2 decimal places.
a 39.114 **b** 7.068 **c** 5.915 **d** 512.715
e 4.259 **f** 12.007 **g** 0.833 **h** 26.8813

11 Round these numbers to the nearest
i tenth **ii** hundredth **iii** thousandth.
a 0.07 **b** 15.9184 **c** 127.9984
d 887.172 **e** 55.144 55 **f** 0.007 49

12 Calculate.
a 13.06 × 100 **b** 208.5 ÷ 100 **c** 1.085 × 1000
d 2487 ÷ 1000 **e** 0.008 ÷ 10 **f** 0.006 19 × 1000
g 45.13 ÷ 1000 **h** 0.000 045 × 100

N1.2 Adding and subtracting negative numbers

Objectives

- Add and subtract with positive and negative numbers

Useful resources

- Number line

Mental starter

During an ice age the sea levels drop. The water was 450 m lower during the last ice age 10 000 years ago. If the present sea level is regarded as 0 metres, how can you write the sea level 10 000 years ago? (−450 m.)
Assuming the sea rose evenly every 1000 years since the last ice age how much did the level rise by every 1000 years? (45 m)
What was the sea level 6000 years ago?
(−450 + 6 × 45 = −180 m)
You may wish to briefly discuss the validity of the assumption.

Introductory activity

Try some basic examples, using a number line to illustrate:

a $4 - 7 = -3$
b $-4 + 10 = 6$
c $-30 + 24 = -6$
d $-17 - 26 = -43$
e $-9 + 4 - 13 = -18$

Encourage students to express general rules:
(i) Adding a negative means take away
(ii) Taking away a negative means add.
Avoid telling students that 'two minuses make a plus' since −3 − 4 looks like two minuses and the answer is not a plus.
Try more complex examples:

a $7 + -4 = \quad 7 - 4 = 3$
b $-3 - -2 = \quad -3 + 2 = -1$
c $9 - -2 = \quad 9 + 2 = 11$
d $-12 + -7 = -12 - 7 = -19$

Encourage students to re-write the question.
Briefly discuss notation that may otherwise confuse:
$4 + (-6) = +4 + (-6) = 4 + -6$

Exercise commentary and misconceptions

Students should not use a calculator.
In questions 1 to 5, some students may be confused by the process of adding a positive number. Point out that this is not a common way of writing a question and that it just means 'add'.
Ensure that if students are using a number line, that they have included zero. Encourage students to think past the ends of their number line without drawing it further.

Plenary

Look at this statement:

$\square - \square + \square = 0.$

The missing numbers are all different, all negative and all prime. What are the missing numbers?
(−2 − −5 + −3 or −3 − −5 + −2 are the only solutions)

N1.2 Adding and subtracting negative numbers

This spread will show you how to:

- Add and subtract with positive and negative numbers

Keywords
Directed number
Negative
Number line
Positive

A number with a plus or minus sign is a **directed number**. You can extend the basic rules of addition and subtraction to include negative numbers.

- Adding a **positive** number counts as addition. Move right along the **number line**.
- Subtracting a positive number counts as subtraction. Move left along the number line.
- Adding a **negative** number counts as subtraction. Move left along the number line.
- Subtracting a negative number counts as addition. Move right along the number line.

For subtracting a negative number, think of reducing an overdraft, or taking ice cubes out of a cold drink.

Example

Work out **a** $-5 + -6$ **b** $+4 - -2$ **c** $-7 - +2$ **d** $-5 + +8$

a Start at -5 on the number line and move 6 places to the left. The answer is -11.

-6
-12 -11 -10 -9 -8 -7 -6 -5 -4

b Start at $+4$ on the number line and move 2 places to the right. The answer is $+6$.

$+2$
3 4 5 6 7

c Start at -7 on the number line and move 2 places to the left. The answer is -9.

-2
-10 -9 -8 -7 -6 -5

d Start at -5 on the number line and move 8 places to the right. The answer is $+3$.

$+8$
-6 -5 -4 -3 -2 -1 0 1 2 3 4

Example

Ben writes: $-5 + -2 = +7$
Is he correct?

Two minuses make a plus. I've got -5 and -2, so the answer must be positive.

No.
Using the number line, you start at -5 and move 2 places to the left.
The correct answer is $-5 + -2 = -7$.

-2
-7 -6 -5 -4

Examiner's tip
Avoid simple rules like 'two minuses make a plus', which can be misleading. Use the number line.

Exercise N1.2

1 Calculate.

a $+7 - +9$ **b** $+5 - +6$ **c** $+8 - +10$
d $-7 + +5$ **e** $-11 + +6$ **f** $-7 + +2$

2 Calculate.

a $-7 + +8$ **b** $-9 + +12$ **c** $-6 + +10$
d $+3 - -6$ **e** $+5 - -7$ **f** $+2 - -3$

3 Calculate.

a $-9 - -4$ **b** $-8 - -6$ **c** $-5 - -1$
d $-6 - -8$ **e** $-5 - -9$ **f** $-3 - -10$

4 Copy and complete these calculations by replacing the boxes with the correct number or sign.

a $-3 + \square = -5$ **b** $+7 \square -5 = +2$
c $\square 8 + \square 5 = +3$ **d** $\square 2 + \square 11 = -13$

5 Calculate.

a $+8 - -14$ **b** $-1 + -11$ **c** $-9 - -7$
d $+3 + -17$ **e** $+8 - -4$ **f** $+13 + -1$
g $+48 - +29$ **h** $-19 + +4$ **i** $+34 + -23$
j $-104 + +43$ **k** $+208 - -136$ **l** $+347 + -298$

6 Calculate.

a $-4.5 + -6.3$ **b** $+2.8 - -3.5$ **c** $+5.6 - -7.9$
d $-9.4 + +8.7$ **e** $-26.5 + -11.7$ **f** $+45.9 - -66.8$

7 Find the balance in these bank accounts after the transactions shown.

a Opening balance £133.45. Deposits of £45.55 and £63.99, followed by withdrawals of £17.50 and £220.

b Opening balance is –£459.77. Deposit of £650, followed by a withdrawal of £17.85.

A negative number represents an overdraft.

8 Find the final temperatures in these science experiments.

a Starting temperature 55 °C. It goes up 32°, then down 100°.

b Starting temperature –15 °C. It goes down 28°, increases by 75°, and then goes down 17°.

c Starting temperature –22 °C. It goes up 12°, then down 2°, then increases by 53°.

N1.3 Multiplying and dividing negative numbers

Objectives

- Multiply and divide with negative numbers

Useful resources

- Number line

Mental starter

A rubber ball is made so that when you drop it, the height of the bounce is always half the height of the previous bounce.

If you drop it from a height of 256 cm, how many bounces will it do before it bounces less than 1 cm? (After the 8th bounce it reaches a height of 1 cm. So after 9 bounces it will bounce less than 1 cm.)

What is the **total** distance travelled by the ball at the moment it hits the ground for the 10th time?

$(256 + 2(128 + 64 + 32 + 16 + 8 + 4 + 2 + 1 + 0.5) = 767$ m.)

Will the ball ever stop bouncing?

Introductory activity

There are some basic rules to recall for multiplication and division:

When you multiply with a positive and a negative you get a negative answer.

If a submarine is 20 m below sea level, you can call its height −20 m. If the submarine dives so that it is 3 times as deep it is now 60 m below or −60 m. So $-20 \times 3 = -60$. (Avoid saying 'a plus and a minus makes a minus' since this is too easily confused with addition and subtraction.) Similarly:

When you divide with a positive and a negative you get a negative answer.

If you are overdrawn at the bank and you owe £80 you could write this as −£80, or 80 below zero. If you halve your debt, you now only owe £40. So $-80 \div 2 = -40$.

Negative × negative gives a positive answer.

Look at this pattern:

$-5 \times 3 = -15$

$-5 \times 2 = -10$

$-5 \times 1 = -5$

$-5 \times 0 = 0$

$-5 \times -1 = ?$

$-5 \times -2 = ?$

What is the pattern in the answers? (Goes up by 5 each time.)

Negative ÷ negative gives a positive answer.

Exercise commentary and misconceptions

Students should use a number line with care – multiplication and division can be thought of as repeated addition and subtraction respectively.

In questions 1 to 5, you might encourage students to think of the numbers as positive first to get the correct digits, then decide whether the answer is positive or negative.

In question 6, some students might turn each calculation into an equation before solving.

Plenary

Split the class into six groups. Challenge each group to devise one question and ensure that each member of the group knows how to answer it.

Assign only one of the following rules to each group: adding a negative; subtracting a negative; multiplying with a negative and a positive; multiplying with two negatives; dividing with a negative and a positive; dividing with two negatives.

Choose one member from each group to explain their example and solution to the rest of the class.

N1.3 Multiplying and dividing negative numbers

This spread will show you how to:

- Multiply and divide with negative numbers

Keywords
Negative
Positive

For multiplication and division, these simple rules tell you the sign of the answer when negative numbers are multiplied or divided.

- **positive number × positive number = positive number**
- **positive number × negative number = negative number**
- **negative number × negative number = positive number**

The same rules apply to division.

Example

Calculate **a** $+4 \times +3$ **b** $-5 \times +4$ **c** $+7 \times -2$ **d** -6×-2 **e** $-5 \times +7$

a $+4 \times +3 = +12$ **b** $-5 \times +4 = -20$ **c** $+7 \times -2 = -14$
d $-6 \times -2 = +12$ **e** $-5 \times +7 = -35$

Example

Calculate **a** $-12 \div +2$ **b** $+50 \div +2$ **c** $+24 \div -8$ **d** $-18 \div -3$ **e** $-48 \div +4$

a $-12 \div +2 = -6$ **b** $+50 \div +2 = +25$ **c** $+24 \div -8 = -3$
d $-18 \div -3 = +6$ **e** $-48 \div +4 = -12$

These examples use both rules.

Example

Calculate **a** $+4 - +3 \times -2$ **b** $\frac{-2 \times +3}{+2 + -4}$
c $+5 - -2 \times +3$ **d** $2(3 \times -4) \div 4(-5 \times 2)$

a $(+4) - (+3) \times (-2)$
$= (+4) - (-6)$
$= +10$

b $\frac{-2 \times +3}{+2 + -4}$
$= \frac{-6}{-2}$
$= +3$

c $(+5) - (-2) \times (+3)$
$= (+5) - (-6)$
$= +11$

d $2(3 \times -4) \div 4(-5 \times 2)$
$= 2(-12) \div 4(-10)$
$= \frac{-24}{-40}$
$= \frac{3}{5}$ Cancel by −8

Carry out multiplication and division before addition and subtraction.

When multiplying or dividing negative numbers, the combination of the signs gives the sign of the answer.

Exercise N1.3

1 Calculate.

a $+5 \times -3$ **b** $+2 \times -9$ **c** $+7 \times -3$
d $-8 \times +7$ **e** $-4 \times +9$ **f** $-6 \times +2$

2 Calculate.

a -4×-4 **b** -2×-8 **c** -3×-5
d -6×-7 **e** -7×-8 **f** -9×-9

3 Calculate.

a $+5 \times -5$ **b** $+4 \times -8$ **c** $-8 \times +9$ **d** $-4 \times +5$
e -3×-10 **f** -7×-7 **g** $+8 \times +2$ **h** $+5 \times -4$
i $-2 \times +9$ **j** -13×-2 **k** $-7 \times +6$ **l** $+12 \times -4$

4 Calculate.

a $-36 \div +12$ **b** $-16 \div +4$ **c** $+28 \div -4$ **d** $+18 \div -9$
e $-38 \div -2$ **f** $-80 \div -16$

5 Calculate.

a $-18 \div +9$ **b** $-20 \div +4$ **c** $-30 \div -6$ **d** $-12 \div -3$
e $-66 \div +3$ **f** $+47 \div -47$ **g** $-80 \div -2$ **h** $+24 \div +6$
i $-45 \div -9$ **j** $-51 \div +3$ **k** $+57 \div -19$ **l** $-81 \div -3$

6 Copy and complete these calculations, replacing the boxes with the correct positive or negative number or sign.

a $-7 \times \square 8 = -56$ **b** $+48 \div \square = -8$
c $\square \div +45 = +1$ **d** $+108 \div \square = -9$

7 Multiply these numbers by 10.

a +45 **b** −15 **c** +6.3 **d** −2.5 **e** −0.073 **f** +0.0092

8 Multiply these numbers by −10.

a +4.9 **b** −6.3 **c** −0.377 **d** +61.97 **e** −14.09 **f** −0.009

9 Divide these numbers by +10.

a −360 **b** +1 **c** −9.8 **d** −0.087 **e** +0.073 **f** −0.0006

10 Divide these numbers by −10.

a +550 **b** −4.8 **c** −52.66 **d** +1560 **e** −0.082 **f** +5.0005

11 Calculate.

a $+18 \div +100$ **b** $+9 \times -3$ **c** $-14 \div +2$ **d** $-3.8 \times +100$

N1.4 Factors and primes

Objectives

- Use the concepts of factors, prime numbers and prime factor decomposition

Useful resources

- OHT of prime factors

Mental starter

Using only the numbers 2, 3, 5, 7, 11 and 13, and the operation ×, make all the integers from 2 to 16. You may use one or more numbers in your answer (for example, just use 7 to make 7). You may use each number once or more than once. Set this challenge to be completed in 90 seconds. Ask what type of numbers have been used (prime). Can **any** integer be written as the product of prime numbers? (Yes.)

Introductory activity

Define a prime number as a number with exactly 2 distinct factors: 1 and itself. (Therefore 1 is not prime.) Ask students to identify the first 10 prime numbers:
(2, 3, 5, 7, 11, 13, 17, 19, 23, 29.)
To write 12 as a product of its prime factors:
The factors of 12 are 1, 2, 3, 4, 6, 12
→ prime factors are 2 and 3
→ $2 \times 2 \times 3 = 12$.
This can be a lengthy method for a large number. Suggest using a factor tree, underlining the prime numbers as you create them. Use an OHT of the first example in the student book.

An alternative method involves systematically dividing the starting number by increasing prime numbers.
Students should know the collective name for these techniques: **prime factor decomposition**.

Exercise commentary and misconceptions

In question 4 encourage students to write the answer out in full, and then use powers.
Question 5 involves finding the product of prime factors for large numbers – encourage students to be systematic and to look for easy factors such as 2, 10, 5 first.
You may need to recap simple tests of divisibility.
Question 6 relates to the volume of a cuboid.

Plenary

Write 10, 100, 1000 as products of the powers of their prime factors. Can you spot a pattern?
What are one million and one billion as a product of their prime factors?
($10 = 2 \times 5$
$100 = 2^2 \times 5^2$
$1000 = 2^3 \times 5^3$
1 million = $10^6 = 2^6 \times 5^6$
1 billion = $10^9 = 2^9 \times 5^9$)

N1.4 Factors and primes

This spread will show you how to:
- Use the concepts of factors, prime numbers and prime factor decomposition

Keywords
Factor
Prime
Prime factor decomposition
Product

You can write a number as a **product** of **factors** in different ways.

- **A prime number is a number with only two factors – itself and 1.**
  $17 = 17 \times 1$

- **Any number greater than 1 can be written as a product of its prime factors. This is called the prime factor decomposition of the number.**
  $56 = 2 \times 2 \times 2 \times 7 = 2^3 \times 7$

- **There is only one prime factor decomposition for any number.**

The factor tree method is a good way to find factors systematically.

Example

Express 84 and 112 as a product of their prime factors.

Start by dividing by the smallest factor.

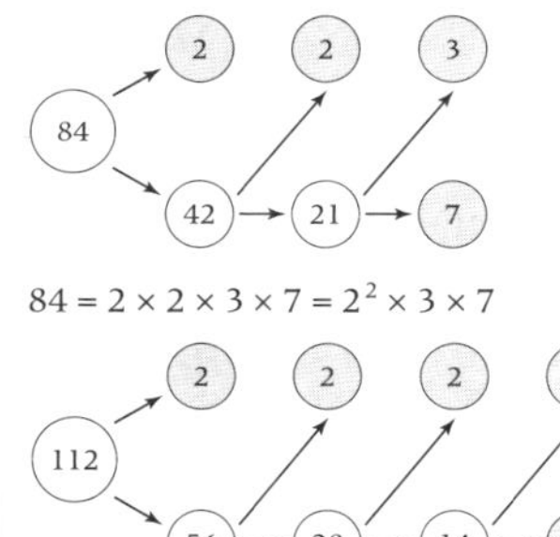

$84 = 2 \times 2 \times 3 \times 7 = 2^2 \times 3 \times 7$

$112 = 2 \times 2 \times 2 \times 2 \times 7 = 2^4 \times 7$

Example

Find the prime factor decomposition of 990.

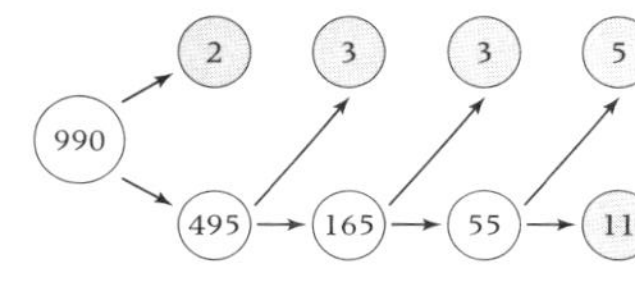

$990 = 2 \times 3^2 \times 5 \times 11$

Examiner's tip
A factor tree helps you keep track of all the prime factors, even if you don't find them in the right order.

Exercise N1.4

1 Copy and complete these calculations to show the different ways that 24 can be written as a product of its factors.

a $24 = \square \times 2$ **b** $24 = 3 \times \square$ **c** $24 = 2 \times 3 \times \square$ **d** $24 = 4 \times \square$

2 Each of these numbers has just two prime factors, which are not repeated. Write each number as the product of its prime factors.

a 77 **b** 51 **c** 65 **d** 91 **e** 119 **f** 221

3 Copy and complete the factor tree to find the prime factor decomposition of 18.

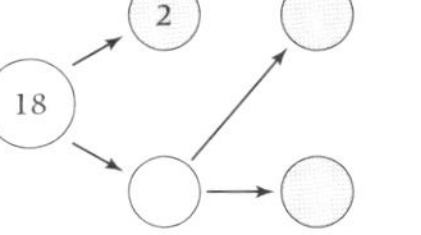

4 For each number, find its prime factors and write it as the product of powers of its prime factors.

a 36	**b** 120	**c** 34	**d** 25
e 48	**f** 90	**g** 27	**h** 60

5 Write the prime factor decomposition for each of these numbers.

a 1052	**b** 2560	**c** 630	**d** 825
e 715	**f** 1001	**g** 219	**h** 289
i 2840	**j** 2695	**k** 1729	**l** 3366
m 9724	**n** 11 830	**o** 2852	**p** 10 179

6 A cuboid is made from 210 small cubes.
The prime factor decomposition of 210 is $2 \times 3 \times 5 \times 7$.
One way of combining the prime factors is $(2 \times 3) \times 5 \times 7 = 6 \times 5 \times 7$.
The dimensions of the cuboid could be $6 \times 5 \times 7$.

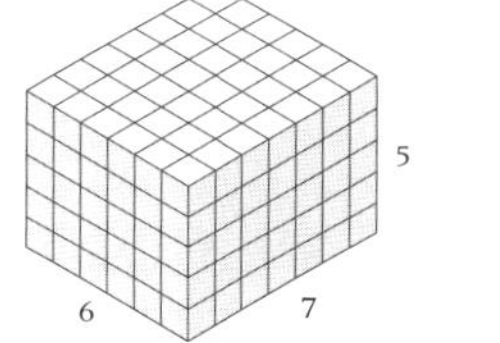

a Find all the ways in which the prime factors of 210 can be combined to make 3 factors.

b Using your answer to part **a**, list the dimensions of all the cuboids that could be made with 210 cubes.

N1.5 Using prime factors: HCF and LCM

Objectives

- Find the highest common factor and least common multiple of two numbers

Useful resources

- OHT of Venn diagram

Mental starter

Elle runs round a 400 m running track in consistent lap times of 60 seconds. Hayley takes 72 seconds. If they both start at the same point and at the same time, how many seconds will pass before Elle overtakes Hayley? At what point on the track will the overtaking take place? How many laps will each of them have completed at that moment?
(360 sec; overtakes at the start point; Elle 6 laps, Hayley 5 laps.)

Introductory activity

Recap HCF and LCM, using the example of 6 and 4. Ask how to deal with large numbers, for example 450 and 945.

First, what are 450 and 945 made up of? Use product of prime factors: $450 = 2 \times 3 \times 3 \times 5 \times 5$
$945 = 3 \times 3 \times 3 \times 5 \times 7$.

Don't write in index form yet as it is easier to see all the prime factors. **What factors have the two numbers got in common?** 3, 3, 5.

Demonstrate that the HCF is the product of all the common factors:
$HCF = 3 \times 3 \times 5 = 45$.

Now express the two numbers using indices:
$450 = 2 \times 3^2 \times 5^2$ and $945 = 3^3 \times 5 \times 7$.

Demonstrate that the LCM is the product of the highest power of each prime factor:
$LCM = 2 \times 3^3 \times 5^2 \times 7 = 9450$.

Show the worked example in the student book. This offers an alternative method to working out HCF and LCM using a Venn diagram. You may need to describe what these are to students who have never met them before.

Make a general point:

- The HCF is the product of the intersection
- The LCM is the product of the union

Exercise commentary and misconceptions

In questions 1 and 2, encourage students to list factors in ascending order.

Questions 3 and 4 involve using a conventional method for finding LCM.

Questions 5 and 6 involve the use of Venn diagrams – encourage students to fill in the intersection first.

Plenary

How can you use a Venn diagram to solve the mental starter?

N1.5 Using prime factors: HCF and LCM

This spread will show you how to:

- Find the highest common factor and lowest common multiple of two numbers

Keywords
HCF (highest common factor)
LCM (least common multiple)
Multiple
Prime
Venn diagram

- The **highest common factor (HCF)** of two numbers is the largest number that is a factor of both of them.

The HCF of 12 and 18 is 6.

- The **least common multiple (LCM)** of two numbers is the smallest number that is a **multiple** of both of them.

The LCM of 12 and 18 is 36.

To find the HCF of two numbers:

- list all the factors of each number
- identify the largest factor that is in both lists.

For example:

Factors of 18 = {1, 2, 3, 6, 9, 18}
Factors of 24 = {1, 2, 3, 4, 6, 8, 12, 24}
HCF of 18 and 24 = 6

To find the LCM of two numbers:

- list the first few multiples of each number
- identify the smallest multiple that is in both lists.

For example:

Multiples of 18 = 18, 36, 54, 72, 90, ...
Multiples of 24 = 24, 48, 72, 96, ...
LCM of 18 and 24 = 72

You can find the HCF and LCM by writing the **prime** factor decomposition for each number in a **Venn diagram**.

Note that there are other methods for finding LCM and HCF.

Example

Find the HCF and LCM of 36 and 28.

Find the prime factor decomposition of each number: $36 = 2^2 \times 3^2$ and $28 = 2^2 \times 7$.

Write these prime factors in a Venn Diagram.

prime factors of 36: 3, 3 | intersection: 2, 2 | prime factors of 28: 7

The common factors are in the middle (the 'intersection').

The HCF of 36 and 28 is the product of the numbers in the intersection: $2 \times 2 = 4$.

The LCM is the product of all of the numbers in the diagram: $2^2 \times 3^2 \times 7 = 4 \times 9 \times 7 = 252$

Exercise N1.5

1 The diagram shows how a student listed all the factors of 36. The lines show how the pairs of factors combine to make 36.

Factors of 36 = 1, 2, 3, 4, 6, 9, 12, 18, 36

a Copy the diagram, and explain why the factor 6 has not been joined to another factor.

b List the factors of 48 in the same way. Draw lines to show the factor pairs.

c The highest common factor (HCF) of 36 and 48 is the largest number that is in both lists. Write down the HCF of 36 and 48.

2 Using the method from question **1**, find the HCF of these pairs of numbers.

a 7 and 8 **b** 4 and 5 **c** 6 and 9
d 14 and 32 **e** 8 and 24 **f** 50 and 70

3 The diagram below shows how a student found the least common multiple (LCM) of 8 and 6.

multiples of 8 = 8, 16, (24), 32, 40, 48 . . .
multiples of 6 = 6, 12, 18, (24), 30, 36 . . .

a List the multiples of 12 and 9 in the same way.

b The least common multiple (LCM) of 12 and 9 is the smallest number that is in both lists.
Write the LCM of 12 and 9.

4 Using the method from question **3**, find the LCM of these pairs of numbers:

a 4 and 5 **b** 12 and 18 **c** 5 and 30
d 12 and 30 **e** 14 and 35 **f** 8 and 20

5 Use the Venn diagram method to find the HCF of 24 and 80.

6 Use the Venn diagram method to work out the HCF and LCM of these pairs of numbers.

a 25 and 120 **b** 16 and 108 **c** 60 and 144
d 42 and 56 **e** 15 and 35 **f** 20 and 110

DID YOU KNOW?
In 1880, John Venn, an English logician, introduced the logic diagrams that now bear his name.

N1 Exam review

Key objectives

- Understand highest common factor, least common multiple, prime number and prime number decomposition
- Multiply and divide by a negative number

Worked solution	Commentary and misconceptions
1	Some students may use a written multiplication technique. Emphasise the wording of this and other similar questions: 'Write down the value of ...', or 'Use a mental method ...'.
a **i** $7.4 \times -3.5 = -25.9$ **ii** $-7.4 \times 3.5 = -25.9$ **iii** $-7.4 \times -3.5 = 25.9$	
b **i** $5.32 \times -10 = -53.2$ **ii** $-53.2 \div 100 = -0.532$ **iii** $-5320 \div -1000 = -5.32$	In part **b**, encourage students to decide whether their answer is sensible – if it isn't they may have moved digits the wrong way.
2	
a **i** $60 = 2 \times 30 = 2 \times 2 \times 5 \times 3 = 2^2 \times 3 \times 5$ **ii** $96 = 4 \times 24 = 2 \times 2 \times 4 \times 6 = 2 \times 2 \times 2 \times 2 \times 2 \times 3 = 2^5 \times 3$	Some students will not break numbers down far enough. Allow students to realise for themselves that it doesn't matter how you split a number up – you should always end up with the same prime factors. Students may prefer to use a different method for finding prime factors – factor trees can be a useful technique, or repeated division.
b prime factors of 60 prime factors of 96	If students use the Venn diagram method, you may wish to use the terms 'intersection' and 'union' to describe the HCF and LCM respectively.

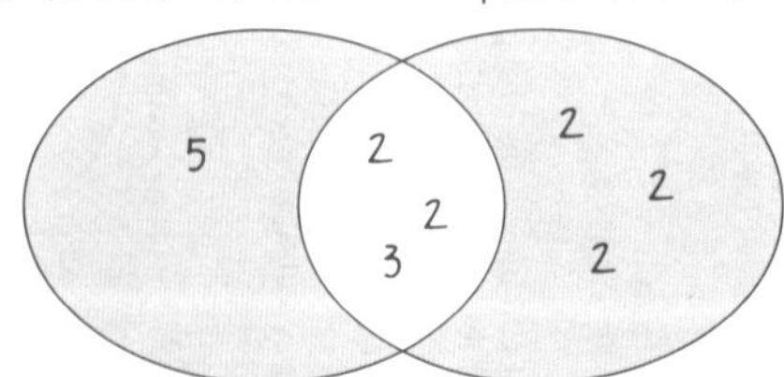

HCF $= 2 \times 2 \times 3 = 12$ (intersection)

c LCM $= 2^5 \times 3 \times 5 = 480$
(product of all the numbers)

S1 Length, area and volume

Objectives

F/H Calculate perimeters and areas of shapes made from triangles and rectangles

F/H Use their knowledge of rectangles, parallelograms and triangles to deduce formulae for the area of a parallelogram, a triangle, and a trapezium, from the formula for the area of a rectangle

F/H Recall the definition of a circle and the meaning of related terms, including chord, tangent, arc, sector and segment

F/H Find circumferences of circles and areas enclosed by circles, recalling relevant formulae

F/H Find the surface area of simple shapes by using the formulae for the areas of triangles and rectangles

Unit overview

This unit consolidates Key Stage 3 knowledge of length and area of simple 2-D shapes before extending to using formulae to calculate the area and perimeter of more complex 2-D shapes and then the surface area of simple 3-D shapes. The area and circumference of circles and semicircles is also covered. This unit provides a basis for further work on 3-D shapes in unit S6.

Prior knowledge

Before your students start this unit they should be able to:

- Measure lengths accurately and use these to find the perimeter of rectangles and triangles
- Calculate the perimeter and area of rectangles and triangles
- Calculate the surface area of cuboids

Differentiation

- **Foundation** focuses on measure, including perimeter and area
- **Foundation Plus** also focuses on measure, including perimeter and area of rectangles, triangles and compound shapes, extending to area and circumference of a circle and then to the volume of cuboids
- **Higher Plus** extends the Higher book to arc length and area of a sector of a circle and to the volume and surface area of more complex 3-D shapes, including pyramids, cones and spheres

S1.1 Area of a rectangle and a triangle

Objectives

- Calculate the perimeter and area of shapes made from rectangles and triangles

Useful resources

OHT of area of a triangle

Mental starter

The area of a particular rectangle is twice the value of the perimeter. Ask: **If the sides are integers what solutions can you find?** Do not let students use a calculator. Most students will try a trial and improvement method – encourage them to list their results in a table. (Solution for sides a, b is that $a = 4b \div (b - 4)$. Integer solutions are 5,20 : 6,12 : 8,8 only.)
For some classes give students a hint, such as one solution has a side of 5, or one solution is a square (rectangle with equal sides).

Introductory activity

Demonstrate that the area of a triangle is half the area of the rectangle. Use the term **perpendicular height** for the height of the triangle and emphasise that this height is always at right angles to the base. You could show an OHT and give students a short exercise with a few simple areas to work out.

Discuss the examples in the student book – the second example contains a compound shape. Ask: **are there any other ways of splitting this shape up to find the area?**

Exercise commentary and misconceptions

Ensure students label units on their answers correctly, as this is often neglected.
In question 2 the triangles are in various orientations, so it is important to identify the perpendicular height.

Some students find it hard to grasp that
$\frac{1}{2} \times$ base $\times$ height
is the same as
(base $\times$ height) $\div$ 2,
or they may think that you have to halve the base **and** the height. Using brackets in the first formula may help.

In question 3 on compound areas, encourage students to organise their stages of working and emphasise that there is not necessarily a right or a wrong way of tackling these problems.

Plenary

Discuss a method for finding the area of a kite – give dimensions.
Encourage different responses – some students may suggest splitting it up into triangles (two or four), and others may suggest drawing a rectangle around it.

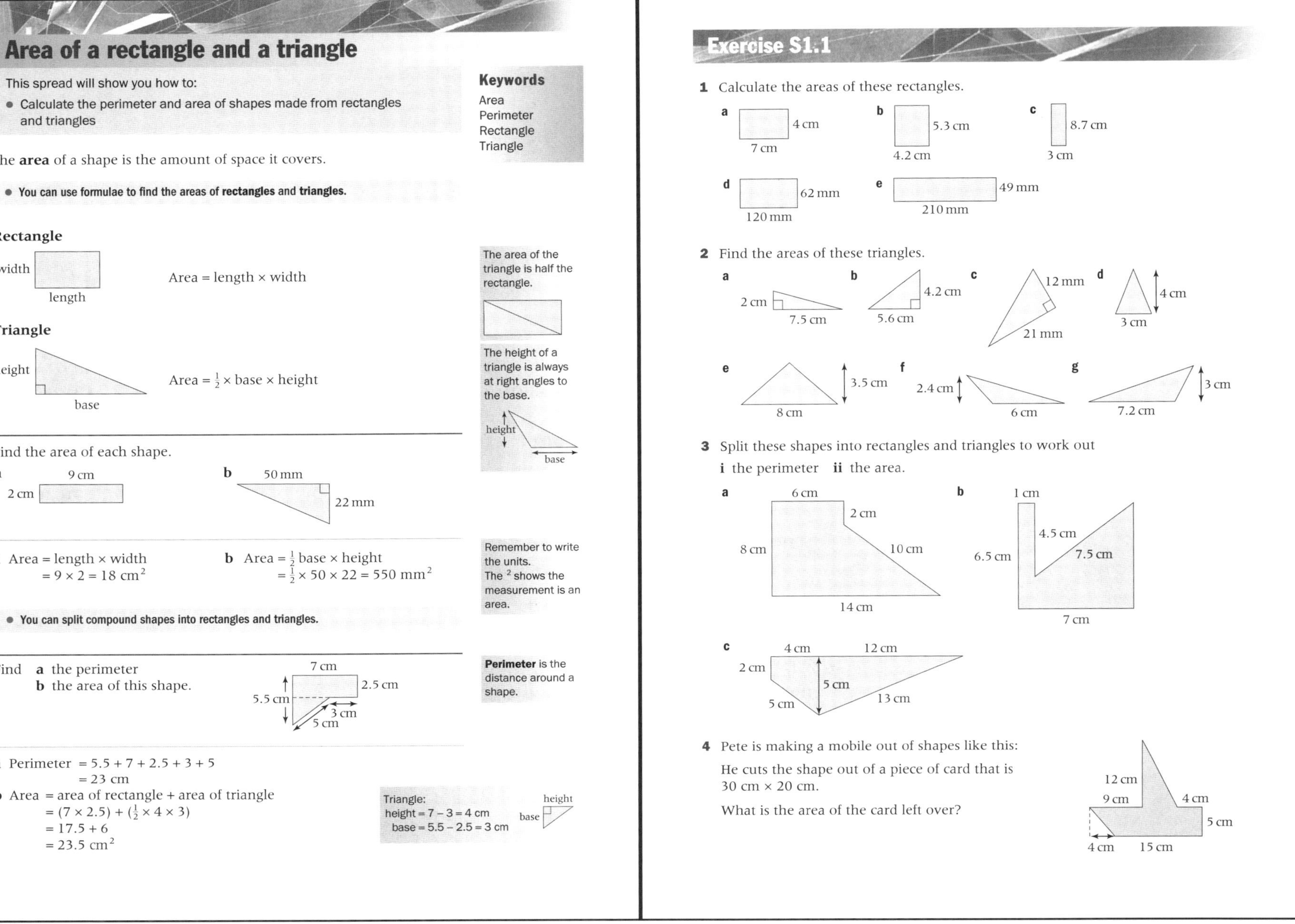

S1.1 Area of a rectangle and a triangle

This spread will show you how to:

- Calculate the perimeter and area of shapes made from rectangles and triangles

Keywords
Area
Perimeter
Rectangle
Triangle

The **area** of a shape is the amount of space it covers.

- **You can use formulae to find the areas of rectangles and triangles.**

Rectangle

Area = length × width

Triangle

Area = $\frac{1}{2}$ × base × height

The area of the triangle is half the rectangle.

The height of a triangle is always at right angles to the base.

Example

Find the area of each shape.

a

b

a Area = length × width
$= 9 \times 2 = 18\text{ cm}^2$

b Area = $\frac{1}{2}$ base × height
$= \frac{1}{2} \times 50 \times 22 = 550\text{ mm}^2$

Remember to write the units. The ^{2} shows the measurement is an area.

- **You can split compound shapes into rectangles and triangles.**

Example

Find **a** the perimeter
b the area of this shape.

Perimeter is the distance around a shape.

a Perimeter = 5.5 + 7 + 2.5 + 3 + 5
= 23 cm

b Area = area of rectangle + area of triangle
$= (7 \times 2.5) + (\frac{1}{2} \times 4 \times 3)$
$= 17.5 + 6$
$= 23.5\text{ cm}^2$

Triangle:
height = 7 − 3 = 4 cm
base = 5.5 − 2.5 = 3 cm

Exercise S1.1

1 Calculate the areas of these rectangles.

2 Find the areas of these triangles.

3 Split these shapes into rectangles and triangles to work out
i the perimeter **ii** the area.

4 Pete is making a mobile out of shapes like this:

He cuts the shape out of a piece of card that is 30 cm × 20 cm.

What is the area of the card left over?

S1.2 Area of a parallelogram and a trapezium

Objectives

- Calculate the area of parallelograms and trapeziums

Useful resources

- OHT of area of a trapezium and parallelogram
- Scissors

Mental starter

This starter builds on work from areas of triangles and rectangles in S1.1

If a triangle has an area of 18 mm^2, find possible integer values for the base and perpendicular height. Restrict solutions to the base $\leqslant$ perpendicular height.

(Answers 1, 36; 2, 18; 3, 12; 4, 9; 6, 6)

Introductory activity

Ask students to cut out a parallelogram, and then cut off the triangle on the end (you will need to show them what you mean by this).

Ask students to reassemble the shapes to form a rectangle. Use this to lead to the formula for the area of a parallelogram.

To demonstrate the area of a trapezium, students could cut out 2 congruent trapeziums, and then lay them side by side to form a parallelogram. Use this to lead to the formula.

You could use an OHT of the area of a parallelogram and a trapezium.

An alternative way of recalling the formula for the area of a trapezium is to imagine it as a rectangle that has been stretched out of shape. The bottom length has been stretched and the top length (parallel) has been compressed. Ask what would happen to the other two lengths (some students may think they would stay the same).

Exercise commentary and misconceptions

In these questions it is important to establish the 'height' – emphasise that in order to do so, students should first establish a base as reference.

Some students will want to split the shapes up – encourage the use of the formula.

Question 3 is rather an awkward shape that will need to be split up into simpler shapes.

Ensure students give units in their answer.

Plenary

Set the problem: you have 24 m of fencing in straight 1m panels. You want to maximise an enclosed area. How can you arrange the fencing as: a triangle; a rectangle; a square; a parallelogram; a trapezium?

Which of your shapes gives the largest area?

Pool results among the class.

S1.2 Area of a parallelogram and a trapezium

This spread will show you how to:

- Calculate the area of parallelograms and trapeziums

Keywords
Area
Parallelogram
Trapezium

You can use the formula for the **area** of a rectangle to find the formula for the area of a **parallelogram**.

This parallelogram can be made into a rectangle.

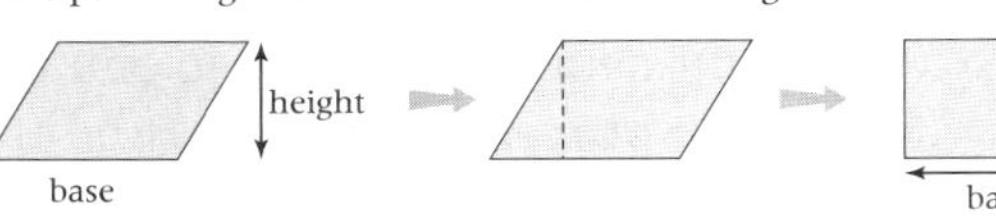

You can cut a triangle from one side and stick it on the other.

The height is at right angles to the base.

- **Area of parallelogram = base × height.**

Two congruent **trapeziums** make a parallelogram.

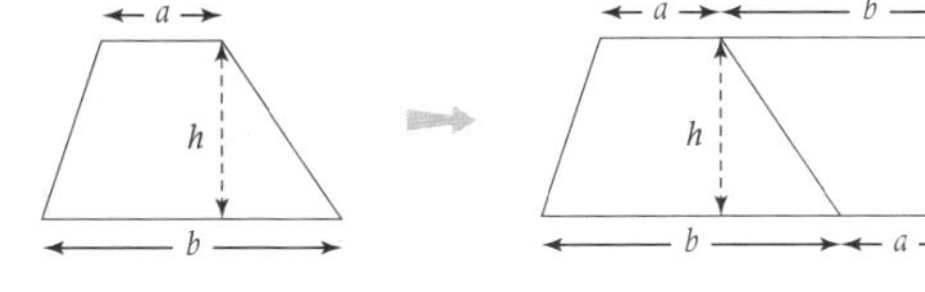

Area of the parallelogram = base × height

$= (a + b) \times h$

- **Area of trapezium** $= \frac{1}{2} \times (a + b) \times h$

The area of a trapezium is half the sum of the parallel sides times the distance between them.

Example

Find the areas of these shapes.

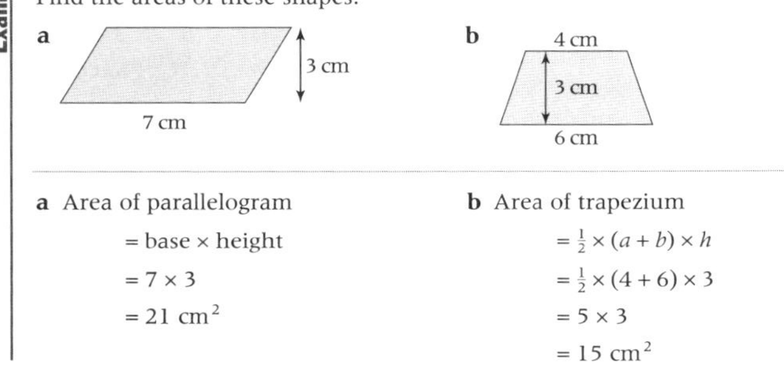

a Area of parallelogram

= base × height

= 7 × 3

= 21 cm^2

b Area of trapezium

$= \frac{1}{2} \times (a + b) \times h$

$= \frac{1}{2} \times (4 + 6) \times 3$

$= 5 \times 3$

$= 15\text{ cm}^2$

Exercise S1.2

1 Find the area of each parallelogram.

a 6 cm, 2.5 cm **b** 5.4 cm, 6.2 cm **c** 3.8 cm, 12 cm

d 4.2 cm, 7.5 cm **e** 2.9 cm, 4.6 cm

2 Find the area of each trapezium.

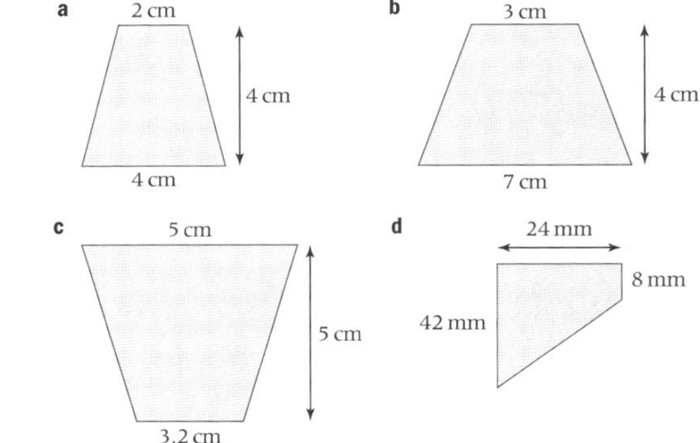

DID YOU KNOW?

The word 'trapezium' originates from the shape made by the ropes and bar of an old-fashioned flying trapeze.

3 Caroline has drawn a sandcastle.
What is the area of her castle and flag?
Start by dividing the shape into parts.

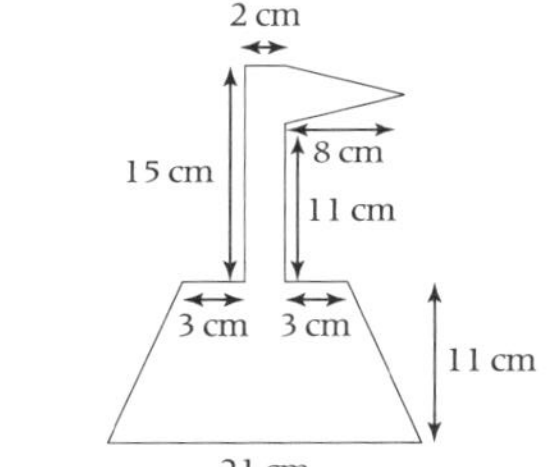

S1.3 Area and circumference of a circle

Objectives

- Use the correct vocabulary to describe the parts of a circle
- Calculate the area and circumference of circles

Useful resources

- Calculators
- Compasses
- String
- OHT of area/circumference of a circle

Mental starter

Recap the terms radius (r), diameter (d), circumference (C) and Area (A).

Ask students to draw a circle of at least 4 cm radius as accurately as possible (preferably using compasses). Then measure the diameter using a ruler (ensure that it passes through the centre).

Estimate the circumference using a piece of string.

Introductory activity

Discuss the formulae for area and circumference of a circle, emphasising that π is actually a number. To rationalise the formula for the area, you could use the approach in the student book, where a circle is split into sectors.

Draw some circles with radii or diameters given and ask students to work out A and C.

Discuss the example in the student book, and reinforce the importance of appropriate accuracy.

Exercise commentary and misconceptions

Emphasise the rule of BIDMAS in the formula for A – indices first then multiply, so the r is squared first.

Encourage students to put the area calculation into their calculator in one go – this avoids students pressing '=' after $\pi \times r$. You may have to recap significant figures, and remind students to give their answers to the accuracy requested.

Students should get into the familiar habit of writing formulae, putting in values, working out answers (down the page).

Ensure that students give appropriate units – linear units for C and squared units for A.

Plenary

Pose the question '**Is it possible to work out the circumference from knowing the area?**'.

Use the example of an area of 75 cm^2. Find the solution without a calculator (using $\pi = 3$, you get $C = 30$).

Alternatively rearrange the formula for area to make r the subject – this can be done before or after 75 m^2 is substituted.

S1.3 Area and circumference of a circle

This spread will show you how to:
- Use the correct vocabulary to describe the parts of a circle
- Calculate the area and circumference of circles

Keywords
Area
Circle
Circumference
Diameter
Radius
Proportion

In a **circle**:
- The **diameter**, d, is the distance across the circle through the centre.
- The **radius**, r, is the distance from the centre to the edge.
- The **circumference**, C, is the perimeter – the distance around the edge.

circumference
radius
diameter

$d = 2 \times r$

The circumference of a circle is in **proportion** to its diameter: $C \approx 3d$

The actual proportion is a decimal. You use a symbol, π (pi).

π is about 3.14

- $C = \pi \times d$ or $C = 2 \times \pi \times r$

You can cut a circle into lots of sectors and lay them out side by side, to make a rectangle.

r
'rectangle'
r
half circumference $= \pi \times r$

Area of 'rectangle' = length × width = $(\pi \times r) \times r = \pi \times r^2$

- Area of a circle $= \pi \times r^2$ or $A = \pi \times r^2$

Example

Find **i** the circumference **ii** the area of each circle.

a $r = 5$ cm

b $d = 32$ mm

a i $C = 2 \times \pi \times r = 2 \times \pi \times 5$
$= 31.415 \ldots$
$= 31.4$ cm (to 3 sf)

ii $A = \pi \times r^2 = \pi \times 5^2$
$= 78.539 \ldots$
$= 78.5\ \text{cm}^2$ (to 3 sf)

b i $C = \pi \times d = \pi \times 32$
$= 100.530 \ldots$
$= 101$ mm (to 3 sf)

ii $A = \pi \times r^2 = \pi \times 16^2$
$= 804.247 \ldots$
$= 804\ \text{mm}^2$ (to 3 sf)

Use the π key on your calculator.

Round your answers to a sensible degree of accuracy. 3 sf is usually good practice.

Exercise S1.3

1 Find the circumferences of these circles.

a 4 cm **b** 38 mm **c** 8 cm **d** 7.5 cm

e 24 mm **f** 42 mm **g** 13.2 cm

Give your answers to 3 sf.

2 Find the circumferences of these circles.

a radius = 12 mm **b** radius = 23 cm **c** diameter = 105 mm
d diameter = 1.2 cm **e** radius = 3.6 cm **f** diameter = 125 cm

3 Find the areas of the circles in question **1**.

4 Find the areas of the circles in question **2**.

5 A circular pond has radius 3 m.
Work out the circumference of the pond.

3 m

6 A round hole has circumference of 44 cm.
Work out the radius of the hole, to 1 decimal place.

r

7 Shamin is cutting out circles for an art project.
She has squares of card that are 4.2 cm wide.

a What is the area of the biggest circle she can cut out?
b What area of card is left?

4.2 cm

S1.4 Area and perimeter of a semicircle

Objectives

- Calculate the length of an arc and the area of a semicircle

Useful resources

- Calculator
- OHT of area/perimeter of a circle

Mental starter

Ask questions that consolidate knowledge of the circumference and area of a circle, using mental methods. For example, estimate the circumference and area of a circle with diameter 12 m (use $\pi = 3$).

Introductory activity

Explain that the word circumference only applies to the whole of a circle. The area of a semicircle is half the area of the circle, but the perimeter of a semicircle is not half the circumference. Discuss the first two examples, emphasising that the perimeter is half the circle plus the diameter.

The last example describes a context.

Exercise commentary and misconceptions

In questions 1 and 2, students should watch for varying units.

Question 5 requires care since a complete integer solution is required, so students will need to round down.

Students may need reminding to add the diameter to get the perimeter of the semicircle.

Accuracy is an important part of the answer to these calculations. If a degree of accuracy is not specified in the question, then suggest an appropriate accuracy – usually 1 decimal place or 3 significant figures.

Plenary

Discuss question 6 – this arch shape is common in exam questions, and students should know how to calculate the perimeter.

S1.4 Area and perimeter of a semicircle

This spread will show you how to:
- Calculate the length of an arc and the area of a semicircle

Keywords
Area
Circumference
Diameter
Semicircle

A **semicircle** is half a circle.

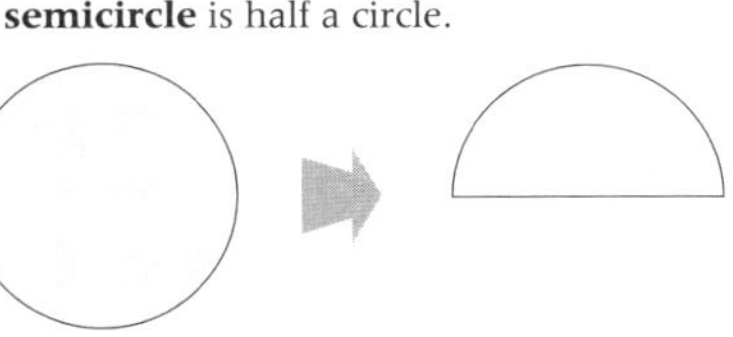

- **Area of a semicircle = $\frac{1}{2}$ × area of whole circle**

Example

Calculate the area of this semicircle.

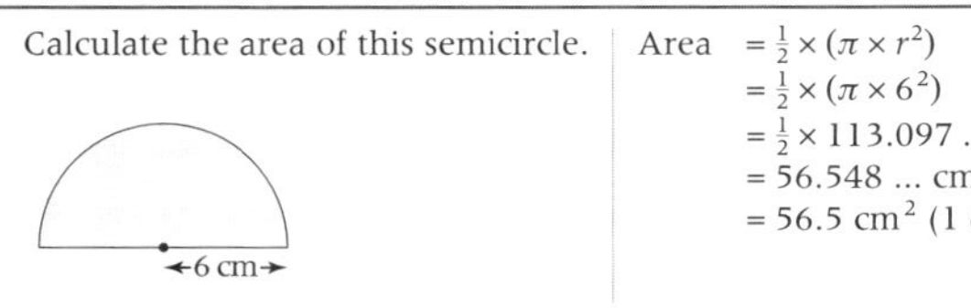

Area $= \frac{1}{2} \times (\pi \times r^2)$
$= \frac{1}{2} \times (\pi \times 6^2)$
$= \frac{1}{2} \times 113.097\ldots$
$= 56.548\ldots\ \text{cm}^2$
$= 56.5\ \text{cm}^2$ (1 dp)

- **Perimeter of a semicircle = $\frac{1}{2}$ × circumference of whole circle + diameter**

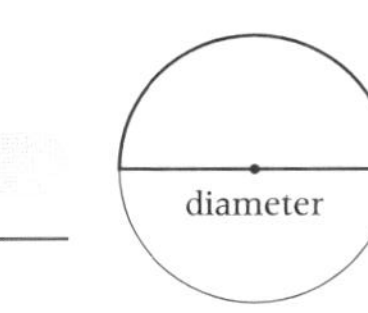

Example

Calculate the perimeter of this semicircle.

6 cm

Circumference of whole circle $= 2 \times \pi \times r$
$= 37.699\ldots$ cm
Perimeter of semicircle $= \frac{1}{2} \times$ circumference of whole circle + diameter
$= (\frac{1}{2} \times 37.699\ldots) + (2 \times 6)$
$= 18.849\ldots + 12$
$= 30.8$ cm (to 1 dp)

Example

A door is shaped in an arch, with a semicircle on top.

Calculate the perimeter of the door, giving your answer to 1 decimal place.

1.5 m
2 m

Perimeter of arch
$= 2\text{ m} + 1.5\text{ m} + 2\text{ m} + \frac{1}{2} \times$ circumference of circle
Circumference $= 2 \times \pi \times r = 1.5 \times \pi = 4.712\ldots$
Perimeter of arch $= 2 + 1.5 + 2 + \frac{1}{2} \times 4.712$
$= 7.9$ m (to 1 dp)

Exercise S1.4

Give your answers to 1 dp.

1 Find the area of each semicircle.

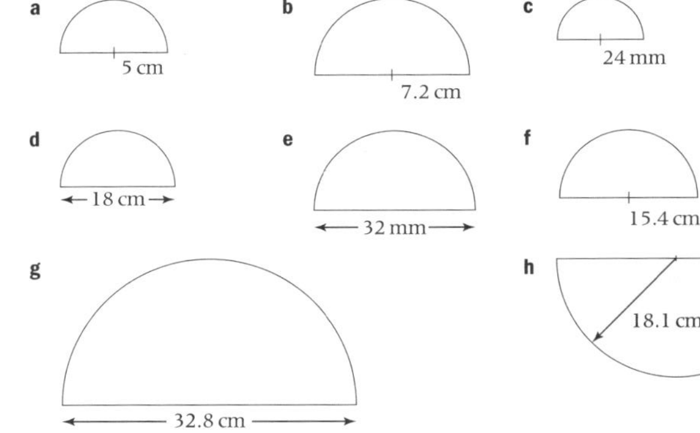

2 Find the area of each semicircle.

a radius = 12 cm **b** radius = 2.3 cm **c** diameter = 12.9 m
d diameter = 22.3 cm **e** radius = 9.5 mm **f** diameter = 3.39 cm

3 Work out the perimeter of each semicircle in question **1**.

4 Work out the perimeter of each semicircle in question **2**.

5 A flowerbed in the park is semicircular.
It has a radius of 2 m.

a Work out the area of the flowerbed.

Percy the park keeper wants to plant flowers that each need an area of 0.3 m^2.

b How many of these flowers can Percy plant in the flowerbed? What space does he have left?

6 Viaduct arches have straight sides 50 m high.
The arch at the top is a semicircle with diameter 8 m.

A spider crawls from ground level on one side, around the arch, and back down the other side.
Work out how far it crawls.

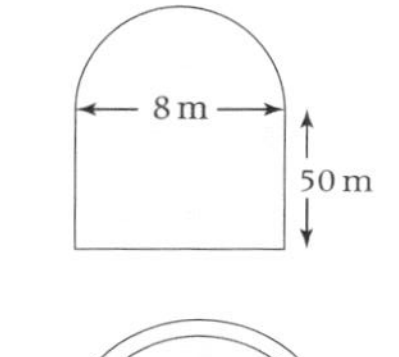

7 A bathroom window is a semicircle with an internal diameter of 80 cm.
Work out the area of the glass in the window.

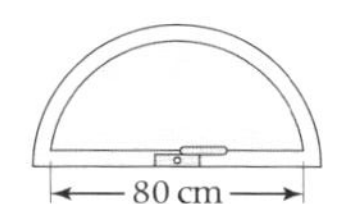

S1.5 Surface area of 3-D shapes

Objectives
- Find the surface area of simple 3-D shapes

Useful resources
- OHT of surface area of prisms
- Cardboard roll

Mental starter
Ask the question: if a cuboid has a surface area of 52 m^2, a base of 12 m^2 and a front face of 6 m^2, what is the area of one of the side faces? (8 m^2)
Students should sketch a cuboid to help visualise the faces. You can extend this starter to ask what the dimensions of the cuboid must be. (Only solution is 2 by 3 by 4)

Introductory activity
Give 2 examples of finding the surface area of a cuboid, first non-calculator, second calculator using π. Use metres in the calculations.
Sketch a right angled triangular prism. Give sides as 6, 8 and 10. Give the length as 12.
Discuss the faces that contribute to the surface.
Remind students of the area of a triangle formula.
(Answer ($24 \times 2 + 120 + 96 + 72 = 336$ m^2).)
Discuss the first example in the student book, and highlight the systematic approach.
Sketch a cylinder with height 10, diameter 8 and a lid and base. Remind students of the circle formulae and discuss what faces make up the cylinder (curved SA and two circles).
Demonstrate the curved surface area by unfolding a cylinder, for example a cardboard roll.
Find the CSA and total SA separately, since exam questions can ask for either. Find an estimate with $\pi = 3$.
(Answer CSA = $\boldsymbol{C} \times h = 24 \times 10 = 240$ m^2
Total SA = $240 + 2 \times 48 = 336$ m^2.)
Discuss the last example in the student book.

Exercise commentary and misconceptions
In question 1, encourage students to find the SA of the 3 different faces then ×2 to reduce working.
In question 2, ensure that students include the top and base (total SA), since the question does not refer to CSA only.
In question 3, encourage students to keep a list of the faces and their areas. A common misconception is that the sloping face is not a rectangle, or that it is the same area as the vertical side face. Also, many students omit to include the vertical side face.

Plenary
A cube has a SA of 54 m^2. Find the dimensions of the cube. Encourage students to sketch the cube and label what they can.
(Answer: 3 by 3 by 3.)
If a cylinder has a curved surface area of 72 cm^2 and a height of 2 cm, find an estimate for the area of the base.
(Answer: $72 = \pi dh \longrightarrow 72 = 3 \times d \times 2 \longrightarrow d = 12 \longrightarrow A = 108$ cm^2.)

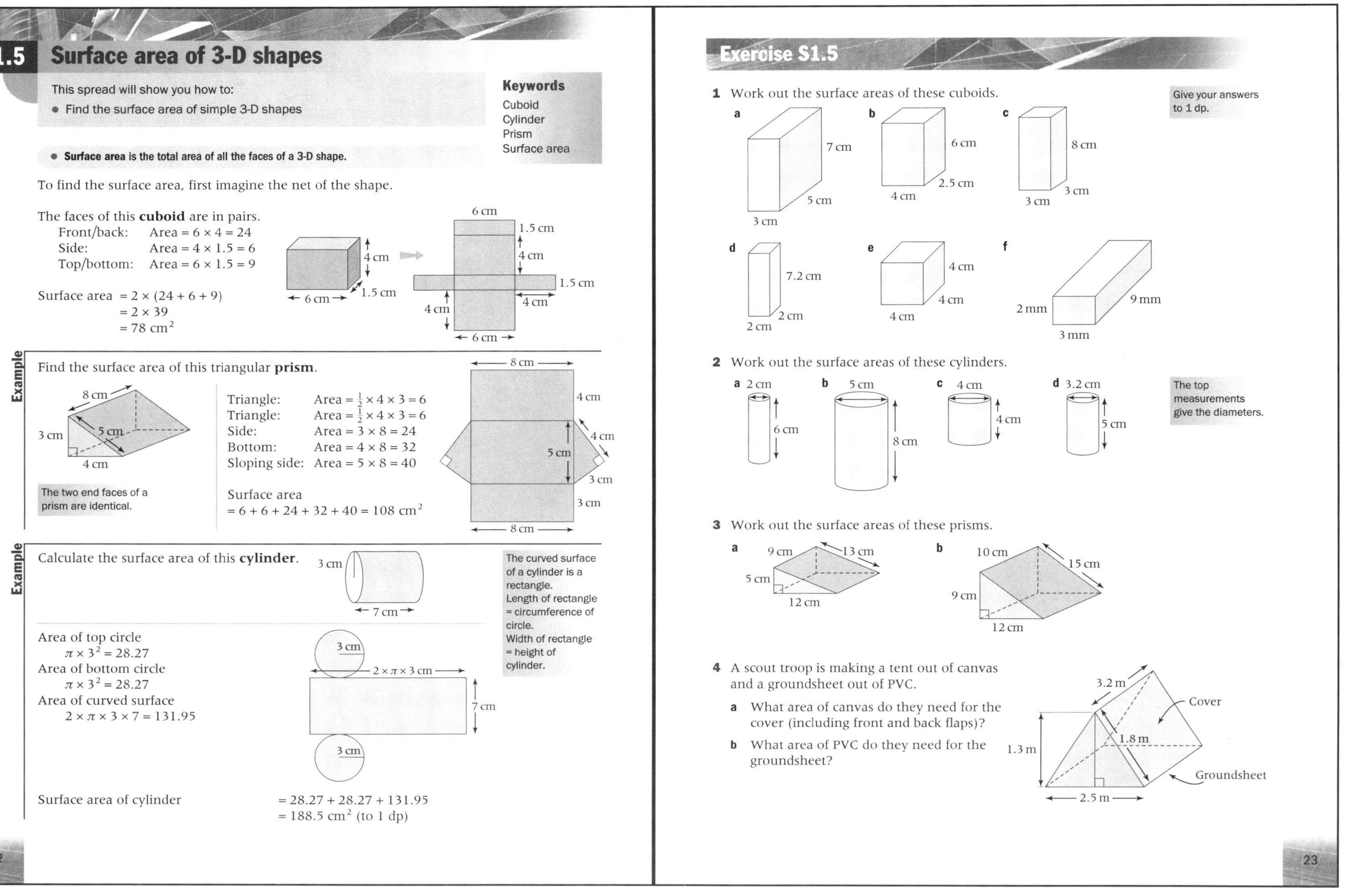

S1.5 Surface area of 3-D shapes

This spread will show you how to:

- Find the surface area of simple 3-D shapes

Keywords
Cuboid
Cylinder
Prism
Surface area

- **Surface area is the total area of all the faces of a 3-D shape.**

To find the surface area, first imagine the net of the shape.

The faces of this **cuboid** are in pairs.

Front/back: Area = 6 × 4 = 24
Side: Area = 4 × 1.5 = 6
Top/bottom: Area = 6 × 1.5 = 9

Surface area = 2 × (24 + 6 + 9)
= 2 × 39
= 78 cm²

Example

Find the surface area of this triangular **prism**.

The two end faces of a prism are identical.

Triangle: Area = $\frac{1}{2}$ × 4 × 3 = 6
Triangle: Area = $\frac{1}{2}$ × 4 × 3 = 6
Side: Area = 3 × 8 = 24
Bottom: Area = 4 × 8 = 32
Sloping side: Area = 5 × 8 = 40

Surface area
= 6 + 6 + 24 + 32 + 40 = 108 cm²

Example

Calculate the surface area of this **cylinder**.

The curved surface of a cylinder is a rectangle.
Length of rectangle = circumference of circle.
Width of rectangle = height of cylinder.

Area of top circle
$\pi \times 3^2 = 28.27$
Area of bottom circle
$\pi \times 3^2 = 28.27$
Area of curved surface
$2 \times \pi \times 3 \times 7 = 131.95$

Surface area of cylinder = 28.27 + 28.27 + 131.95
= 188.5 cm² (to 1 dp)

Exercise S1.5

Give your answers to 1 dp.

1 Work out the surface areas of these cuboids.

2 Work out the surface areas of these cylinders.

The top measurements give the diameters.

3 Work out the surface areas of these prisms.

4 A scout troop is making a tent out of canvas and a groundsheet out of PVC.

a What area of canvas do they need for the cover (including front and back flaps)?

b What area of PVC do they need for the groundsheet?

Exam review

Key objectives

- Calculate perimeters and areas of shapes made from triangles and rectangles
- Find circumferences of circles and areas enclosed by circles
- Solve problems involving surface areas and volumes of prisms and cylinders

Worked solution	Commentary and misconceptions
1	Students may find the combination of information about perimeter and area confusing. Encourage them to read and study the diagram carefully and note down any related information/formulae before attempting the question. Ensure the student knows what they are being asked to find and how this can be achieved using the given information. Some students may need prompting.
Area of square = 16 × Area of triangle $= 16 \times \frac{1}{2} \times 3 \times 2\frac{2}{3}$ $= 64\ \text{cm}^2$ Length of a side of the square $= \sqrt{64}$ $= 8$ cm Perimeter of the square $= 4 \times 8$ $= 32$ cm	Encourage the student to approach such problems step by step. Ensure that the student understands and uses the correct units for length and area.
2	Some students may confuse the terms 'diameter' and 'radius'. Ensure they recall the formula for area of a semicircle correctly, using the correct value for the radius.
Area of a semicircle $= \frac{1}{2} \times \pi r^2$ $= \frac{1}{2} \times \pi \times 7.5^2$ $= 88.35729338$ $= 88.4\ \text{cm}^2$ (3 sf)	Ensure that the student inputs the calculation into their calculator correctly. Some students may need reminding of the order of operations – indices before multiplication. Some students may have difficulty with rounding numbers. Remind them of the rules associated with rounding and that they should avoid rounding during calculations.

A1 Expressions

Objectives

F/H Understand that the transformation of algebraic entities obeys and generalises the well-defined rules of generalised arithmetic

F/H Expand the product of two linear expressions

F/H Manipulate algebraic expressions by collecting like terms, multiplying a single term over a bracket and taking out common factors

F/H Use index notation for simple positive integer powers, and simple instances of index laws

Unit overview

This unit consolidates Key Stage 3 knowledge of the rules of arithmetic and then applies them to algebra. This unit provides an introduction to algebra, preparing the student for more in depth work in further algebra units.

Prior knowledge

Before your students start this unit they should be able to:

- Add, subtract, multiply and divide with integers, using the correct order of operations, including brackets
- Find the factors of integers and the HCF of two integers

Differentiation

- **Foundation** focuses on collecting like terms in simple expressions and substituting numbers into expressions
- **Foundation Plus** extends to factorising and expanding single brackets
- **Higher Plus** extends the Higher book to expressions involving the difference of two squares and algebraic fractions

A1.1 Writing and simplifying expressions in algebra

Objectives

- Use the rules of algebra to write and manipulate algebraic expressions
- Simplify algebraic expressions by collecting like terms
- Use index notation and simple laws of indices

Useful resources

- Number line

Mental starter

If $8^x = 4^y = 2^z$ and x, y, z are positive integers (whole numbers) what possible values can they be?

(2, 3, 6 will be the most likely solution found, but 0, 0, 0 is a trivial solution. Any solution for x: y: z that cancels to the ratio 2 : 3 : 6 is correct.)

Introductory activity

Start with simplifying algebra by adding and subtracting.

Using these examples, ask 'how many different types of term can you see?'
Then ask how the expressions can be simplified.

a $3x + 4x - 6$ (2 types; x and number) $7x - 6$.

b $4y + 1 + 3x - y + 7x$ (3 types; x, y, number)
$3y + 10x + 1$.
Remind students to look at the sign in front of each term to decide if it is + or −.

c $4x^2 - 4x - 2x + 7$ (3 types; x^2, x, number)
$4x^2 - 6x + 7$.
Use a number line if necessary to show $-4 - 2 = -6$.

d $3xy + 5x + 8xy - 2yx$ (Are there 2 or 3 types? Is xy the same as yx? $3 \times 4 = 12$ as does 4×3. Multiplying in either order gives the same result, so $xy = yx$ and there are two types; xy and x.)
$9xy + 5x$.

Next explore simplifying algebra by multiplying.

Use these examples:

a $5 \times 3x$. The numbers will join together, they are the same type (or friends), the x has nothing to join to (no mates!)
$15x$.

b $x \times 5x$. This time the x can be joined, multiplied together. $x \times x = x^2$.
So the answer is $5x^2$.

c $4x \times 3x$. Numbers will multiply and so will the x.
$12x^2$.

Finally look at simplifying algebra by dividing.

Use these examples:

a $12x \div 4 = \frac{12x}{4}$. The 12 can be divided by the 4, but the x has no partner.
$3x$.

b $8xy \div 6 = \frac{8xy}{6}$. The 8 and 6 can cancel.
$\frac{4xy}{3}$.

c $\frac{5y^3}{2y}$. The numbers 5 and 2 don't cancel,
but $\frac{y^3}{y} = y^2$.
$\frac{5y^2}{2}$.

Exercise commentary and misconceptions

Display a number line on the board to assist in +, − with algebra.

Split the lesson into two parts. First remind students how to + and −, then let them try questions 1–4. After 5–10 minutes remind students how to tackle multiplication and division, then they can finish the exercise.

Plenary

Ask the class to come up to the board and take turns in simplifying this expression:

$$\frac{8x^4y^3z^5 \times 2yx^3}{12z^3x^{10}p^2 \times y^4}.$$

$$\left(\text{Answer: } \frac{16x^7y^4z^5}{12x^{10}y^4z^3p^2} = \frac{4z^2}{3x^3p^2}\right)$$

A1.1 Writing and simplifying expressions in algebra

This spread will show you how to:
- Use the rules of algebra to write and manipulate algebraic expressions
- Simplify algebraic expressions by collecting like terms
- Use index notation and simple laws of indices

Keywords
Factorise
Index/indices
Like terms
Product
Simplify

There are rules for writing expressions in algebra
- Do not include the multiplication sign $3 \times p \rightarrow 3p$
- Write divisions as fractions $3 \div p \rightarrow \frac{3}{p}$
- Write numbers first in products $p \times 3 \rightarrow 3p$
- Write letters in products in alphabetical order $4 \times q \times r \times p \rightarrow 4pqr$

- To **simplify** expressions involving addition or subtraction, you collect **like terms** together.

Like terms: $3z, 9z, -4z$ Unlike terms: $5p, 2p^2, 8q$

- To simplify expressions involving multiplication or division, you multiply the numbers then the letters.

You may need to use **indices**.

Example

a Write these using the rules of algebra.
 i $5 \times q \times 3 \times p$
 ii $y \times y \times y \times y \times y$
b Evaluate $3x + 2$ when $x = -4$

a i $5 \times q \times 3 \times p = 15pq$
 ii $y \times y \times y \times y \times y = y^5$
b $3x + 2 = 3 \times (-4) + 2 = -12 + 2 = -10$

Numbers first, letters in alphabetical order

$3x = 3 \times x$

Example

Simplify these algebraic expressions.

a $3p + 9q - 2p + 7q$ b $5q - 7 + q^2$
c $7ab + 3ba$ d $3t \times 4t \times 2t \times 2s$
e $\frac{15ab}{5b}$

a $3p + 9q - 2p + 7q = 3p - 2p + 9q + 7q$
$= p + 16q$
b $5q - 7 + q^2$ cannot be simplified as there are no like terms
c $7ab + 3ba = 7ab + 3ab = 10ab$
d $3t \times 4t \times 2t \times 2s = 48st^3$
e $\frac{15ab}{5b} = \frac{{}^{3}\cancel{15}a\cancel{b}}{\cancel{5}\cancel{b}} = 3a$

'1p' is simply 'p'

'q' and 'q²' are not like terms

Cancelling – divide top and bottom by 5 and by *b*.

Exercise A1.1

1 Write these expressions using the rules of algebra.
 a $5 \times w$ b $6 \div k$ c $y \times y$ d $ab6$ e $k \times k \times 8 \times k$

2 Evaluate these expressions, given that $x = 6$.
 a $3x + 2$ b $10 - x$ c x^2 d $\frac{10x - 16}{2}$ e $3x^2$

3 Simplify these expressions by collecting like terms.
 a $3a + 4b + 8a + 2b$ b $3t + 9 - t + 17$
 c $3x - 4y - 2x - 8y$ d $9p + p^2 + 5p$
 e $10xy + 10yx$ f $6ab + 2ba - ba$

4 Three students tried to simplify $3m + 5$. Which of them did it correctly?

Sara	Paul	Abdul
$3m + 5 = 8m$	$3m + 5 = 15m$	$3m + 5 = 3m + 5$

5 Simplify these expressions.
 a $4m \times 7n$ b $6m \times 2m$ c $\frac{20p}{2}$ d $\frac{14a}{7a}$
 e $2a \times 3b \times 4c$ f $k \times 2k \times 3k$ g $\frac{20ab}{5a}$ h $\frac{45c^2}{5c}$

6 Simplify the expressions in the grid and find the 'odd one out' for each row.

$3p + 2q + p + 5q$	$6p + 3q - 2p + 4q$	$5p - 3q - p + 5q$
$2m \times 3n$	$2 \times n \times m \times 5$	$6mn$
$\frac{24cd}{12c}$	$\frac{2d^2}{d^2}$	$\frac{2d^2}{d}$
$2n - 8$	$3m + 2n - m - 2m$	$3n - 2 - 6 - n$

7 a Write a simplified expression for the
 i perimeter ii area
 of this rectangle

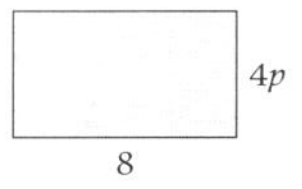

 b What are the measurements of a rectangle with perimeter $6x + 4y$ and area $6xy$?

8 Copy this grid, replacing each expression with its simplified form (where possible).

$3a + 7b - 5a + 2b$	$3a \times 4a$	$\frac{20b}{5}$
$\frac{16ab^2}{8b}$	$2p + 7p^2 + 5p^3 + 8p$	$11abc + 2cab$
$5m - 4$	$3m \times 4m \times 5m$	$\frac{4a}{2a^3}$

A1.2 Expanding single brackets

Objectives

- Multiply a single term over a bracket

Useful resources

- Mini-whiteboards

Mental starter

A rectangle measures 6 cm by $(x + 2)$ cm.
The area is 14 cm^2.
Find the perimeter.
($x = \frac{1}{3}$, perimeter = $16\frac{2}{3}$ cm. You could give students the hint that they should form an equation for area.)

Introductory activity

Explain that when you multiply a bracket by a number, everything inside the bracket is multiplied by the number.

Use these examples:

a $4(3 + 2) = 4 \times 3 + 4 \times 2 = 12 + 8 = 20$
b $5(x + 3) = 5x + 15$
c $2(3x - 5) = 6x - 10$
d $-3(2x + 1) = -6x - 3$ (Remind students $-3 \times 1 = -3$; don't use $+ -3$.)
e $-4x(2x - 1) = -8x^2 + 4$ (Remind students that 'minus × minus gives a positive answer'.)
f $-(4x + 6)$ **What number is the bracket multiplied by?** (−1) (Answer = $-4x - 6$.)

Let students try questions 1 and 2 in Exercise A1.2 before introducing two single brackets in an expression.

Use these examples:

a $2(x + 4) + 3\ (2x - 1) = 2x + 8 + 6x - 3 = 8x + 5$
b $3(x - 1) - 4(2x + 1)$. **What number is the first bracket multiplied by?** (3) **What number is the second bracket multiplied by?** (−4) (It is much easier for students to think of the number −4 as the multiplier rather than multiplying by 4 then subtracting each term.)
($3x - 3 - 8x - 4 = -5x - 7$.)
c $2(x + 6) - 3(x - 1)$. **What number is the first bracket multiplied by?** (2) **What number is the second bracket multiplied by?** (−3)
($2x + 12 - 3x + 3 = -x + 15$.)

Exercise commentary and misconceptions

Try questions 1 and 2 after the first part of the introduction. Then try questions 3–7 after the second part. Watch for errors when multiplying with negatives. Emphasise that the number multiplying the brackets may be positive or negative. Watch for examples where the second bracket is multiplied by −1. If it helps students, encourage them to write the '1' before the bracket.

In questions 5 and 7 students will write formulae that include brackets for area and volume. Then they will form equations. Showing that the expression and the numerical value are the same is often an awkward concept. Write them out on 2 separate pieces of paper. Fix a large equals sign at the front of the class, perhaps held by a student. Bring the expression and the numerical value together from either side. Both sides work out the same, so they are equal. This forms (creates) an equation you can solve and find x.

Question 6b has an infinite set of solutions.

Plenary

Ask various students to come forward and contribute towards removing these brackets.

$$12 + 7x^4py^3(3x^2y + 8py^2 - 1) + 6px^4y^3$$
$$= 12 + 21px^6y^4 + 56p^2x^4y^5 - 7px^4y^3 + 6px^4y^3$$
$$= 12 + 21px^6y^4 + 56p^2x^4y^5 - px^4y^3$$

A1.2 Expanding single brackets

This spread will show you how to:

- Multiply a single term over a bracket

Keywords
Bracket
Expand

A **bracket** in algebra means 'all multiplied by'.

$3(x+5)$ means 'I think of a number, add 5 then **multiply it all** by 3'

To write an expression without brackets, multiply all the terms inside the bracket by the term outside.
This is called **expanding** the bracket.

$3(x+5)$ expanded is $3x+15$

For negative terms, use the rules for multiplying by negative numbers:

- negative term × positive term → negative term $\quad -3(x+5) = -3 \times x + -3 \times 5 = -3x - 15$
- negative term × negative term → positive term $\quad -5(y-8) = -5 \times y + -5 \times -8 = -5y + 40$

Example

Expand each of these.

a $5(m+9)$ b $y(y-7)$
c $3p(2p+7-q)$ d $-4m(m-2)$

a $5(m+9) = 5m + 45$
b $y(y-7) = y^2 - 7y$
c $3p(2p+7-q) = 6p^2 + 21p - 3pq$
d $-4m(m-2) = -4m^2 + 8m$

$y \times y$ is y^2

Be careful with negatives.

Example

Expand and simplify $3(t-2) + 5(2+t)$.

$3(t-2) + 5(2+t) = 3t - 6 + 10 + 5t$
$= 3t + 5t - 6 + 10$
$= 8t + 4$

Collect like terms.

Exam question

A cuboid with a square base of length x has height 1 cm more than the length. Its volume is 230 cm^3.
Show that $x^3 + x^2 = 230$

As the base is square, the length and width must both be x cm.
You are told that the height is 1 cm more than the length, so height is $(x+1)$ cm.
Volume of cuboid = length × width × height
$230 = (x+1) \times x \times x$
$230 = x^2(x+1)$
$230 = x^3 + x^2$ (as requested)

(*Edexcel Ltd., 2003*)

Sketch a diagram to help:

x, x, $x+1$

Exercise A1.2

1 Expand these brackets.

a $4(n+5)$ b $6(b-7)$ c $a(a+3)$
d $a(b-c)$ e $4(2x+3y-4z)$ f $2h(h+9)$

2 Expand these brackets.

a $-3(k+9)$ b $-2(h-5)$ c $-(w-4)$
d $-(t-p)$ e $-k(k+7)$ f $-9(2m-k+4)$
g $-(x^2-x-8)$ h $-2(x^2+3)$ i $-3(1-x)$

Be careful with negatives.

3 Expand and simplify these expressions.

a $3(c+2) + 7(c+8)$ b $4(2x+8) + 5(3x+7)$
c $x(x+8) + x(x+2)$ d $5t(3t+6) + 2t(t+1)$
e $3(x-7) + 4(x-6)$ f $5(2-x) + 7(x-3)$
g $4(m-6) - 2(m+1)$ h $3(g-3) - 7(2g-6)$
i $2(p+5) - (p-4)$ j $(q-4) - (3-q)$

4 Expand and simplify $2x(x+7) + x(9-x) - 3x(2x-7)$.

5 a Using brackets, write a formula for the area of this rectangle.

3

$2x-1$

b Expand the brackets.
c The area of the rectangle is 15 cm^2.
Show that $6x - 18 = 0$.

6 An expression expands to give $24x + 16$.
a What could the expression have been if it involved one pair of brackets?
b What could the expression have been if it involved two single brackets in succession?

7 Write an expression involving brackets for the volume of this cuboid.

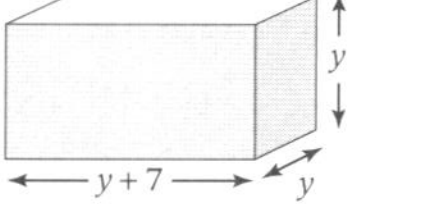

Expand the brackets and simplify your expression.

A1.3 Expanding double brackets

Objectives
- Expand double brackets

Useful resources
- Calculators (for question 4b)

Mental starter
By choosing certain values of x, see whether the following algebraic expressions give the same value. Try $x = 1$, then 5, then 7.

$(x + 2)(x + 4)$
$x^2 + 6x + 8$
$x(x + 6) + 8$
$(x + 3)^2 - 1$
$2(3x + 4) + x^2$.

(If $x = 1$ each expression equals 15; for $x = 5$, expressions = 63; for $x = 7$, expressions = 99.) The class could be split into groups to calculate the different examples. **Which expression is generally the easiest to calculate?** $(x + 2)(x + 4)$ but opinions may vary.

Introductory activity
To multiply 2 brackets together there are various methods commonly used. Students may have seen different methods in previous years.

The students' book uses the FOIL method, but there are several others which students may have used in previous years. It may be helpful to discuss each method for the same example, then discuss the advantages and disadvantages of each method.

Another commonly used method is a multiplication table.

Use a method of choice to expand and simplify

$$\begin{aligned}(3x - 1)(2x + 4) &= 6x^2 + 12x - 2x - 4\\ &= 6x^2 + 10x - 4.\end{aligned}$$

Emphasise the different phrases which can be used to describe the process, for example 'expand', 'multiply out', 'remove the brackets'.

Exercise commentary and misconceptions
In questions 1d and k many students will simply square each term inside the bracket separately. A numerical example such as $(3 + 2)^2$ may help.

$(3 + 2)^2 = 5^2 = 25$,
but $3^2 + 2^2 = 9 + 4 = 13$,
so $(c + 5)^2 \neq c^2 + 5^2$.

Write the expression as $(c + 5)(c + 5)$.

Question 4 may cause some problems and, if so, it may be worth including it in the plenary session.

Because Question 5 involves manipulating equations, with a weaker group it may be best to leave it till a later lesson.

Plenary
Use the plenary to discuss questions 1d and k as mentioned in the exercise commentary if this seems to be a common problem.

In question 4a expand and simply $(a + b)^2$. **What is the link between parts a and b? Do you get the same answer if you calculate $(1.32 + 2.46)^2$ as if you calculate $1.32^2 + 2 \times 1.32 \times 2.46 + 2.46^2$?** (yes; 14.2884.)

A1.3 Expanding double brackets

This spread will show you how to:
- Expand double brackets

Keywords
Expand
FOIL
Product
Simplify

You can **expand** a double bracket in algebra by multiplying pairs of terms.

Each term in the first bracket multiplies each term in the second bracket:

$(p+7)(p+3) \longrightarrow p^2 + 3p + 7p + 21 \longrightarrow p^2 + 10p + 21$

$p^2 + 10p + 21$ is the product of $(p+7)$ and $(p+3)$.

F... Firsts
O... Outers
I... Inners
L... Lasts

Example

Expand and simplify.

$(x+3)(x+4)$

$(x+3)(x+4) = x^2 + 4x + 3x + 12$
$= x^2 + 7x + 12$

F: $x \times x = x^2$
O: $x \times 4 = 4x$
I: $3 \times x = 3x$
L: $3 \times 4 = 12$

Example

Expand and simplify.

a $(x+3)(2x-2)$ **b** $(3x+2)^2$

a $(x+3)(2x-2) = 2x^2 - 2x + 6x - 6$
$= 2x^2 + 4x - 6$

b $(3x+2)^2 = (3x+2)(3x+2)$
$= 9x^2 + 6x + 6x + 4$
$= 9x^2 + 12x + 4$

Use the rules for multiplying negative terms:
O: $x \times -2 = -2x$
L: $3 \times -2 = -6$

Example

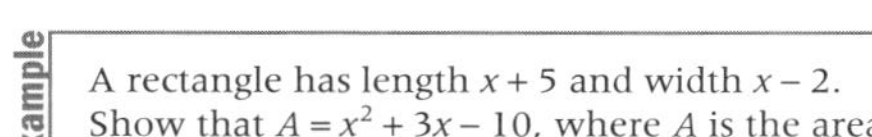

A rectangle has length $x + 5$ and width $x - 2$.
Show that $A = x^2 + 3x - 10$, where A is the area.

Sketch a diagram:

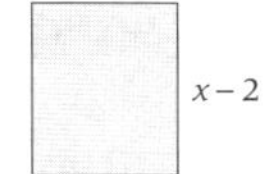

Area of rectangle = length × width
$A = (x+5)(x-2)$
$= x^2 - 2x + 5x - 10$
$A = x^2 + 3x - 10$

Exercise A1.3

1 Expand and simplify these expressions involving double brackets.

a $(x+2)(x+3)$ **b** $(p+5)(p+6)$
c $(w+1)(w+4)$ **d** $(c+5)^2$
e $(x+4)(x-2)$ **f** $(y-2)(y+7)$
g $(t+6)(t-2)$ **h** $(x-2)(x-5)$
i $(y-4)(y-10)$ **j** $(w-1)(w-2)$
k $(p-5)^2$ **l** $(q-12)^2$

2 Expand and simplify.

a $(2x+1)(3x+7)$ **b** $(5p+2)(2p+3)$
c $(3y+4)(2y+1)$ **d** $(2y+6)^2$
e $(5t-4)(2t+4)$ **f** $(5w-1)(3w+9)$
g $(2x+2y)(3x-3y)$ **h** $(3m-4)^2$
i $(2p+5q)(3p-8q)$ **j** $(2m-3n)^2$

3 Write an expression for the areas of this rectangle and square.

a

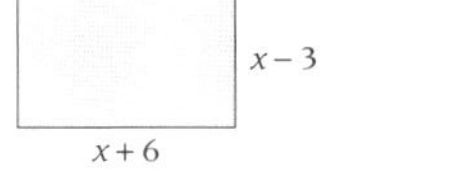

b 2m − 3

4 a Expand $(a+b)^2$.

b Hence, or otherwise, calculate $1.32^2 + 2 \times 1.32 \times 2.68 + 2.68^2$

c Write another calculation that you could work out using this expansion.

5 a The diagram shows a rectangle with an area of 75 cm^2.
Show that $6x^2 + 7x - 78 = 0$.

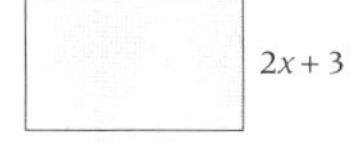

b This triangle also has an area of 75 cm^2. Show that $15x^2 = 14x + 158$.

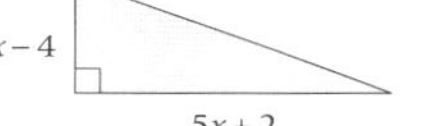

A1.4 Factorising single brackets

Objectives

- Factorise into single and double brackets

Useful resources

- Mini-whiteboards

Mental starter

What is the value of 2×5^2?
(Remind students of the rules of BIDMAS, that is indices/powers before multiplication)
(Answer 50)

The calculation
$2 + 2 \times 3 + 2 \times 3^2 + 2 \times 3^3 + 2 \times 3^4$
can also be written as
$2(1 + 3(1 + 3(1 + 3(1 + 3))))$.
Split the class in half and ask the students in each half to calculate one of the expressions. As soon as each student has the result they should raise their hand.
Why is the second calculation easier and quicker? (242, the brackets mean fewer and easier calculations)

Introductory activity

Expand $3(x + 2) = 3x + 6$.
If you were given $3x + 6$ to start with, could you go backwards and write it with the brackets?

Try $5x + 15$. See how many students get $5(x + 3)$.
What method can be used? What factor is common to the $5x$ and the 15? (5). Write this common factor outside a set of brackets.

Imagine what you must multiply the 5 by to make the answer $5x + 15$.

For example, $8xy + 6x$. **What is common to each part?** (2 and x.) Therefore put $2x$ outside a pair of brackets. **What must you multiply the $2x$ by to make $8xy + 6x$?** ($2x(4y + 3)$.)

For example,
$20x^2 - 10x + 30xy = 10x(2x - 1 + 3y)$. Explain why you have to use a 1 in the bracket.

Try question 1 and question 2 before giving examples of more complex types of factorising.

Sometimes the common factor in each term is more complex. For example,

$2(x + 4) + (x + 4)^2$

The common factor in each term is a bracket itself, $(x + 4)$. So you write down this common factor and follow it with a set of brackets.
$(x + 4)$ ( ).
What is $(x + 4)$ multiplied by to make $2(x + 4)$ and $(x + 4)^2$?
$(x + 4)(2 + (x + 4)) = (x + 4)(6 + x)$. Students should try question 3 before explaining more complex factorising.

$(x + 3)(x + y)$ becomes $x^2 + xy + 3x + 3y$ when expanded. **Is there a single common factor?** (No but it can be factorised (put) into 2 brackets.)

Try $ac + ad + bc + bd$. **Is there a common factor for all four terms?** (No. Try to make 2 brackets out of it.)

The first two terms have a as a common factor and the last two terms have b as a common factor, so $a(c + d) + b(c + d)$.
$(c + d)$ is now a common factor, so $(c + d)(a + b)$.

Try $pq - pr + qs - rs$ ($= (p + s)(q - r)$)

Exercise commentary and misconceptions

Throughout the exercise, encourage students to multiply out to check their solutions as multiplying out is much easier than factorising for most students.

In question 1d, a common mistake is to forget to leave a '1' inside the brackets because the whole term is a common factor.

Question 4, which uses two brackets, is very demanding. Encourage students to first combine pairs of terms.

Although question 6 should be done without a calculator, students may wish to check their work with a calculator.

Plenary

Factorise the numerator and denominator of this fraction. $\dfrac{4x^2y + 6xy}{8y^2x - 2xy}$

Answer $\dfrac{2xy(2x + 3)}{2xy(4y - 2)} = \dfrac{2x + 3}{4y - 2}$

How does this help? Suppose you had to work out its value if $x = 3$ and $y = 1$. **How fast can you do it for the first expression? How fast for the last expression?** (Answer 4.5)

A1.4 Factorising single brackets

This spread will show you how to:

- Factorise into single and double brackets

Keywords
Factorise
Highest common factor (HCF)

The reverse of expanding a set of brackets is called **factorising**.
To factorise an expression, you put brackets in.

Expand

$12(x+2)$ $12x+24$

Factorise

You divide the terms by their **highest common factor (HCF).**

HCF of $12x$ and 24 is 12

$12x+24$ $12(x+2)$

Check your answer by expanding

Example

Factorise fully.

a $6x+9$ **b** $12pq-4pw$ **c** $5x+10x^2-25xy$

a The HCF of $6x$ and 9 is 3.

$6x+9=3(2x+3)$

b $12pq-4pw$
Deal with numbers first, then letters.
HCF of 12 and 4 is 4.
HCF of pq and pw is p.
HCF of $12pq$ and $4pw$ is $4p$.
So, $12pq-4pw=4p(3q-w)$

c The HCF of $5x$, $10x^2$ and $25xy$ is $5x$.

$5x+10x^2-25xy=5x(1+2x-5y)$

$5x \div 5x$ is 1

Example

a Factorise y^2-3y **b** Factorise $(p+q)^2-2(p+q)$

a $y^2-3y=y(y-3)$

b $(p+q)^2-2(p+q)$
$=(p+q)((p+q)-2)$
$=(p+q)(p+q-2)$

Each part in **b** has $(p+q)$ in common

Exercise A1.4

1 Factorise each of these fully by removing common factors.

a $2x+4$ **b** $3y-6$ **c** $12p+36q$
d $25w-5$ **e** $6xy+xw$ **f** $ab-2bc$
g $pqr+qrt-qsw$ **h** $5xy-x$ **i** $2xy+6x$
j $4ab-6a^2$ **k** $25p^2-10p$ **l** $7x+14xy$
m $2ac+4a^2-8a$ **n** $15mn-5m+10m^3$ **o** $6p^4-12p$

2 All three students have completed their factorisations incorrectly.
Explain what they have done wrong.

Clare: $5x+10xy$ $=5x(0+2y)$

Ben: $6pq+3p$ $=3(2pq+1)$

Vicky: $21p+14pq$ $=7p(14+7q)$

3 Factorise these expressions.

a $10(x+y)+13(x+y)$ **b** $(a-b)^2+5(a-b)$
c $6(q+r)-(q+r)^3$ **d** $(pt-w)+6(pt-w)$

4 Factorise fully.

a $ax+bx+ay+by$ **b** $cd+bd+cm+bm$
c $a^2+ab+2a+2b$ **d** $cd+ce-me-md$

First look at what each pair of terms has in common.

5 Write a factorised expression for

a The perimeter of this rectangle

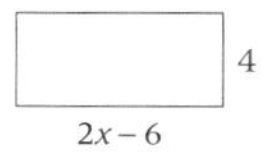

b The perimeter of a square with sides $5b+10$

6 Use factorisation to help you to evaluate these, without a calculator.

a $2\times1.86+2\times1.14$ **b** $3\times5.87-3\times0.37$
c $5.86^2+5.86\times4.14$ **d** $3.32\times6.68+3.32^2$

7 Show that the shaded area of this rectangle is $2(4x+5)$.

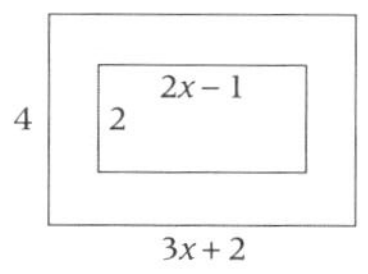

A1.5 Factorising double brackets

Objectives

- Factorise into single and double brackets

Useful resources

- Mini-whiteboards

Mental starter

Ask students to evaluate 3×10^2. (300.) (Remind students of the rules of BIDMAS if necessary, that is indices/powers before multiplication.)

The algebraic expression $3x^2 - 8x - 35$ can also be written as $(3x + 7)(x - 5)$. The two expressions give the same numerical answer for all possible values of x. (Split the class in half and ask each half to calculate one of the calculations with $x = 7$. As soon as a student has the result they should raise their hand.)
Why is the second calculation easier and quicker? (56, the brackets mean fewer and easier calculations.)

Introductory activity

Show different quadratic expressions and ask what they all have in common.
$2x^2 + 3x$, $4x^2 + 4x - 1$, $x^2 + 12$, $-7x^2$ are all quadratics.

Demonstrate how some quadratics can be factorised (put back into brackets).

Look at
$(x + 1)(x + 7) = x^2 + x + 7x + 7 = x^2 + 8x + 7$.

Where did the x^2 come from? The $x \times x$ at the beginning of each bracket.
Where did the 7 come from? The 1×7 at the ends of the brackets.
Where did the 8x come from? The $1x$ and $7x$ made from multiplying out the brackets.

Factorise $x^2 + 3x + 2$ $(x + 1)(x + 2)$
Factorise $x^2 + 2x - 8$ $(x + 4)(x - 2)$
Factorise $3x^2 + 5x$ $x(3x + 5)$. Notice that this has a common factor of 'x', so the answer does not need 2 brackets.

Exercise commentary and misconceptions

Try questions 1–3 before factorising expressions that have a common factor. There are many methods of factorising quadratics with 2 brackets and many students find it difficult. Using a method learnt by rote may not be well recalled. Reinforce where each term in the expanded expression comes from.

Plenary

A quadratic like $x^2 + 6x + 8$ factorises into 2 brackets $(x + 2)(x + 4)$.
What might a cubic like $x^3 + 4x^2 + 5x + 2$ factorise into? (3 brackets.)

Expand $(x + 1)(x + 1)(x + 2)$ and see if this is the factorised form of the cubic expression.

$(x^2 + 2x + 1)(x + 2) = x^3 + 2x^2 + 2x^2 + 4x + x + 2$
$= x^3 + 4x^2 + 5x + 2$ (it works!)

A1.5 Factorising double brackets

This spread will show you how to:

- Factorise into single and double brackets

Keywords
Factor
Factorise
Quadratic

When you expand double brackets, you often get a **quadratic** expression with three terms. Therefore you can factorise a quadratic expression into double brackets.

> A quadratic expression contains a squared term, such as x^2.

$$(x+2)(x+3) = x^2 + 3x + 2x + 6$$
$$= x^2 + 5x + 6$$
$$2 + 3 = 5 \qquad 2 \times 3 = 6$$

- **The two numbers in the brackets multiply to give the number at the end and add to give the number of xs.**

You can use this pattern to help you factorise a quadratic expression. You can check your answer by expanding the brackets.

Example

a Factorise $x^2 + 8x + 15$.
b Factorise $x^2 - 7x - 18$.

a Look for two numbers that multiply to give +15 and add to give +8.
These are +3 and +5.
$x^2 + 8x + 15 = (x + 3)(x + 5)$

> Consider the factor pairs of 15
> 1 and 15
> 3 and 5

b Look for two numbers that multiply to give −18 and add to give −7.

> It can help to write out all the factor pairs.
> Consider the factor pairs of −18:
> −1 and 18
> −18 and 1
> −3 and 6
> −6 and 3
> −2 and 9
> −9 and 2

The two numbers are −9 and +2.
$x^2 - 7x - 18 = (x - 9)(x + 2)$

> You can check by expanding.

Example

Factorise $y^2 - 3y - 28$.

$y^2 - 3y - 28 = (y - 7)(y + 4)$

> Two numbers that multiply to give −28 and add to give −3.
> $-7 \times 4 = -28$
> $-7 + 4 = -3$

Exercise A1.5

1 Factorise each of these into double brackets.

a $x^2 + 6x + 8$	**b** $x^2 + 10x + 21$	**c** $x^2 + 11x + 28$
d $x^2 + 11x + 24$	**e** $x^2 - 8x + 12$	**f** $x^2 - 9x + 18$
g $x^2 - 13x + 36$	**h** $x^2 + x - 12$	**i** $x^2 - 2x - 35$
j $x^2 + 6x - 27$	**k** $x^2 - 14x + 32$	**l** $x^2 + 18x - 40$

2 **a** Write an expression for the perimeter of this triangle.

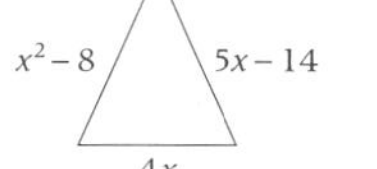

b Factorise your expression.

3 Factorise each of these expressions.

a $x^2 + 6x - 72$	**b** $x^2 - 10x - 24$	**c** $x^2 - 20x + 75$
d $x^2 + 12x - 64$	**e** $x^2 - 64$	**f** $x^2 - 29x + 100$

4 Decide if each of these are single or double bracket factorisations. Factorise each fully.

a $x^2 + 21x + 38$	**b** $5x^2 + 5x + xy$	**c** $x^2 + 22x + 121$
d $x^2 + 7x - 18$	**e** $33 + p^2 + 14p$	**f** $2x^2 + 3xy$

5 Given that the area of this rectangle is 12 cm^2, show that $(x + 1)(x + 8) = 0$.

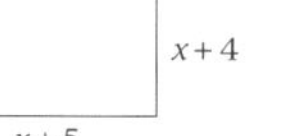

6 Use common factors and double brackets to factorise these expressions fully.

a $2x^2 + 22x + 56$
b $x^3 - 5x^2 - 24x$
c $x^3 - 16x$

> Example:
> $x^3 + 5x^2 + 6x$
> $= x(x^2 + 5x + 6)$
> $= x(x + 2)(x + 3)$

7 Factorise $2.3^2 + 2 \times 2.3 \times 1.7 + 1.7^2$ and use this to show that the calculation results in 16.

A1 Exam review

Key objectives

- Expand the product of two linear expressions
- Manipulate algebraic expressions by collecting like terms, multiplying a single term over a bracket, taking out common factors and factorising quadratic expressions

Worked solution	Commentary and misconceptions
1 a $4x + 2x^2 = 2x(2 + x)$	Students may struggle with the factors of x^2. Remind students that $x^2 = x \times x$. Ensure that students take out all the common factors.
b $(x + 5)(x - 2) = x^2 - 2x + 5x - 10$ $= x^2 + 3x - 10$	Encourage students to expand brackets methodically, ensuring that they don't miss out any terms.
c $3ab^2 \times 4b^3a = 3 \times 4 \times b^2 \times b^3 \times a \times a$ $= 12a^2b^5$	Encourage students to rearrange the expression: numbers first, then letters. Students may find it useful to arrange the expression in alphabetical order.
2 a $P = x + x + 4 + x + x + 4$ $= x + x + x + x + 4 + 4$ $= 4x + 8$ $= 4(x + 2)$.	Encourage students to show all their working. Ensure they give an expression for the perimeter, not the area. Ensure they use the correct rules of arithmetic when manipulating algebraic expressions. Ensure students derive the correct expression for perimeter, including all 4 sides. Encourage them to show the steps of their working, giving their solution in its simplest form. Ensure they collect like terms correctly; common incorrect answers would be $2x + 4$, $x^4 + 8$.
b $P = 54$. $4(x + 2) = 54$ $(x + 2) = \frac{54}{4}$ $(x + 2) = 13.5$ $x = 13.5 - 2 = 11.5$ cm. Length $= x + 4 = 15.5$ cm.	Ensure students substitute $P = 54$ into the expression and rearrange correctly to find x. Ensure students use their value of x to find $x = 4$, the length, in cm.

N2

Decimal calculations

Objectives

F/H Estimate answers to problems involving decimals

F/H Round to a given number of significant figures

F/H Understand where to position the decimal point by considering what happens if they multiply equivalent fractions

F/H Develop a range of strategies for mental calculation

F/H Use standard column procedures for addition, subtraction, multiplication and division of integers and decimals

Unit overview

This unit consolidates Key Stage 3 knowledge of calculating with integers and extends this to calculations involving decimals. Students are encouraged to estimate solutions to calculations and round numbers accurately. Mental and written methods are covered, preparing the students for more complex calculations involving decimals, fractions and percentages in further number units.

Prior knowledge

Before your students start this unit they should be able to:

- Understand and use decimal notation and place value
- Use a range of mental and written methods to add, subtract, multiply and divide with integers

Differentiation

- **Foundation** focuses on whole number calculations
- **Foundation Plus** extends to working with decimals and rounding to 1 dp
- **Higher Plus** extends the Higher book to consider upper and lower bounds of calculations and to calculating with fractions and percentages

N2.1 Approximation and rounding

Objectives

- Round numbers to a given power of ten or number of decimal places
- Round numbers to any number of significant figures

Useful resources

- Place value table

Mental starter

Two athletes wish to be selected to represent their school in athletics at the 100 m race. The first has recorded 3 times of 12, 12.5 and 11.5 seconds. The second has recorded times of 10.5, 12 and 14 seconds. **Which athlete should be chosen to represent the school?** (Means: first athlete 12 and second athlete 12.1666; Best times: first athlete 11.5 and second athlete 10.5. First is more consistent, second is capable of a faster time.)

Introductory activity

Discuss how numbers are rounded when a very precise number is not appropriate.

Discuss this scenario:
A willow tree is 11.4 m high. The birch tree growing next to it is $\frac{4}{7}$ of its height. If you use a calculator to find the height of the birch tree, the display is 6.514285714 ... m. or 6 m, 51 cm, 4 mm, then fractions of a millimetre.

Do you think the height of a tree would be measured to a fraction of a millimetre?
If you use 3 sf what would the height be?
(6.51 m)
What would it be to 3 dp? (6.514 m)
Which do you think is better in this case?

Try a few examples with significant figures:

Round 236.8379 to 2 sf (240)
Round 0.000 040 978 to 3 sf (0.000 041 0)
Remind students which is the first significant number and also where zeros are required.

Now look at the examples in the student book, which offer a mixture of decimal places, significant figures and powers of 10.

Significant figures are useful when you need an approximate answer to a calculation. Rounding numbers to 1 sf **before** the calculation is the general rule. When working out an estimate students **must** write down the numbers they are using in the calculation.

49 books at £19.95 each is approximately
50 × £20 = £1000.

(Discuss appropriate accuracy: In exams students are often asked to give answers to a sensible degree of accuracy. Answers with the same number of significant figures or the same number of decimal places as the question are acceptable. Answers with one less significant figure or decimal place are also acceptable. However the answer should not be more accurate than the question.)

Exercise commentary and misconceptions

Remind students that with dp you begin counting after the decimal point and with sf you begin counting from the first non-zero number. Students might like to put a circle around the first significant figure to identify it.
Many students are unsure when zeros must be added to an answer.
It sometimes helps to read the number out so that the place value is apparent.

Plenary

Round each of these numbers to 4 significant figures.

a 341 587 (341 600)
b 43 562.457 (43 560)
c 590.6853 (590.7)
d 0.004 687 56 (0.004 688)
e 0.00500978 (0.005010)

N2.1 Approximation and rounding

This spread will show you how to:

- Round numbers to a given power of ten or number of decimal places
- Round numbers to any number of significant figures

Keywords
Decimal
Power
Round
Significant

Numbers are **rounded** when it is not appropriate to give an answer that is too precise.

- **Numbers can be rounded:**

to decimal places	4.16 = 4.2 to 1 dp, and 5.663 = 5.66 to 2 dp
to the nearest unit, 10, 100, 1000	32 559 = 33 000 to the nearest thousand

Always check the digit after the one you're rounding to: if it is a 5 or more, round up your final digit.

- When rounding to **significant figures**, count from the first non-zero digit

to 2 sf: 712.4 = 710 and 0.00405 = 0.0041.

to 3 sf: 6.339 = 6.34 and 0.000 000 338 754 = 0.000 000 339.

dp and sf are abbreviations for 'decimal places' and 'significant figures'.

Example

a Round these numbers to 2 dp.
i 34.567 **ii** 3.887 126 **iii** 215.587 54

b Round 323 754.885 to the nearest:
i unit **ii** 10 **iii** 100 **iv** 1000 **v** 10 000

c Round these numbers to 2 sf.
i 39.54 **ii** 217 **iii** 0.000 455 **iv** 12 019 **v** 25.505

a i 34.57 **ii** 3.89 **iii** 215.59
b i 323 755 **ii** 323 750 **iii** 323 800 **iv** 324 000 **v** 320 000
c i 40 **ii** 220 **iii** 0.000 46 **iv** 12 000 **v** 26

Unit, 10, 100, ... are **powers** of ten.

You should always round the original value.

Example

Round 3.447 to **a** 2 dp **b** 1 dp.

a 3.447 = 3.45 to 2 dp **b** 3.447 = 3.4 to 1 dp

Do not round the 2 dp to get the 1 dp answer: use the original value 3.447.

Example

Find approximate answers to:
a 12.3 − 8.9 **b** 76.5 + 184.2 **c** 20 − 14.53

a 12.3 − 8.9 ≈ 12 − 9 = 3
b 76.5 + 184.2 ≈ 80 + 200 = 280
c 20 − 14.53 ≈ 20 − 15 = 5

Rounding to 1 sf is a useful way of finding a quick approximate answer to a calculation.

Exercise N2.1

1 Round these numbers to the nearest 10.
a 28 **b** 32 **c** 50 **d** 209 **e** 776 **f** 23 775

2 Round these decimal numbers to the nearest whole number.
a 5.8 **b** 4.4 **c** 21.67 **d** 39.175 **e** 18.405 **f** 453.66

3 Round these numbers to the nearest 100.
a 205 **b** 173 **c** 52 **d** 734 **e** 1389 **f** 134 545

4 Round these numbers to the nearest 1000.
a 2239 **b** 12 563 **c** 7500 **d** 11 452 **e** 78 466 **f** 155 669

5 Round these numbers to one decimal place (nearest tenth).
a 0.31 **b** 0.73 **c** 0.25 **d** 0.205 **e** 4.55 **f** 105.449

6 Round these whole numbers to two decimal places (nearest hundredth).
a 0.317 **b** 0.455 **c** 15.304 **d** 104.675 **e** 16.445 **f** 0.0036

7 Round these whole numbers to two significant figures.
a 483 **b** 1206 **c** 488 **d** 13 562 **e** 533 **f** 14 511

8 Round these numbers to two significant figures.
a 0.355 **b** 0.421 **c** 0.0566 **d** 0.004 673 **e** 1.357 **f** 0.000 004 152

9 Round these numbers to one significant figure.
a 157 **b** 2488 **c** 4.66 **d** 13.77 **e** 0.000 453 **f** 121 450

10 Use a calculator to work these out.
Write your answers correct to two significant figures.
a 8 ÷ 13 **b** 4 ÷ 7 **c** 5 ÷ 9 **d** 24 × 16 **e** 7.8 × 71 **f** 2093 × 3493

11 By rounding all of the numbers to one significant figure, write a calculation that you could carry out mentally to estimate the answers to these calculations.
a 355 ÷ 21 **b** 39 × 43 **c** 1053 ÷ 92 **d** 4385 + 11 655
e 108 + (2360 ÷ 52)

12 Use mental calculations to work out the value of each of the estimates that you wrote for question **11**.

13 Use a calculator to work out an exact answer for each of the calculations from question **11**.
For each one, write a sentence to say how well the calculator result agrees with the estimated answer that you wrote in question **12**.

N2.2 Mental methods for adding and subtracting decimals

Objectives

- Understand where to place the decimal point in calculations
- Use mental methods to calculate with decimal numbers

Useful resources

- Mini-whiteboards for mental starter

Mental starter

Without using a calculator the numbers 1 to 10 can be added in a way which makes the calculation easier.

$1 + 2 + 3 + 4 + 5 + 6 + 7 + 8 + 9 + 10$

$1 + 9 = 10$, $2 + 8 = 10$, $3 + 7 = 10$, $4 + 6 + = 10$ so there are 4 groups of 10 plus the 10 at the end plus the 5 in the middle.

$4 \times 10 + 10 + 5 = 55$.

Use the same method to add together all the numbers from 1 to 20 (**210**).
Try from 1 to 100 (**5050**).

Introductory activity

Describe a method for working out $2 - 0.34$.

First estimate: $2 - 0.5 = 1.5$

Now multiply **both numbers** by equal powers of 10 to remove decimals. This gives $200 - 34 = 166$.

Then divide by the same powers of 10, so the answer is 1.66. The answer should be checked with the estimate.

Work through the examples in the student book, using terms such as **compensation** where appropriate.

Emphasise the importance of starting with an estimate.

Exercise commentary and misconceptions

Encourage students to experiment with their own methods. Emphasise that there is no single way of doing these calculations, and students should choose and use a method that they feel comfortable with.

Students may need reminding about estimates, and then checking their answer against their estimate.

Plenary

Ask: What is wrong with this way of answering a question?

$7.9 - 2.3 - 3.1 - 1.3$

$\rightarrow 7.9 - 2.3 - 1.8$

$\rightarrow 7.9 - 0.5 = 7.4$

N2.2 Mental methods for adding and subtracting decimals

This spread will show you how to:

- Understand where to place the decimal point in calculations
- Use mental methods to calculate with decimal numbers

Keywords
Compensation method
Estimate

You already know several ways of adding and subtracting whole numbers, such as 36 + 59 or 326 – 138, in your head.
You can use the same techniques with decimal numbers.

- **Start by working out an estimate in your head so that you can check that your final answer is reasonable.**

15.1 – 3.8 ≈ 15 – 4
= 11

Then calculate the exact answer using the compensation method.

15 – 4 = 11
Add 0.1 (for 15.1) = 11.1
Add 0.2 (for 3.8) = 11.3
Answer is 11.3

Or treat the numbers as whole numbers, and adjust the place value afterwards. For 15.1 – 3.8, you could work out 151 – 38 = 113, and adjust this to 11.3.

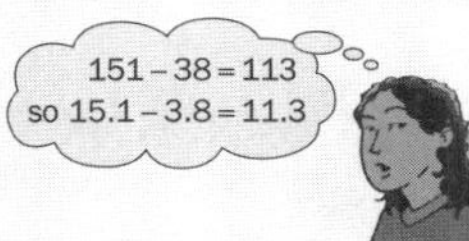

Example

Work these out in your head.

a 4.7 + 9.8 **b** 16.9 – 7.3 **c** 108.4 – 37.1 **d** 6.07 + 5.3 **e** 13.9 – 6.75

a Initial estimate: 5 + 10 = 15.
One method is: 4.7 + 10 = 14.7, and then 14.7 – 0.2 = 14.5.
4.7 + 9.8 = 14.5

b Initial estimate: 17 – 7 = 10.
One method is: 169 – 73 = 169 – 69 – 4 = 96.
16.9 – 7.3 = 9.6

c Initial estimate: 100 – 40 = 60.
One method is to adjust the initial approximation:
60 + 8.4 + 2.9 = 71.3.
108.4 – 37.1 = 71.3

d Initial estimate: 6 + 5 = 11.
Now add the figures after the decimal point (0.07 and 0.3).
6.07 + 5.3 = 11.37

e Initial estimate: 14 – 7 = 7.
The difference between 6.75 and 14 is 7.25, so the difference between 6.75 and 13.9 is 0.1 less.
13.9 – 6.75 = 7.15

There is more than one way to solve each of the these problems.

Exercise N2.2

1 Write the answers to these additions.
a 0.1 + 0.7 **b** 0.2 + 0.3 **c** 0.3 + 0.1
d 0.7 + 0.4 **e** 0.3 + 0.1 **f** 0.5 + 0.5

2 Use your answers to question **1** to write the answers to these additions.
a 5.1 + 0.7 **b** 6.2 + 0.3 **c** 7.3 + 0.1
d 3.7 + 0.4 **e** 11.3 + 0.1 **f** 9.5 + 0.5

3 Work out these in your head and write the answers.
a 4.7 + 5.3 **b** 4.7 + 5.4 **c** 3.6 + 6.7
d 6.8 + 4.3 **e** 7.5 + 8.9 **f** 2.7 + 4.8

4 Work out these in your head and write the answers.
a 3.55 + 4.22 **b** 2.13 + 3.12 **c** 3.18 + 0.42
d 3.72 + 0.18 **e** 1.42 + 0.71 **f** 8.39 + 4.65

5 Work out these in your head and write the answers.
a 3.35 + 0.8 **b** 0.15 + 6.7 **c** 0.7 + 3.88
d 6.92 + 3.5 **e** 5.34 + 6.8 **f** 1.44 + 0.6

6 Write the answers to these subtractions.
a 1 – 0.3 **b** 1 – 0.4 **c** 1 – 0.7
d 4 – 0.6 **e** 11 – 0.8 **f** 15 – 0.9

7 Use your answers to question **6** to write the answers to these subtractions.
a 1.1 – 0.3 **b** 1.1 – 0.4 **c** 1.5 – 0.7
d 4.7 – 0.6 **e** 11.5 – 0.8 **f** 15.3 – 0.9

8 Work out these in your head and write the answers.
a 2.16 – 1.42 **b** 1.51 – 0.46 **c** 6.39 – 4.88
d 15.46 – 8.32 **e** 5.17 – 4.09 **f** 4.29 – 3.65

9 Write the answers to these subtractions.
a 2 – 0.38 **b** 4 – 0.49 **c** 1 – 0.55
d 5 – 0.63 **e** 15 – 0.48 **f** 12 – 0.34

10 Use your answers to question **9** to write the answers to these subtractions.
a 2.1 – 0.38 **b** 4.1 – 0.49 **c** 1.2 – 0.55
d 5.3 – 0.63 **e** 15.5 – 0.48 **f** 12.7 – 0.34

N2.3 Written methods for adding and subtracting decimals

Objectives

- Understand where to place the decimal point in calculations
- Use mental and written methods to calculate with decimal numbers

Useful resources

- Mini-whiteboards

Mental starter

Ask half the class to calculate
$40 + 20 + 10 + 5 + 2.5 + 1.25 + 0.625 + \ldots$ as far as they can in the time available.
Ask the other half to work out $40 \div \frac{1}{2}$.
(Answers 77.5, 78.75, 79.375, 79.6875, 79.84375, 79.921875 or 80.)
What is the answer to the first question getting closer to? Why do you think the answer can be worked out by calculating $40 \div \frac{1}{2}$? **(Begins with 40 and halves each time may be a common answer.)**

What would
$512 + 256 + 128 + 64 + 32 + 16 + 8 + 4 + 2 + 1 + 0.5 + 0.25 + 0.125 + 0.0625 + \ldots$ equal?
($512 \div \frac{1}{2} = 1024$)

Introductory activity

The lesson focuses on written decimal addition and subtraction. Give some examples, outlining the key procedure:

1 Estimate answer
2 Line up decimal points
3 Fill in gaps with zeros
4 Borrow if necessary as you subtract.

You could try these examples:
$34.7 - 7.93 + 1.289 \approx 30 - 8 + 1$
$= 23$
$34.7 - 7.93 + 1.289 = 34.700 - 7.930 + 1.289$
$= 26.770 + 1.289 = 28.059$

Now work through the examples in the student book, emphasising the importance of an estimate.

Exercise commentary and misconceptions

The questions in this exercise focus on using a written method to add and subtract decimals. Some of the questions can be done using a mental method. Encourage students to lay out their work with plenty of space between the digits, and to align the digits accurately. They will be less likely to make mistakes, especially with carrying numbers. Students who have major problems with alignment could use square grid paper initially, but this should be a temporary measure.

Plenary

How far can you keep halving 100 and be sure you have not made an error? Give students a fixed amount of time, then ask them to give their smallest answer that they think is correct. (50, 25, 12.5, 6.25, 3.125, 1.5625, 0.78125, 0.390625, 0.1953125, 0.09765625...)

N2.3 Written methods for adding and subtracting decimals

This spread will show you how to:

- Understand where to place the decimal point in calculations
- Use mental and written methods to calculate with decimal numbers

Keywords
Column
Decomposition
Digit

When adding or subtracting decimal numbers with more than one decimal place, use a written method to ensure accuracy.

Example

Calculate these using a written method.

a 102.773 + 28.47 b 26.44 − 1.105

a Initial estimate: 100 + 30 = 130

```
  1 0 2 . 7 7 3
+   2 8 . 4 7 0
  1 3 1 . 2 4 3
```

b Initial estimate: 30 − 1 = 29

```
  2 6 . 4 ³4̸¹⁰0
−   1 . 1 0 5
  2 5 . 3 3 5
```

You should still use an estimate to help check your answer.

Make sure that you line up the numbers correctly; put the decimal points above each other.

When the numbers in the calculation have different numbers of decimal digits, it can be useful to add extra zeros at the end of the number with fewer digits (shown above in red).

The 'carrying' and **decomposition** of numbers works in the same way as with whole numbers.

- **When you use a written method for adding or subtracting decimals, you should estimate first.**

Example

Here are some incorrect attempts at adding and subtracting with decimals. Try to decide what went wrong in each case.

a 10.03 − 2.55 b 38.53 + 2.474 c 100.773 − 28.782

a Initial estimate: 10 − 3 = 7

```
  1 0 . ⁹0̸¹3
−   2 . 5 5
    8 . 4 8  ✗
```

The top number has not been split up (decomposed) properly. Note that this calculation is actually quite easy to do mentally.

The correct answer is 7.48.

b Initial estimate: 39 + 2 = 41

```
38.53
2.474
6.327  ✗
 1 1
```

In this example the decimal points, and therefore all of the columns, were not properly aligned.

The correct answer is 41.004.

c Initial estimate: 100 − 30 = 70

```
 100.773
− 28.782
 128.011  ✗
```

Here the smallest digit in each column has been subtracted from the bigger one. This seems silly, but it's easy to do when in a hurry!

The correct answer is 71.991

Exercise N2.3

1 Calculate these using a mental method.
a 4.3 + 8.1 b 6.4 + 5.6 c 9.2 + 3.9
d 12.7 + 9.8 e 14.3 + 8.8 f 16.2 + 9.9

2 Work out each of the calculations from question **1**, using a standard written method.
If you do not get the same answers by both methods, check to find your mistake.

3 Some of the calculations below are easy to do mentally.
Others are best performed using a written method.
Find the answer to each, showing your method clearly.
a 5.9 + 7.1 b 0.673 + 1.198 c 94.834 + 106.487
d 16.7 + 28.3 e 36.87 + 2.1 f 17.71 + 3.98

4 Work out these calculations using a standard written method.
Remember to write down an estimate first.
a 31.45 + 108.88 b 182.7 + 59.6 c 81.927 + 16.88
d 104.7 + 98.89 e 57.784 + 103.218 f 61.386 + 40.614

5 Use a calculator to check your answers to question **4**.

6 Work out these using a mental method.
a 8.4 − 6.2 b 9.7 − 0.6 c 17.9 − 2.9
d 7.2 − 2.3 e 15.3 − 6.9 f 17.8 − 14.9

7 Work out the calculations from question **6**, using a standard written method.
Check that you get the same answers by both methods. If not, find and correct your mistake.

8 Work out these calculations using an appropriate method (written or mental). Show your method clearly.
a 5.8 − 3.2 b 16.73 − 8.87 c 9.6 − 3.7
d 109.54 − 17 e 2.37 − 1.4 f 26.25 − 1.98

9 Work out these calculations using a standard written method.
Remember to write an estimate before you do the calculation.
a 21.864 − 7.968 b 104.87 − 85.42 c 417.48 − 57.69
d 24.503 − 16.82 e 19.21 − 18.884 f 102.01 − 90.59

10 Use a calculator to check your answers to question **9**.

11 Work out these calculations using an appropriate method, and show your working clearly.
a 8.6 − 4.5 b 26.4 + 13.8 c 18 − 6.712
d 15.808 − 9.84 e 4.008 − 3.116 f 4.109 − 3.64

N2.4 Mental methods for multiplying and dividing decimals

Objectives

- Multiply and divide decimal numbers
- Understand where to place the decimal point in calculations

Useful resources

- Place value table

Mental starter

Chris has thought of his own ways for remembering difficult times tables. What do you think of his methods? Would you use either of them?

1 $8 \times 7 = ?$ what two numbers come before 8 and 7? Answer 5 and 6. So $8 \times 7 = 56$

2 $8 \times 8 = ?$ eight is an even number, what two even numbers come before eight? 6 and 4. So $8 \times 8 = 64$.

Introductory activity

Given that $12 \times 34 = 408$, then the answers to 1.2×34 or 0.12×0.34 or 0.012×34 can be written immediately with no further calculation.

$1.2 \times 34 \approx 1 \times 34 = 34$

so $1.2 \times 34 = 40.8$ using the same digits and placing the decimal point to give a number close to 34.

$0.12 \times 0.34 \approx 0.1 \times 0.3 = 0.03$ so

$0.12 \times 0.34 = 0.0408$

Show how inverse calculations can also be used.

Given that $0.012 \times 34 = 0.408$

$408 \div 34 = 12$

$40.8 \div 0.34 \approx 40 \div 0.4 = 100$ so

$40.8 \div 0.34 = 120$

Find the value of $0.408 \div 0.0034$ (120)

Given that $238 \times 17 = 4046$

Find the value of $2.38 \times 17\,0000$ (40 460)

Find the value of $404.6 \div 170$ (2.38)

Exercise commentary and misconceptions

Each question in this exercise uses one number fact as the basis for a range of related facts.

Apart from question 6, which asks students to check their answers with a calculator, calculators should not be used.

To tackle the second, third and subsequent parts of a question, make sure that students go back to the original example rather than using an answer they have just found.

Plenary

Give students this fact:

$$\frac{470.89}{21.7} = 21.7$$

What is 217^2? (47 089)

N2.4 Mental methods for multiplying and dividing decimals

This spread will show you how to:

- Multiply and divide decimal numbers
- Understand where to place the decimal point in calculations

Keywords
Estimate
Place value

You already know some methods for mental multiplication and division of whole numbers. For example,

$7 \times 14 = 7 \times (10 + 4) = 7 \times 10 + 7 \times 4 = 70 + 28 = 98$.

You can extend these methods to work with decimal numbers.

- **One basic number fact can be extended to many different decimal calculations. For example, $4 \times 3 = 12$ leads to $4 \times 0.3 = 1.2$, $0.4 \times 3 = 1.2$, $0.3 \times 0.4 = 0.12$, ...**

Start with a basic calculation using whole numbers, and then to adjust the **place value**.

Example

Calculate mentally. **a** 7×19 **b** 0.7×19 **c** 0.007×190

a $7 \times 20 = 7 \times 2 \times 10 = 14 \times 10 = 140$. So, $7 \times 19 = 140 - 7 = 133$.

b The original 7 (in 7×9) has changed to 0.7, so $0.7 \times 19 = 13.3$.
- A rough estimate is another way to get the correct answer. $0.7 \times 19 \approx 1 \times 20$. The answer must be 13.3, since the alternatives (1.33 or 133) are much further from 20.

c Compare this to part **a**. 0.007, not 7, so answer is 1000 times smaller. 190, not 19, so answer is 10 times bigger. So, overall, answer is 100 times smaller. The answer is 1.33.
- An estimate confirms this: $0.007 \times 190 \approx 0.007 \times 200 = 0.007 \times 100 \times 2 = 0.7 \times 2 = 1.4$

Often you know what the digits in the answers are, but you need to decide on the place value. A rough estimate is usually good enough to decide.

Example

Given that $238 \times 17 = 4046$, find the value of
a $2.38 \times 17\,000$ **b** $404.6 \div 170$

a $2.38 \times 17\,000 \approx 2 \times 20\,000$, so the answer must be 40 460.

b The original calculation gives $4046 \div 17 = 238$. $404.6 \div 170 \approx 400 \div 200 = 2$. The answer must be 2.38.

Example

Davinder buys 15 bottles of cola for a party. Each bottle costs 79p. Work out mentally the total cost.

Estimate: $15 \times 80 = 10 \times 80 + 5 \times 80$
$= 800 + 400 = 1200$

$15 \times 79 = 15 \times 80 - 15 \times 1$
$= 1200 - 15 = 1185$

The answer is £11.85.

Exercise N2.4

1 Write the answer to the calculation 4×12.
Use this answer to work out
a 0.4×12 **b** 4×1.2 **c** 0.4×1.2 **d** 40×1.2 **e** 400×0.12

2 Use a mental method to work out 7×13. Then use your answer to work out
a 0.7×13 **b** 7×1.3 **c** 0.7×1.3 **d** 70×1.3 **e** 700×0.13

3 Write the answers to
a 14×3 **b** $64 \div 4$ **c** $35 \div 5$ **d** $208 \div 2$ **e** $48 \div 24$ **f** 4×7

4 Use your answers from question **3** to work out
a 1.4×3 **b** $6.4 \div 4$ **c** $3.5 \div 5$ **d** $20.8 \div 2$ **e** $4.8 \div 2.4$ **f** 0.4×0.7

5 Any multiplication (such as $2 \times 3 = 6$) is part of a larger family of related facts: the equivalent multiplication $3 \times 2 = 6$, and the divisions $6 \div 2 = 3$ and $6 \div 3 = 2$. Write three other related facts for each of these multiplications.
a $7 \times 9 = 63$ **b** $6 \times 8 = 48$ **c** $13 \times 7 = 91$ **d** $15 \times 18 = 270$
e $5 \times 3.5 = 17.5$ **f** $2.4 \times 3.9 = 9.36$

6 You can use your knowledge of place value to extend families of multiplication facts further. For example, starting with $8 \times 7 = 56$, you can write $7 \times 80 = 560$, $56 \div 7 = 8$, $5.6 \div 7 = 0.8$, $5.6 \div 8 = 0.7$, $5.6 \div 80 = 0.07$, and so on.
Use these ideas to write some number facts related to these calculations, and check your answers with a calculator.
a $4 \times 5 = 20$ **b** $7 \times 9 = 63$ **c** $32 \div 8 = 4$ **d** $24 \div 8 = 3$ **e** $81 \div 9 = 9$

7 Given that $5 \times 9 = 45$, calculate
a 0.5×9 **b** 50×90 **c** 0.5×90 **d** 0.5×0.9
e $45 \div 9$ **f** $450 \div 5$ **g** $450 \div 0.9$

8 Given that $38 \times 91 = 3458$, calculate
a 3.8×91 **b** 38×9.1 **c** $3458 \div 91$ **d** 0.38×910
e $345.8 \div 38$ **f** $0.3458 \div 38$

9 Given that $2.91 \times 350 = 1018.5$, write the value of
a 29.1×350 **b** 29.1×3.5 **c** 0.0291×3.5 **d** $101.85 \div 291$
e $10.185 \div 2.91$ **f** $1.0185 \div 350$

10 Given that $4.7 \times 6.3 = 29.61$, write the answer to
a 0.047×630 **b** 47×0.0063 **c** $2961 \div 0.47$

N2.5 Written methods for multiplying and dividing decimals

Objectives

- Use written methods to calculate with decimal numbers
- Understand where to place the decimal point in calculations

Useful resources

- Place value table

Mental starter

8 small cubes are packed into a box to form a large cube. The area of 1 face of the smaller cubes is 70 cm^2. I need to wrap the large cube completely in paper to send in the post. The wrapping paper costs 5p per 10 cm^2.

How much will it cost in pounds to wrap the large cube?

(Area of each face of large cube = 70 × 4 = 280.
Total surface area of large cube = 280 × 6 = 1680.
Cost = 5 × 168 = 840p = £8.40)

Introductory activity

When working with decimals you need to use methods you already know for multiplication and division.
The extra step when working with decimals is that the decimal point is put in after the calculation has been done. The estimate for the calculation helps you decide where to put the decimal point, in the same way as in previous lessons.

For example:
34.27 ÷ 2.3

1 Estimate 30 ÷ 2 = 15
2 Calculate 3427 ÷ 23 = 149 (Demonstrate using methods developed in the class.)
3 Answer to 34.27 ÷ 2.3 = 14.9
This is close to the estimated answer of 15.

Another example:
2.93 × 18.5

1 Estimate 3 × 20 = 60
2 Calculate 293 × 185 = 54 205 (Demonstrate using methods developed in the class.)
3 Answer to 2.93 × 18.5 = 54.205
This is close to the estimated answer of 60.

Exercise commentary and misconceptions

Questions 1–3 and 4–6 form two triplet questions. In the first two questions of each set, students are asked to perform mental calculation with non-decimal numbers, then check their answers with written methods. Finally, they use an estimate to position the decimal point.

Questions 7–14 are in pairs. In the first question the student needs to use the 3-step method of Estimate, Calculate and Answer as before, while the second question asks the student to check the answers using a calculator.

Plenary

Without finding the exact answers, arrange these calculations in descending order of size.

a 2 328 ÷ 97
b 0.02328 ÷ 0.097
c 232.8 ÷ 0.97
d 2.328 ÷ 0.097
e 232.8 ÷ 97
f 23.28 ÷ 0.0097
(**f**, **c**, **d**, **e**, **b**, **a**)

N2.5 Written methods for multiplying and dividing decimals

This spread will show you how to:
- Use written methods to calculate with decimal numbers
- Understand where to place the decimal point in calculations

Keywords
Place value

You already know written methods for multiplying and dividing whole numbers. You can use the same methods with decimal numbers.

- Always start with an estimate.
- Carry out the calculation.
- Use the estimate to position the decimal point in the answer.

Example

Work out 34.27 ÷ 2.3

Rhian uses repeated subtraction.

34.27 ÷ 2.3
– Estimate: 30 ÷ 2 = 15
– Work out 3427 ÷ 23

Use only whole numbers in the calculation

```
 3427
 2300   (100 × 23)
 1127   10 × 23 = 230
  920   20 × 23 = 460
        (40 × 23 = 920)
  207   (9 × 23) = 230 − 23
  207            = 207
    –
```

3427 ÷ 23 = 100 + 40 + 9 = 149
So, 34.27 ÷ 2.3 = 14.9

Tom uses long division.

34.27 ÷ 2.3
– Estimate: 30 ÷ 2 = 15
– Work out 3427 ÷ 23

```
     149
23)3427
   23
   112
    92
    207
    207
      –
```

3427 ÷ 23 = 149
So, 34.27 ÷ 2.3 = 14.9

Rhian and Tom both use a written method to calculate 3427 ÷ 23, and an estimate to position the decimal point in the answer, so that the digits have the correct place values.

Example

Solve 2.93 × 18.5

Rhian uses grid multiplication.

2.93 × 18.5
– Estimate: 3 × 20 = 60
– Work out 293 × 185

×	200	90	3	
100	20000	9000	300	29300
80	16000	7200	240	23440
5	1000	450	15	1465
				54205

293 × 185 = 54205
⇒ 2.93 × 18.5 = 54.205

Tom uses long multiplication.

2.93 × 18.5
– Estimate: 3 × 20 = 60
– Work out 293 × 185

```
   293
 × 185
  1465
 23440
 29300
 54205
```

293 × 185 = 54205
⇒ 2.93 × 18.5 = 54.205

Exercise N2.5

1 Solve these, using a mental method.
a 14 × 7 b 19 × 8 c 21 × 13 d 17 × 19
e 11 × 28

2 Now use a written method to work out the answers for question **1**. Check that you get the same answers with both methods.

3 Use your answers to questions **1** and **2** to write the answers to
a 1.4 × 7 b 19 × 0.8 c 2.1 × 1.3 d 1.7 × 0.019
e 1.1 × 0.28

4 Solve these, using a mental method.
a 320 ÷ 4 b 180 ÷ 15 c 276 ÷ 23 d 357 ÷ 17
e 440 ÷ 20

5 Now use a written method to work out the answers for question **4**. Check that you get the same answers with both methods.

6 Use your answers to questions **4** and **5** to write the answers to
a 3.2 ÷ 4 b 18 ÷ 15 c 2.76 ÷ 2.3 d 0.357 ÷ 1.7
e 0.44 ÷ 2

7 Use a written method to work out these.
a 4.7 × 5.3 b 1.53 × 2.8 c 21.6 × 4.9 d 33.65 × 3.89
e 21.58 × 1.99

8 Use a calculator to check your answers to question **7**.

9 Use a written method to calculate these.
a 34.83 ÷ 9 b 5.425 ÷ 7 c 7.328 ÷ 8 d 451.8 ÷ 60
e 54.39 ÷ 3

10 Use a calculator to check your answers to question **9**.

11 Use a written method to calculate these.
a 58.65 ÷ 17 b 66.4 ÷ 16 c 185.76 ÷ 24 d 7.752 ÷ 1.9
e 3.055 ÷ 1.3

12 Use a calculator to check your answers to question **11**.

13 Use a written method to work out these, giving your answers correct to two decimal places.
a 14.73 ÷ 2.8 b 51.99 ÷ 1.8 c 193.8 ÷ 0.14 d 1013 ÷ 5.77
e 23.78 ÷ 0.83

14 Use a calculator to check your answers to question **13**.

N2 Exam review

Key objectives

- Round to a given number of significant figures
- Estimate answers to problems involving decimals
- Develop a range of strategies for mental calculation
- Understand where to position the decimal point by considering what happens if they multiply equivalent fractions

Worked solution	Commentary and misconceptions
1	Encourage students to estimate answers before attempting calculations. Encourage students to align their written workings by place value to help eliminate errors. Ensure that students round correctly at the end of their calculations.

a $5.635 - 2.91 = 2.725 = 2.7$ (2sf)

b $9.5 + 10.56 = 20.06 = 20$ (2sf)

c $10 \times 4.92 = 49.2$ $0.01 \times 4.92 = 0.0492$

$10.01 \times 4.92 = 49.2 + 0.0492$

$= 49.2492 = 49$ (2sf)

d
1728
− 360
1368
− 1080
288
− 288

$1728 \div 36 = 10 + 30 + 8$

so $17.28 \div 3.6 = 4.8$

Worked solution	Commentary and misconceptions
2	Ensure that students use the information given to perform the calculations. Revision of multiplying and dividing by powers of 10 and place value may help.

a $9.7 = 97 \div 10$ $12.3 = 123 \div 10$

$9.7 \times 12.3 = 11\,931 \div 100 = 119.31$

b $0.97 = 97 \div 100$ $123\,000 = 123 \times 1000$

$0.97 \times 123\,000 = 11\,931 \times 10 = 119\,310$

c $11.931 = 11\,931 \div 1000$, $9.7 = 97 \div 10$

$11\,931 \div 97 = 123$

$11.931 \div 9.7 = 123 \div 100 = 1.23$

A2 Equations and inequalities

Objectives

F/H Use inverse operations

F/H Set up simple equations

F/H Solve simple equations by using inverse operations or by transforming both sides in the same way

H Solve linear equations in one unknown, with integer or fractional coefficients

H Solve linear inequalities in one variable, and represent the solution set on a number line

Unit overview

This unit consolidates Key Stage 3 knowledge of the rules of arithmetic and the rules of algebra introduced in unit A1. It then extends to setting up and solving linear equations using inverse operations, including equations with fractional coefficients and double sided equations. The final spread extends to solving simple inequalities in one variable. This unit prepares students for work on more complex equations and graphs in further algebra units which extend to quadratic equations.

Prior knowledge

Before your students start this unit they should be able to:

- Add, subtract, multiply and divide with integers, decimals and fractions
- Use the rules of algebra with the correct order of operations
- Expand single and double brackets
- Simplify algebraic expressions, including factorisation

Differentiation

- **Foundation** focuses on solving simple one-step linear equations with whole number coefficients and solutions
- **Foundation Plus** extends to solving two-step linear equations and simple linear inequalities
- **Higher Plus** focuses on equations involving fractions

A2.1 Working with inverse operations

Objectives
- Understand and use inverse operations
- Use function machines

Useful resources
- Blank function machines

Mental starter
Write these expressions on the board:

$x^2 + 1$

$5(\sqrt{x} - 2)$

$2(x + 1)^2$

$\frac{3x + 5}{4}$

For each expression, challenge students to find the value of x that makes the expression 50.

Introductory activity
Addition and subtraction are examples of mathematical operations. **Can you think of any others?** (multiplication, division, squaring, square rooting, cubing, cube rooting are the usual suggestions.) Doing things backwards is called the inverse. For example, subtraction is the **inverse** of addition, because if you add four to any number, you must subtract four to get back to the original number.
What do you think is the inverse of division? (multiplication) **And of taking the square root?** (squaring) **And of cubing?** (taking cube root).

A function takes a starting number and changes it into another number.

An example is in $\rightarrow \times 3 \rightarrow +2 \rightarrow$ out. **What answers do you get out if you put these values in? 14, −1, x**
(44, −1, $3x + 2$)

Using the same function, what numbers must have been put in if you get these answers out? 23, p

$\left(7, \frac{p-2}{3}\right)$.

Explain that you are just working the function backwards.

If you have a mathematical expression, you can work out the order in which the operations must be done.

For example, $4x + 9$ means x has been multiplied by 4 and then 9 has been added.
Suppose I added 9 to x and then multiplied by 4, what would the expression look like?
($(x + 9) \times 4$ or $4(x + 9)$)

The method of using inverse operations can be used to solve equations.
Solve the equation $\sqrt{x^2 + 75} = 10$ by inverse operations. **What has been done to x and in what order?**
$x \rightarrow$ squared $\rightarrow +75 \rightarrow$ square rooted $\rightarrow = 10$.

Working backwards:

Square, subtract 75, square root, so
$10 \rightarrow 100 \rightarrow 25 \rightarrow 5$, so $x = 5$.

Exercise commentary and misconceptions
In question 2 you could remind students of the method of multiplying by a decimal to increase by a percentage. The inverse is then to divide by the same decimal.

Question 3 might be useful as an oral exercise.

Most students will not think of the negative square root in question 4.

In question 6 you could remind students that dividing by a half is the same as multiplying by two.

Plenary
A function is in $\rightarrow$ cube it $\rightarrow +6 \rightarrow$ out.
Can you find a number you can put in so that you get the same answer out?
You could give the hint that it is an integer. (Only solution is −2.) Students should realise that as they try larger positive numbers their output gets further away in size from the input.

A2.1 Working with inverse operations

This spread will show you how to:
- Understand and use inverse operations
- Use function machines

Keywords
Inverse
Operation
Variable

- **Addition, subtraction, multiplication and division are all operations.**
- **An operation has an inverse, for example subtraction is the inverse of addition.**

An operation acts on a number or a variable.

The inverse operation changes the variable back to its original value

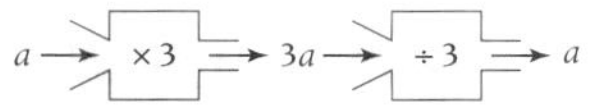

$a \rightarrow [\times 3] \rightarrow 3a \rightarrow [\div 3] \rightarrow a$

x and $2y^2$ are examples of variables.

If two or more operations act on a variable then you do their inverse operations in reverse order to get back to the original value

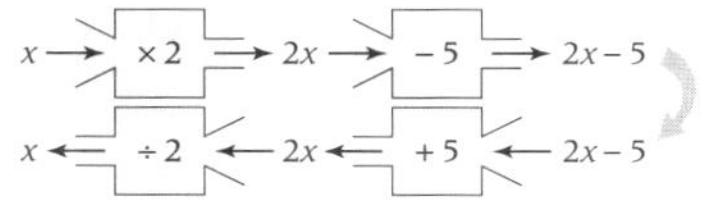

$x \rightarrow [\times 2] \rightarrow 2x \rightarrow [-5] \rightarrow 2x-5$

$x \leftarrow [\div 2] \leftarrow 2x \leftarrow [+5] \leftarrow 2x-5$

The operations are 'multiply by 2' and 'subtract 5'.

The inverse operations are 'add 5' and 'divide by 2'.

Example

Find the value of each letter.

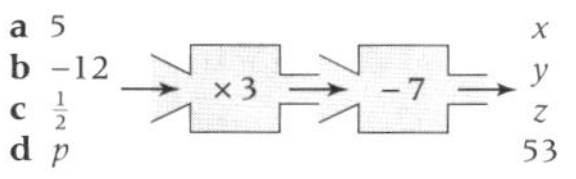

a 5, b −12, c $\frac{1}{2}$, d p $\rightarrow [\times 3] \rightarrow [-7] \rightarrow$ x, y, z, 53

a $5 \xrightarrow{\times 3} 15 \xrightarrow{-7} 8$ so $x = 8$

b $-12 \xrightarrow{\times 3} -36 \xrightarrow{-7} -43$ so $y = -43$

c $\frac{1}{2} \xrightarrow{\times 3} 1\frac{1}{2} \xrightarrow{-7} -5\frac{1}{2}$ so $z = -5\frac{1}{2}$

d $53 \xrightarrow{+7} 60 \xrightarrow{\div 3} 20$ so $p = 20$

To find p, do the inverse operations in the reverse order.

Example

Find the starting number in each case.

a I think of a number, subtract 8 and square root it. The answer is 5.
b I think of a number, multiply by 2, subtract 4 and cube it. The answer is −216.

a $5 \xrightarrow{\text{square}} 25 \xrightarrow{+8} 25 + 8 = 33$ The starting number is 33.

b $-216 \xrightarrow{\text{cube root}} -6 \xrightarrow{+4} -2 \xrightarrow{\div 2} -1$ The starting number is −1.

Undo the operations in the reverse order.

Exercise A2.1

1 Copy and complete these diagrams.

a 20, −8, x, ? $\rightarrow [+4] \rightarrow [+5] \rightarrow$?, ?, ?, 15

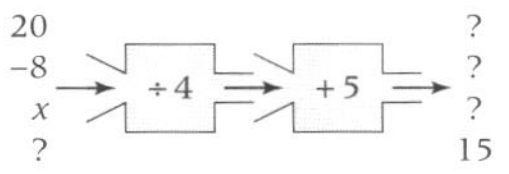

b 5, ?, ?, ? $\rightarrow [\text{Square}] \rightarrow [\times 1.5] \rightarrow$?, 1.5, 13.5, y

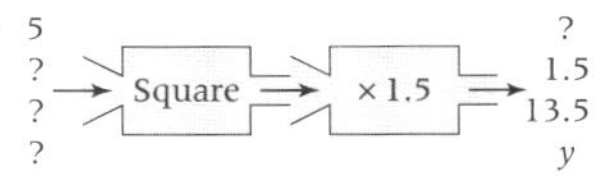

c 5, −7, $\frac{1}{2}$, ? $\rightarrow [?] \rightarrow [-2] \rightarrow$ 18, ?, ?, w

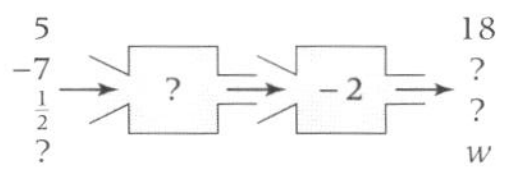

d 5, −2, ?, ? $\rightarrow [\text{Cube}] \rightarrow [\times 2] \rightarrow$?, ?, 128, 2000

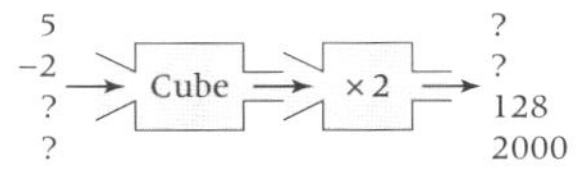

2 Explain why 'adding 10%' cannot be undone using the inverse operation 'subtracting 10%'.

3 In each case, use inverse operations to find the starting number.
a I think of a number, double it and subtract 4. This gives me 7.
b I think of a number, square it, multiply by 3 and get 75.
c I think of a number, add 11, cube root it and get 2.
d I think of a number, halve it, treble it then subtract 6. I get 54.
e I think of a number, multiply it by $\frac{1}{4}$, square root it and get 5.

4 Is there another starting number for question **3b**?
Explain your answer.

5 **a** Write, in order, the operations that have acted on x.

i $x + 3 = 17$ **ii** $2x - 1 = 17$ **iii** $\frac{x+5}{2} = 24$ **iv** $3x^2 = 48$

b Using the inverse operations, in reverse order, find the value of x in each case.

6 Look at this function machine.

Start	$\times 2$	-9	$\div 5$	$+2$	$\div \frac{1}{2}$	Finish

a What value do you end up with if (−3) enters the machine?
b What value did you start with if these values leave the machine?

i 6 **ii** 8.4 **iii** $\frac{1}{4}$

7 **a** Invent your own function machine that uses five operations to convert a value of 5 into 53.
b Use your machine to describe what value you started with if the output is 100.

A2.2 Solving one-sided equations

Objectives

- Set up and solve simple equations

Useful resources

- Mini-whiteboards

Mental starter

Catherine is 10 cm taller than Angela. The sum of both their heights is 3.34 m (or 334 cm).
How tall are they?
(Catherine 172 cm, Angela 162 cm.) Many students will begin by halving 3.34 m.
Encourage them to check their solutions are correct. A simple sketch of the problem may help to explain that you should add or subtract 10 first.

Introductory activity

Many students find the approach to solving equations that works best is 'doing the same thing to both sides' rather than 'moving things over to the other side' (also known as 'change side, change sign'). The method of inverse operations is very quick and simple, though only really useful for equations where the variable only appears once.

It is essential to get the order of operations in the equation correct, before trying the inverse. For example (taken from the student book):

$$\frac{3x-5}{2} = 8.$$

The order is multiply x by 3, subtract 5, divide by 2.

What would the inverse be? (multiply by 2, add 5, divide by 3.)

So $8 \rightarrow 16 \rightarrow 21 \rightarrow 7$ and $x = 7$

Here is another example:
$4(y + 3) = 40$.
What is the order of operations on y? (add 3, multiply by 4.) **What would the inverse be?** (divide by 4, subtract 3.)

So $40 \rightarrow 10 \rightarrow 7$ and $y = 7$
Look at the last example, which involves forming an equation before solving it.

Exercise commentary and misconceptions

In Question 3 remind students that they should simplify the expression for the perimeter before equating it to 30 and then solving the equation to find x.
Many students will forget that the question requires them to find the lengths of the sides, not just x.

Plenary

Individual students can make up 'I think of a number' questions for the whole class. This becomes more of a feat of memory than a mathematical exercise unless you limit the number of operations!

A2.2 Solving one-sided equations

This spread will show you how to:
- Set up and solve simple equations

Keywords
Equation
Operation
Solve

- **An equation is a statement with an equals sign. For example 2x – 4 = 18 is an equation.**
- **To solve an equation, do the same operation to both sides. For example**

2x – 4 = 18	Add 4 to both sides
2x = 22	Divide both sides by 2
x = 11	

This equation is only true when $x = 11$.

Example

Solve these equations.

a $\frac{3x-5}{2} = 8$ **b** $4(y+3) = 40$ **c** $17x = 35$

a $\frac{3x-5}{2} = 8$ Undo the operations in reverse order.
$3x - 5 = 16$ Multiply both sides by 2.
$3x = 21$ Add 5 to both sides.
$x = 7$ Divide both sides by 3.

b $4(y+3) = 40$ Divide both sides by 4.
$y + 3 = 10$ Subtract 3 from both sides.
$y = 7$

or $4(y+3) = 40$ Expand the bracket.
$4y + 12 = 40$ Subtract 12 from both sides.
$4y = 28$ Divide both sides by 7.
$y = 7$

c $17x = 35$ Divide both sides by 17.
$x = \frac{35}{17}$ Change to a mixed number.
$= 2\frac{1}{17}$ Avoid decimals. $\frac{1}{17}$ is a recurring decimal and has to be rounded, so a decimal answer is not exact.

Example

The perimeter of this rectangle is 34 cm. Find x.

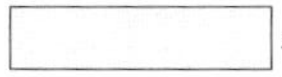

Perimeter $= x + x + 3x + 1 + 3x + 1$
$= 8x + 2$
so $8x + 2 = 34$
$8x = 32$
$x = 4$

Exercise A2.2

1 Solve these one-sided equations.

a $3x - 7 = 8$ **b** $4(x-1) = 20$ **c** $\frac{x}{2} + 8 = 13$

d $\frac{2x-8}{4} = -3$ **e** $2(x^2 + 9) = 68$ **f** $2\left(\frac{x+1}{2} - 3\right) = 8$

g $\frac{3x-5}{2} = 10$ **h** $x^3 - 4 = 12$ **i** $\sqrt{x} - 1 = 9$

2 Copy and complete this crossword by solving the equations in the clues.

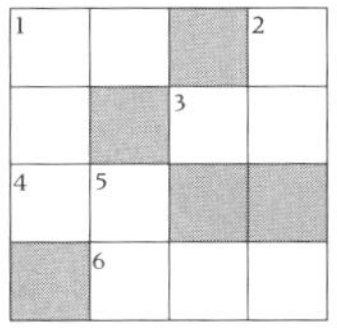

Across
1 $2(x+2) = 30$
3 $\frac{x}{2} - 5 = 15$
4 $\frac{3x-11}{2} = 71$
6 $3(x-100) = 675$

Down
1 $3x + 15 = 330$
2 $x - 10 = 30$
5 $x^2 + 1 = 170$

3 The perimeter of this shape is 30 mm. Find the length of each side.

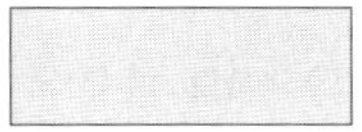

4 This shape is a quadrilateral. Work out the value of x.

The interior angle sum of a quadrilateral is 360°.

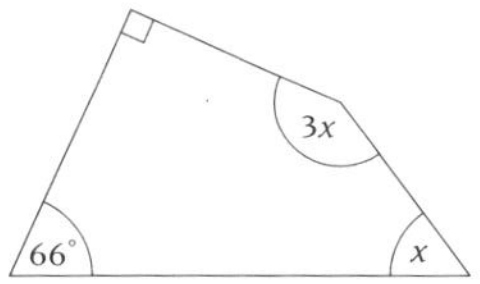

5 a Given that $y = 4x - 8$, work out the value of x when $y = 12$.
b Repeat for $y = -4$.

6 The diagram shows an isosceles triangle. Given that the equal angles are 10 less than double the third angle:

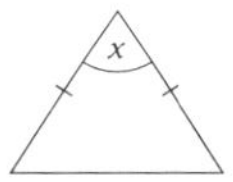

a Write an expression for each equal angle.
b Write an equation connecting the angles.
c Solve the equation, using your answer to find the size of each angle.

A2.3 Solving double-sided equations

Objectives

- Set up and solve simple equations

Useful resources

- Mini-whiteboards

Mental starter

I have x number of goldfish. My brother has half as many as me and my sister has 3 times as many as me. All together we have 63 goldfish. **How many do we each have?** (me = 14, brother = 7, sister = 42.) Can be solved by a trial and improvement method or by forming an equation.

Introductory activity

In the previous lesson the equations had letters on only one side of the equation. The extra step in these questions is to subtract the smaller algebra term from both sides. Then it will become a one-sided equation and you can solve it by using inverse operations as before.
Taking an example from the student book:
$3x + 7 = 5x - 1$.
Which is the smaller algebra term? ($3x$). If you subtract $3x$ from both sides you have $3x + 7 - 3x = 5x - 1 - 3x$ or $7 = 2x - 1$. This is the same as $2x - 1 = 7$ as the two sides can be swapped round. **What must x be?** (4)

Another example from the student book:
$5 - 6y = 2y - 3$.
Which is the smaller algebra term? ($-6y$)
Students may need reminding of the number line; -6 is smaller than 2.
The equation then becomes
$5 - 6y - -6y = 2y - 3 - -6y$
or $5 = 8y - 3$.
What must y be? (1)

Set this contextual problem:
If I was twice Bob's height plus 5 cm and my son was 100 cm taller than Bob, would it be possible to tell how tall Bob is if you knew I was the same height as my son?
(Yes, form an equation. Call Bob's height x cm. Therefore $2x + 5 = x + 100 \longrightarrow x = 95$.)

Discuss the last example in the student book, which involves forming an equation before solving it.

Exercise commentary and misconceptions

In question 1d students need to multiply to remove the brackets first.

In question 2d students must simplify both sides first.

Question 4a needs students to find the angles, not just x.

When forming an equation, make it clear that if you have 2 answers or ways of finding the same thing, you can make them equal.

Plenary

Discuss questions 3 and 4.
Students often find it hard to form equations, and are often much happier solving equations that have been given to them.
Allow students to share their strategies for forming equations with the class.

A2.3 Solving double-sided equations

This spread will show you how to:

- Set up and solve simple equations

Keywords
One-sided
Unknown

When you solve an equation with **unknowns** on both sides

① Subtract the smaller algebra term from both sides, for example

$5x - 3 = 3x + 1$	Subtract $3x$ from both sides.
$2x - 3 = 1$	This is now a **one-sided** equation.

② Solve the one-sided equation by using inverse operations.

$2x - 3 = 1$	Add 3 to both sides.
$2x = 4$	Divide both sides by 2.
$x = 2$	

Example

Solve

a $3x + 7 = 5x - 1$ **b** $5 - 6y = 2y - 3$

a $3x + 7 = 5x - 1$	Subtract 3x from both sides (3x is smaller than 5x).
$7 = 2x - 1$	Add 1 to both sides.
$8 = 2x$	Divide both sides by 2.
$4 = x$	
$x = 4$	
b $5 - 6y = 2y - 3$	Subtract −6y from both sides (−6y is smaller than 2y): Subtracting −6y is the same as adding 6y.
$5 = 8y - 3$	Add 3 to both sides.
$8 = 8y$	Divide both sides by 8.
$y = 1$	

$-6y$ … 0 … $2y$

Example

The diagram shows an isosceles triangle.
Find the length of the equal sides.

$2z - 4$ $4z - 18$

$2z - 4 = 4z - 18$	Write an equation and solve it.
$-4 = 2z - 18$	After subtracting 2z from both sides.
$14 = 2z$	After adding 18 to both sides.
$7 = z$	

Hence, $2z - 4 = 14 - 4 = 10$ and $4z - 18 = 28 - 18 = 10$
The equal sides of the triangle are 10 units long.

Exercise A2.3

1 Solve these equations.

a $2x + 4 = x + 3$ **b** $10 - 3x = 7x - 10$

c $8 - 3x = 5 - 2x$ **d** $4(7 + 2z) = 15 - 8z$

2 In each row of equations, one has a different solution from the other two. Find the odd one out.

a $2a + 7 = 4a + 1$ $6a - 2 = 2a + 6$ $10 + 2a = 7a - 5$

b $10 - 3b = 6b + 1$ $15 - 2b = 14 - b$ $3b - 14 = b$

c $2c + 2 = 4c - 1$ $8c - 7 = 6c - 4$ $1 - 4c = 2 - 8c$

d $2d + d + 8 = 3d + 2d - 4$ $5d + 7d - 3 = 10 - d$ $15 - d - d = 9 - d$

3 In each case, use the information to write an equation and solve it to find the starting number.

a I think of a number, multiply it by 8 and subtract 2. I get the same answer as when I multiply this number by 2 and add 10.

b I think of a number, multiply it by 5 and add 3. I get the same answer as when I multiply it by 2 and subtract it *from* 24.

c Taking double a number from 11 is equal to taking treble that number from 14.

4 In each case, use the information to form an equation and solve it.

a The angles in a triangle are $x°$, $x + 20°$ and $x + 40°$.
Find the angles of the triangle.

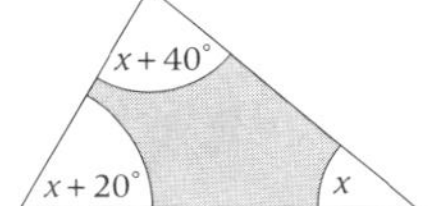

b The perimeters of these two shapes are equal.
What are the dimensions of each shape?

$4x + 3$ $8x$ $4x + 1$

c Abdul is 180 cm tall and Mark is $10x$ cm tall.
Alice is 164 cm tall and Miranda is $9x$ cm tall.
The difference in height between the two boys is equal to the difference in height between the two girls.
How tall are Mark and Miranda?

A2.4 Solving equations with fractions

Objectives

- Solve equations involving fractions

Useful resources

- Mini-whiteboards

Mental starter

If $\frac{12}{x} = y$ and $x^y = 81$ find the values of x and y. ($x = 3$, $y = 4$ is the only solution.)

Students should try to find values of x and y that fit one of the equations and try these values in the other one.

Introductory activity

What happens when you multiply a fraction by a whole number?

For example, $\frac{3}{11} \times 3$? Explain clearly that this must be $\frac{9}{11}$, because when you multiply a fraction by a whole number, you only multiply the top.

If the x is in the denominator of a fraction, the fraction must be cleared first.

For example $\frac{5}{x} = 7$. Multiplying both sides of the equation by x gives $5 = 7x$ and $x = \frac{5}{7}$.

For example $\frac{70}{5x} + 5 = 12$. Multiplying both sides of the equation by $5x$ gives

$70 + 25x = 60x$
$\longrightarrow 70 = 35x$
$\longrightarrow x = 2$.

For example $\frac{21}{x+4} = 5$. Multiplying both sides of the equation by (x + 4) gives $21 = 5(x + 4)$
$\longrightarrow \frac{21}{5} = x + 4$
$\longrightarrow \frac{21}{5} = x - 4$
$\longrightarrow x = \frac{1}{5}$

In this case it would not be advisable to multiply the bracket out.

For example $\frac{16}{x} = \frac{20}{x+1}$.

When both sides are fractions, it is best to cross multiply to clear the fractions $\longrightarrow 16(x + 1) = 20x$
$\longrightarrow 16x + 16 = 20x$
$\longrightarrow 16 = 4x$
$\longrightarrow x = 4$.

Exercise commentary and misconceptions

In question 2, remind students that every term must be multiplied when clearing fractions, or you can combine the number terms before dealing with the fraction.

In question 5, students often get confused between 'divide it into' and 'divide by'.

Question 7 is more challenging than the previous questions. You may need to recap the mean.

Plenary

In the last few lessons, many different types of equations have been solved.
This example has a little bit of everything from the previous 3 lessons. Invite contributions from the whole class to solve it.

$$\frac{3}{2(x-4)} = \frac{2}{1-4x}$$
$$\longrightarrow \frac{3}{2x-8} = \frac{2}{1-4x}$$
$$\longrightarrow \frac{3-12x}{2x-8} = 2$$
$$\longrightarrow 3 - 12x = 4x - 16$$
$$\longrightarrow 3 = 16x - 16$$
$$\longrightarrow 19 = 16x$$
$$\longrightarrow \frac{19}{16} = x$$

A2.4 Solving equations with fractions

This spread will show you how to:

- Solve equations involving fractions

Keywords
Cross multiply
Reciprocal

- **When you solve an equation involving fractions, clear the fractions first.**

You can use the fact that $x \times \frac{1}{x} = 1$ to help you.

$\frac{1}{x}$ is called the **reciprocal** of x.

- **Any non-zero number multiplied by its reciprocal is 1.**

Example

Solve the equation $\frac{15}{x} = \frac{3}{7}$

$\frac{15}{x} = \frac{3}{7}$ Multiply both sides by the common denominator $7x$ to clear the fractions.

$\frac{15}{x} \times 7x = \frac{3}{7} \times 7x$ Cancel common factors.

$15 \times 7 = 3 \times x$, so $105 = 3x$, so $x = 35$

$7x$ is the reciprocal of $\frac{1}{7x}$.

When the equation has just a single fraction on each side you can **cross multiply** to clear the fractions, for example, so

$\frac{15}{x} = \frac{3}{7}$ $15 \times 7 = x \times 3 \longrightarrow 105 = 3x \longrightarrow x = 35$ as above

Cross multiply: LH numerator × RH denominator and vice versa.

Example

Solve these equations.

a $\frac{3}{x} = \frac{4}{9}$ **b** $\frac{12}{p} + 9 = 28$

a $\frac{3}{x} = \frac{4}{9}$ Simple fraction each side, so cross multiply.

$3 \times 9 = 4x$ Divide both sides by 4.

$27 = 4x$

$x = \frac{27}{4} = 6\frac{3}{4}$

b $\frac{12}{p} + 9 = 28$ Subtract 9 from both sides.

$\frac{12}{p} = 19$ Cross multiply ($19 = \frac{19}{1}$).

$12 \times 1 = 19p$ Divide both sides by 19.

$p = \frac{12}{19}$

It is tempting to use cross multiplication straightaway but you cannot cross multiply until you have a simple fraction on each side.

Example

I divide 15 by a certain number and get $\frac{3}{4}$. What is the number? Write an equation and solve it to find the starting number.

If x is the missing number then: $\frac{15}{x} = \frac{3}{4}$ Cross multiply

$15 \times 4 = 3x$, so $3x = 60$, so $x = 20$

Exercise A2.4

1 Solve these equations, using the reciprocal function.

a $\frac{7}{x} = 21$ **b** $15 = \frac{5}{x}$ **c** $\frac{4}{y} = 3$ **d** $\frac{7}{p} = 8$

e $\frac{10}{x} = -2$ **f** $11 = \frac{5}{y}$ **g** $-3 = \frac{7}{y}$ **h** $-\frac{3}{x} = -9$

2 Solve these equations.

a $\frac{5}{x} + 9 = 10$ **b** $\frac{10}{p} + 7 = 8$ **c** $\frac{x}{4} + 3 = 10$ **d** $-2 = 1 + \frac{3}{x}$

3 Solve these equations.

a $\frac{16}{x} + 4 = 2$ **b** $\frac{12}{2y} - 3 = 5$ **c** $\frac{6}{3p} - 1 = 10$ **d** $\frac{15}{2x} + 4 = -2$

4 Solve these equations.

a $\frac{x+1}{3} = \frac{x-1}{4}$ **b** $\frac{2y-1}{3} = \frac{y}{2}$ **c** $\frac{5}{w+5} = \frac{15}{w+7}$ **d** $\frac{3}{x-1} = \frac{9}{2x-1}$

5 Write an equation and solve it to find the starting number in each case.

a I think of a number, divide it into 16 and I get 10. What is my number?

b I think of a number, add 4, divide it into 12 and get 7. What is my number?

c I think of a number, take 3 and divide it into 11. This gives me the same answer as when I take the same number and divide it into 8. What is my number?

6 Use the formula $P = \frac{180}{n+1}$ to find

a the value of P when $n = 4$

b the value of n when $P = 12$

c the value of n when $P = 2\frac{2}{9}$.

7 The means of each set of expressions are equal. Use this information to find x and, hence, the value of each expression.

To find the mean add all the expressions and divide by the number of expressions.

Set 1

$2x - 1$, $3x + 2$, $5x + 4$, $7x$, $6x - 4$, $10 - 2x$

Set 2

$3x - 7$, $5x + 8$, $13 - x$, $2(3x + 1)$, $12 + 4x$

A2.5 Inequalities

Objectives

- Solve simple inequalities, representing the solution on a number line

Useful resources

- Number line

Mental starter

I am thinking of a number.
If I double it, I get an answer that is 1 greater than half my original number.
What's my number?
(Answer $\frac{2}{3}$.)
You could give the clue that it is not an integer.

Introductory activity

Ask how do you read these inequalities? Say them forwards and backwards and give a possible integer value for x in each case.

$x > 2$; $x \leqslant -3$; $1 < x$; $5 \leqslant x$; $-1 < x \leqslant 3$

Inequalities can be drawn on a number line. Draw the 2nd and 4th inequalities from above on the same number line. Use a hollow circle for each then a line and arrow. Is this all we need to do? Students may comment on the problem of showing that −3 and 3 are allowed while −1 is not. Ask students to think about when they use the program WORD and choose the PAGE SETUP option, then go to PAPER SIZE, they are able to select LANDSCAPE or PORTRAIT. Beside each word is a little hollow circle. If you fill in the circle with a dot, does this then mean you select that option? Yes. So likewise with inequalities, if you fill in the hollow circle you are choosing to select (include) that number.

Solving inequalities is just like solving equations, except your answer does not have an =, but an inequality.

Adding or subtracting numbers from either side does not affect the inequality.

Multiplying or dividing by a positive number does not affect the inequality, but multiplying or dividing by a negative number reverses the inequality. This can be shown with a numerical example.

Start with $4 < 6$. Remind students of the relative positions of 4 and 6 on the number line. Add any number to both 4 and 6. **What effect does this have on the relative positions?** (no change.)

Subtract any number from both 4 and 6. **What effect does this have on the relative positions?** (no change.)

Multiply both 4 and 6 by any positive number. **What effect does this have on the relative positions?** (no change.)

Multiply both 4 and 6 by any negative number. **What effect does this have on the relative positions?** (reverses the inequality.)

Discuss the example in the student book.

Exercise commentary and misconceptions

In question 2 where double ended inequalities are used, advise students to set the answer out with the smaller number on the left, the x in the middle and the larger number on the right, then put the inequality signs in place.

In question 4 the most convenient way to deal with minus xs is to add them to both sides. This not only means students do not have to reverse the inequality but also makes future rearranging formulae more straightforward, avoiding the need to multiply through by −1.

Plenary

If $x^2 < 16$ and x is an integer, **what values can x take?** (−3, −2, −1, 0, 1, 2, 3)

If $xy < 7$, $x > 0$, $y > 0$, $x \neq y$ and x and y are both integers, **what pairs of values can x and y take?**

x	1	1	1	1	1	2	2	3	3	4	5	6
y	2	3	4	5	6	1	3	1	2	1	1	1

A2.5 Inequalities

This spread will show you how to:

- Solve simple inequalities, representing the solution on a number line

Keywords
Inequality
Number line
Reverse
Solve

- An **inequality** tells you about two quantities that are unequal.
- You can **solve** an inequality using inverse operations, for example

$3x + 2 > 5$ Subtract 2 from both sides.

$3x > 3$ Divide both sides by 3.

$x > 1$

$x < 3$, $2x > 5$, $x + 1 > 0$ are inequalities.

- **If you multiply or divide by a negative number, an inequality becomes false. For example:**

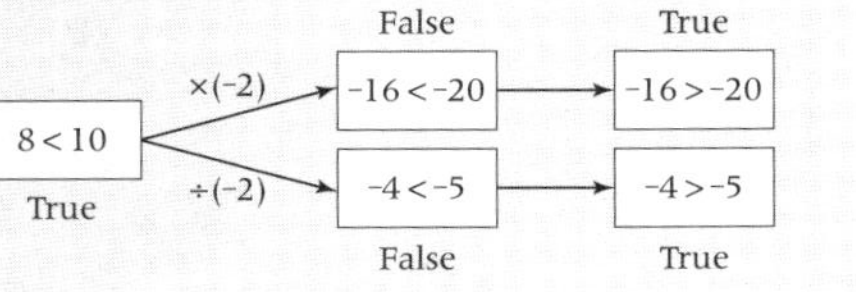

When you multiply or divide an inequality by a negative number you must **reverse** the inequality sign, for example

$7 - 2x \leqslant 3$ Subtract 7 from both sides.

$-2x \leqslant -4$ Multiply by −1 and **reverse** the inequality sign.

$2x \geqslant 4$ so $x \geqslant 2$

You can show the solution of an inequality on a **number line**, for example

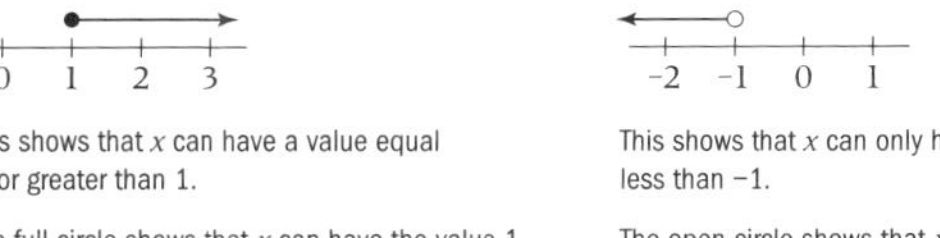

This shows that x can have a value equal to or greater than 1.

The full circle shows that x can have the value 1.

This shows that x can only have values less than −1.

The open circle shows that x cannot have the value −1.

Example

Solve these inequalities and represent the solutions on a number line.

a $5x - 7 \geqslant 3x + 13$ **b** $16 > -4x$

a $5x - 7 \geqslant 3x + 13$ Subtract 3x from both sides.

$2x - 7 \geqslant 13$ Add 7 to both sides.

$2x \geqslant 20$

$x \geqslant 10$

7 8 9 10 11 12

b $16 > -4x$ Divide both sides by 4.

$4 > -x$ Multiply by −1 and reverse sign.

$-4 < x$,

so $x > -4$

−6 −5 −4 −3 −2

Exercise A2.5

1 Use an inequality to represent each of these. For example, I think of a number and it is more than 11: $x > 11$

a I think of a number and it is 3 or below.

b I think of a number and it is between 2 and 8 inclusive.

c I think of a number and it is over −5 but below 12.

2 What inequalities are represented on these number lines?

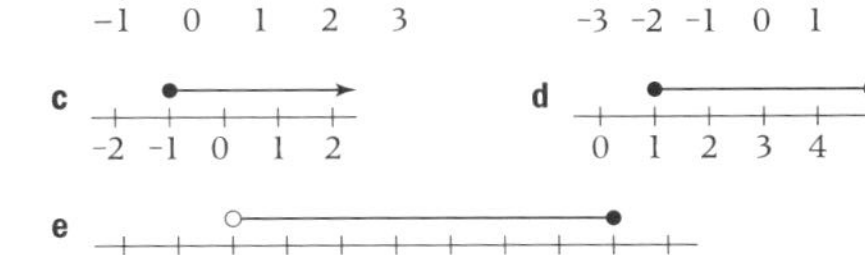

e -9 -8 -7 -6 -5 -4 -3 -2 -1 0 1

3 Is this statement true or false? The inequality $3 \geqslant x$ is represented on this number line.

0 1 2 3 4 5 6

4 Solve these inequalities and represent the solutions on a number line.

a $3x \leqslant 21$ **b** $2x - 5 > 17$ **c** $\frac{p}{2} + 6 \leqslant -2$

d $28 < 7x + 49$ **e** $5y + 3 \leqslant 2y + 5$ **f** $-3y > 9$

g $4(x + 2) \leqslant 16$ **h** $-6x < 30$ **i** $\frac{x}{-5} \geqslant -2$

j $4p - 3 \leqslant 3(p - 2)$ **k** $3(x - 2) < 5(x + 6)$ **l** $6x - 4 \geqslant -2x$

5 **a** The area of this rectangle exceeds its perimeter. Write an inequality and solve it to find the range of values of x.

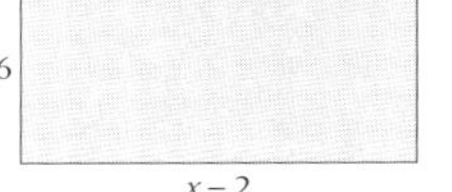

b Given that x is an integer, find the smallest possible value that x can take.

A2 Exam review

Key objectives

- Set up simple equations
- Solve linear equations in one unknown, with integer or fractional coefficients
- Solve simple linear inequalities in one variable, and represent the solution set on a number line

Worked solution	Commentary and misconceptions
1	Students may find inequalities confusing. Encourage them to treat inequalities like equations, recalling the rules for when inequalities are reversed.
a $3x - 2 \leqslant 5x + 4$ $-2 - 4 \leqslant 5x - 3x$ $-6 \leqslant 2x$ $-3 \leqslant x$	
b $-6 < 2x \leqslant 4$ $-3 < x \leqslant 2$ −3 −2 −1 0 1 2	Ensure that students know how to distinguish between < and ⩽, and > and ⩾ in their diagrams.
2	Remind students that each side of the equation must be transformed in the same way. Some students may find rules such as 'swap side, swap sign' helpful.
a $20y - 16 = 18y - 9$ $20y - 18y = -9 + 16$ $2y = 7$	
b $\frac{40 - x}{3} = 4 + x$ $40 - x = 3(4 + x)$	Ensure that students multiply the entire right side of the equation by 3. Encouraging them to use brackets may help.
$40 - x = 12 + 3x$ $40 - 12 = 3x + x$ $28 = 4x$ $7 = x$ or $x = 7$	

01 Collecting data

Objectives

F/H Design an experiment or survey

H Identify possible sources of bias and plan to minimise it

F/H Design and use data collection sheets and two-way tables

H Select and justify a sampling scheme and a method to investigate a population, including random sampling

F/H Identify the modal class for grouped data

F/H Calculate the mean, range and median of continuous data

F/H Calculate the mean for large data sets with grouped data

H Find the median, quartiles and interquartile range for large data sets

Unit overview

This unit consolidates Key Stage 3 knowledge of questionnaires and data collection. It extends from designing a survey and collecting data to calculating averages and spread of data sets. This unit provides a basis for work covered in further data units, including analysis of data.

Prior knowledge

Before your students start this unit they should be able to:

- Use appropriate data collection methods
- Record data efficiently
- Add, subtract, multiply and divide with integers and decimals
- Understand and use percentages

Differentiation

- **Foundation** focuses on identifying and collecting data from a variety of sources, including grouped data, and using two-way tables
- **Foundation Plus** has a similar content to the Foundation book but set in more challenging contexts
- **Higher Plus** extends the Higher book to include stratified sampling

D1.1 Designing a survey

Objectives

- Design a questionnaire, recognising bias and taking steps to avoid it

Useful resources

- OHT illustrating a questionnaire and data collection sheets

Mental starter

A sample is to be chosen from an alphabetical list of names.

- Amy suggests choosing every 10th name.
- Beth suggests choosing the first 20 names.
- Caitlin suggests choosing her friends' names.
- Denise suggests choosing names at random.

Discuss which is likely to give the least biased sample.

Introductory activity

When writing questions for a questionnaire, there are several points which students should consider.

- Questions must be very clear about what they mean (unambiguous). **Why is the question 'Do you listen to a lot of music?' not very clear?** (Are all kinds of music included? What is meant by a lot?)
- The questions must have a clear set of possible answers (responses). The question 'How long do you spend listening to pop music in an average day?' has the responses: less than half an hour, more than an hour, between three and four hours. **Why are the set of responses not clear?** (Between half an hour and an hour not included, more than an hour includes between three and four hours, some people may listen for more than four hours, some people may not listen to music at all.)
- Questions must not be biased (leading), trying to make you answer a certain way. **What is wrong with the question 'Do you agree that most pop music is rubbish?'?** (Can be difficult to disagree with the person asking the question.)
- Questions must not be too personal, for example you can't ask for someone's exact income or weight.

Some exam questions will ask why a particular question is poor, and then ask you to write a better question; other exam questions just ask you to write a question on a particular subject, but in either case you need to remember the points above.

Exercise commentary and misconceptions

When deciding if a question is appropriate, students should check it against the criteria. When rewriting the question ensure the list of responses do not miss out any possible answer, remember to include 'other' if appropriate.

In question 2a, students could either ask respondents to check which of a list of sports they like, or ask them to put a list of sports in their order of preference. They need to decide whether Sally is interested in participation in sports or in watching sports.

Plenary

Describe this question from a questionnaire: Most people have stolen something in their lives. Have you ever taken something you should not have and how did you feel about it?

What is wrong with this question?
(This is not an easy question to rewrite. Ask students to try and design a single question that relates to some part of the original question.)

D1.1 Designing a survey

This spread will show you how to:

- Design a questionnaire, recognising bias and taking steps to avoid it

Keywords
Biased
Data
Questionnaire
Survey

Organisations carry out statistical surveys to collect data that help them plan for the future.

You can use a **questionnaire** and collect **data** in a **survey**.

You have to choose suitable questions for a questionnaire.

Suitable questions
- can be answered yes or no
- ask for facts.

Unsuitable questions
- may be vague
- are leading or **biased**
- could be embarrassing.

Questions must have responses that
- cover all possible answers
- do not overlap or have gaps.

Do you have an MP3 player?
How many CDs do you own?
none ☐ 1–10 ☐ 11–20 ☐ over 20 ☐

Do you listen to a lot of music?
Do you agree that Coldplay is the best band in the world?
How cool are you?

Where do you buy your CDs?
internet ☐ store ☐ other ☐
How much do you spend on CDs each month?
£0 to £14.99 ☐ £15 to £29.99 ☐
£30 and up ☐

Examiner's tip
The techniques described in this unit will be useful for your statistical coursework task.

Example

A recently restyled breakfast radio show conducted this survey.

1 What is your opinion of the new breakfast show?
Fantastic ☐ Good ☐
2 How long do you listen to the show?
10 min ☐ 1 hour ☐ 1–2 hours ☐

a Comment on these questions.
b Write two questions for the survey to find out if listeners like the new show and for how long they listen.

a Question **1** does not include all possible responses, for example if you do not like the new show or don't think it's an improvement.
In question **2** there are gaps, between 10 min and 1 hour, and an overlap: 1 hour appears twice.

b **1** What is your opinion of the new breakfast show compared to the old one?
Much better ☐ Better ☐
Neither better nor worse ☐ Not as good ☐
2 How much of the breakfast show do you listen to?
all of it ☐ over an hour ☐
half an hour to 1 hour ☐ less than half an hour ☐

Exercise D1.1

In this exercise you can find out if any questions you write 'work' by testing them out on groups of people in your class. The larger the sample the more reliable the results.

1 Katy is doing a survey to find out how often people go to the cinema and how much they spend. She writes this question:

How many times a month do you go to the cinema?

a What is wrong with this question?
b Write an improved question to find out how often people visit the cinema.
c Write a question to find out how much people spend when they go to the cinema.

2 Sally put this question in a questionnaire:

Do you agree that tennis is the best sport?

a **i** What is wrong with this question?
ii Write a better question to find out the favourite sport. Include some response boxes.

Sally also wants to find out how often people play sport.

b Design a question for Sally to use. Include some response boxes.

3 James wants to know which flavour crisps he should stock in the school tuck shop. He asks this question:

Do you prefer plain or ketchup flavoured crisps?

a What is wrong with this question?
b Think about what crisps you and your friends like and design a better question for James to use. You should include some response boxes.
c James also put this question in his questionnaire:

How many times have you visited the tuck shop?
Once ☐ Lots of times ☐

i Write two things that are wrong with this question.
ii Design a better question for James to use. Include some response boxes.

4 Merlin wants to find out how far people would travel to see their favourite band perform. He writes this question:

How far would you travel to see your favourite band?
less than 1 mile ☐ 5–10 miles ☐ any distance ☐

a **i** What is wrong with this question?
ii Design a better question for Merlin to use. Include some response boxes.

Merlin also wants to find out how much people would pay for a ticket to see their favourite band.

b Design a question for Merlin to use. Include some response boxes.

D1.2 Collecting data – choosing a sample

Objectives

- Design a questionnaire, recognising bias and taking steps to avoid it
- Understand the concept of random sampling

Useful resources

- A population of objects e.g. straws

Mental starter

Describe this scenario: Government figures say that 15% of boys between the ages of 14 and 16 smoke more than 20 cigarettes a week. A teacher wants to know if this is true. He carried out a survey at a mixed school asking 5% of all the pupils, he chose the students randomly. He asked 'Do you smoke?' 3% of those he asked said 'YES'. **What is wrong with the survey?**

(Should only have asked boys; should only ask those aged 14 to 16; the question was not specific enough, did not mention 20 a day; students are unlikely to answer honestly to a teacher, the survey needed to be confidential; it may be a small school and 5% of the school may be too few people; students at the school might not be typical of pupils all over the country.)

Introductory activity

Discuss the meaning of the term 'population': The term 'population' refers to all things in the group you are investigating. If you are conducting a survey about drivers in the UK, your population is 'all people who drive in the UK'. The term 'sample' refers to a group taken from the population that you will investigate so that you can make claims about the whole population.

When picking a sample you must try to ensure that it is a fair spread of people (it represents the population) and is picked randomly from the population. There are a number of different methods used for sampling, some of the most common are listed here, each one has advantages and disadvantages. (Ask students to comment on these methods.)

- Give a number to each member of the population and then use random numbers to choose your sample.
- Telephone people randomly taking their number from the phone book.
- Post questionnaires to people randomly, taking their addresses from the electoral register.
- Stand at a particular place on the high street and ask members of the public.
- List all members of the population in alphabetical order and pick say every 10th one.
- Ask members of your family and friends.

Exercise commentary and misconceptions

The main theme here is the idea that a sample must be representative. A biased sample is often caused when the surveyor makes their life easier by taking an easy sample.
Some students find it hard to understand the concept of random sampling. The term 'random sampling' could be used to describe the sample that gives every member of the population the same chance of being picked. In practice a purely random sample is very hard to achieve with a large population.

Plenary

I want to choose a random sample of 140 students from my school. There are seven year groups and four tutor groups in each year. So I decide to pick five students at random from each tutor group. **Comment on this method**.
(It will do well at giving a representative spread of ages. You may not get a good spread of males/females. The 6th form might not be very big compared to the other years, and yet we are picking the same number from each year group no matter how big the year group is.) **What might be more fair?** (You could pick 10% of all the pupils in the school by picking randomly, 10% of the girls in year 7, 10% of the boys in year 7, 10% of the girls in year 8 and so on throughout the school.)

D1.2 Collecting data – choosing a sample

This spread will show you how to:

- Design a questionnaire, recognising bias and taking steps to avoid it
- Understand the concept of random sampling

Keywords
Bias
Population
Random
Sample

It may be time-consuming, too costly, too long or too impractical to collect data from everyone. In these cases you should ask a representative **sample**.

You must choose the sample so that it is not biased. For example, a survey of preferred music using a sample of friends is biased as friends are more likely to have similar opinions.

- A sample should represent a whole **population**.

One way of avoiding **bias** is to use a **random** sample.

The population is the group of people or items being surveyed

- **In a random sample each member of the population has the same chance of being included.**
- **Methods for choosing a random sample include:**
  - **taking names out of a hat**
  - **giving everyone a number and using a calculator or random number tables to pick numbers.**

The larger the sample the more reliable the results

Example

James carries out a survey to find out if people in his town enjoy sport.

He stands outside a football ground and surveys people's opinions as they go in to watch a match.

Write two reasons why this is not a good sample to use.

People who watch football usually enjoy sport.
More men than women go to watch football so the survey could be gender biased.

Example

A train company carried out a survey about a local rail service.

They telephoned 100 people from a page of the telephone directory to answer a questionnaire on the rail service.

Write three reasons why this sample could be unrepresentative.

Only people who have a land-line telephone (and are not ex-directory) can be included in the sample.
Only people on one page of the telephone directory are included in the sample.
Some people may not be at home when they are phoned.

Exercise D1.2

1 Katy is doing a survey to find out how often people go to the cinema and how much they spend. She stands outside a cinema and asks people as they go in.

Write a reason why this sample could be biased.

2 Sally wants to find out how often people play sport. Sally belongs to an athletics club. She asked members in her athletics club.

How could this sample be biased?

3 James wants to know which flavour crisps he should stock in the school tuck shop.

a He asks his mum, dad, auntie and uncle.

Explain why this is not a good sample to use.

b He asks only Year 11 at his school.

Explain why this sample could be biased.

c Describe how James could take a sample of 50 people. (There are 1000 people in his school.)

4 Merlin wants to find out how far people would travel to see their favourite band perform.

a He asks all his friends.

Write a reason why this sample could be biased.

b He goes into town one Saturday morning and asks anyone listening to music on a MP3 player.

How could this sample be biased?

5 Jenny carries out a survey to find out the most popular band. She asks 10 of her friends – all girls.

How could this sample be biased?

6 Wayne carries out a survey to find out the most popular car colour. He stands on a street corner and notes the colour of the first 15 cars that pass by.

Write a reason why this sample could be biased.

7 Lisa wants to find out how people travel to work.

a She asks people at a bus stop one morning.

How could this sample be biased?

b She opens the telephone directory at a random page and phones everyone on that page.

How could this sample be biased?

D1.3 Designing a data collection sheet – two way tables

Objectives
- Design and use data collection sheets and two-way tables

Useful resources
- Blank two-way tables

Mental starter
The table shows information about students in a school.

	Left handed	**Right handed**
Girls	12	48
Boys	8	28

What fraction of the girls are left-handed?
What percentage of the total are boys?

Introductory activity
A data collection sheet collects data from a questionnaire or survey. As well as recording the data, you can analyse the relationship between two sets of data by recording them in a two-way table.

As a class activity, you could use a two-way table of gender and some other characteristic of the students, such as tutor group. Each student can record his/her position with a tally in the table.

		Tutor group		
		Ms Jones	Mr Khan	Mrs MacMillan
Gender	Male			
	Female			

You can then use the table to work out some percentages, such as the percentage of girls in the class who are in Mr Khan's group.

Exercise commentary and misconceptions
The class could be split into pairs or small groups and each group assigned one of the questions 1–7. They could then design the questions and data collection sheet, use the rest of the class as subjects and then analyse the results.

Plenary
Why can't you have a three way table?
(There are only two directions on a flat piece of paper.)

Give examples of occasions when collecting data when you would favour a questionnaire over a data collection sheet or visa versa.

D1.3 Designing a data collection sheet – two-way tables

This spread will show you how to:
- Design and use data collection sheets and two-way tables

Keywords
Data
Two-way table

- You can use a data collection sheet to collect **data** from a questionnaire or experiment.
- You can use a **two-way table** to collate the two sets of results.

Example

Two questions on a questionnaire are:

'Are you male or female?' and 'How old are you?'

Design a two-way table to collect this data.

	Under 10	10–19	20–29	30–40	40+	Total
Male						
Female						
Total						

- You can use data in a two-way table to find other results.

Example

The table gives information about Key Stage 4 students at a school.

	Boys	Girls
Year 10	68	117
Year 11	89	126

a Work out the percentage of Key Stage 4 students in Year 10 who are boys.
b Work out the percentage of Key Stage 4 students who are girls.

Find the totals in the table.

	Boys	Girls	Total
Year 10	68	117	**185**
Year 11	89	126	**215**
Total	**157**	**243**	**400**

There are 400 students at Key Stage 4 $(68 + 117 + 89 + 126)$.

a There are 68 Year 10 boys: $\frac{68}{400} \times 100 = 17\%$

b There are 243 $(117 + 126)$ girls altogether: $\frac{243}{400} \times 100 = 60.75\%$

Exercise D1.3

1 Katy is doing a survey to find out how often people go to the cinema and how much they spend.
Design a suitable data collection sheet in the form of a two-way table that she could use.

2 Sally wants to find out how often people play sport.
She wants to divide the results into those from males and those from females.
Design a suitable data collection sheet in the form of a two-way table that she could use.

3 James wants to know which flavour crisps he should stock in the school tuck shop.
He also wants to know which year groups prefer which flavours.
Design a suitable data collection sheet in the form of a two-way table that he could use.

4 Merlin wants to find out how far people would travel to see their favourite band perform and how much they would spend on a ticket to watch them.
Design a suitable data collection sheet in the form of a two-way table that he could use.

5 Jenny carries out a survey to find out people's favourite band and how many of that band's CDs they own.
Design a suitable data collection sheet in the form of a two-way table that she could use.

6 Wayne carries out a survey to find the most popular car colour and the most popular make of car.
Design a suitable data collection sheet in the form of a two-way table that he could use.

7 Lisa wants to find out how people travel to work and how long it usually takes them.
Design a suitable data collection sheet in the form of a two-way table that she could use.

8 The table gives information about the number of students in Years 7–9 that attended a school disco.

	Year 7	Year 8	Year 9
Boys	42	58	96
Girls	78	104	122

a How many students attended the disco?
b Work out the percentage of students that were
i Year 8 girls **ii** in Year 7 **iii** boys.

D1.4 Averages and spread

Objectives

- Find an average and a measure of spread for a data set

Useful resources

- (Random) number generator (eg on calculator)

Mental starter

Write down a set of numbers, for example 3.2, 9.1, 9.5, 3.7, 5.2

Ask students to find the mean, median and range using a mental method. (6.14, 5.2, 6.3)

Introductory activity

Discuss this scenario:

15 students of the same age at different schools were asked how many homeworks they were set in an average week.

The results were: 3, 6, 7, 7, 8, 10, 11, 12, 13, 14, 15, 16, 17, 18, 20. Find the mode, median, mean, range, lower quartile, upper quartile and interquartile range.

- **MO**de = **MO**st common result (first 2 letters the same) (7).
- **M**edian = **M**iddle result (both words have 6 letters) (12).
- Mean = Add up all the results and divide by the total frequency (number of results) (11.8. Note that a decimal result is permissible for a mean).
- Range = difference between the highest and lowest result (20 – 3 = 17).
- Lower quartile (LQ) = The result $\frac{1}{4}$ of the way from the bottom (7).
- Upper quartile (UQ) = The result $\frac{1}{4}$ of the way from the top (16).
- Inter quartile range = Difference between UQ and LQ (16 – 7 = 9).

Discuss how both the range and the interquartile range give rough ideas of how spread out the results are. The interquartile range is the more sophisticated measurement because it does not get altered by one extreme result at the top or bottom end, but the range does.

Discuss the example in the student book.

Exercise commentary and misconceptions

Students often forget to put data in order before finding the median, range or interquartile range. None of the data in question 1 is in order.

All parts of question 4 refers back to question 1g.

Plenary

Why can't you represent all data in a list of numbers like the example used at the beginning of class?

Sometimes data is not numbers, but words, for example, colours.

Data may be too long, for example, a list of 5000 numbers.

The individual results may not be known, you may have them in groups, for example, five people aged 0–10 years old.

Encourage other reasons.

D1.4 Averages and spread

This spread will show you how to:

- Find an average and a measure of spread for a data set

Keywords
Average
Interquartile range
Lower quartile
Mean
Measure of spread
Median
Mode
Range
Upper quartile

You can summarise data using an **average** and a **measure of spread**.

- **An average is a single value.**
  There are three types of average:
  - **the mode is the value that occurs most often**
  - **the median is the middle value when the data are arranged in order**
  - **the mean is calculated by adding all the values and dividing by the number of values.**

- Spread is a measure of how widely dispersed the data are.
  Two measures of spread are:
  - the range
  - the **interquartile range** (IQR).

  If there are one or more extreme values the IQR is a better measure of spread than the range.

An extreme value is a value well outside the range of the rest of the data

- **Range = highest value – lowest value**
- **IQR = upper quartile – lower quartile**

Lower quartile $= \frac{1}{4}(n + 1)$th value

Upper quartile $= \frac{3}{4}(n + 1)$th value

Example

Louise collected data on the number of times her friends went swimming in one month.

4 7 22 1 6 2 1 5 6 6 4

Work out the: **a** range **b** mode **c** mean **d** median **e** interquartile range.

In order the data are: 1 1 2 4 4 5 6 6 6 7 22

a Range = 22 – 1 = 21

b Mode = 6

c Mean = 5.8
$(4 + 7 + 22 + 1 + 6 + 2 + 1 + 5 + 6 + 6 + 4) \div 11 = 64 \div 11 = 5.8$

The mean does not have to be an integer even if all the data values are integers.

d There are 11 values.
Median $= \frac{11+1}{2} =$ 6th value = 5

If there are n values Median value $= (\frac{n+1}{2})$th value.

e Interquartile range = upper quartile – lower quartile

Lower quartile $= (\frac{11+1}{4})$th value
$= (\frac{12}{4})$th value
= 3rd value
= 2

Upper quartile $= (\frac{3(11+1)}{4})$th value
= 9th value
= 6

IQR = 6 – 2 = 4

Exercise D1.4

1 For these sets of numbers work out the
i range **ii** mode **iii** mean
iv median **v** interquartile range

a 5, 9, 7, 8, 2, 3, 6, 6, 7, 6, 5
b 45, 63, 72, 63, 63, 24, 54, 73, 99, 65, 63, 72, 39, 44, 63
c 97, 95, 96, 98, 92, 95, 96, 97, 99, 91, 96
d 13, 76, 22, 54, 37, 22, 21, 19, 59, 37, 84
e 89, 87, 64, 88, 82, 88, 85, 83, 81, 89, 90
f 53, 74, 29, 32, 67, 53, 99, 62, 34, 28, 27, 27, 27, 64, 27
g 101, 106, 108, 102, 108, 105, 106, 109, 103, 105, 107, 104, 104, 105, 105

2 For the set of numbers in question **1e**, explain why the interquartile range is a better measure of spread to use than the range.

3 For the set of numbers in question **1f**, explain why the mode is not the best average to use.

4 **a** Subtract 100 from each of the numbers in question **1g** and write down the set of numbers you get.
b For your set of numbers in **a**, work out the
i range **ii** mode **iii** mean
iv median **v** interquartile range
c Compare your answers for the measures of spread in part **b i** and **v** and **1g i** and **v**.
What do you notice?
d Add 100 to your answers for the measures of average in part **b ii**, **iii** and **iv**.
Compare these answers to the answers you got in question **1g**.
What do you notice?
e Give a reason for what you noticed in parts **c** and **d**.

5 A scientist takes two sets of measurements from her experiment.
Her results are:

Set A:	0	99	99	100	100	100	100	100	101	101	200
Set B:	0	0	99	99	100	100	100	101	101	200	200

a For each set of measurements, work out the
i range **ii** mode **iii** mean
iv median **v** interquartile range
b Discuss whay you notice about the measurements and your answers to part **a**.

D1.5 Mean of two combined data sets

Objectives

- Find the mean of two combined data sets

Useful resources

- Mini-whiteboards for mental starter

Mental starter

Write these values on the board:

23 26 29 31 34 25 39
42 43 64 71

Ask students to find:

- the mean and the range
- the median and the interquartile range.

Introductory activity

Describe this scenario: Suppose a class has 10 students. Each student sits a test marked out of 20. The six girls have a mean mark of 11, and the mean mark for the four boys is 16. Find the mean mark for the whole class.

Mean = total of all marks ÷ number of students.

The mean for the girls is their total number of marks divided by the number of girls, so the total number of marks for the girls is $11 \times 6 = 66$ marks.

What is the total number of marks for the boys? ($16 \times 4 = 64$.)

The mean for the class is

$$\frac{\text{total marks}}{\text{number of students}} = \frac{66 + 64}{10} = 13$$

Is the class mean the average of the means for the boys and the girls? (average of the 2 means is $\frac{11 + 16}{2} = 13.5$, not the same as the class mean.)

Exercise commentary and misconceptions

The concept that the mean value can be taken as the value for each member of the group is not an easy one to grasp.
Encourage students to start each problem by first identifying the variable. Then find the total number or amount for the combined data set.

All the questions depend on recalling that the mean for a group is the total for that group divided by the number in the group.

Plenary

Ask the questions:
Which average is not always a number? (Mode, for example, most common colour.) **Which average is always one of the data results?** (Mode.) **Which average cannot always be found?** (There may not be a mode if each result appears only once, also there will not be a median or mean if the data is not numerical.)

D1.5 Mean of two combined data sets

This spread will show you how to:

- Find the mean of two combined data sets

Keywords
Mean

- **Mean** = $\frac{\text{Sum of all values}}{\text{Number of values}}$

You can combine two data sets to form one larger data set by finding the mean of all the data.

Example

32 students, 12 boys and 20 girls, in class 8Z sat a maths test.
The boys' mean mark was 63%.
The girls' mean mark is 78%.
Work out the mean mark for class 8Z.

Boys: Total sum of marks $63 \times 12 = 756$
Girls: Total sum of marks $78 \times 20 = 1560$
Total sum of marks for boys and girls $756 + 1560 = 2316$
Mean mark of all students
$2316 \div 32 = 72.375\% = 72\%$ to nearest whole mark.

Example

50 students answered a survey question about time spent on the internet one evening.
30 of the students were boys and 20 were girls.

The mean time spent on the internet by all 50 students was 18 minutes.
The mean time spent on the internet by the 30 boys was 24 minutes.

Work out the mean time spent on the internet by the 20 girls.

Total time spent on internet by all 50 students:
$50 \times 18 = 900$ minutes

Total time spent on the internet by the 30 boys:
$30 \times 24 = 720$ minutes

Total time spent on the internet by the 20 girls:
$900 - 720 = 180$ minutes

Mean time spent on internet by the 20 girls:
$180 \div 20 = 9$ minutes

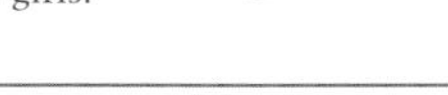

Example

There are 13 boys and 16 girls in a class.

In a test, the mean mark for the boys was p.

In the same test, the mean mark for the girls was q.
Find an expression for the mean mark of all 29 students.

$$\text{Mean} = \frac{13p + 16q}{29}$$

Exercise D1.5

1 An athletics club has 100 members, 60 boys and 40 girls.

The mean time the boys spent training one day was 86 minutes.
The mean time the girls spent training one day was 72 minutes.

Work out the mean time spent training on one day for all 100 members of the athletics club.

2 There are 120 students in Year 11 at St Edmunds school.
75 are girls and 45 are boys.

The mean time spent on homework each week for boys is 5.2 hours.
The mean time spent on homework each week for girls is 8.6 hours.

Work out the mean time spent on homework for all 120 students in Year 11 at St Edmunds school.

3 Of the students in Year 13 at St Edmunds school, 60 boys and 20 girls have passed the driving test.

The mean number of driving lessons that all 80 students had before passing the test was 19.75.
The mean number of driving lessons for the boys was 12.

Work out the mean number of driving lessons for the girls.

4 The mean mark in a statistics test for class 10Z was 84%.

There are 32 students in the class, 12 of whom are girls.
The mean mark in the test for these girls was 93%.

Work out the mean mark in the statistics test for the boys.

5 Thirty boys and girls were asked how many times they had visited the cinema in the past year.

The average number of times was 5.4.
The average number of times for the 12 boys that were asked was 2.5.

Work out the average number of times the girls in the group visited the cinema in the past year.

6 A fitness test was taken by 25 girls and 52 boys.

The average fitness score for the girls was 6.4, and the average fitness score for the boys was 9.2.

Work out the average fitness score for the whole group.

D1 Exam review

Key objectives

- Design an experiment or survey
- Design and use two-way tables for discrete and grouped data
- Calculate mean, range and median of small data sets with discrete data

Worked solution | **Commentary and misconceptions**

1

Some students may find the use of algebra in this question confusing. Encourage them to treat the letter symbols like any number. It may help to work through a numerical example before students attempt this question.
Ensure that students recall the correct formulae linking mean time and total time and know how to use it.

Total time spent watching television by the 30 boys:

$30 \times a = 30a$ hours

Total time spent watching television by the 20 girls:

$20 \times b = 20b$ hours

Total time spent watching television by the 50 children:

$30a + 20b$ hours

Mean time spent watching television by the 50 children:

$$\frac{30a + 20b}{50} = \frac{3a + 2b}{5} \text{ hours}$$

Ensure that students give the required solution. They are asked to find the mean number of hours, not the actual number of hours.

2

The aim of this question is to encourage students to work methodically and record results clearly. This is a suggested solution but obviously individual students' work may vary.
Ensure that students understand the value of using two-way tables to represent split data.

	Salt & vinegar	Ready salted	Cheese & onion	Other	Total
Females					
Males					
Total					

Fractions, decimals and percentages

Objectives

F Understand equivalent fractions, simplifying a fraction by cancelling all common factors

F/H Order fractions by rewriting them with a common denominator

F/H Add and subtract fractions by writing them with a common denominator

F/H Multiply and divide a given fraction by an integer, by a unit fraction and by a general fraction

F/H Understand and use unit fractions as multiplicative inverses

H Distinguish between fractions with denominators which are represented by terminating decimals, and other fractions which are represented by recurring decimals

F Convert simple fractions of a whole to percentages of the whole and vice versa

F/H Perform short division to convert a simple fraction to a decimal

H Convert a recurring decimal to a fraction and vice versa

Unit overview

This unit extends from whole number calculation, covered in unit N2, to calculations with fractions, decimals and percentages. Students are introduced to the three forms and shown how they relate to each other and are used in calculations. This unit provides the knowledge required to solve problems involving fractions, decimals and percentages covered in further number units.

Prior knowledge

Before your students start this unit they should be able to:

- Understand and use fraction notation
- Recognise simple fractions and the proportion of a whole they represent
- Cancel fractions to their simplest form
- Understand and use decimal notation
- Understand and use place value
- Convert between simple fractions, decimals and percentages
- Find the LCM and HCF of two integers

Differentiation

- **Foundation** focuses on simple fractions, decimals and percentages, including addition and subtraction of fractions and converting between forms
- **Foundation Plus** extends to multiplying and dividing with fractions
- **Higher Plus** extends the Higher book to using fractions, decimals and percentages in problem solving

N3.1 Ordering fractions

Objectives

- Compare and order fractions by rewriting them with a common denominator

Useful resources

- Mini-whiteboards

Mental starter

Ask students to find the lowest common multiple of two numbers, for example 18 and 24.
Extend to three numbers, for example 12, 18 and 30.

Introductory activity

Remind students of LCM referring to the mental starter. **What numbers will 6 and 8 both divide into?** (24, 48, 72, 96, 120, 144, 168, 192, 216, 240)

These are all common multiples of 6 and 8, but 24 is the lowest (least) common multiple (LCM). **What is the LCM of 4 and 3?** (12)

If John scores $\frac{5}{6}$ in a French test and $\frac{7}{8}$ in a German test, which language is he best at? Hard to say because the tests are out of different numbers. Change the fractions so they have a common denominator; the least common multiple is the easiest. The fractions will still remain equivalent as long as you multiply top and bottom. ($\frac{20}{24}$ and $\frac{21}{24}$ so better at German.)

When putting a list of fractions into order of size it can be awkward to change them all to a common denominator. Split them into two groups first, less than half and more than half.

Put these fractions in ascending order $\frac{4}{7}$ $\frac{2}{7}$ $\frac{2}{3}$ $\frac{3}{8}$ $\frac{2}{5}$ $\frac{11}{21}$.

$\frac{3}{8}$ and $\frac{2}{5}$ are both less than $\frac{1}{2}$, $\frac{1}{2} = \frac{1}{2}$ and $\frac{4}{7}$, $\frac{2}{3}$ and $\frac{11}{21}$ are all more than $\frac{1}{2}$.

$\frac{3}{8} = \frac{15}{40}$, $\frac{2}{5} = \frac{16}{40}$ so $\frac{3}{8}$ is less than $\frac{2}{5}$. $\frac{4}{7} = \frac{12}{21}$, $\frac{2}{3} = \frac{14}{21}$, $\frac{11}{21} = \frac{11}{21}$, so the order is $\frac{11}{21}$, $\frac{4}{7}$, $\frac{2}{3}$.

The complete order is $\frac{3}{8}$, $\frac{2}{5}$, $\frac{2}{4}$, $\frac{11}{21}$, $\frac{4}{7}$, $\frac{2}{3}$.

Remind students of the meaning of ascending and descending.

Exercise commentary and misconceptions

In questions 1 and 8 the easiest way to find LCMs is to list the first few multiples of each number, and select the smallest number which occurs in both (all) lists, but some students may be familiar with the prime factor method.

In questions 3–6 it is good practice for students to write out the original fraction and show how it is being altered, showing multiplication working top and bottom. This will lead to less errors.

The commonest mistake in questions 10 and 11 is to confuse ascending and descending order. Some students may want to change the fractions to decimals to compare them. This is fine as a check on their work, but they do need to master the method of equivalent fractions.

Plenary

Spanners used for tightening nuts can be measured in inches or mm. The inch measurements are

(in ascending order) $\frac{1}{4}$, $\frac{5}{16}$, $\frac{3}{8}$, $\frac{7}{16}$, $\frac{1}{2}$, $\frac{9}{16}$, $\frac{5}{8}$, $\frac{11}{16}$, $\frac{3}{4}$, $\frac{7}{8}$, $\frac{15}{16}$, 1, $1\frac{7}{16}$.

Give these to students in a random order and ask them to order them in **descending** order. Why might it be important to know how the spanners compare to each other? (When using a spanner, if one is a little too big, you need to try the next size down.)

N3.1 Ordering fractions

This spread will show you how to:

- Compare and order fractions by rewriting them with a common denominator

Keywords
Ascending
Common denominator
Denominator
Descending
Least common multiple (LCM)
Numerator

Which fraction is bigger $\frac{2}{5}$ or $\frac{5}{12}$?

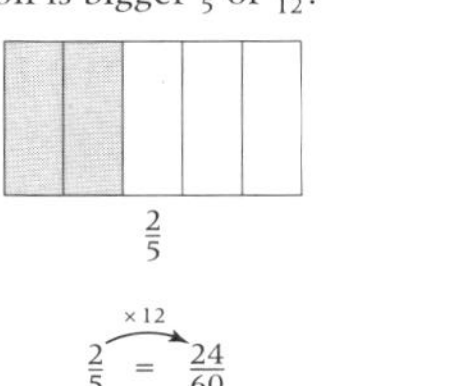

$\frac{2}{5}$

$\frac{5}{12}$

$$\frac{2}{5} \xrightarrow{\times 12} \frac{24}{60}$$

$$\frac{5}{12} \xrightarrow{\times 5} \frac{25}{60}$$

So $\frac{5}{12}$ is bigger than $\frac{2}{5}$.

- **In the fraction $\frac{2}{5}$**
  - **The top number, 2, is the numerator.**
  - **The bottom number, 5, is the denominator.**
  - **The common denominator of $\frac{2}{5}$ and $\frac{5}{12}$ is 60, the LCM of 5 and 12 (the original denominators).**

The **least common multiple (LCM)** is the lowest number that two (or more) numbers will divide into exactly.

To compare two or more fractions:

- Find the common denominator.
- Work out the equivalent fractions.
- Write the fractions in ascending or descending order.

Ascending – going up
Descending – going down

Example

Write these in ascending order. $\frac{7}{8}$ $\frac{5}{6}$ $\frac{3}{4}$

$8 = 2^3$
$6 = 2 \times 3$
$4 = 2^2$

LCM of 8, 6 and 4 $= 2^3 \times 3$
$= 24$

$\frac{7}{8} = \frac{21}{24}$ Multiply numerator and denominator by 3.

$\frac{5}{6} = \frac{20}{24}$ Multiply numerator and denominator by 4.

$\frac{3}{4} = \frac{18}{24}$ Multiply numerator and denominator by 6.

$\frac{18}{24} < \frac{20}{24} < \frac{21}{24}$

In ascending order the fractions are: $\frac{3}{4}, \frac{5}{6}, \frac{7}{8}$

Exercise N3.1

1 Write the least common multiple of each pair of numbers.

a 2 and 4 **b** 2 and 5 **c** 3 and 8 **d** 4 and 6
e 4 and 10 **f** 7 and 5 **g** 15 and 20 **h** 20 and 30

2 The diagram shows that $\frac{1}{2} = \frac{2}{4}$
Draw diagrams to show that

a $\frac{1}{2} = \frac{3}{6}$ **b** $\frac{2}{3} = \frac{4}{6}$
c $\frac{3}{5} = \frac{9}{15}$ **d** $\frac{3}{4} = \frac{15}{20}$

3 Write each fraction as an equivalent fraction with a denominator of 60.

a $\frac{1}{3}$ **b** $\frac{1}{4}$ **c** $\frac{2}{3}$ **d** $\frac{2}{5}$

4 Write each fraction as an equivalent fraction with a denominator of 24.

a $\frac{3}{4}$ **b** $\frac{1}{3}$ **c** $\frac{3}{8}$ **d** $\frac{5}{12}$

5 Rewrite each fraction with the denominator shown.

a $\frac{2}{3} = \frac{\ }{30}$ **b** $\frac{3}{7} = \frac{\ }{42}$ **c** $\frac{7}{9} = \frac{\ }{45}$ **d** $\frac{5}{8} = \frac{\ }{40}$

6 Write the fractions in each pair as equivalent fractions with a common denominator. Say which fraction in each pair is larger.

a $\frac{1}{5}$ and $\frac{3}{10}$ **b** $\frac{2}{3}$ and $\frac{3}{4}$ **c** $\frac{2}{5}$ and $\frac{1}{3}$ **d** $\frac{7}{10}$ and $\frac{2}{3}$

7 Draw diagrams (like the ones in question **2**) to illustrate your answers to question **6**.

8 Find the least common multiple of each set of numbers.

a 2, 3 and 5 **b** 3, 4 and 6 **c** 2, 3 and 8 **d** 2, 4 and 7

9 Rewrite each set of fractions with a common denominator.

a $\frac{1}{2}, \frac{2}{3}$ and $\frac{3}{4}$ **b** $\frac{1}{5}, \frac{3}{4}$ and $\frac{7}{20}$ **c** $\frac{1}{8}, \frac{7}{12}$ and $\frac{2}{3}$ **d** $\frac{2}{3}, \frac{3}{4}$ and $\frac{2}{7}$

10 Write each set of fractions in ascending order. Show your working.

a $\frac{2}{3}, \frac{1}{5}$ and $\frac{2}{15}$ **b** $\frac{1}{4}, \frac{2}{5}$ and $\frac{7}{20}$ **c** $\frac{3}{7}, \frac{3}{8}$ and $\frac{5}{14}$ **d** $\frac{2}{3}, \frac{5}{6}$ and $\frac{2}{7}$

11 Write each set of fractions in descending order. Show your working.

a $\frac{2}{5}, \frac{1}{2}, \frac{3}{10}$ and $\frac{1}{4}$ **b** $\frac{1}{4}, \frac{3}{20}, \frac{4}{5}$ and $\frac{1}{10}$ **c** $\frac{2}{5}, \frac{3}{8}, \frac{3}{4}$ and $\frac{17}{40}$ **d** $\frac{5}{6}, \frac{11}{24}, \frac{7}{12}$ and $\frac{5}{8}$

N3.2 Adding and subtracting fractions

Objectives

- Understand the terms equivalent and improper fractions and mixed numbers
- Add and subtract fractions

Useful resources

- Mini-whiteboards for mental starter

Mental starter

Challenge students to add and subtract decimals mentally. For example:
5.96 – 2.31
4.37 + 1.52
6.8 – 4.74
7.2 + 3.815

Introductory activity

Ask the questions:
What is £5 + 4 pints? Can't be done because the two measures are not the same. **What is 3m + 100 cm?** (4 m or 400 cm) Can be done because the measurements can be changed so that they are both in the same units. Conclude that you must change fractions so they have the same denominator before you can add them.

What is $\frac{3}{5} + \frac{1}{5}$? ($\frac{4}{5}$). Imagine a pizza sliced into 5 pieces (fifths). You eat 3 pieces ($\frac{3}{5}$) and a friend eats 1 piece ($\frac{1}{5}$). **How much has been eaten?** ($\frac{4}{5}$). But suppose you eat $\frac{2}{5}$ of one pizza and $\frac{3}{10}$ of a second pizza. **How much have you eaten altogether?**

($\frac{2}{5} + \frac{3}{10} = \frac{4}{10} + \frac{3}{10} = \frac{7}{10}$)?

Remind students that improper (top-heavy) fractions should be changed into mixed numbers.

Exercise commentary and misconceptions

Questions 1–5 add or subtract fractions that already have a common denominator.

Questions 6–11 involve fractions with different denominators.

Questions 12 and 13 change improper fractions to mixed numbers and vice versa.

Question 14 adds and subtracts mixed numbers. With mixed numbers some students will convert to improper fractions first, others will keep them as mixed numbers. Both methods have drawbacks. Converting to improper numbers can involve very large numbers. Keeping them as mixed fractions can involve having to interpret a negative fraction. Let students try both methods if possible.

Plenary

After winning the lottery, a philanthropist donates $\frac{1}{2}$ of the money to her local hospital,
$\frac{1}{4}$ to mend a church roof,
$\frac{1}{8}$ to build a new half pipe (skating ramp),
and $\frac{1}{16}$ to plant new trees.
What fraction was left? ($\frac{1}{16}$).

N3.2 Adding and subtracting fractions

This spread will show you how to:
- Understand the terms equivalent and improper fractions and mixed numbers
- Add and subtract fractions

Keywords
Common denominator
Improper fraction
Mixed number

- You can only add and subtract fractions if they have common denominators.

  $\frac{2}{8}+\frac{5}{8}=\frac{7}{8}$
- If the fractions have different denominators, change them to equivalent fractions with the same denominator, then add.

  $\frac{11}{12}-\frac{1}{3}=\frac{11}{12}-\frac{4}{12}=\frac{7}{12}$
- If your answer is an improper fraction, change it to a mixed number.

  $\frac{3}{4}+\frac{2}{5}=\frac{15}{20}+\frac{8}{20}=\frac{23}{20}=1\frac{3}{20}$
- Cancel any common factors in the numerator and denominator.

  $\frac{4}{5}-\frac{3}{10}=\frac{8}{10}-\frac{3}{10}=\frac{5}{10}=\frac{1}{2}$

$\frac{23}{20}$ is an **improper fraction** – its numerator is larger than its denominator.

$1\frac{3}{20}$ is a **mixed number** – a whole number and a fraction.

Example

Calculate **a** $\frac{1}{2}+\frac{1}{3}$ **b** $\frac{3}{4}-\frac{1}{5}$ **c** $\frac{1}{4}+\frac{5}{6}$

a $\frac{1}{2}+\frac{1}{3}=\frac{3}{6}+\frac{2}{6}=\frac{3+2}{6}=\frac{5}{6}$ The lowest common denominator is 6

b $\frac{3}{4}-\frac{1}{5}=\frac{15}{20}-\frac{4}{20}=\frac{15-4}{20}=\frac{11}{20}$ The lowest common denominator is 20

c $\frac{1}{4}+\frac{5}{6}=\frac{3}{12}+\frac{10}{12}=\frac{3+10}{12}=\frac{13}{12}=1\frac{1}{12}$ The lowest common denominator is 12

A common mistake is to simply add or subtract the numerators and denominators.

Example

Both of these students calculated incorrectly.
Say what they did wrong and find the correct answer.

a $\frac{2}{3}+\frac{3}{4}$

Jodi wrote

$\frac{2}{3}+\frac{3}{4}=\frac{2+3}{3+4}=\frac{5}{7}$ ✗

b $\frac{7}{8}-\frac{2}{3}$

Abdul wrote

$\frac{7}{8}-\frac{2}{3}=\frac{7-2}{8-3}=\frac{5}{5}=1$ ✗

a $\frac{2}{3}+\frac{3}{4}=\frac{8}{12}+\frac{9}{12}$
$=\frac{17}{12}$
$=1\frac{5}{12}$

Jodie added the numerators and denominators instead of finding the common denominator first.

b $\frac{7}{8}-\frac{2}{3}=\frac{21}{24}-\frac{16}{24}$
$=\frac{5}{24}$

Abdul subtracted the numerators and denominators instead of finding the common denominator first.

a The answer $\frac{5}{7}$ cannot be correct. Both $\frac{2}{3}$ and $\frac{3}{4}$ are bigger than $\frac{1}{2}$, so the answer must be greater than 1.

b $\frac{7}{8}$ is smaller than 1, so when $\frac{2}{3}$ is taken away the answer must be less than 1.

Exercise N3.2

1 The diagram shows that $\frac{1}{4}+\frac{1}{4}=\frac{1+1}{4}=\frac{2}{4}=\frac{1}{2}$
Draw diagrams to show that
a $\frac{1}{3}+\frac{2}{3}=\frac{2+1}{3}=\frac{3}{3}=1$ **b** $\frac{1}{5}+\frac{3}{5}=\frac{1+3}{5}=\frac{4}{5}$

2 Work out **a** $\frac{1}{5}+\frac{2}{5}$ **b** $\frac{1}{4}+\frac{3}{4}$ **c** $\frac{2}{7}+\frac{3}{7}$ **d** $\frac{3}{8}+\frac{5}{8}$

3 Draw diagrams (like the ones from question **1**) to illustrate your answers to question **2**.

4 Work out **a** $\frac{2}{3}-\frac{1}{3}$ **b** $\frac{4}{5}-\frac{1}{5}$ **c** $\frac{5}{6}-\frac{1}{6}$ **d** $\frac{9}{10}-\frac{3}{10}$

5 The diagram shows that $\frac{1}{3}+\frac{1}{4}=\frac{4}{12}+\frac{3}{12}=\frac{7}{12}$
Draw diagrams to show that
a $\frac{1}{3}+\frac{1}{2}=\frac{2}{6}+\frac{3}{6}=\frac{5}{6}$
b $\frac{3}{5}+\frac{3}{10}=\frac{6}{10}+\frac{3}{10}=\frac{9}{10}$

+ =

6 Calculate **a** $\frac{1}{5}+\frac{1}{10}$ **b** $\frac{2}{3}+\frac{1}{6}$ **c** $\frac{2}{5}+\frac{3}{20}$ **d** $\frac{1}{8}+\frac{1}{4}$

7 Draw diagrams (like the ones from question **5**) to illustrate your answers to question **6**.

8 Work out **a** $\frac{3}{8}-\frac{1}{4}$ **b** $\frac{5}{6}-\frac{1}{3}$ **c** $\frac{4}{5}-\frac{1}{20}$ **d** $\frac{3}{8}-\frac{1}{16}$

9 Work out **a** $\frac{4}{7}-\frac{1}{3}$ **b** $\frac{4}{5}-\frac{2}{3}$ **c** $\frac{8}{9}-\frac{5}{6}$ **d** $\frac{3}{5}-\frac{1}{4}$

10 The diagram shows what happens when you add fractions with a total that is greater than 1.
In this example, $\frac{3}{4}+\frac{4}{5}=\frac{15}{20}+\frac{16}{20}=\frac{15+16}{20}=\frac{31}{20}=1\frac{11}{20}$

+ =

Draw diagrams to find the answers to **a** $\frac{3}{5}+\frac{1}{2}$ **b** $\frac{5}{6}+\frac{3}{4}$

11 Calculate **a** $\frac{2}{3}+\frac{1}{2}$ **b** $\frac{4}{5}+\frac{1}{2}$ **c** $\frac{1}{3}+\frac{4}{5}$ **d** $\frac{4}{7}+\frac{1}{2}$

12 Change these improper fractions into mixed numbers.
a $\frac{5}{4}$ **b** $\frac{9}{5}$ **c** $\frac{13}{8}$ **d** $\frac{17}{4}$

13 Change these mixed numbers to improper fractions.
a $1\frac{3}{4}$ **b** $1\frac{7}{16}$ **c** $1\frac{5}{9}$ **d** $2\frac{4}{7}$

14 Calculate
a $2\frac{3}{5}+1\frac{1}{3}$ **b** $2\frac{3}{4}+1\frac{5}{6}$ **c** $4\frac{3}{7}+3\frac{1}{2}$ **d** $5\frac{4}{9}+2\frac{3}{7}$
e $3\frac{3}{5}-2\frac{1}{4}$ **f** $2\frac{1}{2}-1\frac{3}{4}$ **g** $3\frac{3}{4}-1\frac{4}{5}$ **h** $7\frac{3}{7}-2\frac{1}{2}$

N3.3 Multiplying and dividing fractions

Objectives

- Understand unit fractions and use them as multiplicative inverses
- Multiply and divide fractions

Useful resources

- Mini-whiteboards

Mental starter

A police chief has to allocate her officers to different duties. A quarter in the office, half that many on car patrol, three times as many on foot patrol as car patrol and the rest on investigative work.
If she puts 2250 officers on foot patrol, how many will there be on all the other duties?
(2250 foot; 750 cars; 1500 office; 6000 total; 1500 investigate.)

Introductory activity

Discuss what students already know about multiplying and dividing fractions.

Here are some examples you can use that give students an idea of why the methods work:

$\frac{1}{4} \times \frac{1}{2}$. This must be halving $\frac{1}{4}$. So you must get $\frac{1}{8}$. Notice the 4 and 2 have been multiplied.

$\frac{1}{4} \times \frac{3}{2}$. This is still multiplying by a half, but this time there are 3 times as many halves. So you will get $\frac{3}{8}$ instead of just $\frac{1}{8}$. Notice that the numerators and denominators have both been multiplied.

$3 \div \frac{1}{4}$. How many quarters are there in 3?
There are 12. Notice that $3 \times 4 = 12$. You can turn the second fraction upside down (invert it) and then multiply so
$3 \div \frac{1}{4} = 3 \times \frac{4}{1} = 12$.
This works for more complex fractions as well.

$\frac{2}{5} \times \frac{4}{3} = \frac{8}{15}$ (multiply numerators and denominators).

$\frac{2}{3} \div \frac{5}{7}$ (the dividing fraction is inverted, then used to multiply) $\frac{2}{3} \times \frac{7}{5} = \frac{14}{15}$.

When multiplying or dividing fractions, the first step is to change any mixed numbers into improper fractions.

$$(1\tfrac{3}{5})^2 \div 7 = \tfrac{8}{5} \times \tfrac{8}{5} \div \tfrac{7}{1}$$
$$= \tfrac{64}{25} \div \tfrac{7}{1}$$
$$= \tfrac{64}{25} \times \tfrac{1}{7}$$
$$= \tfrac{64}{175}.$$

Exercise commentary and misconceptions

Strangely, multiplying and dividing fractions is easier than adding and subtracting. Most students find the methods easier to understand from examples rather than a full explanation of how and why these methods work.

Questions 1–3 help students to understand the methods used in later questions.

Questions 4–8 provide basic practice at multiplying and dividing with fractions.

In question 10 students must change to improper fractions first. Students may be tempted to deal with the integer parts separately from the fraction parts as they did when adding and subtracting.

Plenary

The ancient Egyptians only used what are called unit fractions, $\frac{1}{2}, \frac{1}{4}, \frac{1}{5}, \frac{1}{10} \ldots$
If they wanted any other fraction, they had to add unit fractions together. They always used as few unit fractions as possible in the sum. For instance, they had no way of writing $\frac{3}{4}$ as a single fraction. They wrote $\frac{1}{2} + \frac{1}{4}$, not $\frac{1}{4} + \frac{1}{4} + \frac{1}{4}$.
How could they write $\frac{7}{10}$ and $\frac{5}{6}$?
($\frac{1}{2} + \frac{1}{5}$ and $\frac{1}{2} + \frac{1}{3}$.)

N3.3 Multiplying and dividing fractions

This spread will show you how to:
- Understand unit fractions and use them as multiplicative inverses
- Multiply and divide fractions

Keywords
Multiplicative inverse
Reciprocal
Unit fraction

The diagram shows the multiplication

$\frac{2}{3} \times \frac{3}{4} = \frac{6}{12} = \frac{1}{2}$

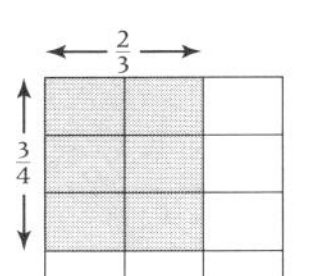

You get the same result if you multiply the numerators together and multiply the denominators together.

- **To multiply fractions, multiply the numerators and then the denominators, then cancel any common factors.**

Example

Find a $\frac{2}{3} \times \frac{4}{5}$ b $\frac{4}{9} \times \frac{3}{5}$ c $\frac{7}{4} \times \frac{5}{2}$

a $\frac{2}{3} \times \frac{4}{5} = \frac{2 \times 4}{3 \times 5} = \frac{8}{15}$

b $\frac{4}{9} \times \frac{3}{5} = \frac{4 \times 3}{9 \times 5} = \frac{12}{45} = \frac{4}{15}$

c $\frac{7}{4} \times \frac{5}{2} = \frac{7 \times 5}{4 \times 2} = \frac{35}{8} = 4\frac{3}{8}$

- Multiplying by a **unit fraction** is the same as dividing by its denominator. For example, multiplying by $\frac{1}{5}$ is the same as dividing by 5.

$$10 \times \frac{1}{5} = \frac{10}{5} = 2 \text{ and } 10 \div 5 = 2$$

- Dividing by a unit fraction is the same as multiplying by its denominator.

$$10 \div \frac{1}{2} = 10 \times 2 = 20$$

A unit fraction has a numerator of 1.

The **multiplicative inverse** of an integer is its **reciprocal**.
For example, the reciprocal of 5 is $\frac{1}{5}$.

You combine these two ideas when you divide by a fraction.

For example,

$10 \div \frac{5}{2} = 10 \div 5 \div \frac{1}{2}$ — $\div 5$ is the same as $\times\frac{1}{5}$

$= 10 \times \frac{1}{5} \times 2$ — $\div\frac{1}{2}$ is the same as $\times 2$

$= 10 \times \frac{2}{5}$

$= \frac{20}{5} = 4$

- **To divide by a fraction, multiply by its multiplicative inverse.**

Example

Find a $\frac{3}{7} \div 5$ b $\frac{5}{12} \div \frac{3}{4}$ c $3\frac{1}{2} \div 2\frac{1}{5}$

a $\frac{3}{7} \div 5 = \frac{3}{7} \times \frac{1}{5} = \frac{3}{35}$

b $\frac{5}{12} \div \frac{3}{4} = \frac{5}{12} \times \frac{4}{3} = \frac{20}{36} = \frac{5}{9}$

c $3\frac{1}{2} \div 2\frac{1}{5} = \frac{7}{2} \div \frac{11}{5} = \frac{7}{2} \times \frac{5}{11} = \frac{35}{22} = 1\frac{13}{22}$

The **multiplicative inverse** of a fraction is the original fraction 'turned upside down'. The inverse of $\frac{3}{5}$ is $\frac{5}{3}$.

Change mixed numbers to improper fractions first.

Exercise N3.3

1 Write the reciprocal (multiplicative inverse) of

a 4 b 6 c 10 d 12

2 Write the multiplicative inverse of

a $\frac{1}{5}$ b $\frac{1}{9}$ c $\frac{1}{2}$ d $\frac{1}{3}$

3 Rewrite each of these divisions as multiplications.

a $8 \div 5$ b $6 \div 4$ c $9 \div 5$ d $17 \div 3$

For example, you can write $7 \div 5$ as $7 \times \frac{1}{5}$.

4 Copy and complete these sentences.

a Dividing a number by 4 is the same as multiplying the number by ___

b Multiplying a number by $\frac{1}{2}$ is the same as dividing the number by ___

c Dividing a number by $\frac{1}{3}$ is the same as multiplying the number by ___

5 Draw diagrams, like the one on page 78, to show that

a $\frac{1}{4} \times \frac{1}{2} = \frac{1}{8}$ b $\frac{3}{4} \times \frac{1}{3} = \frac{3}{12} = \frac{1}{4}$ c $\frac{2}{5} \times \frac{2}{3} = \frac{4}{15}$ d $\frac{1}{4} \times \frac{3}{5} = \frac{3}{20}$

6 Calculate these, giving your answers in their simplest form.

a $\frac{3}{4} \times \frac{1}{5}$ b $\frac{2}{3} \times \frac{2}{9}$ c $\frac{2}{7} \times \frac{1}{4}$ d $\frac{5}{16} \times \frac{4}{5}$

e $\frac{5}{9} \times \frac{4}{7}$ f $\frac{7}{8} \times \frac{2}{21}$ g $\frac{4}{5} \times \frac{3}{13}$ h $\frac{8}{35} \times \frac{7}{24}$

7 Calculate, giving your answers as fractions in their simplest terms.

a $3 \div 4$ b $6 \div 8$ c $4 \div 5$ d $8 \div 10$

e $3 \div 7$ f $9 \div 5$ g $24 \div 7$ h $22 \div 8$

For example, $13 \div 4 = 13 \times \frac{1}{4} = \frac{13}{4} = 3\frac{1}{4}$.

8 Calculate

a $\frac{5}{8} \div 4$ b $\frac{3}{4} \div 6$ c $\frac{2}{3} \div 7$ d $\frac{3}{16} \div 9$

e $4 \div \frac{1}{5}$ f $6 \div \frac{2}{3}$ g $5 \div \frac{2}{5}$ h $11 \div \frac{3}{7}$

9 Calculate

a $\frac{1}{8} \div \frac{3}{8}$ b $\frac{1}{5} \div \frac{4}{5}$ c $\frac{1}{14} \div \frac{3}{7}$ d $\frac{1}{10} \div \frac{2}{5}$

e $\frac{2}{3} \div \frac{3}{4}$ f $\frac{5}{8} \div \frac{7}{9}$ g $\frac{1}{12} \div \frac{3}{8}$ h $\frac{2}{7} \div \frac{3}{4}$

10 Calculate

a $1\frac{1}{2} \times \frac{3}{4}$ b $2\frac{3}{4} \times \frac{2}{5}$ c $1\frac{1}{5} \times 2\frac{1}{2}$ d $1\frac{2}{3} \times 1\frac{4}{5}$

e $2\frac{7}{8} \div \frac{3}{5}$ f $4\frac{1}{4} \div 3\frac{1}{2}$ g $5\frac{3}{8} \div 2\frac{3}{4}$ h $9\frac{1}{3} \div 2\frac{1}{4}$

N3.4 Converting fractions to decimals

Objectives

- Distinguish between recurring and terminating decimals
- Convert fractions to decimals and percentages and vice versa

Useful resources

- Number line

Mental starter

If you work out $\frac{1}{3}$ as a decimal you get 0.333333 or $0.\dot{3}$ (show this by short division). **What does that imply about $\frac{2}{3}$ and $\frac{3}{3}$?** This leads to $\frac{3}{3} = 0.99999999$ or $0.\dot{9}$ which is the same as 1. Discuss why these are the same number.

Work out $4 \div 3$ and $5 \div 3$ as decimals. (1.33333 ..., 1.66666 ...) **What does this suggest about the value of 1.99999 ... or $1.\dot{9}$?** (It equals $6 \div 3 = 2$)

Introductory activity

Discuss how all fractions can be changed into decimals.

Some change into decimals that stop (terminate) such as $\frac{2}{25} = 0.08$.

Some change into recurring decimals such as $\frac{5}{6} = 0.83333$ ... or $0.8\dot{3}$.

How can you tell which of these types a fraction will change into? Simplify the fraction fully before you start. **Which numbers are easy to divide by?** (10, 100, 2, 5 ...) All these have only twos and fives as prime factors. If the denominator of a fraction is made up of only twos and fives when written as a product of prime factors, then the fraction will terminate, otherwise it will recur. For example:

$\frac{26}{80} = \frac{13}{40}$

$40 = 2 \times 2 \times 2 \times 5$

so $\frac{13}{40}$ will terminate (= 0.325).

$\frac{5}{18}$

$18 = 2 \times 3 \times 3$

so $\frac{5}{18}$ recurs (= 0.277777 ...).

If you want to change a fraction into a decimal without a calculator, a good method is short division. Make sure that the number you are dividing by goes on the outside of the division sign.

To change a fraction into a percentage, first change it into a decimal, then multiply by 100. A decimal has a denominator of 1, and a percentage is a fraction with a denominator of 100, so you are multiplying the fraction by $\frac{100}{100}$ or 1.

$\frac{5}{8} = 0.625 = 62.5\%$

$\frac{2}{3} = 0.6666 = 66.7\%$

Exercise commentary and misconceptions

Questions 1–8 are in pairs: the first question of each pair changes fractions to decimals and the second question changes these decimals to percentages.

In question 7 the decimals are recurring – encourage students to use the dot notation and challenge them to think of how it is used if more than 1 figure recurs.

To help with question 11, remind students to write the denominators as the product of prime factors.

To help with question 12, encourage students to do $1 \div 13$ as long division, and to think about the remainder each time.

Plenary

There are some decimals that do not terminate or recur. **What must they do?** (Keep going but not repeat.)

Four of the following numbers are like this, which are they? $\sqrt{2}$, 0.3^2, π, $\frac{4}{5}$, e, $\frac{7}{13}$, $\sqrt{9}$, $\sqrt{11}$ ($\sqrt{2}$, π, e, $\sqrt{11}$).

You may need to briefly discuss the exponential constant e = 2.7182 ...

These are called irrational numbers.

N3.4 Converting fractions to decimals

This spread will show you how to:
- Distinguish between recurring and terminating decimals
- Convert fractions to decimals and percentages and vice versa

Keywords
Denominator
Numerator
Recurring
Terminating

- **To convert a fraction to a decimal divide the numerator by the denominator.**

Example

Write these fractions as decimals. **a** $\frac{5}{8}$ **b** $\frac{5}{9}$ **c** $\frac{1}{7}$

a $\frac{5}{8} = 5 \div 8 = 8\overline{)5.000}$ = 0.625

b $\frac{5}{9} = 5 \div 9 = 9\overline{)5.000\ldots}$ = 0.555 ...

c $\frac{1}{7} = 1 \div 7 = 7\overline{)1.000000000\ldots}$ = 0.142857142 ...

0.625 is a **terminating** decimal

0.555 ... = $0.\dot{5}$ is a **recurring** decimal. The dot over the 5 shows the recurring digit.

$0.\dot{1}4285\dot{7}$ is a recurring decimal. The dots show the recurring group of digits.

To decide if a fraction will be a terminating or a recurring decimal, look at the denominator.

- If the only factors of the denominator are 2 and/or 5 or combinations of 2 and 5 then the fraction will be a terminating decimal.
- If the denominator has any factors other than 2 and/or 5 then the fraction will be a recurring decimal.

Example

Say whether these fractions are terminating or recurring decimals.

a $\frac{9}{20}$ **b** $\frac{4}{30}$ **c** $\frac{7}{16}$ **d** $\frac{11}{13}$

a denominator = $20 = 2 \times 2 \times 5 \rightarrow \frac{9}{20}$ is a terminating decimal
b $\frac{4}{30} = \frac{2}{15}$ denominator = $15 = 3 \times 5 \rightarrow \frac{2}{15}$ is a recurring decimal
c denominator = $16 = 2^4 \rightarrow \frac{7}{16}$ is a terminating decimal
d denominator = $13 \rightarrow \frac{11}{13}$ is a recurring decimal

Simplify the fraction.

To convert a fraction to a percentage
- write it as a decimal
- multiply the decimal by 100%.

$100\% = \frac{100}{100} = 1$ so you are multiplying by 1 which does not change the value of the decimal.

Example

Write as percentages **a** $\frac{5}{8}$ **b** $\frac{5}{9}$ **c** $\frac{1}{7}$

a $\frac{5}{8} = 0.625 = 0.625 \times 100\% = 62.5\%$
b $\frac{5}{9} = 0.\dot{5} = 0.\dot{5} \times 100\% = 55.\dot{5}\% = 55.6\%$ to 1 dp
c $\frac{1}{7} = 0.\dot{1}4285\dot{7} = 0.\dot{1}4285\dot{7} \times 100\% = 14.\dot{2}8571\dot{4}\% = 14.3\%$ to 1 dp

Exercise N3.4

You should be able to do all of these mentally.

1 Write a decimal equivalent for each of these fractions.
a $\frac{1}{2}$ **b** $\frac{3}{4}$ **c** $\frac{2}{5}$
d $\frac{1}{10}$ **e** $\frac{1}{5}$ **f** $\frac{1}{4}$

2 Convert each of the decimals from question **1** to a percentage.

3 Convert each fraction to a percentage, using a written method. Show your working.
a $\frac{5}{8}$ **b** $\frac{4}{5}$ **c** $\frac{7}{8}$
d $\frac{3}{5}$ **e** $\frac{3}{8}$ **f** $\frac{1}{8}$

You should practise using a written method here, even if you can do these mentally.

4 Use a calculator to check your answers to question **3**.

5 Use a calculator to convert these fractions to decimals.
a $\frac{1}{16}$ **b** $\frac{7}{25}$ **c** $\frac{7}{125}$
d $\frac{3}{40}$ **e** $\frac{7}{16}$ **f** $\frac{1}{32}$

6 Convert the decimal answers from question **5** to percentages.

7 Use a written method to convert each fraction to a decimal. Use the 'dot' notation to represent recurring decimals.
a $\frac{1}{3}$ **b** $\frac{1}{6}$ **c** $\frac{2}{3}$
d $\frac{1}{7}$ **e** $\frac{1}{9}$ **f** $\frac{5}{6}$

DID YOU KNOW?
Blaise Pascal's mechanical calculator, invented in 1645, used dials and gears to add and subtract positive numbers and was based on the decimal system.

8 Convert your decimal answers from question **7** to percentages.

9 Use a calculator to check your answers to questions **7** and **8**.

10 Use an appropriate method to convert these fractions to decimals.
a $\frac{3}{7}$ **b** $\frac{3}{16}$ **c** $\frac{17}{80}$
d $\frac{5}{9}$ **e** $\frac{4}{25}$ **f** $\frac{5}{7}$

11 State whether each of these fractions will give a recurring decimal or a terminating decimal. Explain your answers.
a $\frac{1}{25}$ **b** $\frac{3}{20}$ **c** $\frac{4}{11}$
d $\frac{1}{126}$ **e** $\frac{1}{125}$ **f** $\frac{1}{128}$

12 Shula says, 'I used my calculator to change $\frac{1}{13}$ to a decimal, and I got the answer 0.07692308. There is no repeating pattern, so the decimal does not recur.' Explain why Shula is wrong.

N3.5 Converting decimals and percentages to fractions

Objectives

- Distinguish between recurring and terminating decimals
- Convert fractions to decimals and percentages and vice versa

Useful resources

- Number line

Mental starter

Ask students to convert these simple decimals to fractions:

0.25, 0.1, 0.5, 0.75, 1.75

Extend to these simple recurring decimals:

$0.\dot{3}$, $0.\dot{6}$, $1.\dot{3}$

Introductory activity

To change between a decimal and percentage you need to think what a percentage is: a fraction with a denominator of 100.

23% is $\frac{23}{100}$ which is 0.23. So to change a percentage to a decimal divide by 100.

To change percentages into fractions, remember that they are already fractions of 100. For example $16\% = \frac{16}{100} = \frac{8}{50} = \frac{4}{25}$ (cancel down).

To change terminating decimals to fractions, remember what place value they are in; tenths, hundredths, thousandths.

For example $0.3 = \frac{3}{10}$.

$0.21 = \frac{21}{100}$.

$0.306 = \frac{306}{1000} = \frac{153}{500}$ after cancelling.

Recurring decimals are the most difficult to change into fractions. The method is best explained using the three examples in the student book.

To change $0.\dot{4}$ into a fraction

$10 \times 0.\dot{4} = 4.\dot{4}$

$1 \times 0.\dot{4} = 0.\dot{4}$

$9 \times 0.\dot{4} = 4$

$0.\dot{4} = \frac{4}{9}$

To change $0.\dot{5}\dot{6}$ into a fraction

$100 \times 0.\dot{5}\dot{6} = 56.\dot{5}\dot{6}$

$1 \times 0.\dot{5}\dot{6} = 0.\dot{5}\dot{6}$

$99 \times 0.\dot{5}\dot{6} = 56$

$0.\dot{5}\dot{6} = \frac{56}{99}$

To change $0.\dot{3}1\dot{5}$ into a fraction

$1000 \times 0.\dot{3}1\dot{5} = 315.\dot{3}1\dot{5}$

$1 \times 0.\dot{3}1\dot{5} = 0.\dot{3}1\dot{5}$ $999 \times 0.\dot{3}1\dot{5}$

$= 315$ $0.\dot{3}1\dot{5} = \frac{315}{999} = \frac{35}{111}$

Exercise commentary and misconceptions

In questions 1–6 remind students to simplify fractions by cancelling where possible.

In questions 8–10 remind students of the method for converting recurring decimals.

Plenary

If 6 people share a prize equally amongst themselves, what percentage do they each receive? Give the answer as a decimal percentage, then as a mixed fraction percentage. ($100\% \div 6 = 16.66\ldots\% = 16\frac{2}{3}\%$.)

N3.5 Converting decimals and percentages to fractions

This spread will show you how to:

- Distinguish between recurring and terminating decimals
- Convert fractions to decimals and percentages and vice versa

Keywords
Denominator
Recurring
Terminating

- **To convert a percentage to a fraction, divide the percentage by 100.**

  For example,

  $43.7\% = \frac{43.7}{100} = 0.437$

- **To convert a terminating decimal to a fraction:**

  1 Write the decimal as a fraction with denominator 10, 100, 1000, ..., according to the number of decimal places. For example, $0.45 = \frac{45}{100}$

  2 Simplify the fraction.

  $\frac{45}{100} = \frac{9}{20}$

2 decimal places so denominator is 100.

Example

Convert these to fractions **a** 0.306 **b** 45% **c** 32.5% **d** 0.52

a $0.306 = \frac{306}{1000} = \frac{153}{500}$

b $45\% = \frac{45}{100} = \frac{9}{20}$

c $32.5\% = 0.325 = \frac{325}{1000} = \frac{65}{200} = \frac{13}{40}$

d $0.52 = \frac{52}{100} = \frac{26}{50} = \frac{13}{25}$

The next example shows how to convert **recurring** decimals to fractions. You multiply by a power of 10, then subtract one lot of the original decimal to produce a fraction.

Example

Write as fractions **a** $0.\dot{4}$ **b** $0.\dot{5}\dot{6}$

a

$10 \times 0.\dot{4} = 4.\dot{4}$ one recurring figure so multiply decimal by 10

$10 \times 0.\dot{4} - 1 \times 0.\dot{4} = 4.\dot{4} - 0.\dot{4}$ subtract one lot of the original decimal.

$9 \times 0.\dot{4} = 4$

$0.\dot{4} = \frac{4}{9}$

b

$100 \times 0.\dot{5}\dot{6} = 56.\dot{5}\dot{6}$ two recurring figures so multiply decimal by 100

$100 \times 0.\dot{5}\dot{6} - 1 \times 0.\dot{5}\dot{6} = 56.\dot{5}\dot{6} - 0.\dot{5}\dot{6}$ subtract one lot of the original decimal.

$99 \times 0.\dot{5}\dot{6} = 56$

$0.\dot{5}\dot{6} = \frac{56}{99}$

Exercise N3.5

1 Convert these percentages to decimals.

a 43% **b** 86% **c** 94% **d** 45.5% **e** 3.75% **f** 105%

2 Convert these decimals to fractions. You should be able to do these mentally.

a 0.5 **b** 0.25 **c** 0.2 **d** 0.125 **e** 0.75 **f** 0.9

3 Convert these decimals to fractions.

a 0.51 **b** 0.43 **c** 0.413 **d** 0.719 **e** 0.91 **f** 0.871

4 Convert these percentages to fractions.

a 49% **b** 53% **c** 73% **d** 81% **e** 37% **f** 19%

5 Convert these decimals to fractions. Give your answers in their simplest form.

a 0.32 **b** 0.55 **c** 0.44 **d** 0.155 **e** 0.64 **f** 0.265

6 Convert these percentages to fractions. Give your answers in their simplest form.

a 55% **b** 62% **c** 84% **d** 65% **e** 72% **f** 18.5%

7 Which of these fractions will make recurring decimals? Explain your answer.

$\frac{22}{25}$ $\frac{17}{20}$ $\frac{8}{11}$ $\frac{2}{5}$

8 Write each of these recurring decimals using 'dot' notation.

a 0.111 ... **b** 0.555 ... **c** 0.75555 ... **d** 0.346346346 ...

e 0.7656565 ...

9 Use the method shown on page 82 to convert these recurring decimals to fractions.

a $0.\dot{2}$ **b** $0.\dot{6}$ **c** $0.\dot{2}\dot{5}$ **d** $0.\dot{2}\dot{7}$ **e** $0.\dot{5}4\dot{5}$ **f** $0.\dot{6}0\dot{5}$

10 Convert these percentages to fractions.

a $52.\dot{2}\%$ **b** $5.\dot{5}\%$ **c** $45.\dot{2}\dot{7}\%$ **d** $8.\dot{3}\dot{5}\%$ **e** $66.\dot{6}\%$ **f** $8.\dot{2}0\dot{5}\%$

11 Find a fraction equal to the recurring decimal $0.\dot{0}12345678\dot{9}$, giving your answer in its simplest form. Show your working.

N3 Exam review

Key objectives

- Order fractions by rewriting them with a common denominator
- Add and subtract fractions by writing them with a common denominator
- Distinguish between fractions which are represented by terminating decimals, and fractions which are represented by recurring decimals
- Convert a recurring decimal to a fraction and vice versa
- Multiply and divide a given fraction by an integer, by a unit fraction and by a general fraction

Worked solution	Commentary and misconceptions
1	Ensure that students are able to cancel fractions to their simplest form correctly.
a **i** $60\% = \frac{60}{100} = \frac{6}{10} = \frac{3}{5}$ **ii** $0.52 = \frac{52}{100} = \frac{26}{50} = \frac{13}{25}$	
b	Some students may be tempted to convert the forms to fractions before completing the calculation. Discourage them from doing this as it leads to more steps of calculation. Ensure that students give their solutions in their simplest form.
i $0.36 + 0.61 = 0.97$ $0.97 = \frac{97}{100}$ **ii** $33\% - 20\% = 13\% = \frac{13}{100}$	
iii $\frac{5}{7} \div \frac{10}{9} = \frac{5}{7} \times \frac{9}{10} = \frac{45}{70} = \frac{9}{14}$	Ensure that students understand how to use fractions as multiplicative inverses.
2	
a 0.067 0.56 0.6 0.605 0.65	Students may find ordering similar looking numbers confusing. Encouraging them to examine the place value of each digit may help.
b −10 −6 −4 2 5 **c** $\frac{2}{5}$ $\frac{1}{2}$ $\frac{2}{3}$ $\frac{2}{4}$	Students may find it useful to convert to decimals or percentages. Encourage them to use the method they feel most comfortable with.

S2

Angles and circles

Objectives

F/H Distinguish between lines and line segments; use parallel lines, alternate angles and corresponding angles

F/H Understand the consequent properties of parallelograms and a proof that the angle sum of a triangle is 180 degrees

F/H Understand a proof that the exterior angle of a triangle is equal to the sum of the interior angles at the other two vertices

F/H Explain why the angle sum of a quadrilateral is 360 degrees

F/H Recall the definition of a circle and the meaning of related terms, including chord, tangent, arc, sector and segment

H Prove and use the facts that the angle subtended by an arc at the centre of a circle is twice the angle subtended at any point on the circumference, the angle subtended at the circumference by a semicircle is a right angle, that angles in the same segment are equal, and that opposite angles of a cyclic quadrilateral sum to 180 degrees

H Understand that the tangent at any point on a circle is perpendicular to the radius at that point

H Understand and use the fact that tangents from an external point are equal in length

Unit overview

This unit consolidates Key Stage 3 knowledge of angles. It extends from angles on lines to angles in polygons and circles, including circle theorems. This unit provides a basis for further work on problems involving angles in units S4 and S8.

Prior knowledge

Before your students start this unit they should be able to:

- Understand and use angle notation
- Understand and use angles on straight lines and at a point
- Recall the angle properties of triangles and quadrilaterals
- Understand the concept of parallel and perpendicular lines
- Understand and use the definitions of parts of circles

Differentiation

- **Foundation** focuses on angles on a straight line and at a point, using a protractor to measure angles. It extends to angles in triangles, including triangles on a coordinate grid
- **Foundation Plus** extends to angles in scalene triangles and quadrilaterals, including both interior and exterior angles
- **Higher Plus** focuses on angles in circles and extends the Higher book to the alternate segment theorem

S2.1 Angles in straight lines

Objectives

- Use parallel lines, alternate angles, corresponding angles and interior angles

Useful resources

- Rulers

Mental starter

State the facts:
Angles k and j are both multiples of 10.
Angle $3j$ is vertically opposite angle k.
Angles $2j$ and $(k - 20)$ are next to each other (adjacent) on a straight line.
Angle $(k - j + 10)$ is a right angle.
What value do k and j have?
($j = 40$; $k = 120$.)

Drawing a sketch may help to reinforce the relationship between the angles. Trial and improvement is one way to solve the problem although it can also be solved by simultaneous equations.
You could give the hint that the answers are multiples of 20.

Introductory activity

Draw two pairs of parallel lines horizontally with a line intersecting. Identify the **corresponding** angles and show that the four angles all match up (correspond) in the same position.
Identify the **alternate** angles that lie inside the parallel lines – two angles are the same if they are on the other (alternate) side of the line.
You may use the letters F and Z when helping students to spot these angles, but the terms 'F angle' and 'Z angle' are not commonly acceptable as a full explanation.

Exercise commentary and misconceptions

In questions 2, 3 and 4 it is easiest (though not essential) to find the angles in alphabetical order.

In question 2 remind students that they must give angle facts or reasons for their answers.

Rules for vertically opposite angles, straight line angles and angles at a point could also be used in this exercise. There is often more than one way of finding the angles correctly.
Remind students that the diagrams are not drawn to scale so no assumptions can be made about the size of an angle.

Plenary

What is the sum of the two angles on the same side of the line between 2 parallel lines? (180°) These are sometimes called allied angles or co-interior angles, although these terms are not often used and do not appear in any exam board vocabulary. However, the rule is often useful.

The bearing of B from A is 030°. The bearing of C from B is 110°. **What is angle ABC?** (100°)

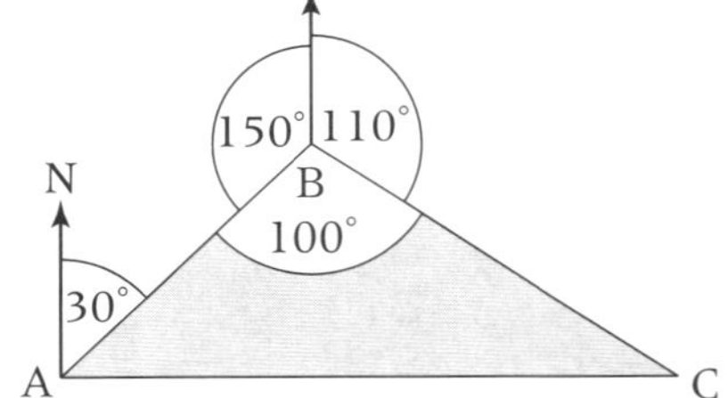

S2.1 Angles in straight lines

This spread will show you how to:
- Use parallel lines, alternate angles, corresponding angles and interior angles

Keywords
Alternate
Corresponding
Interior
Parallel
Supplementary
Vertically opposite

- Angles are formed when two lines cross, as at this crossroads.

$a = c$ and $b = d$

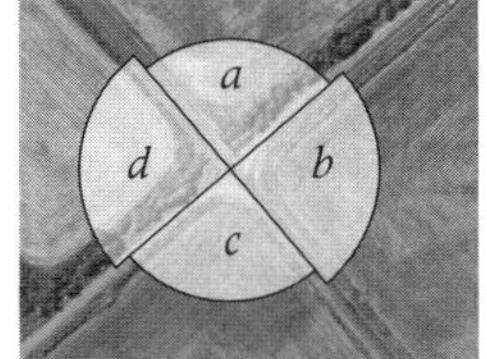

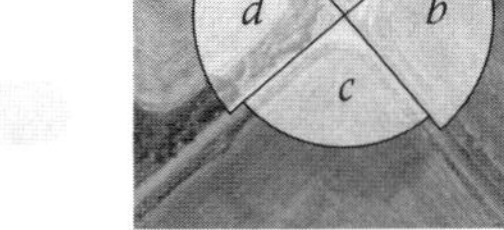

- **Vertically opposite** angles are equal.

Parallel lines are lines that never cross.

You need to remember these angle facts for parallel lines.

- **Alternate** angles are equal.

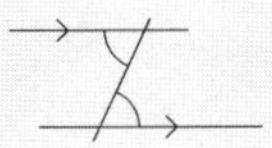

These are sometimes called Z angles.

- **Corresponding** angles are equal.

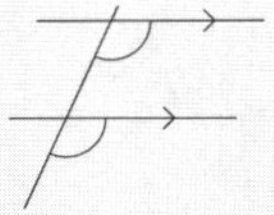

These are sometimes called F angles.

- **Interior** angles are **supplementary**. $a + b = 180°$

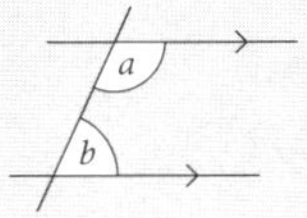

Supplementary angles add up to 180°.

When you work out angles, you should always say which angle fact you are using.

Example

Find the missing angles. Give reasons for your answers.

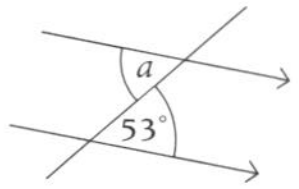

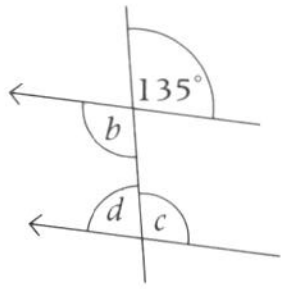

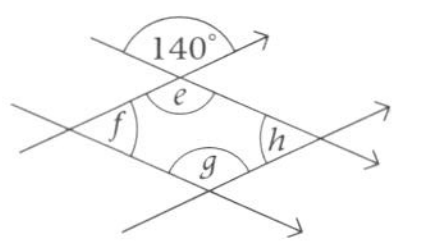

$a = 53°$ alternate angles

$b = 135°$ vertically opposite
$c = 135°$ corresponding
$d = 45°$ interior angles

$e = 140°$ vertically opposite
$f = 40°$ interior angles
$g = 140°$ interior angles
$h = 40°$ interior angles

Exercise S2.1

You may be able to find more than one pair in each diagram.

1 Copy the diagrams.
 i Use colour to show alternate angles in each diagram.
 ii Use another colour to show corresponding angles in each diagram.

a b 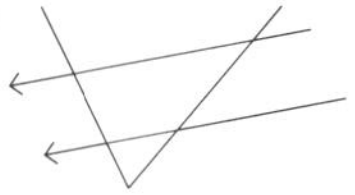 c

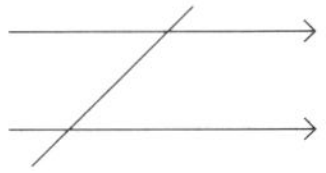

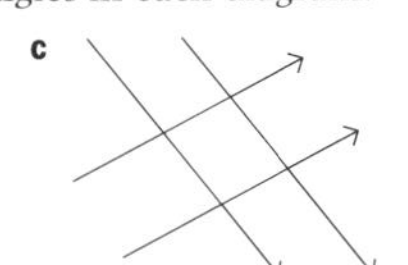

2 Find the missing angles in each diagram.
Write down which angle fact you are using each time.

a b 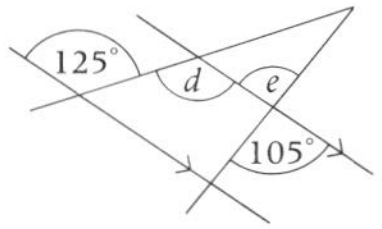 c

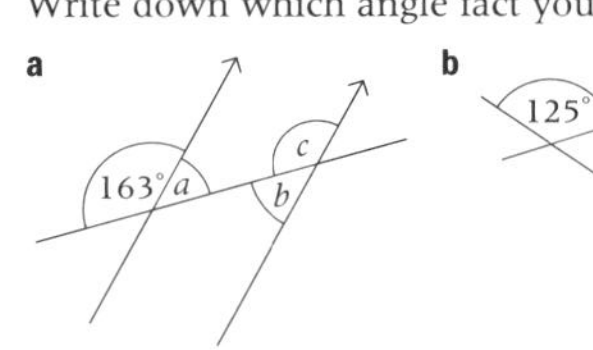

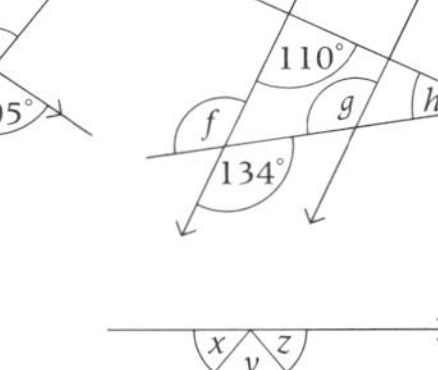

3 **a** Angles x, y and z are on a straight line. Write down the value $x + y + z$.
 b Use alternate angles to work out the two missing angles in the triangle.
 c Use your answers to **a** and **b** to show that angles in a triangle add up to 180°.

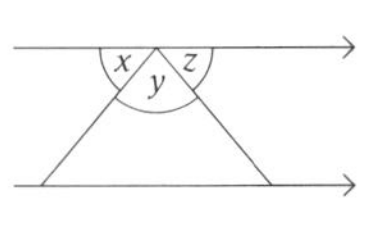

4 **a** Work out the missing angles in this diagram.
 b Describe the quadrilateral formed between the pairs of parallel lines.

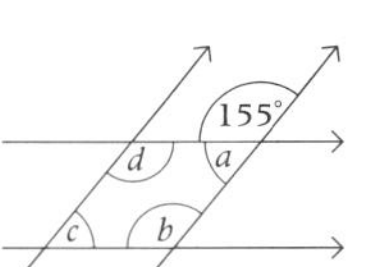

5 Work out the missing angles.

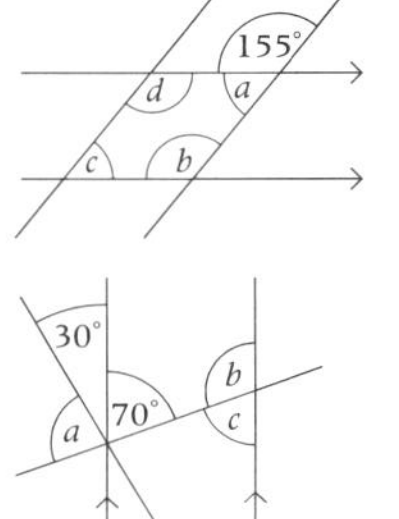

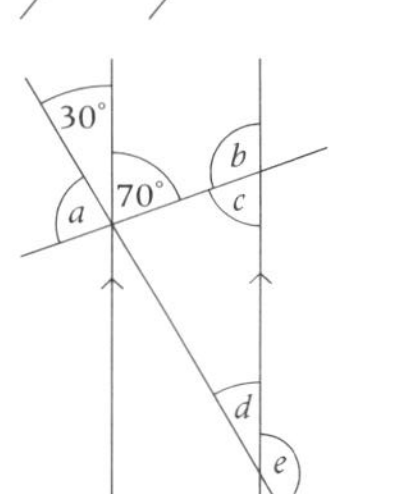

S2.2 Angles in polygons

Objectives

- Calculate and use the interior and exterior angles of polygons

Useful resources

- Rulers
- Chalk

Mental starter

Demonstrate how an isosceles trapezium can be split into a rectangle and 2 isosceles triangles.
Draw such a shape and give some of the angles.
Ask students to work out the missing angles.
Extend to the interior angles of an irregular quadrilateral.

Introductory activity

Draw a chalk regular pentagon or regular hexagon large enough for students to walk around outside. Extend (produce) each side to make the exterior angles. If a student walks around the perimeter of the polygon in a complete cycle they will have turned through 360°and turned at each exterior angle. So the exterior angles sum to 360°.

Each exterior angle is the same size. **What is each one?** (360 ÷ 5 = 72° or 360 ÷ 6 = 60°)

Will the sum be 360° for all polygons? (Yes, the angles will still sum to 360°, and each angle will be 360° ÷ number of sides, because you still have to go all around the shape by turning at each exterior angle.)

What about irregular polygons? (Exterior angles still sum to 360°, but are not equal.)

How can you find the sum of the interior angles of a polygon?

Discuss with students before taking the example of a pentagon.

How many triangles can you split a pentagon into? (3) Each contains 180°. So 3 × 180 = 540°. Ensure students do not cross over the diagonals. The easiest way is to draw all the lines from the same vertex.

What is the relationship between interior and exterior angles at a vertex? (Straight line, so 180°.)

Exercise commentary and misconceptions

Question 1 deals with regular polygons, but questions 2 and 3 deal with irregular polygons.

In question 1 students may spot patterns in their results. Encourage them to write them down or explain them verbally.

In questions 2 and 3 students could form an equation.

Plenary

The sum of the interior angles of a polygon is 1800°. **How many sides does it have?** (There must be 10 triangles so 12 sides.) Students could use patterns found in their table from question 1 to help them.

How can you work out the number of sides in a polygon if you know the sum of its interior angles? (Divide the sum by 180 and add 2.)

S2.2 Angles in polygons

This spread will show you how to:
- Calculate and use the interior and exterior angles of polygons

Keywords
Exterior angle
Interior angle
Polygon
Vertex

A **polygon** is a closed shape with three or more straight sides.

- **A regular polygon has all sides the same length and all interior angles equal.**

A square is a regular quadrilateral.

The **interior angles** are inside the polygon.

The **exterior angles** are made by extending each side in the same direction.
Exterior angles are outside the polygon.

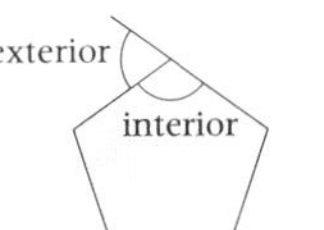

- **For any polygon the exterior angle sum = 360°.**
- **For a regular polygon with n sides, each exterior angle = 360° ÷ n.**

A pentagon divides into 3 triangles. Angles in a triangle add up to 180°. So interior angle sum of a pentagon = 3 × 180 = 540°

You can divide any polygon into triangles by drawing diagonals from a **vertex** (corner).

The number of triangles is always two less than the number of sides.

- **For a polygon with n sides the interior angle sum = $(n - 2) \times 180°$.**

Example

In a regular octagon find

a an interior angle **b** an exterior angle.

a An octagon has 8 sides and divides into 6 triangles.
Interior angle sum is 6 × 180° = 1080°
Each interior angle is 1080° ÷ 8 = 135°.

b Each exterior angle is 360° ÷ 8 = 45°.

Example

An irregular hexagon has angles 108°, 92°, 120°, 134°, 115° and x.

Find the size of angle x.

Any hexagon has 6 sides and divides into 4 triangles.

Sum of interior angles is 4 × 180° = 720°

Sum of the 5 given angles is
108° + 92° + 120° + 134° + 115° = 569°

So x = 720° − 569° = 151°

Exercise S2.2

1 **a** Copy the table of regular polygons.
b Draw each polygon and divide it into triangles by drawing diagonals from a vertex.
c Complete the table.

Shape	△	□	⬠	⬡	heptagon	octagon
Number of sides						
Number of triangles the shape splits into						
Sum of the interior angles in the shape						
Size of one interior angle						
Size of one exterior angle						

2 Find the missing angles in these quadrilaterals.

a
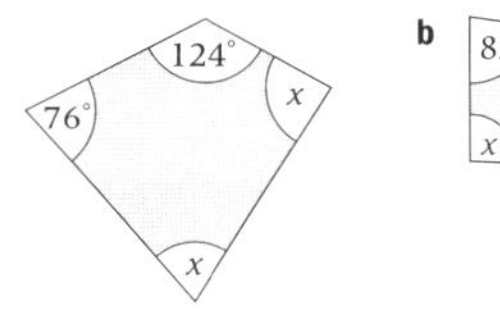

b
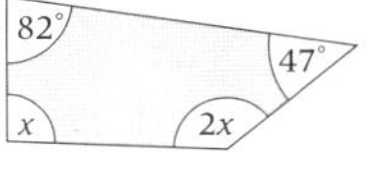

c
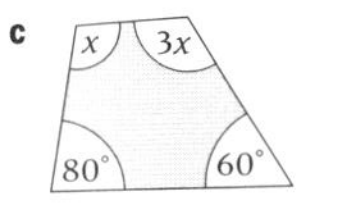

3 Find the missing angles in these irregular polygons.

a
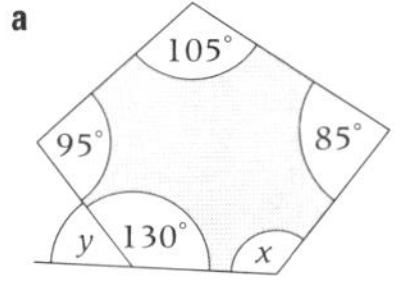

b
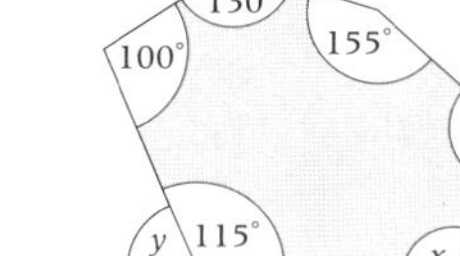

4 **a** Find the exterior angle marked x in each triangle.

i **ii** **iii**

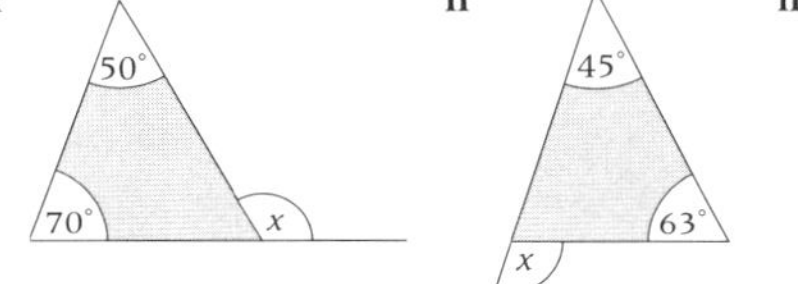

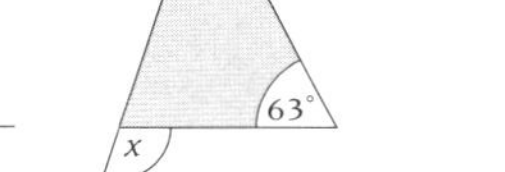

b Use your answers to part **a** to help you copy and complete this statement.
The exterior angle of a triangle is equal to________________

5 Use a diagram to explain why the interior angle sum of a quadrilateral is 360°.

S2.3 Circle theorems

Objectives

- Recognise the parts of a circle, using correct vocabulary to describe them
- Calculate angles in a circle by using the circle theorems

Useful resources

- OHT of circle theorems
- Geometry tool for drawing circles and polygons

Mental starter

Challenge students to mentally work out the interior angle of a pentagon, hexagon, octagon, nonagon, decagon and dodecagon.

Introductory activity

Draw a circle and mark two points on the circumference. Join both points to a third point on the circumference. Join them to a different point on the circumference. Join them to the centre of the circle.

The angle you make at the circumference is always the same, but the angle in the centre is larger. **How much larger? How would you describe the relationship?** Encourage the class to make suggestions. Conclude that the relationship is 'twice as big'.

Use these three examples, drawing the appropriate diagrams.

- If the angle at the centre of a circle is 140°, find the angle at the circumference.
- If the angle at the centre of a circle is 220°, find the angle at the circumference.
- If the angle at the circumference of a circle is 30°, find the angle in the same segment.

Exercise commentary and misconceptions

Students should sketch the diagrams and fill in any angles they can find, using any previous knowledge as well as angle theorems.

Students need to remember to give all the reasons for their answers.

Check that students do not take measurements since diagrams are not drawn to scale.

Students will often make invalid assumptions. You cannot assume a line passes through the centre unless the centre is labelled on the line.

Plenary

Discuss all the circle terminology that students should be familiar with, including chords, segments, tangents and arcs.

Discuss circle facts, including area and circumference.

S2.3 Circle theorems

This spread will show you how to:
- Recognise the parts of a circle, using correct vocabulary to describe them
- Calculate angles in a circle by using the circle theorems

Keywords
Arc
Chord
Diameter
Segment

A straight line joining two points on the circumference is a **chord**.

A chord divides the circle into two **segments**.

An **arc** is the part of the circumference that joins two points.

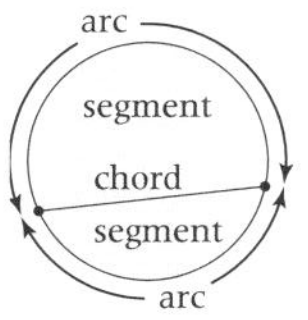

arc
segment
chord
segment
arc

The **circumference** is the distance around the outside of a cicle.

The **diameter** is the longest chord.

- **The angle at the centre of a circle is double the angle at the circumference from the same arc.**

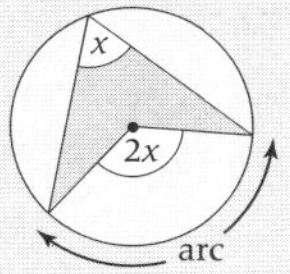

x
$2x$
arc

- **Angles from the same arc in the same segment are equal.**

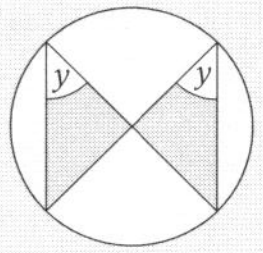

y y

Example

Find the missing angles. Give a reason for each answer.

a	a, 140°	**a** = 70°	angle at the centre is double the angle at the circumference
b	52°, b	**b** = 360° – 104° = 256°	angles at a point
c	220°, c	**c** = 110°	angle at the centre is double the angle at the circumference
d, e	d, e, 30°, 65°	**d** = 30° **e** = 65°	angles on same arc are equal angles on same arc are equal
f	52°, f	**f** = 104°	angle at the centre is double the angle at the circumference

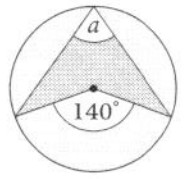

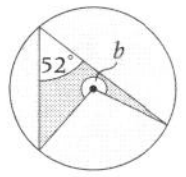

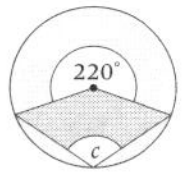

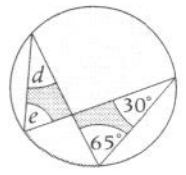

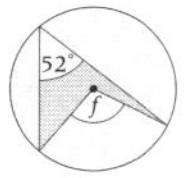

The angle at the centre is a reflex angle.

Exercise S2.3

1 Find the missing angles.
Give a reason for each answer.

a a, 126°
b b, 92°
c 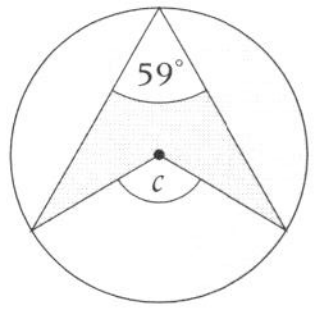 59°, c
d 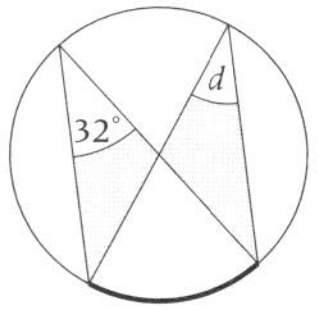 32°, d
e e, 78°
f f, 206°
g 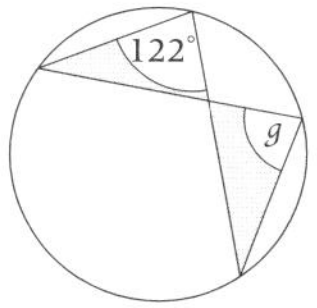 122°, g
h 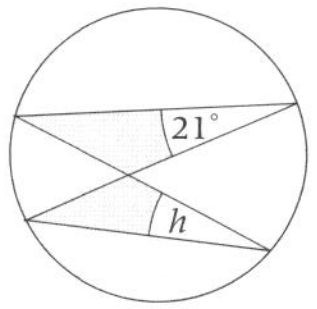 21°, h
i i, 280°
j 38°, j, 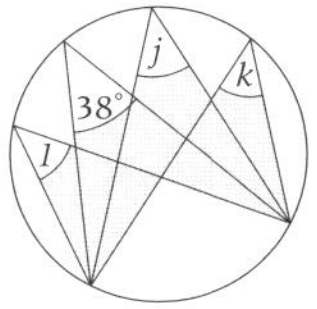 k, 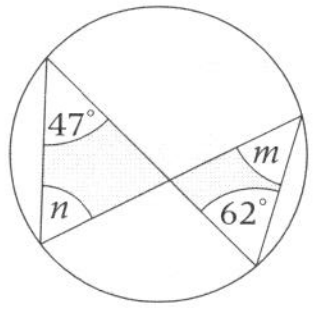 l
k 47°, n, m, 62°
l 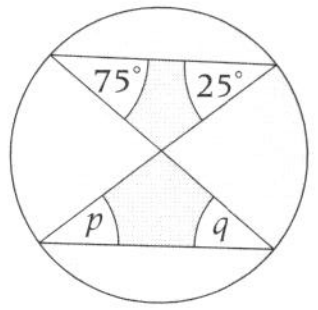 75°, 25°, p, q
m r
n 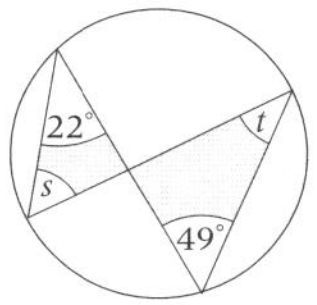 22°, s, t, 49°
o u, 45°, 54°, v
p w
q x
r y

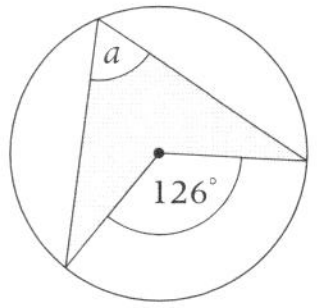

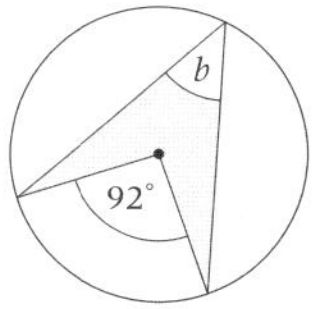

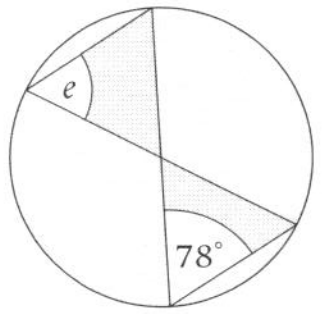

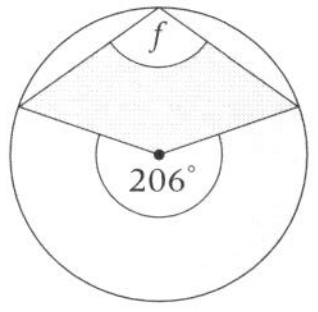

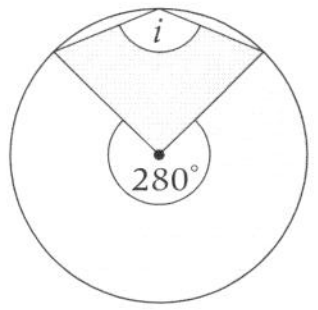

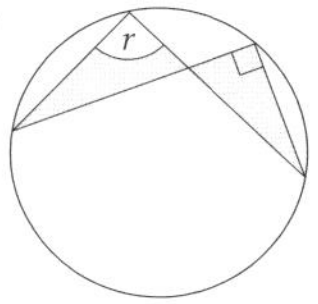

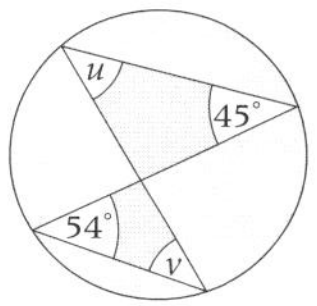

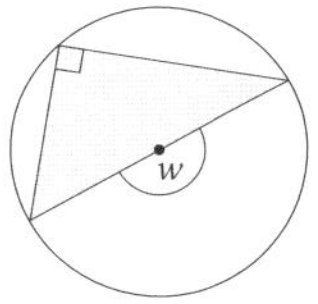

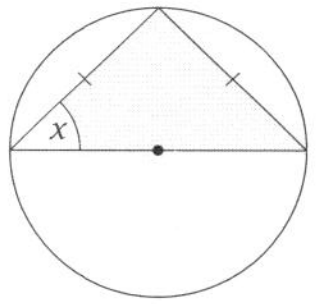

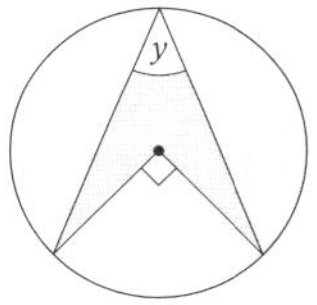

S2.4 More circle theorems

Objectives

- Calculate angles in a circle by using the circle theorems

Useful resources

- Compasses, angle measurer, ruler
- Geometry tool

Mental starter

Read out a list of some different types of triangle – equilateral, obtuse-angled isosceles, right-angled isosceles, acute-angled isosceles, obtuse scalene, right-angled scalene, acute-angled scalene. **If each vertex of a triangle is on the circumference of a circle, which type of triangle can have one of its sides passing through the centre of the circle?**
Split the class into groups and assign two types of triangle to each for each group to investigate. (Only right-angled triangles will pass through the centre.)

Introductory activity

Ask students to draw circles of differing size. They should draw in the diameter and then join the two ends of the diameter to a third point on the circumference to form a triangle.
What sort of triangle is being made?
(Right-angled.) Emphasise that any triangle drawn inside a semicircle is always right-angled.

Instruct students to draw a quadrilateral inside a circle with each vertex touching the circumference – try this a number of times. Measure the angles and look for a relationship between them – remember you may be up to 2 or 3 degrees out in your measurements. (Opposite angles sum to 180°.) These quadrilaterals go round inside a circle – they are **cyclic quadrilaterals**.

Draw a cyclic quadrilateral and label one angle 82°. Find the opposite angle. (98°)

Draw angles in opposite segments and label them 64° and c. Draw an angle in the centre of the circle labelled d. Find c (116°) and d (128°).

Exercise commentary and misconceptions

Students should fill in any angles they can find, remembering to use any previous knowledge of angles.
Ensure no students take measurements since diagrams are not drawn to scale.
You cannot assume a line passes through the centre unless the centre is labelled on the line.
Students need to remember to give all the reasons for their answers.
Previous rules from S2.3 may be used.

Plenary

Do isosceles trapeziums fit inside (inscribe) a circle? (Yes, opposite angles sum to 180°, reflectional symmetry and allied angles.)

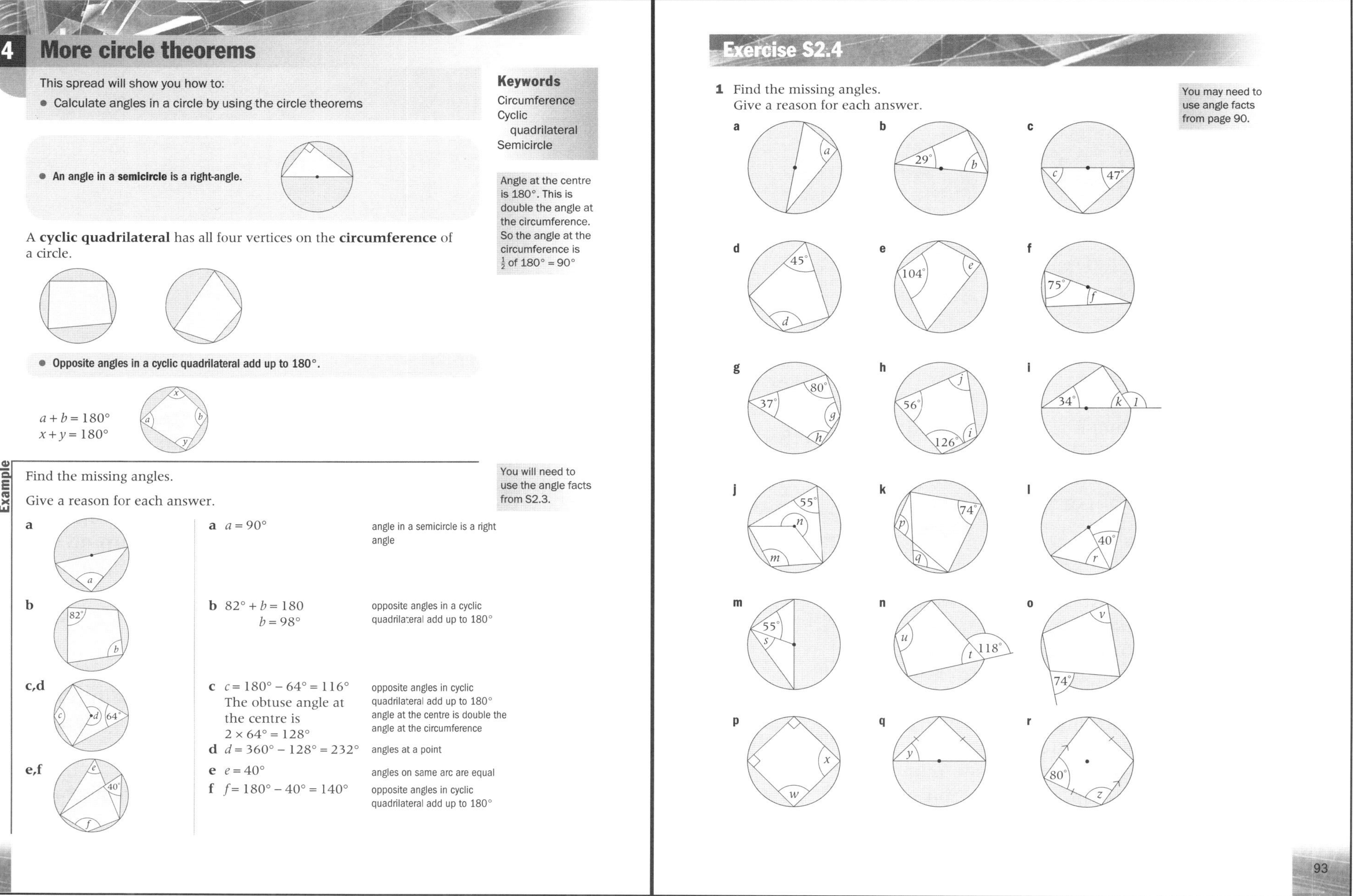

S2.4 More circle theorems

This spread will show you how to:

- Calculate angles in a circle by using the circle theorems

Keywords
Circumference
Cyclic quadrilateral
Semicircle

- **An angle in a semicircle is a right-angle.**

Angle at the centre is 180°. This is double the angle at the circumference. So the angle at the circumference is $\frac{1}{2}$ of $180° = 90°$

A **cyclic quadrilateral** has all four vertices on the **circumference** of a circle.

- **Opposite angles in a cyclic quadrilateral add up to 180°.**

$a + b = 180°$
$x + y = 180°$

Example

Find the missing angles.

Give a reason for each answer.

You will need to use the angle facts from S2.3.

	Working	Reason
a	$a = 90°$	angle in a semicircle is a right angle
b	$82° + b = 180$ $b = 98°$	opposite angles in a cyclic quadrilateral add up to 180°
c	$c = 180° - 64° = 116°$ The obtuse angle at the centre is $2 \times 64° = 128°$	opposite angles in cyclic quadrilateral add up to 180° angle at the centre is double the angle at the circumference
d	$d = 360° - 128° = 232°$	angles at a point
e	$e = 40°$	angles on same arc are equal
f	$f = 180° - 40° = 140°$	opposite angles in cyclic quadrilateral add up to 180°

Exercise S2.4

1 Find the missing angles.
Give a reason for each answer.

You may need to use angle facts from page 90.

S2.5 Tangents to circles

Objectives

- Apply the tangent theorems to circle problems

Useful resources

- Compasses, angle measurers and rulers
- Geometry tool

Mental starter

A pentagon has the following angles measured clockwise from the top: 60°, 120°, 90°, 90°, 120°. **Can you draw a single line to split it into a triangle and a quadrilateral? Do these two figures have special names?** (Equilateral triangle and rectangle.) **Does the pentagon have any lines of symmetry?** (1) **What are the two congruent shapes made by this line of symmetry?** (Trapeziums.)

Introductory activity

Ask students to visualise these scenarios:

Imagine a bicycle. **If you draw a line from the hub of the wheel to the point of the ground where the wheel touches the ground, what angle does this line make with the ground?** (90°). The line is a radius and the ground is a tangent.
Use this scenario to highlight the key point: the angle between the tangent and the radius at a point is a right angle.

Imagine a rubber band round a pulley wheel. If you pull the band a short distance away from the wheel, the two lines of the band are tangents to the wheel and are of equal length. It doesn't matter how far you pull the band away.
Use this scenario to highlight the key point: two tangents drawn from a point to a circle are equal.
Discuss the example in the student book.

Exercise commentary and misconceptions

Students should fill in any angles they can find, remembering to use any previous knowledge of angles.
Ensure no students take measurements since diagrams are not drawn to scale.
You cannot assume a line passes through the centre unless the centre is labelled on the line.
Students need to remember to give all the reasons for their answers. Previous rules from S2.3 and S2.4 may be used.

Plenary

Relate this topic to the subject of forces – discuss students' knowledge of circular motion, and encourage students to use real-life examples such as a fairground ride.

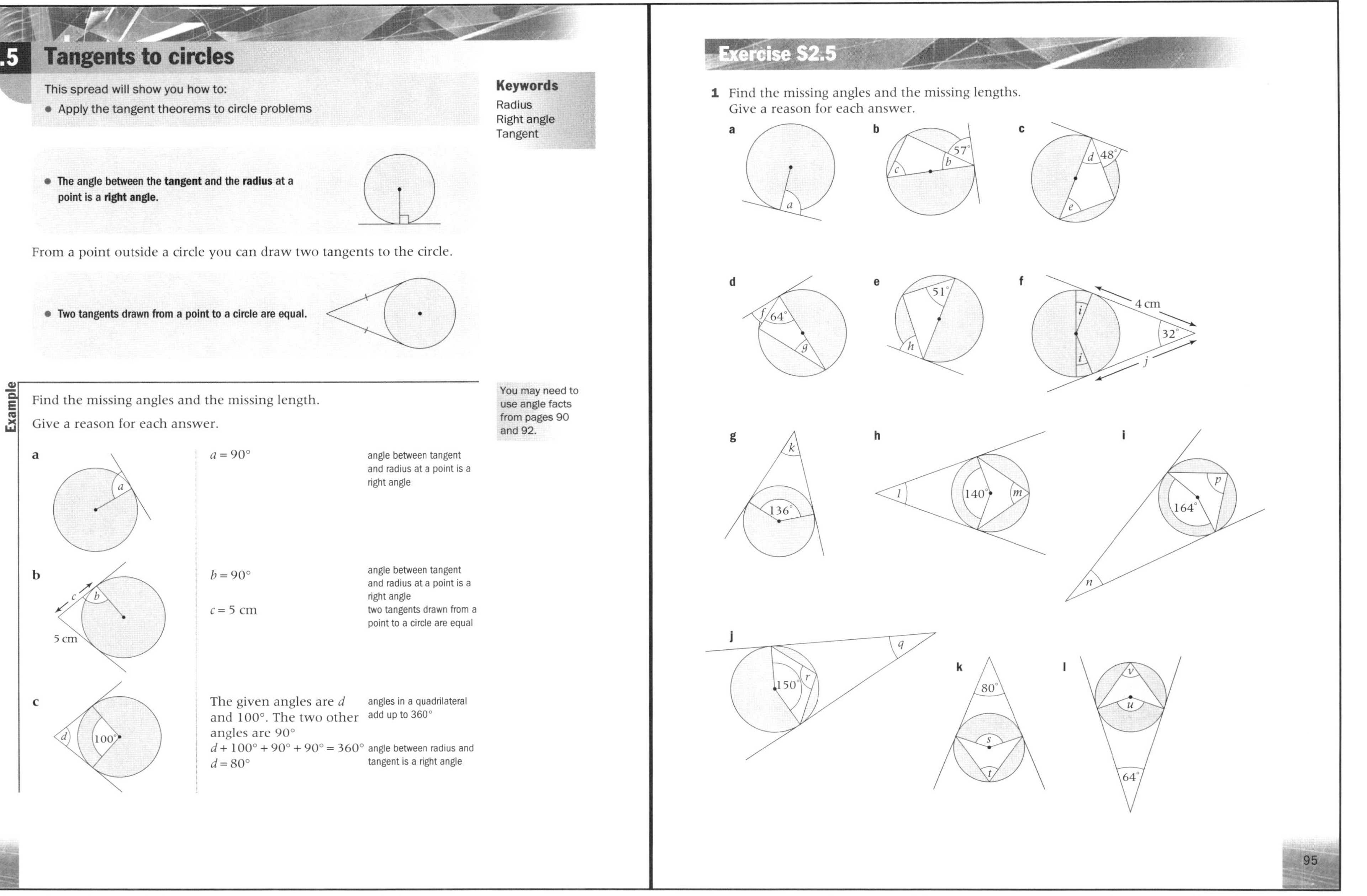

S2.5 Tangents to circles

This spread will show you how to:

- Apply the tangent theorems to circle problems

Keywords
Radius
Right angle
Tangent

- The angle between the **tangent** and the **radius** at a point is a **right angle**.

From a point outside a circle you can draw two tangents to the circle.

- **Two tangents drawn from a point to a circle are equal.**

Example

Find the missing angles and the missing length.

Give a reason for each answer.

a $a = 90°$ — angle between tangent and radius at a point is a right angle

b $b = 90°$ — angle between tangent and radius at a point is a right angle

$c = 5$ cm — two tangents drawn from a point to a circle are equal

c The given angles are d and 100°. The two other angles are 90° — angles in a quadrilateral add up to 360°

$d + 100° + 90° + 90° = 360°$ — angle between radius and tangent is a right angle

$d = 80°$

You may need to use angle facts from pages 90 and 92.

Exercise S2.5

1 Find the missing angles and the missing lengths.
Give a reason for each answer.

S2 Exam review

Key objectives

- Use parallel lines, alternate angles and corresponding angles
- Calculate and use the sums of the interior and exterior angles of polygons
- Recall the definition of a circle and the meaning of related terms, including chord, tangent, arc, sector and segment
- Understand that the tangent at any point on a circle is perpendicular to the radius at that point
- Prove and use the facts that the angle subtended by an arc at the centre of a circle is twice the angle subtended at any point on the circumference, the angle subtended at the circumference by a semicircle is a right angle, that angles in the same segment are equal, and that opposite angles of a cyclic quadrilateral sum to 180°

Worked solution	Commentary and misconceptions
1 Each exterior angle is $360° \div 6 = 60°$	Students should recall that for a regular polygon with n sides, exterior angle = $360° \div n$.
2 **a** FE is a tangent and AC passes through the midpoint. Therefore, ACE = 90°. So ACB = 90° − 63° = 27° **b** ABC = 90° (angle in a semicircle) So BAC = 180° − (90° + 27°) = 63°	Students should recall the relevant theorems for angles in circles and state them to explain their workings.

A3 Sequences

Objectives

F Generate terms of a sequence using term-to-term and position-to-term definitions of the sequence

F/H Generate common integer sequences

F/H Use linear expressions to describe the *n*th term of an arithmetic sequence, justifying its form by reference to the activity or context from which it was generated

Unit overview

This unit consolidates Key Stage 3 knowledge of simple number patterns and extends this to more general sequences and the rules from which they are generated. Students are required to recall the rules of algebra learnt in units A1 and A2 and use this knowledge to describe and generate sequences. Understanding and using linear expressions is important in this unit, preparing the students for work involving more complex algebraic formulae in further algebra units.

Prior knowledge

Before your students start this unit they should be able to:

- Understand and use fraction notation
- Add, subtract, multiply and divide with integers, decimals and fractions
- Use the rules of algebra with the correct order of operations
- Expand single and double brackets
- Simplify algebraic expressions, including factorisation
- Recognise simple number patterns
- Understand and recognise square numbers
- Use tables to organise numerical results

Differentiation

- **Foundation** focuses on simple sequences, finding terms of a sequence using simple position-to-term rules and describing sequences and patterns using words
- **Foundation Plus** extends to using algebraic expressions to describe simple sequences and patterns
- **Higher Plus** extends the Higher book to more complex quadratic sequences and then to solving quadratic equations and quadratic graphs

A3.1 Number patterns

Objectives
- Generate common sequences and describe how number patterns are formed

Useful resources
- OHT of pattern sequences

Mental starter
Use only the even numbers from 0 to 10. Multiply any two of these even numbers together, not using a number twice. **What numbers can you make?** (8, 12, 16, 20, 24, 32, 40, 48, 60, 80.)

Introductory activity
Spotting the next number in a pattern can be difficult. First look at the difference between successive numbers.

For example 7, 12, 19, 29, 43 (going up by +5, +7, +10, +14). Going up by 2 extra, then 3 extra, then 4 extra each time. **Next two terms?** (62, 87.)

Next look to see if you are multiplying or dividing each time. For example, 4, 12, 36, (multiplying by three). **Find next two terms**. (108, 324.)

Look to see if you are using the previous term to find the next term. For example, 3, 4, 7, 11, 18, (add previous two terms to find next term). **Find next two terms**. (29, 47.)

Look to see if it is a special sequence, for example square numbers, cube numbers, triangle numbers, prime numbers.
What are the next two terms of 1, 4, 9, 16, 25, 36, ...? (49, 64.)
What are the next two terms of 1, 8, 27, 64, 125, 216, ...? (343, 512.)
What are the next two terms of 1, 3, 6, 10, 15? (21, 28.)
What are the next two terms of 2, 3, 5, 7, 11, 13, 17, ...? (19, 23.)
Discuss the examples in the student book, which extend to tiling patterns.

Exercise commentary and misconceptions
Students will need to be familiar with multiples and powers of a number. Using first and second differences can help establish the rules. Point out that you are finding the rule for getting the next term from the previous term, but it is difficult to find the answer for any term, say the 124th term! What would be needed is a rule for finding a term using only the position it is in.

Plenary
Divide the square number sequence by the cube number sequence, one term at a time.
($\frac{1}{1}, \frac{4}{8}, \frac{9}{27}, \frac{16}{64}, \frac{25}{125}, \ldots = \frac{1}{1}, \frac{1}{2}, \frac{1}{3}, \frac{1}{4}, \frac{1}{5}, \ldots$)

Can you see why it works out this way? (Each term is made by squaring on top and cubing underneath, this means multiplying twice on top and multiplying 3 times underneath. Cancelling leaves a single term in the denominator.)

A3.1 Number patterns

This spread will show you how to:

- Generate common sequences and describe how number patterns are formed

Keywords
Sequence
Term

- A **sequence** is a set of numbers that follow a pattern, for example

  5, 9, 13, 17, 21, ... are the first five **terms** of a sequence that goes up in 4s

  3, 6, 12, 24, 48, ... are the first five terms of a sequence that doubles

  1, 4, 9, 16, 25, ... is the sequence of square numbers
  (1 × 1, 2 × 2, 3 × 3 and so on)

  1, 8, 27, 64, 125, ... are the cube numbers
  (1 × 1 × 1, 2 × 2 × 2, 3 × 3 × 3 and so on)

Examiner's tip
The techniques learned in this unit will be useful for your Using and Applying Mathematics coursework task.

You can find sequences in patterns, for example,

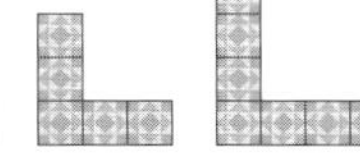

Number of tiles is: 3, 5, 7, ...
The next diagram would need 9 tiles.

Example

Write the next two terms in each sequence.

a 4, 5, 7, 10, 14, 19, ... **b** 0.6, 0.7, 0.8, 0.9, ...

a 4, 5, 7, 10, 14, 19, ... The terms increase by 1, then 2, then 3.
so the next two terms are 19 + 6 = 25 and 25 + 7 = 32.

b 0.6, 0.7, 0.8, 0.9, ... The terms increase by 0.1. The next two terms are 0.9 + 0.1 = 1.0 and 1.0 + 0.1 = 1.1.

You may be tempted to continue 0.7, 0.8, 0.9, 0.10, ... but 0.10 is less than 0.9!

Example

Name each of these sequences.

a 1, 3, 5, 7, 9, ... **b** 25, 36, 49, 64, ...

a These are the odd numbers. **b** These are the square numbers, starting at 5 × 5 (or 5^2).

Example

How many tiles would be in the tenth diagram in this pattern?

Diagrams	1	2	3	4
Number of tiles	1	5	9	13

A table is a good way of organising results so that you can see what is happening.

The number of tiles increases by 4 each time. If you continue this, you get 1, 5, 9, 13, 17, 21, 25, 29, 33, **37**.
The tenth diagram would have 37 tiles.

Exercise A3.1

1 Copy each sequence and add the next two terms.

a 4, 9, 14, 19, 24, __, __ **b** 100, 93, 86, 79, 72, __, __
c 1, 2, 4, 7, 11, __, __ **d** 9, 99, 999, 9999, 99 999, __, __
e 1, 1, 2, 3, 5, 8, __, __ **f** 54, 27, 13.5, 6.75, __, __

2 Copy and fill in the missing numbers in each sequence.

a 4, __, 10, __, 16, ... **b** 4, __, __, 32, 64, ...
c 95, __, __, __, 87, ... **d** 1, __, 27, 64, __, ...

3 Write the first five terms of each of these well-known number patterns.

a Multiples of 3 **b** Powers of 2
c Prime numbers **d** Square numbers over 100

4 The triangular numbers form a sequence.

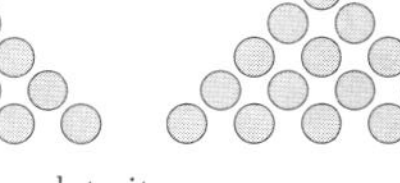
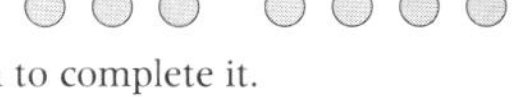
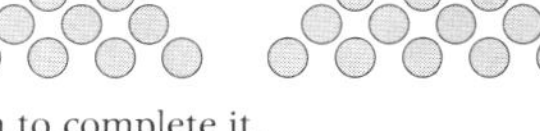
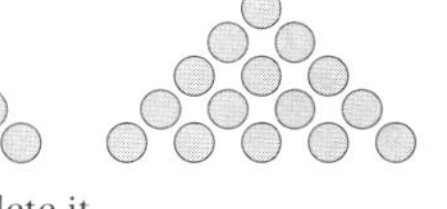
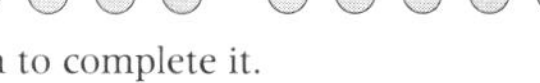

a Copy the table and use the diagram to complete it.

Diagrams	1	2	3	4	5
Number of dots	1	3			

b Hence, write the first 10 triangular numbers.

c Why do you think the square numbers (1, 4, 9, 16, 25, ...) got their name?

d How did the cube numbers get their name? Write the first five cube numbers.

5 Find the tenth term in each of these number patterns.

a (1 × 2), (2 × 3), (3 × 4), (4 × 5), ...
b $\frac{1}{2}, \frac{2}{3}, \frac{3}{4}, \frac{4}{5}, \ldots$
c (5 × 2), (5 × 4), (5 × 8), (5 × 16), ...
d (1 × 1), (4 × 8), (9 × 27), (16 × 64), ...

6 Look at this number pattern.

$7^2 = 49$
$67^2 = 4489$
$667^2 = 444\,889$

a Write the next two lines in the pattern.
b What is $66\,666\,666\,667^2$?
c What is $\sqrt{4\,444\,444\,444\,888\,888\,889}$?

A3.2 Generating sequences

Objectives
- Use position-to-term rules to write sequences
- Find and describe the *n*th term of a sequence, using this to find other terms

Useful resources
- Mini-whiteboards

Mental starter
Ask the students to stand up. Starting with 100, go around the class asking students for the next term in the sequence you describe: count down in 7s, in square numbers, in steps of 0.95.

Introductory activity
If the rule for finding the terms of a sequence is 'position × 5 – 2', **what are the first four terms?** (3, 8, 13, 18.)

The position is often given the letter 'n', so this rule would be $5n - 2$.

Find the first 4 terms if the rule is $3n - 5$. (−2, 1, 4, 7.)

A rule is often referred to as 'the nth term of T_n or $T(n)$'. For example $T_n = 2n^2$. **Find the first four terms.** (2, 8, 18, 32.)
Discuss the example in the student book.

Exercise commentary and misconceptions
Questions 1 and 2 are fairly straightforward, requiring students to generate the first five terms of a sequence.

In question 3 students may not realise that they need only calculate the third term (or the tenth term), not the whole sequence.

In question 4 students may need you to point out that the shapes in the flow chart have a meaning, if they are not familiar with flow charts.

Plenary
When something oscillates it goes back and forward. When something gets further and further away it diverges, when it gets closer and closer it converges. **How would you describe the sequences created by these *n*th terms?**

(i) $5 \times (-2)^n$
(−10, 20, −40, 80, −160, 320, ...)
oscillating and diverging

(ii) $\frac{n}{n+1}$
($\frac{1}{2}, \frac{2}{3}, \frac{3}{4}, \frac{4}{5}, \frac{5}{6}, \ldots$)
converging on 1.

A3.2 Generating sequences

This spread will show you how to:

- Use position-to-term rules to write sequences
- Find and describe the *n*th term of a sequence, using this to find other terms

Keywords
*n*th term
Position
Position-to-term rule
Sequence
Term

- **You can use a position-to-term rule to write a sequence.**

For example, this is a sequence using the position-to-term rule

Multiply by 5 and add 2

The first 5 terms are:

Postion	1	2	3	4	5
Term	$1 \times 5 + 2 = 7$	$2 \times 5 + 2 = 12$	$3 \times 5 + 2 = 17$	$4 \times 5 + 2 = 22$	$5 \times 5 + 2 = 27$

You can write a position-to-term rule in algebraic notation.
For example, T_n represents the ***n*th term** of the sequence

$$T_n = 3n - 2$$

The rule connects the **position** of a **term** in a sequence to its value.

To find T_1, the first term, put $n = 1$ in the rule.

Hence, $T_1 = 3 \times 1 - 2 = 1$
$T_2 = 3 \times 2 - 2 = 4$
$T_3 = 3 \times 3 - 2 = 7$
$T_4 = 3 \times 4 - 2 = 10$

The sequence $T_n = 3n - 2$ is 1, 4, 7, 10, ...

Example

Find the first five terms of the sequences with these position-to-term rules.

a $T_n = 5n + 7$

b $T_n = 3n^2$

a $T_1 = 5 \times 1 + 7 = 12$
$T_2 = 5 \times 2 + 7 = 17$
$T_3 = 5 \times 3 + 7 = 22$
$T_4 = 5 \times 4 + 7 = 27$
$T_5 = 5 \times 5 + 7 = 32$, ...
Hence, the sequence with $T_n = 5n + 7$ is 12, 17, 22, 27, 32, ...

You may notice a pattern that allows you to generate the terms quickly, for example this sequence goes up in 5s.

b $T_1 = 3 \times 1^2 = 3$
$T_2 = 3 \times 2^2 = 12$
$T_3 = 3 \times 3^2 = 27$
$T_4 = 3 \times 4^2 = 48$
$T_5 = 3 \times 5^2 = 75$, ...
Hence, the sequence with $T_n = 3n^2$ is 3, 12, 27, 48, 75, ...

Remember 'BIDMAS' ... the power comes before multiplication, so square before multiplying by 3.

Exercise A3.2

1 Write the first five terms of each sequence.

a $T_n = 8n + 2$ **b** $T_n = 5n - 4$ **c** $T_n = 7n$
d $T_n = 10 - 2n$ **e** $T_n = n^2 - 3$ **f** $T_n = 2n^2$

2 Generate the first five terms of each sequence.

a $T_n = 9 - n$ **b** $T_n = (n + 1)(n + 2)$ **c** $T_n = \frac{1}{n}$
d $T_n = (-n)^3$ **e** $T_n = (2n - 1)(n + 1)(n - 3)$ **f** $T_n = n^4$

3 Here are four position-to-term rules.

$T(n) = n^2$ $T_n = 18 + 2n$ $T_n = n^3$ $T_n = n(n + 5)$

Write rules that give

a two sequences with an identical first term
b two sequences whose third term is 24
c two sequences whose tenth term is greater than 100.

4 **a** Generate all the terms and the formula for T_n, the *n*th term of the sequence described by the flow chart.

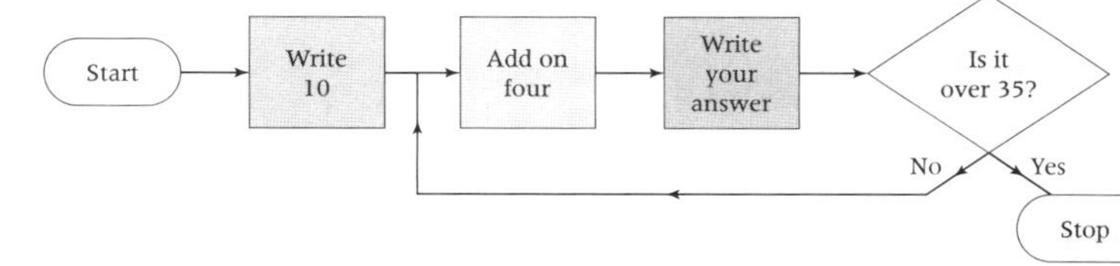

b Design your own flow chart for the sequence 2, 5, 10, 17, 26, 37, 50.

5 Write the algebraic position-to-term rule for

a **i** Two sequences with identical 1st terms
ii Two sequences with identical 5th terms

b If two sequences have identical 1st terms and identical 5th terms, do they have to be identical sequences? Discuss your answer.

A3.3 Finding the *n*th term

Objectives

- Find and describe the *n*th term of a sequence, using this to find other terms

Useful resources

- Graph paper

Mental starter

Beginning with the number 1, double the answer to get 2, then keep doubling to get as large a number as possible without making a mistake. Give the class 2 minutes. **Who has the highest correct answer in the class?** (1, 2, 4, 8, 16, 32, 64, 128, 256, 512, 1024, 2048, 4096, 8192, 16 384, 32 768, 65 536, 131 072, 262 144, 524 288, 1 048 576, 2 097 152, 4 194 304, 8 388 608, 16 777 216, 33 554 432, 67 108 864.)

Introductory activity

Explain that a sequence that goes up or down by the same number is called a linear sequence. Suppose you have the sequence 5, 8, 11, 14, 17, ...

You could say the rule is +3, this would help you to work out the 6th term from the 5th term by adding 3. But it would not be very helpful in working out the 257th term. You need a rule for working out any term if you know what **position** it is in. This sequence has a rule of 'position × 3 + 2'

If you want to find the term in the 7th position $7 \times 3 + 2 = 23$.
257th term is $257 \times 3 + 2 = 773$.

The rule is normally called the '*n*th term' or 'T_n' and instead of writing 'position × 3 + 2' you use the letter n to stand for the position so '$3n + 2$'. How can you find the rule for any term just from the sequence?

Look at the sequence

5	8	11	14	17

Because the sequence goes up by 3 each time, the rule will begin with $3n$.

The positions n are

1	2	3	4	5

but if you try $3 \times n$ for each term, you are always 2 below the sequence

3	6	9	12	15

so the *n*th term, T_n, is $3n + 2$.

Look at the sequence

14	8	2	−4	−10

Because the sequence goes down by 6 each time, the rule will begin with $-6n$.

The positions n are

1	2	3	4	5

but if you try $-6 \times n$ for each term you are always 20 below the sequence

−6	−12	−18	−24	−30

so the *n*th term, T_n, is $6n + 20$.

Discuss the examples in the student book.

Exercise commentary and misconceptions

Encourage students to use the table format in the students' book.

Some students may notice that the number that is put onto the end of the *n*th term formula is the number that would come just before the first term in the sequence. For example in −1, 4, 9, 14, 19 the term before −1 would be −6 and the *n*th term is $5n - 6$. This is not a good way to learn how to find the *n*th term since students may easily apply this wrongly, but for those students who understand this topic it is a convenient short cut.

Plenary

Discuss question 6, particularly asking students to describe their graphs. What would happen if the sequence was continued indefinitely? Encourage students to think in terms of approaching a limiting value.

A3.3 Finding the *n*th term

This spread will show you how to:

- Find and describe the *n*th term of a sequence, using this to find other terms

Keywords
Formula
Linear
*n*th term

- **In a linear sequence the terms go up or down by the same amount.**

  12, 15, 18, 21, 24, ... goes up in 3s, so it is linear
  100, 95, 90, 85, 80, ... goes down in 5s, so it is linear

You can find the ***n*th term** of a linear sequence by comparing the sequence to the times table to which it is related. For example the sequence 10, 17, 24, 31, 38, ... is connected to the 7× table, hence:

Position	1	2	3	4	5
Multiples of 7	7	14	21	28	35
Sequence term	10	17	24	31	38

The terms go up in 7s.

By comparing the sequence to the 7× table, you can see that you have to add 3 to the multiples of 7 to get the terms of the sequence. Hence *n*th term T_n = Position × 7 + 3

$T_n = 7n + 3$

You can use the *n*th term **formula** to find any term of the sequence quickly, for example the 100th term of the above sequence is
$T_{100} = 7 \times 100 + 3$ so $T_{100} = 703$

Example

Find the *n*th term and, hence, the 100th term of

a 2, 8, 14, 20, 26, ... **b** 25, 20, 15, 10, 5, ...

a Compare the sequence to the multiples of 6.

Position	1	2	3	4	5
Multiples of 6	6	12	18	24	30
Term	2	8	14	20	26

This sequence goes up in 6s, so it must be connected to the 6 times table.

First term = 2 = 6 − 4
Subtract 4 from the multiple of 6 to get each term.
$T_n = 6n - 4$
Hence, $T_{100} = 6 \times 100 - 4 = 596$

b Compare this sequence to the multiples of −5.

Position	1	2	3	4	5
Multiples of 5	−5	−10	−15	−20	−25
Term	25	20	15	10	5

If a sequence goes down use negative multiples.

On a number line, to get from −5 to 25, you need to add 30.

+5 +25
−5 0 5 10 15 20 25

First term = 25 = 30 − 5
Second term = 20 = 30 − 10
Add 30 to the multiple of −5 to get each term.
$T_n = -5n + 30$ or $T_n = 30 - 5n$
Hence, $T_{100} = 30 - 5 \times 100 = -470$

Exercise A3.3

1 Find a formula for the *n*th term, T_n, of each sequence.

a 4, 9, 14, 19, 24, ... **b** 1, 3, 5, 7, 9, ...
c 10, 12, 14, 16, 18, ... **d** 1, 1.5, 2, 2.5, 3, ...
e −4, −2, 0, 2, 4, ... **f** 1, 2, 3, 4, 5, ...
g The multiples of 13 **h** Counting up in 10s, starting from 4
i 10, 8, 6, 4, 2, ... **j** 100, 95, 90, 85, 80, ...
k 50, $49\frac{3}{4}$, $49\frac{1}{2}$, $49\frac{1}{4}$, ...
l Counting down in multiples of 4 from 75

2 Write five linear sequences of your own that have a third term of 15. Find a formula for the *n*th term of each one, and use this to find the 100th term in each case.

3 Are these true or false?

a The 50th term of 2, 5, 8, 11, 14, ... is more than 150.
b The 50th term of 5, 9, 13, 17, 21, ... is even.
c The 100th term of 1000, 990, 980, 970, 960, ... is negative.

Find the position-to-term rule for each sequence.

4 Here are some terms of a sequence. In each case find the formula for the *n*th term, T_n.

a The 5th term is 20, the 6th term is 28 and the 7th term is 36.
b The 100th term is 302, 101st term is 305 and the 102nd term is 308.
c The 153rd term is 260, the 154th term is 262 and the 155th term is 264.

5 How could you find the *n*th term formula for these fractional sequences? See if you can work out what it would be in each case.

a $\frac{3}{7}, \frac{5}{10}, \frac{7}{13}, \frac{9}{16}, \frac{11}{18}, \ldots$
b $\frac{10}{30}, \frac{12}{27}, \frac{14}{24}, \frac{16}{21}, \frac{18}{18}, \ldots$
c $\frac{7}{1}, \frac{8}{4}, \frac{9}{9}, \frac{10}{16}, \frac{11}{25}, \ldots$
d $\frac{1}{11}, \frac{8}{9}, \frac{27}{7}, \frac{64}{5}, \frac{125}{3}, \ldots$

Find separate formulae for the numerator and denominator.

6 **a** What is the *n*th term formula for $\frac{1}{1}, \frac{1}{2}, \frac{1}{3}, \frac{1}{4}, \frac{1}{5}, \ldots$?
b Plot this sequence on a graph, with the term number on the *x*-axis and the term on the *y*-axis.
c What happens if you continue this sequence indefinitely?
d Investigate sequences of your own that behave in a similar way.

A3.4 Describing patterns

Objectives

- Explain how the formula for the nth term of a sequence works

Useful resources

- OHT of pattern sequences
- 1 cm square grid paper for plenary

Mental starter

In a fence each panel needs a post on either side. Suppose a fence panel is 6 feet wide and a post is 3 inches wide. **How would you make a straight fence that is 44 feet long?**
(Use 7 panels (42 feet) and 8 posts (2 feet).)
You may need to give the hint that there are 12 inches in a foot.

Introductory activity

Sometimes you don't have a sequence of numbers, but a sequence of diagrams.

Look at the triangle sequence in the student book. Recapping the previous lesson, show how you can use a table to find the nth term. Extend to the cube sequence, which offers a 3-D example.

In another example pentagons are joined together to form a chain. When 4 pentagons are joined you need 17 lines. How many lines for n pentagons?

To form the sequence of numbers you need to imagine 1 pentagon, then 2, 3 and so on. The sequence is 5, 9, 13, 17. Number of lines $= 4n + 1$.

This can also be done by recognising that each extra pentagon drawn needs four lines, so the formula must start with $4n$, but you need to add 1.
Extend to the hexagon example in the student book.

Exercise commentary and misconceptions

The questions require students to find the formula connecting the number of objects in the diagrams. Encourage students to use the table format in the students' book as this makes it easier to find the sequence, but they also need to be able to visualise the development of the patterns to explain why the formulae work.

Plenary

Draw a 4 by 4 square. There are 30 squares inside it. **Can you find all of them?**
How many in a 3 by 3 square? (14)
Draw a 10 by 10 square. **How long might it take to count all the squares inside?**
Fortunately there is a formula for finding the number of squares inside. In an n by n square grid there are

$\frac{n(n+1)(2n+1)}{6}$ squares inside. How many for a 10 by 10?
$(10 \times 11 \times 21 \div 6 = 2130 \div 6 = 385$ squares.)

A3.4 Describing patterns

This spread will show you how to:

- Explain how the formula for the nth term of a sequence works

Keywords
Formula
nth term
Pattern

- **You can describe a pattern using a formula.**

This pattern is made of triangles.

Number of triangles (n)	1	2	3	4
Number of arrows	3	5	7	9

+2 +2 +2

Number of arrows = 2 × number of triangles +1
so ***n*th term** of pattern = $2n + 1$
This works because each extra triangle needs two arrows and you need one arrow at the start.

Examiner's tip
The situations described on this page are particularly relevant to coursework tasks.

This pattern is made from cubes.
The outside faces of the cubes in each block are painted.

How many faces are painted each time?

Block number (n)	1	2	3	4
Number of painted faces	6	10	14	18
4× table	4	8	12	16

The terms go up in 4s.

In the nth block,
number of painted faces = 4 × block number + 2
so, nth term = $4n + 2$

Each cube in the pattern has 4 painted faces and the 2 end faces are painted.

Example

Find a formula connecting the number of edges and the number of hexagons and explain why it works.

Make a table.

Number of hexagons (n)	1	2	3	4
Number of edges	6	11	16	21
5× table	5	10	15	20

The terms go up in 5s.

Number of edges = 5 × number of hexagons + 1
$E = 5n + 1$

Each hexagon needs five edges to join it to the previous one, and there is one edge to start the whole pattern off.

Exercise A3.4

1 For each pattern, there is a formula. Explain why each formula works.

a 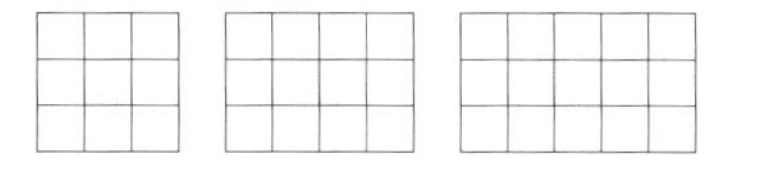

$E = 3s + 1$ where E = number of edges, s = number of squares

b

$W = B + 4$ where W = number of white tiles, B = number of coloured tiles

c

$M = L(L + 1)$ where M = number of nails and L = width of rectangle

2 Derive your own formulae for these patterns and explain why they work.

a Connect the number of coloured tiles (B) and the number of white tiles (W).

b i Connect the number of white circles (W) with the number of coloured circles (B).

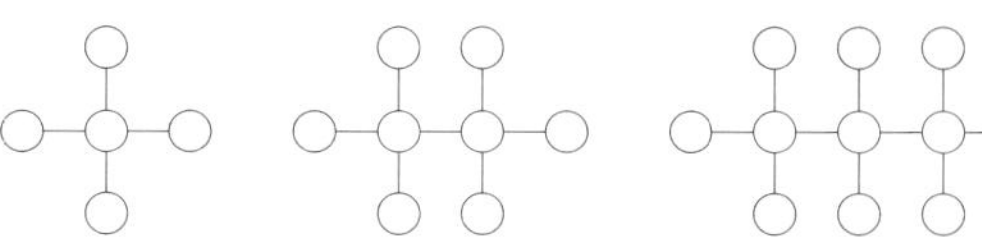

ii Use the same pattern to connect the number of white circles (W) with the number of lines (L).

c Find a formula for

i the number of presents (P) swapped at a party with n people, if all the guests give one another a present

ii the number of handshakes (H) at a party with n people, if all the guests shake hands with each other.

Find the first 3 or 4 terms and make a table.

3 a A square jigsaw puzzle has n^2 pieces. It has C corner pieces, E edge pieces and M middle pieces. Write a formula for C, E and M in terms of n and show that your formula accounts for all the pieces.

b Extend to a rectangular puzzle with m pieces in its length and n pieces in its width.

A3.5 Quadratic sequences

Objectives

- Use quadratic expressions to describe the nth term of a sequence

Useful resources

- OHT of table in introductory activity

Mental starter

Go around the class asking individual students for terms of the sequence of square numbers. Challenge students to go as far as possible using mental methods: (1, 4, 9, 16, 25, 36, 49, 64, 81, 100, 121, 144, ...)

Introductory activity

Sometimes an nth term (T_n) is more complex than say $4n + 2$. For example $n^2 + 1$ gives the sequence 2, 5, 10, 17, ...

How can you find the nth term when it is more complex? Look at this sequence:
2, 8, 18, 32, 50. This example is given in the student book.
Notice that the amount it increases by changes by the same amount each time: +6, +10, +14, +18, so the **increases** go up by four each time. That means n^2 is in the formula.

Position	1	2	3	4	5
Square number	1	4	9	16	25
Term	2	8	18	32	50

What is the link between these numbers?
(From the sequence values you can see that you need to double n^2. So nth term = $2n^2$.)
Discuss the other example in the student book.

Exercise commentary and misconceptions

In question 2 using a table of position, square number and term will help students to 'spot' the link and therefore the nth term.

In question 3 students should discover that half the second difference tells you what to multiply n^2 by, but watch for students getting the subtraction wrong.

This exercise can take quite a time to finish!

Plenary

The sequence 1, 4, 27, 256, 3125, ... has an nth term. **Which one of the following is it?**

a $1 + 3(n - 1)^2$
b n^n
c $\dfrac{n^2 + 1}{n + 1}$.

Split the class into 3 groups to investigate ...

(**a** generates 1, 4, 13, 28 ...,
b generates 1, 4, 27, 256, 3125 ...,
c generates 1, $\frac{5}{3}$, $\frac{10}{4}$, ...
so answer is **b**.)

Can you find the 10th term?
(10^{10} = 10 000 000 000 = ten billion.)

A3.5 Quadratic sequences

This spread will show you how to:

- Use quadratic expressions to describe the nth term of a sequence

Keywords
Constant
Quadratic sequence
Second difference

In a quadratic sequence the difference between terms increases.
For example, the square numbers:

Term	1,	4,	9,	16,	25,	36...
Difference	3	5	7	9	11	

However, if you work out the differences a second time they are **constant**.

Constant means 'staying the same'.

Term	1,	4,	9,	16,	25,	36,
1st difference	+3	+5	+7	+9	+11	
2nd difference	+2	+2	+2	+2		

- If a sequence has a constant **second difference** it is a **quadratic sequence**.

So this sequence is a quadratic sequence.

	3,	6,	11,	18,	27...
1st difference	+3	+5	+7	+9	
2nd difference	+2	+2	+2		

Compare the sequence to the square numbers.

Position	1	2	3	4	5
Square numbers	1	4	9	16	25
Term	3	6	11	18	27

Notice that you need to add 2 to the square numbers to get the sequence

$T_n = n^2 + 2$

This is the position-to-term rule for the sequence.

Example

Find a formula for each of these quadratic sequences.

a 0, 3, 8, 15, 24, ... **b** 2, 8, 18, 32, 50, ...

a Make a table.

Position	1	2	3	4	5
Square numbers	1	4	9	16	25
Term	0	3	8	15	24

The terms are 1 less than the square numbers, so $T_n = n^2 - 1$

Learn the square numbers.

b Make a table.

Position	1	2	3	4	5
Square numbers	1	4	9	16	25
Term	2	8	18	32	50

The terms are double the square numbers so $T_n = 2n^2$

Examiner's tip
These are difficult and unlikely to be examined on. However, you may encounter quadratic sequences in your coursework.

Exercise A3.5

1 Generate the first five terms of each of these quadratic sequences.

a $T_n = n^2 + 5$ **b** $T_n = n^2 - 2$ **c** $T_n = 3n^2$
d $T_n = n^2 + 2n$ **e** $T_n = 2n^2 - 1$ **f** $T_n = 3n^2 + 1$

2 Find the formula for T_n, the nth term of these sequences.

a 4, 7, 12, 19, 28, ... **b** −2, 1, 6, 13, 22, ...
c 10, 40, 90, 160, 250, ... **d** 3, 12, 27, 48, 75, ...
e 2, 6, 12, 20, 30, ... **f** $\frac{1}{2}$, 2, $4\frac{1}{2}$, 8, $12\frac{1}{2}$, ...

3 The value of the second difference can tell us something about the formula for the nth term of a quadratic sequence.

a Copy and complete this table.

nth term	First five terms	First differences	Second differences
$2n^2$	2, 8, 18, 32, 32, 50, ...	6, 10, 14, 18, ...	4, 4, 4, ...
$3n^2$	3, 12, _, _, _, _, ...		
$4n^2$	4, _, _, _, _, _, ...		
$5n^2$	5, _, _, _, _, _, ...		
$10n^2$	10, _, _, _, _, _, ...		

b What do you notice about the second difference and how it is connected to the quadratic formula for each sequence?

c Use your findings to help you to find the nth term for these sequences.

i 6, 24, 54, 96, 150, ... **ii** 1, 7, 17, 31, 49, ...
iii 4, 14, 30, 52, 80, ...

4 a Using differences, find a formula connecting the height of each rectangle (h) with the number of tiles (T).

Start by making a table.

Rectangle height (h)	1	2	3	4
Number of tiles	2			
Square numbers	1			

b Explain why your formula works.

5 a Write a formula for the area, A, of each rectangle, connecting it to the pattern number, n.

b Explain **i** why your formula works
ii why your formula is quadratic.

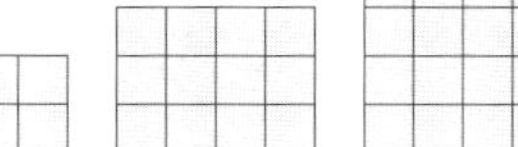

A3 Exam review

Key objectives

- Generate common integer sequences
- Generate terms of a sequence using term-to-term and position-to-term definitions of the sequence
- Use linear expressions to describe the *n*th term of an arithmetic sequence

Worked solution | **Commentary and misconceptions**

1

Students struggling with the concept of algebra may have problems with finding and using the rules of sequences. Approaching questions logically and systematically will help. Encourage students to use a table to record results.

a

Some students may need prompting to see the link between the sequence and the 2 × table.

Position	1	2	3	4	5	6
Term			5	7	9	11
Difference			2	2	2	

Position	1	2	3	4	5	6
2 x table	2	4	6	8	10	12
Difference			5	7	9	11

Term = 2 x position − 1
So, nth term = 2n − 1

b 1st term = 2 x 1 − 1 = 2 − 1 = 1
2nd term = 2 x 2 − 1 = 4 − 1 = 3

Ensure that students use the correct order of operations: multiplication before addition and subtraction.

2

Students struggling with the concept of algebra may have problems with finding and using the rules of sequences. Approaching questions logically and systematically will help. Encourage students to use a table to record results. Some students may need prompting to see the link between the sequence and the 6 × table.

Pattern	1	2	3
6 x table	6	12	18
Number of sticks	6	12	18

Number of sticks = Pattern number x 6
So, m = 6n

Ensure that students give a formula as their solution, not an expression.

Displaying and interpreting data

Objectives

F/H Draw and produce, using paper and ICT, diagrams for continuous data including, scatter graphs, and stem-and-leaf diagrams

H Draw and produce, using paper and ICT, box plots and histograms for grouped continuous data

F/H Appreciate that correlation is a measure of the strength of the association between two variables

F/H Distinguish between positive, negative and zero correlation using lines of best fit

F/H Interpret a wide range of graphs and diagrams and draw conclusions

H Compare distributions and make inferences, using shapes of distributions and measures of average and spread, including median and quartiles

Unit overview

This unit uses the knowledge of data and data collection from unit D1 and extends to representing and interpreting data using a variety of graphs and diagrams. It provides a basis for work involving representing and interpreting data in further data units.

Prior knowledge

Before your students start this unit they should be able to:

- Plot points in all four quadrants of the coordinate grid
- Use tables to record results
- Draw and interpret linear graphs
- Understand and use percentage notation
- Calculate a given percentage of an amount
- Use the language of probability
- Calculate the averages and range of a data set

Differentiation

- **Foundation** focuses on displaying data using a range of diagrams and graphs
- **Foundation Plus** extends to using scatter graphs
- **Higher Plus** extends the Higher book to include cumulative frequency diagrams

D2.1 Scatter diagrams

Objectives
- Draw and use scatter diagrams
- Understand the concept of correlation

Useful resources
- 2 mm graph paper

Mental starter
A skateboarder has the idea that the taller you are, the longer the skateboard you need to achieve a maximum performance. He comes up with the formula

$$length\ of\ skateboard = \frac{4 \times height}{5} - 80.$$

Board length and height are measured in cm. **If this formula were true, what would be the difficulties in skateboarders using it?** (Someone 1 m tall has a skateboard 0 cm long! People just over 1 m will have boards too small to fit their feet on. Shops will have to sell boards of so many different lengths.) In actual fact most boards are between 78 and 83 cm with small children finding the slightly smaller board easier.

Introductory activity
Discuss what students know about scatter diagrams.
In which of these situations would you draw a scatter diagram?

- Comparing the height of a person and the length of his/her arm
- Comparing the age of a car and its engine size
- Counting the number of siblings for all students in a tutor group.

Only the first one. A scatter diagram is used to look for a connection (correlation) between two variables (things that are being measured). The first one is well suited to a scatter diagram because you would expect there to be a relationship between the two variables. You could draw a diagram for the second, but it would be meaningless because the age of a car does not depend on how large an engine it has. In the third example you are measuring frequency (the number of people who have 0, 1, 2, 3 or more siblings) and would use a frequency diagram such as a bar chart or line graph to illustrate this.

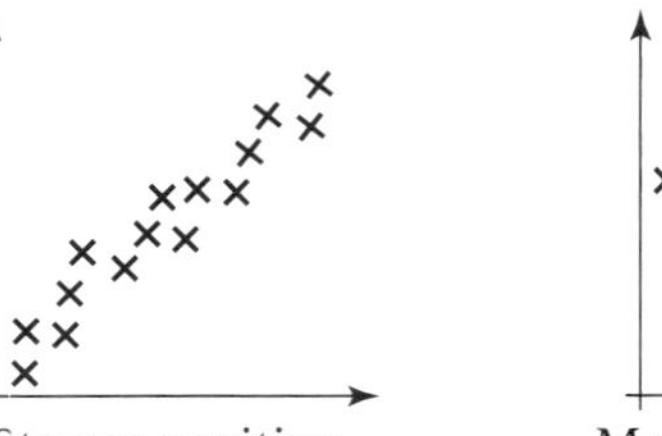

Strong positive correlation

Moderate negative correlation

Weak positive correlation

No correlation

What can you say about the type of correlation in each of these scatter diagrams?

Exercise commentary and misconceptions
When students draw scatter diagrams, encourage them to make them as large as possible, using a whole page for each diagram if necessary. Many students will tend to draw them far too small. Students should try to design scales that are easy to use. Some students will use the given values as the scale, instead of finding the values on a regular scale. Remind students that they do not have to include the origin in the diagram, so for question 1 the scales should go from 130 to 160 cm and from 30 to 55 kg.

It is important that students get into the habit of describing fully the relationship between the variables, rather than just writing 'There is a strong positive correlation' for example. This is asked for in part **c** of each question.

Plenary
Why is the statement 'figures show that the amount of carbohydrates you eat affects your metabolism' not very helpful? It tells you there is a connection (correlation) but does not tell you if your metabolism gets faster or slower if you eat more carbohydrates. In this case you need to say whether the correlation is positive or negative.

D2.1 Scatter diagrams

This spread will show you how to:

- Draw and use scatter diagrams
- Understand the concept of correlation

Keywords
Correlation
Scatter diagram

- **A scatter diagram shows you the relationship between two numerical variables.**

If there is a close relationship, or **correlation**, between the variables, the points will lie roughly on a straight line.

Height / Age of child — If the line slopes upwards there is **positive correlation**

Value / Age of car — If the line slopes downwards there is **negative correlation**

House number / Age — If the points are widely scattered, there is **no correlation.**

Examiner's tip
The diagrams described in this unit may be useful for your statistical coursework task.

Example

Sally recorded the circumference and weights of eight pumpkins.

Circumference, cm	142	124	136	140	139	128	132	135
Weight, kg	28	21	25	26	26	23	23	24

a Draw a scatter diagram of these data.
b Describe the correlation shown.
c Describe the relationship between the circumference and weight of these pumpkins.

a

Weight (kg): 20, 21, 22, 23, 24, 25, 26, 27, 28, 29
Circumference (cm): 120, 125, 130, 135, 140, 145

b There is positive correlation.

c The correlation shown suggests that larger pumpkins weigh more.

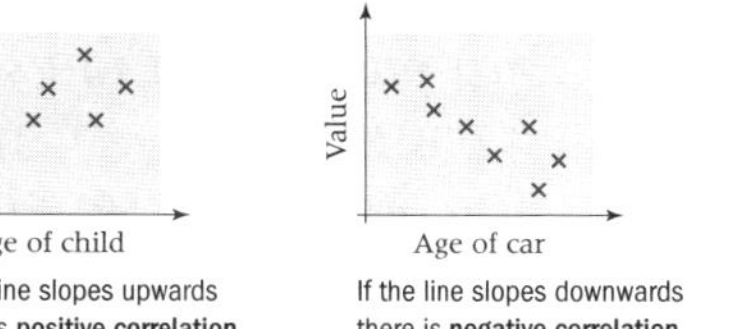

Exercise D2.1

1 Kevin recorded the heights and weights of eight boys.

	Adam	Ben	Carl	Don	Evan	Fred	Gavin	Henry
Height, cm	157	136	142	140	138	152	156	160
Weight, kg	47	32	36	37	33	39	42	51

a Draw a scatter diagram of these data.
b Describe the correlation shown.
c Describe the relationship between the height and weight of these boys.

You will need to use the graphs you draw in this exercise for Exercise D2.2.

2 Iain recorded the percentages achieved in Statistics and Mathematics exams by 10 students.

Statistics %	78	82	74	75	93	70	66	62	77	89
Mathematics %	70	76	61	70	89	65	59	58	73	82

a Draw a scatter diagram of these data.
b Describe the correlation shown.
c Describe the relationship between the percentages achieved in Statistics and Mathematics for these students.

3 Louise recorded information about the average number of minutes per day spent playing computer games and the reaction times of nine students.

Minutes per day spent playing computer games	40	60	75	40	35	20	80	50	45
Reaction time, seconds	5.2	4.3	3.9	5.5	6.0	7.2	3.6	4.8	5.0

a Draw a scatter diagram of these data.
b Describe the correlation shown.
c Describe the relationship between the average number of minutes per day spent playing computer games and reaction time.

4 Bob caught nine fish during one angling session.
He recorded information about the weights and lengths of the fish he caught in this table.

Weight, g	500	560	750	625	610	680	600	650	580
Length, cm	30	32	50	44	39	48	40	45	36

a Draw a scatter diagram of these data.
b Describe the correlation shown.
c Describe the relationship between the weights and lengths of the fish.

D2.2 Using scatter diagrams

Objectives
- Draw and use lines of best fit

Useful resources
- OHT of scatter graph
- Rulers

Mental starter
The table shows the number of burglaries recorded and the number of dishwashers owned in England and Wales.

Burglaries (millions)	1.39	1.23	1.21	1.17	1	0.9	0.83
Dishwashers owned (millions)	6.9	7.4	7.6	8	10.8	11.2	12
Year	1993	1994	1995	1996	1997	1998	1999

Can you cut the number of burglaries by increasing the sales of dishwashers? (No; as burglaries have gone down, the number of dishwashers owned has gone up, but that does not meant that they are having an effect on each other.) A correlation between two variables does not mean that one is causing the other.

Introductory activity
Recap the previous lesson on scatter diagrams. Look at the example in the student book, which relates to the circumference and weight of pumpkins. How could you draw a line through the crosses to give an idea of the trend? 'Lines of best fit' should pass through the centre of the crosses, have roughly the same number of crosses above and below, not be extended past the crosses at either end, and not be forced in the direction of the origin (0, 0).

What is wrong with extending the line to predict other results outside the range of the data?

You do not know that the relationship will continue to be governed by a straight line, so your result will be unreliable.

How reliable is predicting a result within the range of the data? The correlation is only moderate, so your estimate is not totally reliable.

Exercise commentary and misconceptions
Encourage students to use common sense when deciding if the correlation means one measure is causing the other to change.
Explain that extending the line of best fit slightly outside the data may be fine, but results are unreliable.
Estimates should be rounded to a sensible degree of accuracy, especially if the correlation is not strong.

Plenary
In the 1930s most people did not believe that smoking seriously damaged your health, in fact they thought it actually improved your health! If a scatter diagram showing the age at which people died and the number of cigarettes they smoked per week had been produced, do you think this would have convinced people of the dangers?

What might the scatter diagram look like? (Strong negative correlation.)

D2.2 Using scatter diagrams

This spread will show you how to:
- Draw and use lines of best fit

Keywords
Correlation
Line of best fit
Prediction
Scatter diagram

- **You can use a scatter diagram that shows correlation to predict other results.**

To predict results you first draw a **line of best fit**.

- A line of best fit should follow the trend of the plotted points

The line of best fit can slope downwards. It doesn't have to start at (0, 0).

Example

a Draw a line of best fit for the data Sally collected on the circumference and weight of eight pumpkins in D2.1.
b Use the line to predict
i the weight of a pumpkin, circumference 137 cm
ii the circumference of a pumpkin that weighs 22 kg.

a

Weight (kg)
29 28 27 26 25 24 23 22 21 20
120 125 130 135 140 145
Circumference (cm)

Draw a vertical line from 137 cm up to the graph, then across to the vertical axis

b i The weight of a pumpkin with a circumference 137 cm is 25 kg.
ii The circumference of a pumpkin weighing 22 kg is 127 cm.

It would not be sensible to use this graph to predict the weight of a pumpkin that is 120 cm around as this is outside the range of the collected data plotted in the graph.

Exercise D2.2

1 Look at the graph you drew in Exercise D2.1 question **1**.

a Draw a line of best fit for the data Kevin collected on the height and weight of eight boys.

b Use the line to predict

i the weight of a boy 145 cm tall

ii the height of a boy who weighs 45 kg.

2 Look at the graph you drew in Exercise D2.1 question **2**.

a Draw a line of best fit for the data Iain collected on Statistics and Mathematics results.

b Use the line to predict

i the percentage in Mathematics for Imogen who achieved 80% in Statistics

ii the percentage in Statistics for Katherine who achieved 80% in Mathematics.

c Give a reason why it would not be sensible to use this graph to predict the percentage achieved in the Mathematics exam for a person who achieves 46% in the Statistics exam.

3 Look at the graph you drew in Exercise D2.1 question **3**.

a Draw a line of best fit for the data Louise collected on time spent playing computer games and reaction time.

b Use the line to predict

i the reaction time of Jim, who spends on average 70 minutes per day playing computer games

ii the time per day spent playing computer games for Ellis, who has a reaction time 6.4 seconds.

c Give a reason why it would not be sensible to use this graph to predict the reaction time of Tom who spends on average 180 minutes per day playing computer games.

4 Look at the graph you drew in Exercise D2.3 question **4**.

a Draw a line of best fit for the data Bob collected on the weight and length of fish.

b Use the line to predict

i the length of a fish that weights 700 g

ii the weight of a fish that is 42 cm long.

D2.3 Stem-and-leaf diagrams

Objectives

- Draw stem-and-leaf diagrams, using them to find the median and range of data sets

Useful resources

- OHT of stem-and-leaf diagram
- 5 mm square grid paper

Mental starter

Write these values on the board:
28 32 21 53 60 24 37 49 35 44 13
Ask students to find the mean, median and range.

Introductory activity

These are the ages of 16 people at a squash club: 23, 34, 27, 19, 18, 35, 42, 25, 43, 9, 22, 32, 39, 34, 30, 41 (the 9 year old is very talented!). This data can be arranged numerically, but there is a more sophisticated way called a stem-and-leaf diagram.

First make the 'stem'. This is made of the most significant digits in the data, so for this example it will be the 'tens' digits. They range from 0 to 4 in this example, so write them in ascending order down the page.

Next put the less significant digits beside the stem digits in the order in which they were written in the data list.

Now copy the diagram, but this time put the 'leaf' digits in order.

Finally give the key for the diagram, so that someone reading this table knows that the tens are on the left and the units are on the right.

0	9
1	8 9
2	2 3 5 7
3	0 2 4 4 5 9
4	1 2 3

Key: | 1 | 8 | stands for 18 years old

Describe this scenario:.

Because the data is now in order, it is easy to find the median (31), the lower quartile (22.5), the upper quartile (37) and the interquartile range (14.5).
Discuss the example in the student book, which relates to a set of test results.

Exercise commentary and misconceptions

Encourage students to check that the number of results in the question always match the number of results in their table/diagram.
Mistakes are very easy to make so encourage students to use a pencil. Students will only get full marks if the diagram **is** ordered.
In order to save time, students will be tempted to put them in order straight away by scanning the list. Encourage students not to do this; it is likely to lead to errors. Ensure students record duplicate results properly, that they keep the data in columns and that they **always** include a key, even if the question does not ask for it. Grid paper can help to keep the digits aligned.

Plenary

The weights in tonnes of a sample of cars from the USA are 3.5, 2.7, 1.8, 3.9, 2.8, 2.7, 4.3.
A sample of cars from the UK are 2.3, 0.9, 1.9, 2.2, 1.5, 3.1, 0.7. Put the results in a 'back-to-back stem-and-leaf diagram'.

This can be done by having two students draw the 'back-to-back stem-and-leaf' on the board by standing 'back to back'.
What can you say about the weights of cars in the two countries? (American cars tend to be heavier.)

D2.3 Stem-and-leaf-diagrams

This spread will show you how to:

- Draw stem-and-leaf diagrams, using them to find the median and range of data sets

Keywords
Interquartile range
Lower quartile
Median
Range
Stem-and-leaf
Upper quartile

You can represent small amounts of data on a **stem-and-leaf** diagram. A stem-and-leaf diagram shows all the data and the overall trend.

- **You can use a stem-and-leaf diagram to find statistics such as the median.**

When you draw a stem-and-leaf diagram choose the stem according to the data.

For example, for heights given to nearest cm, data will be in the 100s: 165, 154, 178, ... so in the stem you use

15 |
16 |
17 |

Example

These are the percentages for a Statistics test taken by 23 students.

58 54 78 66 67 40 45 38 58 73 51 49
47 53 41 36 59 64 52 43 39 80 37

a Draw a stem-and-leaf diagram for these data.

b Describe the trend.

c For these data find the

i range

ii median

iii lower quartile and upper quartile

iv interquartile range.

a

Stem	Leaves
3	8 6 9 7
4	0 5 9 7 1 3
5	8 4 8 1 3 9 2
6	6 7 4
7	8 3
8	0

Use the tens digit as the stem

Use the units digit as the leaves

Now put the stem and leaves in order

Stem	Leaves
3	6 7 8 9
4	0 1 3 5 7 9
5	1 2 3 4 8 8 9
6	4 6 7
7	3 8
8	0

Key: 7 | 3 stands for 73%

Always give a key

b Most students scored less than 60%.

c i Range = 44% — highest value – lowest value: 80 – 36 = 44

ii Median = 52% — data is in order so count up to find 12th value: $\frac{1}{2}(23 + 1) = 12$

iii Lower quartile (LQ) 41%
Upper quartile (UQ) 64% — lower quartile is the 6th value in the 1st quarter of data: $\frac{1}{4}(23 + 1) = 6$ then count back 6 values to find upper quartile

iv Interquartile range (IQR) 23% — IQR = UQ – LQ 64 – 41 = 23

Exercise D2.3

For these data sets

a Draw a stem-and-leaf diagram to show the information. (Remember to include a key.)

b Describe the trend.

c Find the

i range **ii** median **iii** lower quartile and upper quartile
iv interquartile range.

1 Percentage achieved in a Statistics test

78 82 74 45 69 75 93 54 61 70 48 66
62 51 77 59 51 89 81 52 63 71 59

2 Minutes per day spent playing computer games

40 26 75 84 33 39 28 66 67 71 80
37 52 47 63 49 41 44 58 69 43
73 55 59 43 61 38 29 30 77 60

3 Time taken, in minutes, to solve a crossword puzzle

12 24 21 16 8 9 3 31 18 27 35
41 26 12 17 6 5 19 29 32 37 40
15 22 10 33 11 7 20 27 29 34

4 Weight in kg of Year 10 boys

47 51 63 39 42 57 36 37 49 32 60 54
56 45 52 48 61 58 56 59 70 66 56

5 Height in cm of a group of Year 10 girls

153 147 160 146 162 158 159 171 149 152 150 163
172 167 165 155 155 157 154 168 150 172 152

6 IQ scores of a Year 10 tutor group in a comprehensive school

101 112 125 109 98 107 108 117 121 116 94
91 105 106 114 118 126 131 92 88 129
89 116 103 108 127 110 117 104 119 133

7 Time taken to the nearest minute to change a flat tyre

10 17 3 19 22 27 16 9 6 30 23 21
12 21 9 25 23 18 33 8 32 15 11

D2.4 Interpreting stem-and-leaf diagrams

Objectives

- Draw and use back-to-back stem-and-leaf diagrams

Useful resources

- OHT of table in introductory activity
- 5 mm square grid paper

Mental starter

Show this frequency table on the board:

Value	1	2	3	4	5
Frequency	29	32	17	15	4

Ask students to find the median.
Ask students to find the mean.

Introductory activity

Discuss this scenario, relating to the reading habits of senior citizens and students: A large sample of each group was interviewed and asked 'How many hours a week, on average, do you spend reading? Include only books, not magazines or newspapers.'
Imagine a stem-and-leaf diagram has been drawn and the following information gleaned:

Number of hours	Students	Senior citizens
Minimum	0	3
LQ	17	7
Median	20	10
UQ	24	20
Maximum	34	30
Interquartile range	7	13

What does the difference in the median tell you about who reads more? (Students' median is higher so on average they read more.)

What does the difference in the interquartile ranges tell you about the variation (spread) of the number of hours that each group spends reading? (Seniors' interquartile range is larger so there is more variation (spread) in the number of hours they spend reading than for students.)

Discuss how to show two sets of information on a stem-and-leaf diagram (refered to in D2.3 plenary). Discuss the back-to-back stem-and-leaf diagram in the example in the student book.

Exercise commentary and misconceptions

Students may need reminding of the main features of a stem-and-leaf diagram, especially the key. An example of a back-to-back stem-and-leaf diagram can be found in the plenary of D2.3. Students must interpret the meaning of the median and IQR with reference to the question. Just stating the values, and that they are larger or smaller than each other, is not sufficient. Encourage students to use the words 'variation' or 'spread' and talk about the specific variable that is being measured in the question.

In questions 2 and 3 the 'stems' will contain two digits.

Plenary

What do you think is the most common word used in the UK for a computer password? (The answer is 'password'.) **What would be a good way to choose a password that is easy to remember and difficult for others to guess?** (Discuss; numbers or words or a mixture of both, such as 5hape which uses 5 instead of S.)

D2.4 Interpreting stem-and-leaf diagrams

This spread will show you how to:

- Draw and use back-to-back stem-and-leaf diagrams

Keywords
Average
Interquartile range (IQR)
Median
Measure of spread
Stem-and-leaf diagram

You can represent two data sets of the same variable on a back-to-back **stem-and-leaf** diagram.
The data sets will share a common stem.

- **You can compare the two data sets using an average and a measure of spread.**

The range measures the spread of all the data.

Range = highest value − lowest value

The **interquartile range** (**IQR**) measures the spread of the middle half of the data.

IQR = upper quartile − lower quartile

Example

Draw a back-to-back stem-and-leaf diagram to show the test results of a group of girls and a group of boys.

Girls 52 34 58 46 41 57 47 35 49 47 54

Boys 61 43 47 56 59 39 58 69 52 46 54

Compare the performances of the boys and the girls.

Girls	Stem	Boys
5 4	3	9
9 7 7 6 1	4	3 6 7
8 7 4 2	5	2 4 6 8 9
	6	1 9

Key: 1 | 4 | 3 stands for 41% for girls, 43% for boys

There are 11 girls.

Median = $\frac{(11+1)}{2}$th result
= 6th result
= 47

Range = 58 − 34
= 24

Lower quartile = $\frac{(11+1)}{4}$th value
= 3rd result
= 41

Upper quartile = 54
IQR = 54 − 41
= 13

There are 11 boys.

Median = $\frac{(11+1)}{2}$th result
= 6th result
= 54

Range = 69 − 39
= 40

Lower quartile = $\frac{(11+1)}{4}$th value
= 3rd result
= 46

Upper quartile = 59
IQR = 54 − 41
= 13

1 The boys' average is 7 marks higher than the girls' average.
2 The IQR is the same for the boys and the girls.
3 The range of the boys' marks is greater than that of the girls.

Exercise D2.4

For these data sets

a Draw a back-to-back stem-and-leaf diagram. (Remember to include a key.)

b For both data sets in each question work out the

i median **ii** interquartile range **iii** range.

c Use your answers to part **b** to make comparisons between the two data sets.

1 Percentages achieved in two tests

Test A: 38 37 62 45 42 55 56 61 49 52
47 58 43 51 44 56 41 44 53

Test B: 65 72 57 79 66 48 53 54 41 75
56 63 69 72 53 44 57 61 70

2 Height to the nearest cm of a sample of men and a sample of women

Men: 176 183 184 172 168 175 183 159 169 174
160 180 178 167 182 188 171 178 158 165
177 169 167

Women: 157 148 151 167 174 165 169 158 153 155
161 158 155 172 156 162 166 149 178 154
152 150 162

3 IQ scores of two classes, X and Y

X: 105 123 131 117 118 104 98 96 103 112 110
117 126 129 123 109 108 115 99 89 121 134
106 105 122 124 116 110 118 115 130

Y: 118 119 104 121 126 118 109 97 114 129 130
107 116 87 93 128 121 118 113 103 102 114
107 131 99 106 124 132 119 126 114

4 Reaction times to the nearest tenth of a second by a sample of boys and a sample of girls

Boys: 4.2 5.7 3.2 3.8 6.4 3.8 6.1 5.9 5.3 5.6 3.6 4.4
5.2 3.2 3.8 5.8 4.7 4.5 6.2 6.8 7.1 6.6 7.2

Girls: 4.4 4.6 5.2 4.3 6.7 7.2 8.0 4.0 8.2 7.7 6.3 7.6
4.8 5.9 5.2 7.4 7.3 6.2 6.5 5.6 6.6 6.3 5.5

5 Times taken to the nearest minute for a sample of children to complete two jigsaw puzzles

Puzzle P: 13 17 19 10 8 22 31 11 24 27
6 37 18 12 29 14 8 17 9

Puzzle Z: 28 21 29 15 12 9 32 17 18 11
19 16 8 33 24 14 25 17 23

D2.5 Box plots

Objectives

- Draw box plots

Useful resources

- 2 mm graph paper

Mental starter

Ask students to find $\frac{1}{2}$, $\frac{1}{4}$ and $\frac{3}{4}$ of these amounts:
30, 46, 35, 116, 440, 208.

Introductory activity

To draw a box plot or a box and whisker diagram you need five bits of information about your data: the minimum value, the LQ, the median, the UQ and the maximum value.

These are the heights of 10 trees measured to the nearest metre: 4, 4, 6, 6, 7, 7, 8, 15, 16, 20.

Minimum value = 4
LQ = 6,
median = 7
UQ = 15,
maximum value = 20.

Ask students to draw the box plot, emphasising that the scale must be drawn first. Sketch the desired result on the board, and go around the class checking students' attempts.

Exercise commentary and misconceptions

The data used in this exercise are those from exercise D2.3, so the students should have already found the five values they need for each question.

Ensure students use an exact scale for the axes and don't just label the axes with the points they are trying to represent.

Students should use a reasonable and easy to read scale. Many students are tempted to draw very small box plots; encourage a larger rather than smaller diagram. Students need not include zero on their axes.

Plenary

Discuss: What do box plots help you to interpret? (The median and two measures of spread: the IQR and the range.) **What other statistics might it be helpful to know from the data?** (Mode and mean.) **Can these be found from a box plot?** (no.)

D2.5 Box plots

This spread will show you how to:

- Draw box plots

- You can represent data sets on a **box plot**.
- To draw a box plot you need the **median**, the **upper** and **lower quartiles**, and the highest and lowest values.

Keywords
Box plot
Box and whisker
Interquartile range (IQR)
Lower quartile
Median
Upper quartile

Example

A group of 15 students took a science test.

These are their results.

52, 62, 71, 46, 41, 49, 36, 57, 60, 80, 41, 40, 79, 39, 64

a Find the median, the range and the upper and lower quartiles.
b Draw a box plot to represent these results.
c Find the interquartile range.

A box plot is sometimes called a **box and whisker diagram**.

a First arrange the data in order.

36, 39, 40, 41, 41, 46, 49, 52, 57, 60, 62, 64, 71, 79, 80

Median $= \frac{(15+1)}{2}$th value
= 8th value
= 52

Lower quartile $= \frac{(15+1)}{4}$th value
= 4th value
=41

Range = 80 − 36
= 44

Upper quartile $= \frac{3}{4}(15+1)$th value
= 12th value
= 64

b

Lower quartile
Median
Upper quartile
35 40 45 50 55 60 65 70 75 80 85
Test marks %
Lowest mark
Highest mark

Use graph paper and remember to scale and label the axis.

c Interquartile range = 64 − 41 = 23

Exercise D2.5

The data sets for questions **1** to **7** were used in Exercise D2.3 to draw stem-and-leaf diagrams.

Use the values you found for the median, lower quartile and upper quartile (or use the data to find them again) and draw a box plot to show the information for each of these data sets.

1 Percentage achieved in a Statistics test

78 82 74 45 69 75 93 54 61 70 48 66
62 51 77 59 51 89 81 52 63 71 59

2 Minutes per day spent playing computer games

40 26 75 84 33 39 28 66 67 71 80
37 52 47 63 49 41 44 58 69 43
73 55 59 43 61 38 29 30 77 60

3 Time taken, in minutes, to solve a crossword puzzle

12 24 21 16 8 9 3 31 18 27 35
41 26 12 17 6 5 19 29 32 37 40
15 22 10 33 11 7 20 27 29 34

4 Weight in kg of Year 10 boys

47 51 63 39 42 57 36 37 49 32 60 54
56 45 52 48 61 58 56 59 70 66 56

5 Height in cm of a group of Year 10 girls

153 147 160 146 162 158 159 171 149
152 150 163 172 167 165 155 155 157
154 168 150 172 152

6 IQ scores of a Year 10 tutor group in a comprehensive school

101 112 125 109 98 107 108 117 121
116 94 91 105 106 114 118 126 131
92 88 129 89 116 103 108 127 110
117 104 119 133

7 Time taken to the nearest minute to change a flat tyre

10 17 3 19 22 27 16 9 6 30 23 21
12 21 9 25 23 18 33 8 32 15 11

D2 Exam review

Key objectives

- Draw and produce, using paper and ICT, diagrams for continuous data, including scatter graphs and stem-and-leaf diagrams, and box plots for grouped continuous data
- Draw lines of best fit by eye, understanding what these represent
- Distinguish between positive, negative and zero correlation using lines of best fit

Worked solution | **Commentary and misconceptions**

1 a

4	9 9
5	0 2 2 2 3 4 5 5 8 8 9
6	0 0 1 1 1 3 9
7	0 2 2

Key: 5 | 1 stands for 51 minutes

Students should always include a key with stem-and-leaf diagrams.

b **i** Median = 58 minutes

ii lower quartile = 52 minutes

upper quartile = 61 minutes

inter quartile range = 61 − 52 = 9 minutes

iii range = 72 − 49 = 23 minutes

2

Number of Pages	80	90	100	105	115	130	140	140	160	170
Weight (g)	160	180	180	210	230	270	270	290	320	300

Encourage students to order data before drawing a graph.

a, c

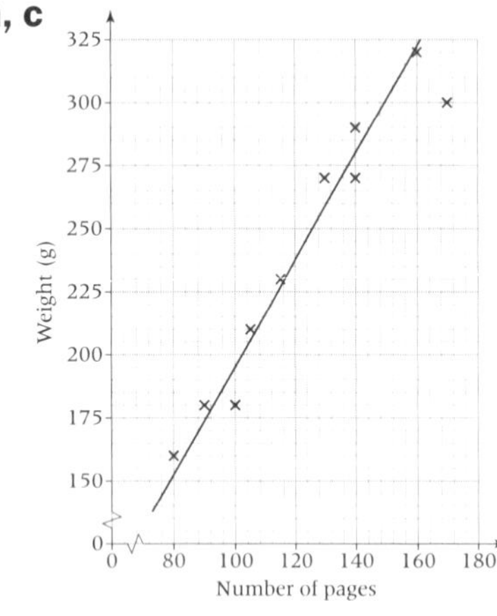

b There is a positive correlation between the number of pages and the weight of a book.

Ensure that students draw a suitable line of best fit.

d **i, ii**

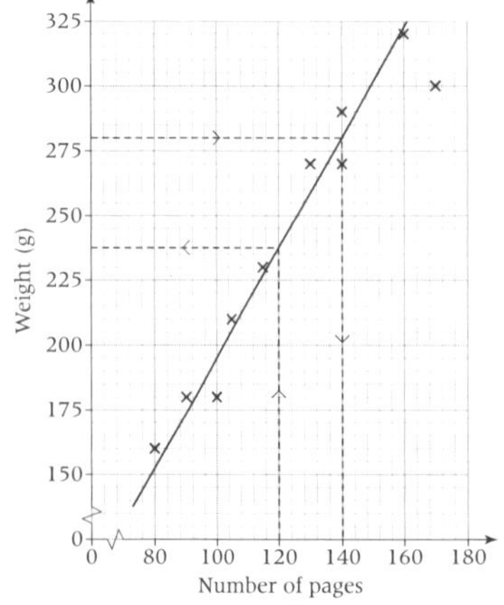

Ensure students use their graph correctly to find the corresponding values.

A4 Straight-line graphs

Objectives

H Understand that the form $y = mx + c$ represents a straight line and that m is the gradient of the line and c is the value of y-intercept

F/H Plot graphs of functions in which y is given explicitly in terms of x, or implicitly

H Explore the gradients of parallel lines and of straight lines perpendicular to each other

H Find the equation of a given straight line in the form $y = mx + c$

H Find the gradient of straight lines given by equalities of the form $y = mx + c$ (where values are given for m and c)

Unit overview

This unit consolidates the Key Stage 3 knowledge of graphs. It extends the knowledge of algebra gained in previous units to graphical representations of linear equations. This unit provides a basis for solving equations graphically, covered in unit A7 and for work on graphs featured in unit A8.

Prior knowledge

Before your students start this unit they should be able to:

- Plot points in all four quadrants of the coordinate grid
- Use and understand the vocabulary associated with straight-line graphs
- Draw lines in the coordinate grid, given several points on the line
- Understand and use fraction notation
- Add, subtract, multiply and divide with integers and fractions
- Use the rules of algebra with the correct order of operations
- Solve linear equations in one unknown

Differentiation

- **Foundation focuses** on plotting points and straight lines given a table of values. Finding the equation of vertical and horizontal lines is also covered
- **Foundation Plus** extends to interpreting the gradient and y-intercept of straight-line graphs
- **Higher Plus** extends the Higher book to implicit functions, perpendicular and parallel lines and solving inequalities graphically

A4.1 Straight-line graphs

Objectives

- Recognise that linear equations have straight-line graphs
- Plot straight-line graphs, given a linear equation

Useful resources

- 2 mm graph paper
- Graph plotting tool

Mental starter

Which integers between -10 and 10 are neither even nor prime? (Note, negatives can be odd, even or prime in exactly the same way as positive numbers. 0 is even, 1 is not prime, so list is -1, 1, -9, 9.) **What answers can you make by adding two of these numbers? For example, $-1 + 9 = 8$.** (-10, -9, -8, -1, 0, 1, 8, 9, 10)

Introductory activity

Imagine you have a rule for changing one number into another number.

Starting number $\rightarrow \times 2 \rightarrow -1 \rightarrow$ end number. Call the starting number x and the end number y. You can write this rule as $y = 2x - 1$.

Think of some possible starting numbers (x) and the matching y numbers. Ask students to give a variety of examples: decimal, fraction, negative, standard form. How can you show a picture of the possible results?

Draw a set of axes from -10 to 10 in the x and y directions. Give students these instructions:

Select a set of values to use for x and complete a table.

Plot points. Extend line through points. Label the line.

Go around the class and check graphs. Explain that equations like $y = 2x - 1$ are called linear equations.

Discuss the example in the student book, which relates to an implicit equation.

Exercise commentary and misconceptions

Question 1 should be done without trying to draw the graphs.
$y = 7$, and $x = -2$ may cause problems because there is only one algebraic term. Students may need reminding of lines parallel to the axes.

If students calculate the values of y wrongly, or label the axes incorrectly, they will not be able to draw straight lines in questions 2, 3 and 5. In question 4 students should not draw the graph.

Question 6 could be used in the plenary session as a starting point for a discussion on the intersection of lines.

Plenary

Discuss question 6 **or**
How could you draw the line for $8x - 4000y = 0$?

Rearrange to $8x = 4000y \rightarrow x = 500y$.

Mark an x-axis scale in units of 500.

x	0	500	1000
y	0	1	2

A4.1 Straight line graphs

This spread will show you how to:

- Recognise that linear equations have straight-line graphs
- Plot straight-line graphs, given a linear equation

Keywords
Axes
Linear
Plot

- The graphs of equations such as $y = 2x + 1$ (with no x^2 or higher powers) are straight lines. For example,

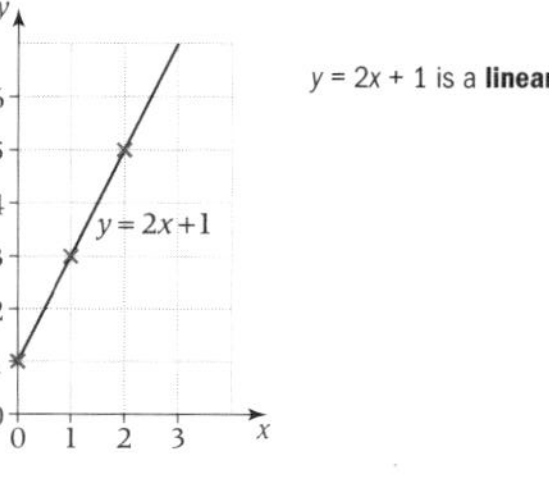

The graph of a linear equation is a straight line.

- You need three points to plot a straight line.

For example, to draw the graph of $y = 3x - 2$, first make a table of x and y values.

x	1	2	3
y	1	4	7

when $x = 1$, $y = (3 \times 1) - 2 = 1$

Draw x- and y-**axes** and **plot** the three points.

Two points are enough to fix a straight line but, as a check, work out a third point.

For an equation such as '$2x + 3y = 12$' it is a little harder to find the points.

$2x + 3y = 12$ is an **implicit** equation – you can't calculate the value of y directly from the equation.

Example

Plot the graph of $2x + 3y = 12$.

x	**0**	6	**3**
y	4	**0**	2

Draw the x and y axes and plot the points.

When the equation is in implicit form, as here, putting $x = 0$ and $y = 0$ gives you two points. For the third, choose an x-value between the first and second values.

Exercise A4.1

1 Which of these equations have straight line graphs?

$y = 2x + 3$	$y = 7 - 3x$	$y = x^2 + 1$	$y = 5x$
$y = 7$	$y = 2x - x^3$	$2x + 7y = 8$	$x = -2$

2 **a** For each equation, copy and fill in the table of values.

i $y = 3x + 2$

x	0	1	2
y			8

ii $y = 2x - 4$

x	1	2	3
y	−2		

iii $2x + 5y = 10$

x	0		1
y		0	

b Draw x- and y-axes from −8 to 8 and plot the graphs.

3 I have a mobile phone. Each month, I pay £5 line rental. For every hour I then spend on the phone, I pay £7.

a Copy and complete this table of values to show my total phone bill for different lengths of time spent on the phone.

x (hours on phone)	1	2	3	4	5
y (total bill £)	12				

b Plot a graph to show hours against total bill.

c Use your graph to find the approximate cost if I spend 3 hours and 15 minutes on the phone one month.

d What is the equation of the graph you have drawn?

4 **a** The point (2, 5) lies on the graph $y = 2x + 1$. Does (3, 8) lie on this graph? Explain your answer.

b The point (2, 5) lies on the graph $y = 2x + 1$. Name another point that lies on this line.

5 **a** Plot the graphs $y = 3x - 1$ and $y = 3x + 2$ on the same axes.
b Explain why there is no point that lies on both of these graphs.
c Name the equation of another line that would have no points in common with these two.

6 Would you prefer to get £3 per week pocket money, with 20p for every chore done (such as washing up) or £5 per week pocket money with 15p for every chore done? Use line graphs to report on your favoured option and to find how many chores you would need to do to receive the same amount under both options.

A4.2 More straight-line graphs

Objectives

- Recognise and understand the form of equations corresponding to horizontal, vertical and diagonal line graphs

Useful resources

- 2 mm graph paper
- Graph plotting tool

Mental starter

Give students the equation of a graph, for example:

$y = 4x - 3$, $y = x^2$, $y = x^3$, $y = 2x + 1$, $y = 1 - x$.

Ask students to sketch the shape of the graph on mini-whiteboards.

Introductory activity

Draw axes from −10 to 10. Draw a vertical line through $x = 4$. Label coordinates of 3 points on the line. **What do you notice about the coordinates? What would be a good name for this line?** ($x = 4$). Draw a vertical line through $y = -7$. Label three points on the line. **What do you notice about the coordinates? What would be a good name for this line?** ($y = -7$). Notice that $x = c$ lines go vertically, the opposite way to the x-axis, and the $y = c$ lines are horizontal. **What are the equations of the axes themselves?** (x-axis has equation $y = 0$ and y-axis has equation $x = 0$.) Label the axes with their equations.

Discuss the examples in the student book, focusing particularly on how to find the intersection point of two straight lines.

Exercise commentary and misconceptions

In question 4, students should be encouraged to plot the graphs for part **a** if necessary, then try the other parts of the question without plotting the graphs.

Most students will need to plot graphs for question 6.

Plenary

The lines $y = 5$ and $x = 5$ intersect at a point. Call this point A.

The lines $y = -3$ and $x = -3$ intersect at point B. **What is the equation of the line joining A and B?** ($y = x$).

To help explain the result, write down the coordinates of A and B and some other points along the line joining them. **What is the connection between the x and y value?** (They are always the same, so call the line $y = x$.)

A4.2 More straight line graphs

This spread will show you how to:

- Recognise and understand the form of equations corresponding to horizontal, vertical and diagonal line graphs

Keywords
Diagonal
Horizontal
Intersect
Vertical

A straight line can be **diagonal**, **vertical** or **horizontal**.

The x-coordinate of every point on this vertical line is 2.

The y-coordinate can have any value.

The equation of the line is $x = 2$

The y-coordinate of every point on this horizontal line is 3.

The x-coordinate can have any value.

The equation of the line is $y = 3$

- **Horizontal lines have equations of the form $x = c$.**
- **Vertical lines have equations of the form $y = c$.**

c stands for a number.

Example

Give three points that would lie on each of these lines
a $y = 5$ **b** $x = -2$.

a $y = 5$
Since the y-coordinate is 5, possible points are (1, **5**), (2, **5**) and (17, **5**)
b $x = -2$
Since the x-coordinate is −2, possible points are (**−2**, 1), (**−2**, 2) and (**−2**, 11)

Example

Where do the graphs $x = 4$ and $y = -1$ intersect?

When lines **Intersect** they cross.

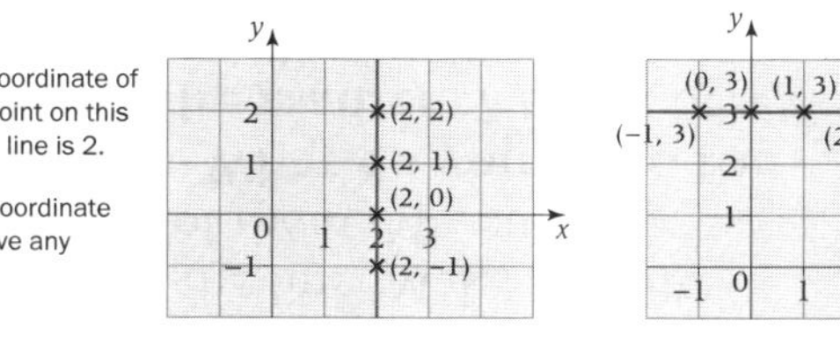

All points on the line $x = 4$ have x-coordinate 4.

All points on the line $y = -1$ have y-coordinate −1.

The lines intersect at (4, −1).

Exercise A4.2

1 **a** Copy and complete the table, deciding if each equation is that of a horizontal, vertical or diagonal line, or none of these.

$x = 9$ $y = 2x - 1$ $x = -0.5$ $y = 7$ $y = x^2 + x$

Horizontal	Vertical	Diagonal	None of these

b Add an equation of your own to each column in the table.

2 Match each line with its equation.

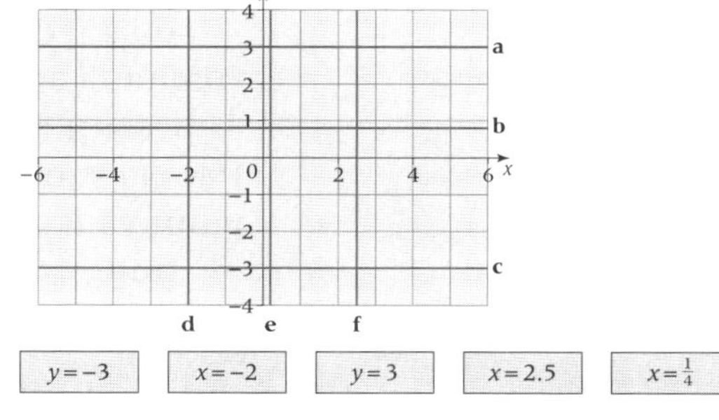

$y = -3$ $x = -2$ $y = 3$ $x = 2.5$ $x = \frac{1}{4}$ $y = \frac{3}{4}$

3 On one set of axes labelled from −6 to +6, plot these graphs.

a $x = 5$ **b** $y = 2$ **c** $x = 1.6$ **d** $y = -3$ **e** $y = 1$ **f** $x = -1\frac{1}{4}$

4 Where do these graphs intersect? Plot the graphs if you need to.

a $x = 5$ and $y = 2$ **b** $x = 4$ and $y = -3$
c $x = -2$ and $y = 9$ **d** $y = -4$ and $x = -2$

5 **a** Give the equations of four lines which, when plotted, form the sides of a rectangle.
b Repeat part **a** for a square.
c Repeat part **a** for an isosceles right-angled triangle.

6 **a** Which point with integer coordinates fits these clues?

Above $y = -1$, below $y = 3$, below $y = 2x + 1$,
above $y = 2 - x$ and left of $x = 2$.

b Write your own clues to describe the point (3, 4).

A4.3 Gradients and intercepts

Objectives

- Find the gradient and y-axis intercept of straight-line graphs

Useful resources

- 2 mm graph paper
- Graph plotting tool

Mental starter

Give students the equation of a graph, for example:
$y = 5$, $y = -2$, $x = 1$, $x = 4$, $x = 0$, $y = 0$.
Ask students to sketch the shape of each graph on mini-whiteboards.

Introductory activity

Discuss: the gradient of a line is its steepness. How do you find it?

Draw axes from −12 to 12 in both directions. Plot the points (4, 9) and (−1, −11). Draw and extend a line through these points. Pick any two clear points on the line. Make a right-angled triangle out of them. Measure the height and the base. **What do you do to the numbers to find the gradient?** (Can't be +, − or × because you would get different answers if you picked different points along the line – you should get the same answer everywhere on the same line.) Must be divide, but which way? Sketch a very steep line that has a base of 2 and a height of 20. If you did base ÷ height, you get 0.1. Height ÷ base gives 10. Compare this to your original line (base ÷ height = 0.25. Height ÷ base gives 4). Steeper lines must have steeper gradients.

So $gradient = \frac{height}{base}$. How many units does your line go up for every 1 unit moved along in the x-direction? (4 units). This is the meaning of gradient: the number of units you move up every time you move 1 unit in the x-direction.

Draw a line through the point (0, 0) and (−2, 8) on the same axes as the first example. What is the gradient? (Answer looks like 4.) But the two lines are not going in the same direction, so do not have the same gradient. Use negative gradients for lines that slope down and positive for those that slope up. Explain that the y intercept is the number where the line cuts the y-axis.

A possible acronym for recalling the gradient formula is TUBA. Referring to the fraction you use to find gradient, Top is Up and Bottom is Across.

Exercise commentary and misconceptions

Question 3 is the key question which links gradient and intercept to the equation of the line.

Plenary

What is the gradient of $y = 5$? (zero) **What is the gradient of $x = -2$?** (infinity or undefined) This can be established by looking at steeper and steeper lines, and by looking at flatter and flatter lines.

A4.3 Gradients and intercepts

This spread will show you how to:

- Find the gradient and y-axis intercept of straight line graphs

Keywords
Coefficient
Constant
Gradient
Rise
Run
y-axis intercept

- The **gradient** of a straight line tells you how steep it is.

To work out the gradient find how many units the line **rises** for each unit it **runs** across the page.

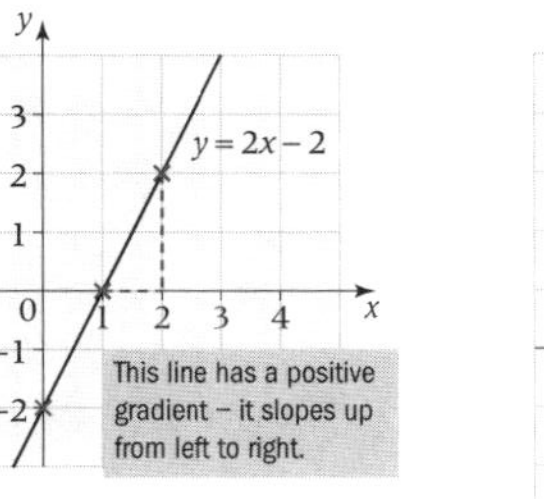

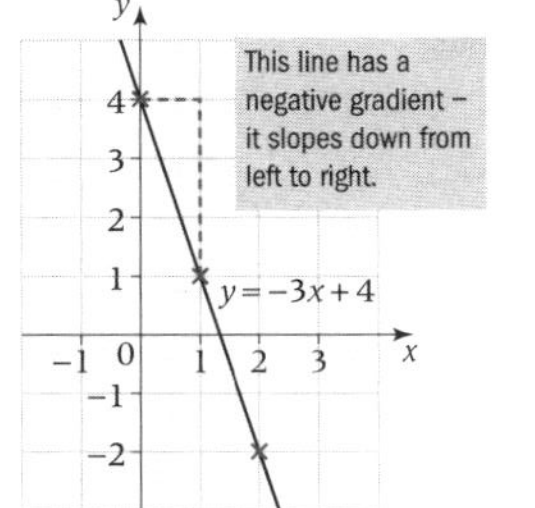

For the line $y = 2x - 2$, **gradient** $= \frac{\textbf{rise}}{\textbf{run}} = \frac{2}{1} = 2$

For the line $y = -3x + 4$, **gradient** $= \frac{\textbf{rise}}{\textbf{run}} = \frac{3}{-1} = -3$

- **The gradient is the coefficient of x (the number of xs) in the equation of the line.**

The **intercept** is the distance from the origin to where the line cuts the y-axis.

The line $y = 2x - 2$ cuts the y-axis at $(0, -2)$.
The y-intercept is -2.
The line $y = -3x + 4$ cuts the y-axis at $(0, 4)$.
The y-intercept is 4.

- **The intercept is the constant term (the number) in the equation of the line.**

Example

Find the gradient and intercept of the lines

a $y = 3x - 4$ **b** $y = \frac{1}{2}x + 5$

a gradient = 3, intercept = −4 **b** gradient = $\frac{1}{2}$, intercept = 5

If the equation is not in the form $y = ...$, rearrange it first, for example

$3x + 2y = 12 \Rightarrow 2y = -3x + 12 \Rightarrow y = -\frac{3}{2}x + 6$

Now you can see that the gradient is $-\frac{3}{2}$ and the intercept is 6.

Example

Find the gradient and intercept of the lines

a $x + y = 4$ **b** $2x - 5y = 10$

a $x + y = 4$
$y = 4 - x$
$y = -x + 4$
gradient = −1, intercept = 4

b $2x - 5y = 10$
$-5y = 2x + 10$ Divide both sides by −5
$y = \frac{2}{5}x - 2$
gradient = $\frac{2}{5}$, intercept = −2

Exercise A4.3

1 Write the gradient and intercept of each line in the diagram.

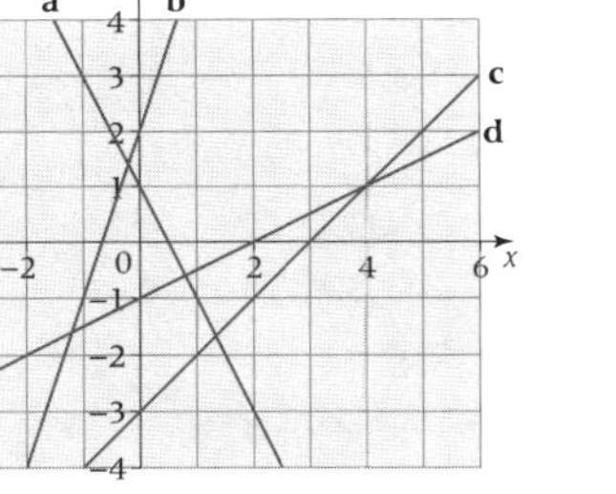

2 On a pair of axes labelled from −8 to +8, draw lines with these key characteristics.

a positive gradient

b y-intercept = 3

c gradient = $\frac{1}{4}$

d negative gradient and y-intercept = −2

e y-intercept = 5 and gradient = 3

f y-intercept = 1 and gradient = $\frac{2}{3}$

3 **a** Plot the line $y = 3x - 2$ on a pair of axes.

b Write its gradient and intercept.

c What do you notice about the connection between the gradient and intercept and the equation of the line?

d Repeat for the line $y = 2 - \frac{1}{2}x$.

4 The points (2, 1) and (5, 3) are shown.

a What is the gradient of the line joining these two points?

b Explain how you could have found the gradient without counting squares and only using the coordinates given.

c What is the gradient of a line joining (2, 9) to (31, 67)?

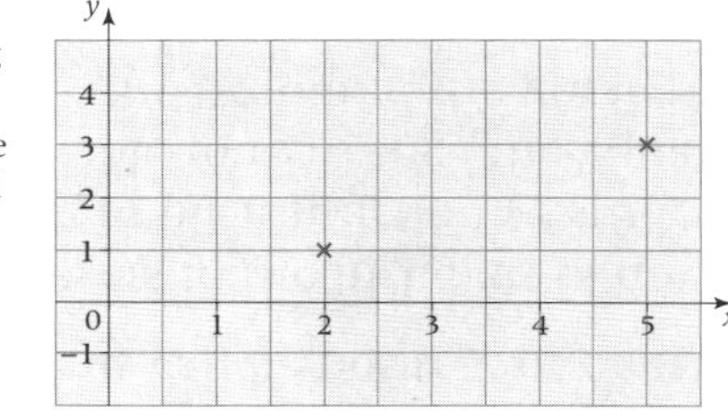

DID YOU KNOW?

To warn drivers of steep sections of road, gradient is shown on road signs as a percentage or ratio (see N4). This road has a gradient of one metre up for every seven metres along.

A4.4 The equation $y = mx + c$

Objectives

- Recognise that equations of the form $y = mx + c$ have straight-line graphs

Useful resources

- Graph plotting tool

Mental starter

Helen cycles up hill for 2 km. She rises in height by 400 m and the ride takes her 8 min. Fran cycles up hill for 3 km. She rises in height by 300 m and the ride takes her 16 min. **Who is the fitter rider? How did you decide?** (Helen. One solution is to look at what each rider can achieve in the same amount of time, say 8 min.)

Introductory activity

Draw axes from −10 to 10 in both directions. Draw the lines $y = 2x - 1$ and $y = -3x + 2$ taking x values at 0, 2, 4.

Find the gradient of each line. (2 and −3) Find the y-intercept of each line (−1 and 2). **How are the values of the gradient and the y-intercept related to the equations of the lines?** (y = gradient × x + y-intercept.) Explain that the normal symbol for the gradient is m. The value of the y-intercept normally uses the symbol c. So the formula
y = gradient × x + y-intercept
becomes $y = m \times x + c$ or $y = mx + c$.

Notice that m can be positive or negative as can c. When c is negative you write − instead of +.

What can you tell about the lines $y = 5x$ and $y = 5x - 4$? They are parallel since the gradients are both 5.

Exercise commentary and misconceptions

In question 1 students may need reminding that gradients can be positive or negative.

Check answers to the fifth part of question 2, which is an implicit function so students will need to be able to rearrange the equation into the form $y = mx + c$ to find the gradient and y-intercept.

Encourage students to draw a sketch to help them in questions 4 and 5, but they should not need to plot the graphs.

Plenary

What are the gradient and intercept of the line $2y = 5x + 6$? (Rearrange to $y = 2.5x + 3$, gradient 2.5, intercept = 3.)

Do the lines $y = 5x - 4$ and $10y - 50x + 2 = 0$ intersect? Do not attempt to draw them. Find their gradients.
First line $m = 5$.

Rearrange second line to
$10y = 50x - 2 \longrightarrow y = 5x - 0.2$, $m = 5$.
Lines have same gradient so they are parallel and they do not intersect.

A4.4 The equation $y = mx + c$

This spread will show you how to:

- Recognise that equations of the form $y = mx + c$ have straight-line graphs

Keywords
Coefficient
Constant
Gradient
Intercept
Parallel
$y = mx + c$

You can find the **gradient** and **intercept** of a line from its equation.

- **You can write the equation of any straight line in the form $y = mx + c$.**

For example
$x + y = 5 \Rightarrow y = -x + 5$
$4x - y = 6 \Rightarrow 4x = 6 + y \Rightarrow y = 4x - 6$

- **The gradient of the line $y = mx + c$ is m.**
- **The y-axis intercept of the line $y = mx + c$ is c.**

m is the **coefficient** of x.

c is the **constant** (number).

$y = mx + c$ — m is the gradient, c is the y-axis intercept

Example

What are the gradient and intercept of each of these lines?

a $y = 2x + 5$ **b** $y = 1 - 3x$

a Compare the equation to $y = mx + c$.
$m = 2$ and $c = 5$
Hence the gradient = 2 and the intercept = 5.

b In this equation $m = -3$ and $c = 1$.
Hence the gradient = −3 and the intercept = 1.

- **Parallel lines have the same gradient.**

Example

Find the equation of a line parallel to $y = 4x + 5$.

The line $y = 4x + 5$ has gradient 4.

A line that is parallel to it will have the same gradient.

So, $y = 4x + 1$ is parallel to the line $y = 4x + 5$.

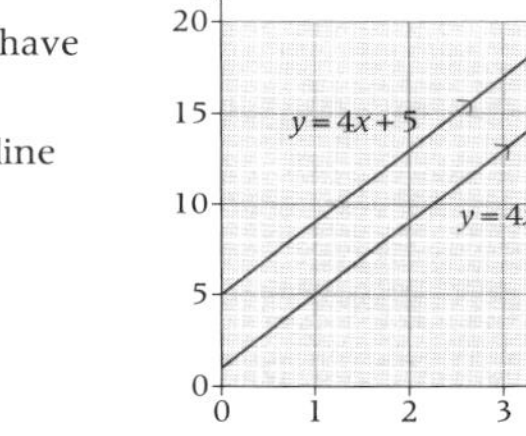

Many other equations are possible.

Exercise A4.4

1 Match each line with its equation.

$y = 4x - 2$ | $y = 3x + 1$ | $y = x$ | $y = 2 - 4x$

a b c d

2 Copy and complete the table.

Equation	Gradient	Direction	Intercept
$y = 4x + 3$			
$y = 3x + 4$			
$y = 9x - 2$			
$y = 4x - 5$			
$2y = 8x + 6$			

3 For each graph

i write its gradient and intercept

ii write the equation of the line.

a b c d

4 Write the equation of a straight line that is

a parallel to $y = 3x + 3$

b parallel to $y = 7 - 2x$ and cuts the y-axis at (0, 3)

c a mirror image, in the y-axis, of $y = 3x + 1$.

5 **a** A straight line passes through (0, 4) and (2, 10). What is its equation?

b A straight line is parallel to $2y = 6x - 9$ and goes through (1, 7). What is its equation?

A4.5 Finding the equation of a straight-line graph

Objectives

- Find the equation of a line joining several points

Useful resources

- Graph plotting tool

Mental starter

Here are three equations:

$y = 2x + 1$
$y = 5x - 3$
$y = \frac{1}{2}x + 3$

Here are three coordinates:
(1, 3), (2, 7), (−2, 2).
Ask students to identify which coordinates are on which straight line.

Introductory activity

Recap from previous lessons that students have:

- Drawn a line from the equation.
- Identified lines like $x = 3$ and $y = -4$.
- Found the gradient and y-intercept from both the graph and the equation.
- Found the equation from the graph.

Now we are going to find the equation of a line when we know the gradient and the coordinates of one point on the line.

For example, a line has gradient 4 and passes through (2, 7). The equation must be $y = 4x + c$. It passes through (2, 7) so $x = 2$ and $y = 7$ must fit in the equation and $7 = 4 \times 2 + c \longrightarrow c = -1$.
Equation is $y = 4x - 1$.

We can also find the equation of the line that passes through two points.

For example, a line passes through (2, 6) and 4, 9). Drawing a sketch, the gradient is
$\frac{9-6}{4-2} = \frac{3}{2}$, so equation is $y = \frac{3}{2}x + c$.
(2, 6) is on line so $6 = \frac{3}{2} \times 2 + c$
$\longrightarrow c = 3$ and equation is $y = \frac{3}{2}x + 3$.
Discuss the examples in the student book.

Exercise commentary and misconceptions

A sketch for question 4 will help students decide between positive and negative gradients.

Question 5 involves an implicit function.

Plenary

A line has equation $y = ax + b$, where a and b are numbers. It crosses the y-axis at A and the x-axis at B. **What are the coordinates of the points A and B?**

(Answer: A(0, b) B($\frac{-b}{a}$, 0).)
A specific example may help to explain why the x-intercept is at $\frac{-b}{a}$.

A4.5 Finding the equation of a straight line graph

This spread will show you how to:

- Find the equation of a line joining several points

Keywords
Equation
Gradient
Intercept
Parallel
$y = mx + c$

- If you know the **gradient** of a line and the y-axis **intercept** you can write the equation of the line.

Example

What is the equation of a line with gradient 9 passing through (0, 5)?

Gradient = 9 and intercept = 5,
so the equation of the line is $y = 9x + 5$.

Remember, $y = mx + c$

- If you know the gradient and a point on the line you can find the equation of the line.

Example

What is the equation of a line with gradient 8 that passes through the point (2, 7)?

Gradient = 8, so equation is $y = 8x + c$.
The line goes through (2, 7) so,

$7 = 8 \times 2 + c$ — Put $x = 2$ and $y = 7$ in the equation $y = 8x + c$
$7 = 16 + c$
$c = -9$

The equation of the line is $y = 8x - 9$.

- If you know two points on a line you can find the equation of the line.

Example

Find the equation of the line joining (1, 2) and (4, 3).

Gradient = $\frac{\text{rise}}{\text{run}} = \frac{1}{3}$

Equation of line is $y = \frac{1}{3}x + c$

The line goes through (1, 2) so substitute 1 for x and 2 for y in $y = \frac{1}{3}x + c$.

$2 = \frac{1}{3} \times 1 + c$
$2 = \frac{1}{3} + c \rightarrow c = \frac{5}{3}$

The equation is $y = \frac{1}{3}x + \frac{5}{3} \rightarrow 3y = x + 5$

Check using the point (4, 3):
When $x = 4$, $3y = 4 + 5 = 9 \rightarrow y = 3$

Exercise A4.5

1 Copy and complete the table.

Gradient	Intercept	Equation
3	5	
5	−2	
−2	7	
$\frac{1}{2}$	9	
$-\frac{1}{4}$	−3	
0	4	
1	0	

2 **a** Which of these lines will pass through the point (2, 8)?

$y = 4x$ $\quad y = 2x + 3$ $\quad y = 12 - 2x$ $\quad y = 7x - 5$ $\quad y = 5x - 1$

b Write the equations of two lines that pass through (1, 4).

3 Find the equations of the nine lines described in the table.

a Gradient of 7 and intercepts y-axis at (0, 5)	**b** Gradient of $\frac{1}{2}$ and passes through (0, 3)	**c** Parallel to a line with gradient 4 and passing through (3, 8)
d Gradient of 3 and passing through (4, 7)	**e** Gradient of −2 and cutting through (4, −3)	**f** Parallel to $y = \frac{1}{4}x - 1$ and passing through (0, −2)
g Passing through (0, 1) and (1, 5)	**h** Passing through (0, 2) and (5, 7)	**i** Passing through the midpoint of (1, 7) and (3, 13) with a gradient of 8

4 **a** What is the gradient of the line joining (0, 5) to (12, 41)?
b What is the equation of the line joining (0, 5) to (12, 41)?
c Repeat **a** and **b** for the points (3, 10) and (5, 6).

5 Where does the line $2y = 9x - 5$ cross
a the y-axis
b the x-axis
c the line $4y = x + 24$?

6 Find the equation of this line in the form $ax + by = c$, where a, b, c are values.

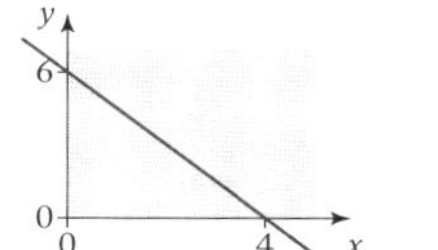

Exam review

Key objectives

- Plot graphs of functions in which y is given explicitly in terms of x, or implicitly
- Understand that the form $y = mx + c$ represents a straight line and that m is the gradient of the line and c is the value of y-intercept
- Find the equation of a given straight line in the form $y = mx + c$

Worked solution | **Commentary and misconceptions**

1 a

Encourage students to write values in a table before plotting the graph. Encourage them to consider the range of values they include. Students should interpret the equations, in terms of y-intercept and gradient, so they will know what shape to expect.

Table of values for $y = 2x - 1$:

x	−2	−1	0	1	2
y	−5	−3	−1	1	3

Table of values for $y = 2x + 2$:

x	−2	−1	0	1	2
y	−2	0	2	4	6

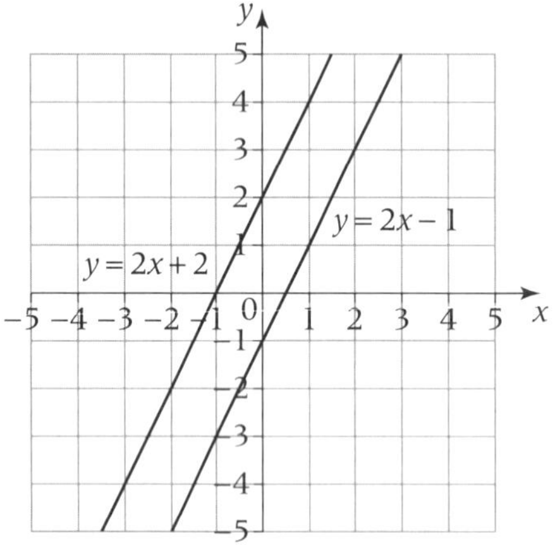

Students should label their graph clearly and choose a suitable range of values to plot.

b Both lines have gradient 2.
Therefore they are parallel and will never cross.

Students who have plotted their graphs incorrectly may struggle with this question. A good understanding of gradients and parallel lines should, however, lead to a correct answer.

2 The straight line passing through D and C is parallel to the straight line passing through A and B and therefore has gradient 2.
The y-intercept of the straight line passing through D and C is 6.
Therefore the line has equation $y = 2x + 6$.

This question tests students' understanding of the form of the equation of a straight-line graph, including the gradient and y-intercept. Some students may need prompting that the two lines are parallel and of the implication this has on gradient.

D3 Relative frequency

Objectives

F Understand and use the probability scale

F List all outcomes for single events, and for two successive events, in a systematic way

F Use fact that probability of not happening is 1 – probability of happening

F/H Identify different mutually exclusive outcomes and know that the sum of the probabilities of all these outcomes is 1

H Know when to add or multiply two probabilities: if A and B are mutually exclusive, then the probability of A or B occurring is $P(A) + P(B)$

F/H Design and use two-way tables for discrete and grouped data

F/H Understand and use estimates or measures of probability from theoretical models, or from relative frequency

Unit overview

This unit extends the knowledge of data collection and interpreting gained in units D1 and D2 to processing and representing data. The unit begins by introducing students to the probability scale and calculating probabilities for single events before proceeding to mutually exclusive events and theoretical probabilities. This unit prepares students for further work on calculating probabilities in later data units.

Prior knowledge

Before your students start this unit they should be able to:

- Add, subtract, multiply and divide with whole numbers and fractions
- Understand fraction, decimal and percentage notation
- Convert between fractions, decimals and percentages

Differentiation

- **Foundation** introduces probability, looking at both experimental results and theoretical models
- **Foundation Plus e**xtends to expected frequency and relative frequency and using two-way tables
- **Higher Plus** extends the Higher book to estimating results and independent events

D3.1 Probability

Objectives

- Understand and use the probability scale
- Calculate and explain the probability of an outcome of an event

Useful resources

- Number line

Mental starter

Write a selection of numbers on the board including fractions, decimal fractions, positive and negative numbers, and numbers greater than 1.
Ask students to sort these into two groups, those that can represent probabilities and those that cannot. Discuss choices.

Introductory activity

Recap probability from Key Stage 3 work.
Referring to the mental starter, emphasise that probability is a number between 0 and 1.
Discuss the formula for the probability of an outcome.
Discuss the scenario:
I get to work in one of four different ways depending on the weather, how tired I am, and other factors. I kept a record for 180 days and found that the number of times I used a particular form of transport was

Method of transport	Walk	Cycle	Car	Bus
Frequency	9	126	18	27

What is the probability that I cycle to work? ($\frac{126}{180} = \frac{7}{10}$).

What is the probability that I swim to work? (Impossible, so probability = 0.)
When I get to work I either have a cup of tea, or a cup of coffee or I go straight to the classroom. The probability I have a cup of tea is 0.6, the probability I go straight to the classroom is 0.36. **What is the probability I have a cup of coffee?** (1 – 0.6 – 0.36 = 0.04.)

Exercise commentary and misconceptions

For questions 1–4, remind students that the probability of an outcome not happening is always:
1 – the probability that it does happen.

In question 3 the answers should be left as fractions, as they form recurring decimals, but in question 4 decimals can be used.

In questions 5, 6 and 7 encourage students to form equations to solve for missing values.

Students should not use percentages or ratios for probabilities unless the question does so.

Plenary

There are a number of coloured balls in a bag, each ball has a number on it. The probability of selecting a particular ball is listed in the table.

	Red	Blue	Green
Number 1	0	0.3	0.2
Number 2	0.1	0.15	0.05
Number 3	0.2	0.08	0.02

What is the probability of selecting a red ball? (0.1 + 0.2 = 0.3.)

What is the probability of selecting a ball with 2 on it? (0.1 + 0.15 + 0.05 = 0.3.)

What is the probability of selecting a ball that is red or has a 2 on it?
(0.1 + 0.2 + 0.15 + 0.05 = 0.5 not 0.6.)
Discuss.

D3.1 Probability

This spread will show you how to:

- Understand and use the probability scale
- Calculate and explain the probability of an outcome of an event

Keywords
Event
Outcome
Probability

- **Probability** is a measure of how likely an **outcome** of an **event** is.
- Probability is measured on a scale from 0 to 1.

0 ———————————— 1
Outcome: cannot happen — Outcome: certain to happen

Throwing a dice is an event. Getting a 6 is an outcome.

For all other outcomes the probability is a fraction between 0 and 1.

You can write probability as a fraction, a decimal or a percentage.

- **Probability of an outcome** = $\dfrac{\text{number of favourable outcomes}}{\text{total number of outcomes}}$

Example

There are 30 students in Class 10Z, 18 girls and 12 boys.
All of the students are aged 14 or 15 and all own a mobile phone.
A student is chosen at random from Class 10Z. What is the probability that the student chosen

a is aged 10 **b** owns a mobile phone **c** is a girl **d** is a boy?

a This outcome cannot happen. All the students are 14 or 15, so P(aged 10) = 0.
b This outcome is certain to happen. All the students own a mobile phone, so P(owns mobile) = 1.
c There are 18 girls and a total of 30 students, so P(girl) = $\frac{18}{30}$.
d There are 12 boys and a total of 30 students, so P(boy) = $\frac{12}{30}$.

P(aged 10) is a short way of writing 'the probability that a student is aged 10'.

The probability that a girl is chosen or that a boy is chosen is certain to happen as the students are all either girls or boys.

Notice that $\frac{18}{30}$ (P(girl)) + $\frac{12}{30}$ (P(boy)) = 1 (P(certainty))

All possible outcomes are accounted for.

- **Sum of probabilities of all possible outcomes = 1**

Example

A spinner has circles, squares and triangles on its face.
The table gives the probabilities of landing on circle, square and triangle.
Work out the value of x.

Outcome	Circle	Square	Triangle
Probability	0.25	0.625	x

Total probability = 1 $\quad 0.25 + 0.625 + x = 1$

$0.875 + x = 1$

So, $x = 0.125$

Exercise D3.1

1 The probability that a girl chosen at random in Class 10Y has a cat is 0.47.
What is the probability that a girl chosen at random in Class 10Y does not have a cat?

2 The probability that a boy chosen at random in Class 10X does not wear glasses is 0.35.
What is the probability that a boy chosen at random in Class 10X does wear glasses?

3 A dish contains 5 orange, 3 lemon and 10 strawberry sweets.
One sweet is chosen at random.
What is the probability that the sweet is

a orange **b** banana **c** strawberry **d** lemon **e** not lemon?

4 A bag contains 7 green, 8 blue and 10 red marbles.
Nine of the marbles are large, 16 are small.
One marble is chosen at random.
What is the probability that the marble is

a green **b** not green **c** large **d** small **e** red **f** not blue?

5 Four girls compete to become head girl.
The probabilities of being chosen are shown in the table.

Amy	Beth	Cathy	Debs
0.14	0.32	0.27	x

Work out the value of x.

6 Five boys want to be captain of the cricket team.
The probabilities of their being chosen are shown in the table.

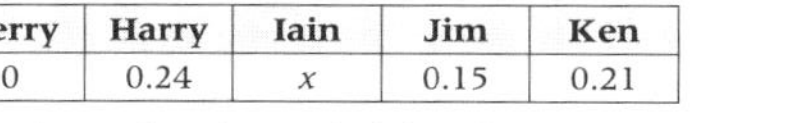

Gerry	Harry	Iain	Jim	Ken
0	0.24	x	0.15	0.21

a Explain what the probability that Gerry is chosen is 0 means.

b Work out the value of x.

7 Four teams are left in a football competition.
The probabilities of their winning the competition are shown in the table.

City	United	Rovers	Rangers
0.18	0.22	x	$2x$

a Explain the chances of Rangers winning compared to the chances of Rovers winning.

b Work out the value of x.

D3.2 Mutually exclusive outcomes

Objectives

- Calculate the probability of mutually exclusive outcomes

Useful resources

- Spinners and dice

Mental starter

More than half the people who play the lottery select some of their numbers by using birthdays, so they use numbers between 1 and 31. **If you used numbers between 32 and 49 only (49 is the highest number possible) would you be more likely to win the jackpot?** (No, you will still have the same chance of winning, but if you do win you are likely to share the jackpot with fewer people.)

Introductory activity

The term 'mutually exclusive events' when applied to a number of different outcomes means that the events are all separate from each other, only one of them can happen. For example, if you play a game of football, you could win, lose or draw (could include match abandoned!). **Are these events mutually exclusive?** (Yes, only one of them could happen.)

If you roll a dice you could roll an even number, a prime number, or a factor of 6. **Why are these events not mutually exclusive?** (Each of these events includes outcomes that are also in the other events (even numbers are 2, 4 and 6; prime numbers are 2, 3, 5; factors of 6 are 1, 2, 3 and 6).)

Can you think of two events that are mutually exclusive when rolling a dice? (Examples are odd and even **or** 'below 3' and 'above 2' **or** prime and 1.) Only mutually exclusive events can be added together.

The probability my train in the morning is early is 0.3, the probability it is on time is 0.4. **What is the probability it is late?** (0.3, because the three events of early, on time and late are mutually exclusive and together include all possible outcomes.)
Discuss the example in the student book.
Highlight the key points.

Exercise commentary and misconceptions

In question 1, students should first give their answers in fractions out of 20, then cancel (if possible). It is not advisable to use decimals in any answer unless the question uses them.

For question 2 students may need to be reminded of the definition of a prime number (exactly two factors so '1' is not prime) and a multiple.

Plenary

How many possible outcomes are there from throwing a coin once? (2: H or T.)
How many outcomes from throwing a coin twice? (4: HH, HT, TH, TT.)
Fill in this table.

Number of throws	1	2	3	4	5	6
Number of outcomes	2	4	(8)	(16)	(32)	(64)

Encourage students to look for a pattern in the results (outcomes × 2 = next number of outcomes) rather than write out all the outcomes.

Can you tell what the probability is of throwing heads on 10 consecutive occasions? (Ten throws, $2^{10} = 1024$ outcomes, so probability of heads on 10 consecutive occasions = $\frac{1}{1024}$.)

D3.2 Mutually exclusive outcomes

This spread will show you how to:
- Calculate the probability of mutually exclusive outcomes

Keywords
Mutually exclusive
Outcome

- **Two or more outcomes are mutually exclusive if they cannot happen at the same time.**

For example,

When you roll a dice, the outcomes 'an even number' and 'a 3' can both happen, but not at the same time.

When you flip a coin, the outcomes 'head' and 'tail' can both happen, but not at the same time.

So the outcomes are mutually exclusive.

- When two or more outcomes are mutually exclusive the addition rule (sometimes called the OR rule) is used to find their probability.
- For two outcomes, A and B, the probability of A or B happening is the sum of P(A) and P(B).

P(A or B) = P(A) + P(B)

- **Probability of an event not happening = 1 – Probability an event happens**
- **For an event A, P(not A) = 1 – P(A) or P(A) = 1 – P(not A)**

Example

A spinner has 12 equal sides: five green, four blue, two red and one white.
The spinner is spun.

a What is the probability that the spinner lands on

i green or white **ii** green or blue **iii** blue or white

iv blue or red or white **v** not green?

b Why are the answers to parts **iv** and **v** the same?

a i P(green) = $\frac{5}{12}$ P(white) = $\frac{1}{12}$ P(green or white) = $\frac{5}{12} + \frac{1}{12} = \frac{6}{12}$

ii P(green) = $\frac{5}{12}$ P(blue) = $\frac{4}{12}$ P(green or blue) = $\frac{5}{12} + \frac{4}{12} = \frac{9}{12}$

iii P(blue) = $\frac{4}{12}$ P(white) = $\frac{1}{12}$ P(blue or white) = $\frac{4}{12} + \frac{1}{12} = \frac{5}{12}$

iv P(blue) = $\frac{4}{12}$ P(red) = $\frac{2}{12}$ P(white) = $\frac{1}{12}$

P(blue or red or white) = $\frac{4}{12} + \frac{2}{12} + \frac{1}{12} = \frac{7}{12}$

v P(not green) = 1 – P(green) = $1 - \frac{5}{12} = \frac{7}{12}$

b The possible outcomes are green, blue, red and white.
Part **v**, P(not green), is the same as part **iv**, P(blue or red or white).

Exercise D3.2

1 A spinner has 20 equal sides: 7 have circles, 5 have pentagons, 4 have squares, 3 have triangles, 1 has a rectangle.
The spinner is spun.
What is the probability that the spinner lands on

a a circle **b** not a circle **c** a rectangle

d not a rectangle **e** a circle or a square **f** a pentagon or a square

g a triangle or a square or a rectangle

h a circle or a pentagon or a triangle **i** not a square

2 A nine sided dice has one digit (1, 2, 3, 4, 5, 6, 7, 8, 9) on each of its sides.
The dice is rolled.
What is the probability that the dice lands

a on the number 7 **b** not on the number 7

c on an odd number **d** on a multiple of 3

e on a multiple of 5 **f** not on a multiple of 5

g on a multiple of 4 **h** not on a multiple of 4

i on a prime number?

3 The table shows information about the type of pet owned by students in Class 10A. No student owns more than one pet.

Pet	Cat	Dog	Hamster	Fish	No pet
Number of students	7	8	2	4	9

a How many students are there in the class?

b One student is chosen at random from the class.
Work out the probability that the student chosen will own

i a cat **ii** a cat or a dog **iii** a dog or a fish

iv a cat or a hamster **v** no pets **vi** a pet.

4 The table shows information about the number of driving lessons students in Class 12C had in February one year.

Number of driving lessons	0	1	2	3	More than 3
Number of students	5	2	8	12	3

One student is chosen at random from the class.
Work out the probability that the student chosen had

a 1 driving lesson **b** 1 or 2 driving lessons

c 2 or more driving lessons **d** 2 or fewer driving lessons.

D3.3 Probability and expectation

Objectives

- Use two-way tables to calculate probabilities

Useful resources

- OHT of two-way tables

Mental starter

The seedings of the winners of the men's singles at Wimbledon for the past 20 years are 1, 1, 4, 1, U, 1, 1, 1, 1, 5, 2, 1, 1, 12, 6, 3, 3, 3, 11, 4. U stands for unseeded. **What seeded player is most likely to win based on the past records?** (Number one seed.) **Is the chance that the number one seeded player wins greater than 0.5?** (No, only nine times out of last 20 years.)

Introductory activity

Discuss this scenario:

A random sample of men and woman who have passed their driving test were asked if they passed first time.

	Men	Woman
Passed first time	15	12
Failed first time	9	4

Find the probability that a person chosen at random passed first time.

P(person passed first time) = $\frac{15}{40} = \frac{3}{8}$.

Find the probability that a woman chosen at random passed first time.

P(woman passed first time) = $\frac{12}{16} = \frac{3}{4}$.

Were men or woman more successful at passing first time?

P(man passed first time) = $\frac{15}{24} = \frac{5}{8}$ compared to P(woman passed first time) = $\frac{6}{8}$.

So women from this survey were more successful.

Based on this survey, how many men from a group of 160 men would you expect to have passed first time?
($\frac{5}{8} \times 160 = 100$.)

Introduce the term **two-way table** and discuss the first example in the student book. Write down the formula for expected number, which is the key point for this lesson. Discuss the second example.

Exercise commentary and misconceptions

It may be worth going through question 1 orally first, to ensure that all students understand how to complete the table. Students should then be able to complete questions 2 and 3.

Plenary

Is the sample data from the introduction a reliable way of making a conclusion about whether men or women are the more careful drivers?
No, for 2 main reasons. Firstly passing first time does not necessarily mean you are a more or less careful driver. Secondly, the sample may be biased. **In what ways do you think it might be biased?** (Possibly all from one place, all at same time, small sample.)

D3.3 Probability and expectation

This spread will show you how to:
- Use two-way tables to calculate probabilities

Keywords
Expected number
Two-way table

- **Probabilities can be found from information given in a two-way table.**

Example

Each of the students in Class 10Z went abroad last year.
The two-way table shows some information about the countries visited.

	Europe	America	Rest of the world	Total
Girls	13			18
Boys		2		12
Total	22	5		

a Complete the table.

b One student is chosen at random from Class 10Z. Write down the probability that the student

i visited Europe **ii** is a boy who visited America.

a

	Europe	America	Rest of the world	Total
Girls	13	**3**	**2**	18
Boys	**9**	2	**1**	12
Total	22	5	**3**	**30**

Use the totals to find the missing numbers.
3 + 2 = 5
13 + 9 = 22

b i Probability that the student visited Europe = $\frac{22}{30}$
ii Probability that the student is a boy who visited America = $\frac{2}{30}$

You can use the probability of a particular outcome happening to calculate the **expected number** of times it will occur.

- **Expected number = Total number of outcomes × Probability of a particular outcome happening**

Example

The foreign countries visited by the students in Class 10Z is typical of the foreign countries visited by the students in Year 10 at the same school.
There are 240 students in Year 10 at the school.
How many students would you expect to visit

a Europe **b** America **c** America or the Rest of the world?

a P(Europe) = $\frac{22}{30}$ so expected number: $240 \times \frac{22}{30} = 176$
b P(America) = $\frac{5}{30}$ so expected number: $240 \times \frac{5}{30} = 40$
c P(America or Rest of the world) = P(America) + P(Rest of the world)
$= \frac{5}{30} + \frac{3}{30} = \frac{8}{30}$

Expected number: $240 \times \frac{8}{30} = 64$

There are 240 students, so total number of outcomes is 240.

Exercise D3.3

1 The two-way table shows where each of the 32 students in Class 10Y spent their Easter holiday.

	France	Spain	UK	Total
Girls	3			18
Boys		8		
Total	6	12		32

a Copy and complete the table.

b One student is chosen at random from Class 10Y. Find the probability that the student

i visited France **ii** did not visit France
iii is a boy **iv** is a boy who visited Spain
v visited France or Spain **vi** is a girl who did not visit France.

2 The two-way table shows some information about the favourite activity of 50 students.

	Orienteering	Paintballing	Quadbiking	Total
Girls	11		4	23
Boys		8		
Total	16			50

a Copy and complete the table.

b One student is chosen at random. Find the probability that the student's favourite activity is

i paintballing **ii** not paintballing
iii paintballing or orienteering **iv** quadbiking or paintballing.

c These students are a sample chosen from a larger group of 400 students. How many of the larger group would you expect

i to prefer orienteering **ii** to be girls?

3 The two-way table shows some information about the preferred subject of the 120 students in Year 10.

	Science	Humanities	Other subjects	Total
Girls	29		32	
Boys		3		
Total	56	21		120

a Copy and complete the table.

b One student is chosen at random.
Find the probability that the student

i is a girl **ii** prefers humanities
iii prefers humanities or science **iv** is a boy who prefers science.

c This year group is typical of the whole school.
There are 600 students in the school.
How many of the whole school would you expect to prefer

i science **ii** humanities?

D3.4 Theoretical and experimental probability

Objectives

- Calculate theoretical probabilities and relative frequencies
- Understand the concepts of experimental and theoretical probability

Useful resources

- One drawing pin for each pair of students.
- 2 mm graph paper

Mental starter

Write down sums and products of fractions, for example:

$\frac{2}{3} + \frac{1}{4}$

$\frac{2}{5} + \frac{3}{8}$

$\frac{4}{9} \times \frac{3}{10}$

$\frac{7}{12} \times \frac{6}{11}$

Introductory activity

The task is to estimate the number of times a drawing pin will fall on its head if it is dropped 1500 times.

Split students into pairs and give each pair one drawing pin. Ask the students to drop the drawing pin 10 times and record the number of times it falls on its head and on its side. Then ask them to drop it a further 10 times and record the total number of times it falls on its head or on its side from the 20 drops so far. Ask them to keep repeating this as far as they can in 5 minutes.

Write this table for students to copy and complete:

Total number of drops	10	20	30	40
Number of heads				
Number of sides				

Remind students to include the previous results each time they fill in the next column. Encourage students to work collaboratively.

Ask each pair to calculate an estimate for the probability of a drawing pin landing on heads. Point out that the more trials you base your answer on, the more reliable your estimate is. The answer can be given as a fraction, or worked out on the calculator as a decimal. Students should write their answer both ways. Now ask how many heads they would expect if the pin is dropped 1500 times. (1500 × probability.)

How could you get a more reliable estimate for the number of heads we would expect? (Estimate the probability from the combined class's results.)

If you had been asked how many times you would expect this coin to come down heads in 1500 throws, would you have done a similar experiment? (No, because each time a coin is thrown, heads and tails are equally likely because the coin is unbiased.)

The drawing pin is an example of experimental probability; the coin of theoretical probability.

Exercise commentary and misconceptions

Encourage students to leave probabilities as fractions rather than use decimals. Question 4 will be a problem if they round the recurring decimal.

Students need to establish in questions 4 and 5 whether the initial number of trials is to be included.

Check that students understand '4 or more' and 'not 5' in question 5.

Plenary

Draw a graph of how the estimated probability changed as you carried out more trials in the introductory experiment. Students will need to convert their results to decimals first. Discuss what this shows. (More trials produce a more reliable estimate.)

D3.4 Theoretical and experimental probability

This spread will show you how to:
- Calculate theoretical probabilities and relative frequencies
- Understand the concepts of experimental and theoretical probability

Keywords
Bias
Experimental
Fair
Theoretical

- You can calculate the **theoretical** probability when an event is **fair** or unbiased.

For example:
A fair coin is thrown.
Probability of a fair coin landing on heads = $\frac{1}{2}$

one head, two outcomes

An unbiased dice is rolled.
Probability of an unbiased dice landing on 2 = $\frac{1}{6}$

one '2', six outcomes

- Theoretical probability is based on equally likely outcomes.
- Theoretical probabilities are calculated on the assumption that the number of favourable outcomes and the number of possible outcomes are as expected.

Example

A fair spinner with 8 sides, three yellow, two red, two blue and one white, is spun.
Work out the probability that the spinner lands on

a yellow **b** red or blue **c** not white.

a P(yellow) = $\frac{3}{8}$ **b** P(red or blue) = $\frac{4}{8}$ **c** P(not white) = $1 - \frac{1}{8} = \frac{7}{8}$

- **You use experimental probability when the event is unfair or biased or when the theoretical outcome is known.**

Example

A biased coin is thrown 200 times. It lands on heads 140 times.

a Estimate the probability of this coin landing on heads on the next throw.
b Estimate the number of heads you expect to get when the coin is thrown 500 times.

a P(Head) = $\frac{140}{200}$ **b** Estimated expected number = $500 \times \frac{140}{200} = 350$

Example

Tom carries out a survey about the number of people in his village who are left-handed.
He asks 40 people and five of them are left-handed.

a Estimate the probability that a person in the village is left-handed.
b 192 people live in his village. Estimate how many are left-handed.

a P(left-handed) = $\frac{5}{40}$ **b** Estimated number = $192 \times \frac{5}{40} = 24$

- **The closer experimental probability is to theoretical probability the less likely it is that there is bias.**

Exercise D3.4

1 The probability that a biased dice will land on a six is 0.35.
The dice is rolled 400 times.
Estimate the number of times the dice will land on a six.

2 The probability that a biased coin will land on tails is 0.24.
The coin is thrown 500 times.
Estimate the number of times the coin will land on tails.

3 A biased coin is thrown 80 times. It lands on heads 50 times.
Estimate the probability that this coin will land on heads on the next throw.

4 A biased four-sided dice is rolled 120 times. The table shows the outcomes.

Score	1	2	3	4
Frequency	22	34	44	20

a Explain why it is twice as likely that the dice will land on 3 as on 1.
b The dice is rolled once more. Estimate the probability that it lands on 2.
c The dice is rolled a further 300 times.
How many of those times would you expect the dice to land on 4?

5 A biased dice is rolled 100 times. The table shows the outcomes.

Score	1	2	3	4	5	6
Frequency	7	20	32	17	13	11

a The dice is rolled once more. Estimate the probability that the dice will land on

i 6 **ii** 3 **iii** 1 or 2 **iv** 4 or more **v** a number that is not 5.

b The dice is going to be rolled a further 400 times.
How many times would you expect the dice to land on

i 2 **ii** 4 or 5?

6 There are 240 students in Year 10 at Endeavour School.
a In a survey of 30 students from Year 10, 13 owned an MP3 player.
How many of the whole of Year 10 would you expect to own an MP3 player?
b In a survey of 40 students from Year 10, 19 said they liked peaches.
How many of the whole of Year 10 would you expect to like peaches?
c In a survey of 48 students from Year 10, 5 were left-handed.
How many of the whole of Year 10 would you expect to be left-handed?

D3.5 Relative frequency

Objectives

- Calculate theoretical probabilities and relative frequencies

Useful resources

- Results from last session's drawing pin experiment

Mental starter

Ask students to **estimate** the decimal equivalent of these fractions:

$\frac{144}{200}, \frac{79}{150}, \frac{217}{300}, \frac{470}{700}, \frac{642}{900}$

Introductory activity

Recap the experiments with the drawing pins and the conclusion that increasing the number of trials improved the estimate of the probability (or relative frequency) of a pin landing on its head.

Emphasise that the terms 'relative frequency' and 'experimental probability' mean the same.

Discuss the first example.
Emphasise that the best estimate includes all the results.
Discuss the second example, and use it to highlight how you can assess bias.

Exercise commentary and misconceptions

Students should be aware that randomness means that experiments are not likely to give the exact expected result.

It is difficult to generalise about the exact number of trials you need to do before you can draw a conclusion from your results. 30 is not enough and 100 is acceptable, but the more trials, the more reliable your conclusions.

In question 5 make sure that students are cumulating the results, not calculating the probability for each set of 10 flips separately.

Plenary

The number of goals a premiership footballer has scored in his last 10 matches is 0, 0, 2, 3, 0, 0, 5, 0, 0, 3. Based on these results you could argue that he is most likely to score no goals with a probability of 0.6. **Does this mean you would not want him in your team?**
(No, just looking at this answer ignores his other scores. He actually has a mean score of 1.3 goals per match, this would make him one of the best players in the premiership.)

D3.5 Relative frequency

This spread will show you how to:

- Calculate theoretical probabilities and relative frequencies

Keywords
Bias
Relative frequency

- **Experimental probability is also known as relative frequency.**
- **Relative frequency is the proportion of successful trials in an experiment.**
- **The more trials that are carried out the more reliable the estimate of probability.**

Example

Rachael has a mixed colours packet of seeds.
She plants 10 seeds each week for 7 weeks.
The table shows the number of purple flowers that grew in each group of seeds.

Week	1	2	3	4	5	6	7
Number of purple flowers	4	6	5	7	4	6	5

a Work out the relative frequency of a purple flower.
b Find the best estimate of the probability of getting a purple flower.

a

Week	1	2	3	4	5	6	7
Number of purple flowers	4	6	5	7	4	6	5
Relative frequency	$\frac{4}{10}$	$\frac{10}{20}$	$\frac{15}{30}$	$\frac{22}{40}$	$\frac{26}{50}$	$\frac{32}{60}$	$\frac{37}{70}$

For each successive week, find the total number of purple flowers and the total number of seeds planted.

b The best estimate includes all the results. P(purple flower) = $\frac{37}{70}$

You can compare the relative frequency of an outcome with the theoretical probability. If they are quite different, the experiment may be biased.

Example

Dan suspects that a particular dice has a bias towards the number 3.
Dan rolls the dice 30 times and gets these results

4	3	3	3	6	5	1	3	2	5
1	3	4	5	3	3	6	2	5	4
6	3	1	3	6	5	3	2	4	3

a What is the relative frequency of rolling a 3?
b Is the dice biased toward 3? Explain your answer.

a The number 3 is rolled 11 times. Relative frequency = $\frac{11}{30}$

b Theoretical probability of rolling a '3' is P(3) = $\frac{1}{6}$
In 30 rolls expected number of 3s = $30 \times \frac{1}{6} = 5$
5 is not close to 11 so the dice does appear to be biased towards 3.

Exercise D3.5

1 Jim carries out an experiment.
He throws a coin 320 times.
The coin lands on tails 114 times.
Is the coin fair? Explain your answer.

2 A spinner has 10 equal sides, 5 black and 5 red.
Dave carries out an experiment.
He spins the spinner 280 times.
The spinner lands on black 133 times.
Is the spinner fair? Explain your answer.

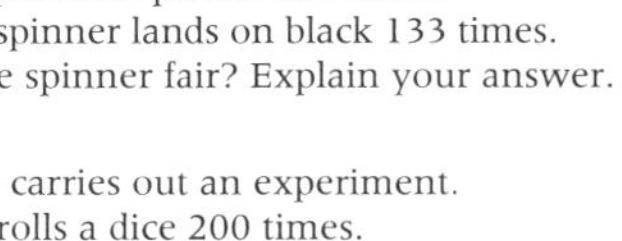

3 Jane carries out an experiment.
She rolls a dice 200 times.
The table shows the outcomes.

Outcome	1	2	3	4	5	6
Frequency	32	34	35	31	35	33

Is the dice fair? Explain your answer.

4 Tom has a four-sided dice.
He rolls the dice 100 times.
The table shows the outcomes.

Outcome	1	2	3	4
Frequency	18	44	19	19

Is the dice fair? Explain your answer.

5 Clara suspects that a coin is biased. She flips the coin and notes how many heads she gets in each group of 10 flips.
Clara flips the coin 100 times in total.
The table shows her results.

Group of 10 flips	1	2	3	4	5	6	7	8	9	10
Number of heads	4	3	4	2	5	4	4	3	3	2
Relative frequency										

a Copy the table and complete for the relative frequency.

b Write the best estimate of the probability of the coin landing on heads.

c Is the coin biased? Explain your answer.

D3 Exam review

Key objectives

- Understand and use the probability scale
- Understand and use estimates or measures of probability from theoretical models, or from relative frequency
- Identify different mutually exclusive outcomes and know that the sum of the probabilities of all these outcomes is 1

Worked solution

Commentary and misconceptions

1

This question tests the students' understanding of experimental and theoretical probability.

a

Ensure that students understand how to calculate probabilities given the numbers in the table.

i $P(2) = \frac{10}{100} = \frac{1}{10}$

ii $P(6) = \frac{23}{100}$

iii $P(2 \text{ or } 6) = P(2) + P(6)$

$= \frac{1}{10} + \frac{23}{100}$

$= \frac{33}{100}$

Some students may need prompting that these two events are mutually exclusive and that therefore the probabilities need to be added.

b No.
For a fair dice, all the probabilities should be $\frac{1}{6}$.
In a practical experiment you expect some variation but the range of results in this case is too large to suggest that the dice is fair.

Students should show an understanding of the expected probabilities involved in rolling a fair dice.
Ensure that they appreciate that from this experiment they cannot be certain that the dice is either fair or unfair but they can express an opinion.
Justification of their answer must be given.

2

Ensure that students understand how to use two-way tables before attempting the question.

a

	France	Germany	Spain	Total
Female	2	23	9	34
Male	15	2	9	26
Total	17	25	18	60

b $P(\text{Germany}) = \frac{25}{60} = \frac{5}{12}$

Encourage students to perform this single calculation instead of using the separate male and female totals and adding the two probabilities.

Proportionality

Objectives

F/H Perform short division to convert a simple fraction to a decimal

F/H Convert simple fractions of a whole to percentages of the whole and vice versa

F/H Use percentages to compare proportions

F/H Use ratio notation, including reduction to its simplest form and its various links to fraction notation

F/H Divide a quantity in a given ratio

H Solve problems and word problems, including those involving ratio and proportion, repeated proportional change and reverse percentages, inverse proportion, surds, measures and conversion between measures, and compound measures defined within a particular situation

H Check and estimate answers to problems

H Select and justify appropriate degrees of accuracy for answers to problems

Unit overview

This unit consolidates the knowledge of calculating with fractions, decimals and percentages gained in units N1, N2 and N3. The concept of proportionality and ratio is introduced and then used within problem-solving situations, laying the basis for work on more complex problems involving ratio and proportion and percentage change in units N7 and N8.

Prior knowledge

Before your students start this unit they should be able to:

- Add, subtract, multiply and divide with integers, fractions, decimals and percentages
- Convert between fractions, decimals and percentages
- Calculate the proportion of an amount as a fraction or a percentage
- Calculate amounts after a percentage change
- Understand and use a range of units of measure
- Use formulae from mathematics and other subjects
- Convert between common imperial and metric units

Differentiation

- **Foundation** introduces proportion in the context of simple problems. Reading tables and exchange rates are also included
- **Foundation Plus** extends to problems involving compound measures, including speed
- **Higher Plus** extends the Higher book to more complex problems involving inverse proportion, repeated proportional change and ratio

N4.1 Introducing proportion

Objectives
- Describe and calculate proportions, using fractions, decimals or percentages
- Understand direct proportion and ratio

Useful resources
- Mini-whiteboards
- Number lines

Mental starter
Go around the class counting on in steps of $\frac{1}{32}$. Ensure that students give the simplified version of the fraction:
$\frac{1}{32}, \frac{1}{16}, \frac{3}{32}, \frac{1}{8}, \ldots$

Introductory activity
Remind students that the phrase '$\frac{3}{5}$ of 85' can be written '$\frac{3}{5} \times 85$' or 'of means multiply'.

The multiplication can be done in different ways, depending on the numbers being used.

$\frac{3}{5}$ of $85 = \frac{3}{5} \times 85 = 3 \times 17 = 51$. This method works when the numbers cancel and the arithmetic is easy!

Discuss the methods used in the first example.

Here is an example of showing one quantity as a proportion of another quantity:

I have 600 duck eggs, but 72 are past their 'sell by date'. **What percentage is unsaleable?**
$\frac{72 \div 6}{600 \div 6} = \frac{12}{100} = 12\%$. Always write as a fraction first, then convert to a percentage.

Discuss the second and third examples.

Exercise commentary and misconceptions
Questions 1 and 2 are easier if the students are familiar with the technique of cancelling fractions.

In question 3 encourage students to calculate estimates. Also encourage simple short cuts with percentages: to find 95% of 400 g it is easier to calculate 5% of 400 g then subtract this amount from 400 g.

Plenary
I have £60 and I decide to spend a proportion of it. **Do all these four statements mean the same? (a) I spend $\frac{1}{4}$ (b) I spend 0.25 (c) I spend and save the money in the ratio 1 : 4 (d) I spend 25%**. (All are the same except the ratio in which I spend $\frac{1}{5}$.) Note that you can change between fractions, decimals and percentages quite easily, but ratio is always a little more complex.

N4.1 Introducing proportion

This spread will show you how to:

- Describe and calculate proportions, using fractions, decimals or percentages
- Understand direct proportion and ratio

Keywords
Proportion

Proportions can be described and calculated using fractions, decimals or percentages.

There are 12 students in Class 3A.
$\frac{2}{3}$ of them are girls.
Number of girls = $\frac{2}{3} \times 12 = 8$

1000 ml of Quango contains 100 ml of fruit juice.
Proportion of fruit juice = $\frac{100}{1000}$
$= \frac{1}{10}$

Example

Find **a** $\frac{3}{5}$ of 85 **b** 28% of 360 **c** 120% of 45 **d** $\frac{7}{8}$ of 86

a $\frac{3}{5_1} \times 85^{17} = \frac{3}{1} \times 17 = 51$

Cancel the 5s before multiplying.

b Estimate: 30% of 400 = 120
By long multiplication, 28 × 36 = 1008
So 28% of 360 = 100.8

c 120% of 45 = (100% of 45) + (20% of 45)
20% of 45 = 45 ÷ 5 = 9 ➡ 120% of 45 = 45 + 9 = 54

Work out 20% of 45 and add it.

d $\frac{1}{8}$ of 86 = 86 ÷ 8 = 10 + $\frac{6}{8}$ = $10\frac{3}{4}$ ➡ $\frac{7}{8}$ of 86 = 86 − $10\frac{3}{4}$ = 76 − $\frac{3}{4}$ = $75\frac{1}{4}$

Work out $\frac{1}{8}$ of 86 and then subtract.

Example

What proportion is **a** 4 of 32 **b** 5 of 35 **c** 12 of 100?
Write your answers as percentages.

a $\frac{4}{32} = \frac{1}{8}$
so 4 is $\frac{1}{8}$ of 32
$\frac{1}{8} = \frac{1}{8} \times 100\%$
$= 12\frac{1}{2}\%$

b $\frac{5}{35} = \frac{1}{7}$
so 5 is $\frac{1}{7}$ of 35
$\frac{1}{7} = \frac{1}{7} \times 100\%$
$= 14.3\%$

c $\frac{12}{100} = \frac{3}{25}$
so 12 is $\frac{3}{25}$ of 100
$\frac{3}{25} = \frac{3}{25} \times 100\%$
$= 12\%$

Example

A 1 kg bag of 'Grow Up' fertiliser contains 45 grams of phosphate. A 500 gram packet of 'Top Crop' fertiliser contains 20 grams of phosphate.
What is the proportion of phosphate in each fertiliser?

Proportion of phosphate in 'Grow Up' = $\frac{45}{1000} \times 100\%$
$= 4.5\%$

Proportion of phosphate in 'Top Crop' = $\frac{20}{500} \times 100\%$
$= 4\%$

Percentages are easy to compare.

Exercise N4.1

You should be able to do all of these mentally.

1 One tenth ($\frac{1}{10}$) of the weight of a soft drink is sugar. Find the amount of sugar in these weights of drink.
a 750 g **b** 45 g **c** 1 kg **d** 1250 g

2 Three fifths ($\frac{3}{5}$) of the volume of a fruit cocktail is orange juice. Find the amount of orange juice in these volumes of fruit cocktail. You should show all of your working.
a 150 ml **b** 380 ml **c** 2 litres **d** 280 cm^3

See Example 1 part **a**.

3 Calculate
a 120% of 50 g **b** 90% of 40 mm
c 95% of 400 g **d** 80% of 39 km

Try these mentally. See Example 1 part **c**.

4 What proportion is
a 5 of 50 **b** 6 of 80 **c** 9 of 45 **d** 15 of 80?

5 Write your answers from question **4** as percentages.

6 Find these proportions, giving your answers as
i fractions in their lowest terms **ii** percentages.
a 3 out of every 20 **b** 4 parts in a hundred
c 8 out of 20 **d** 64 in every thousand

7 Find these.
a 38.3% of 192 mm **b** $\frac{3}{8}$ of £840
c 19% of £52.00 **d** $\frac{7}{8}$ of 960 kg

8 A 250 ml glass of fruit drink contains 30 ml of pure orange juice. What proportion of the drink is pure orange juice? Give your answer as
a a fraction in its lowest terms **b** a percentage.

9 Antifreeze contains 10% concentrated antifreeze; the rest is water.
a How much concentrate is contained in 2 litres of antifreeze?
b How much antifreeze can you make with 300 ml of concentrate?
c How much water would you add to 200 ml of concentrate?

10 Samantha wins £4500 in a competition. She gives $\frac{1}{3}$ to her mother and $\frac{1}{5}$ to her sister.
a How much does she keep? Show your working.
b What proportion of the prize money does she give away?
Give your answer as a fraction and as a percentage.

N4.2 Direct proportion

Objectives

- Understand direct proportion and ratio
- Solve problems involving proportion and proportional change

Useful resources

- Mini-whiteboards

Mental starter

Decribe this scenario:

A jeweller can make seven necklaces in two days. How many complete necklaces can he make in:
8 days; 2 weeks; 5 weeks?

Introductory activity

Remind students that two variables are in direct proportion if the ratio between them stays the same as the actual values vary.

For example, £300 is shared between Ben and Laura in the ratio 2 : 3. **Who gets more?** (Laura) **How many parts has the money been split into?** (5 parts). Each part represents $\frac{1}{5}$, Ben gets 2 parts, Laura 3 parts.
1 part is £300 ÷ 5 = £60.

So Ben gets 2 × £60 = £120.
Laura 3 × £60 = £180.
Check £120 + £180 = £300 (the total amount).

Discuss how you can use a ratio when you don't know the overall total amount. For example, juice and water are mixed in the ratio 3 : 5. **If 96 ml of juice are used, how much water will be added?** The 96 ml represents three parts of the drink. So one part is 96 ÷ 3 = 32 ml. Water is five parts of the drink, so there are 5 × 32 ml = 160 ml of water. Emphasise that the ratio 96 : 160 simplifies to 3 : 5.

The student book gives three methods of solving the problem 'A pipe 2.5 m long weighs 35 kg. How much would 5.5 m weigh?'

Discuss the three methods. The unitary method and the algebraic method always work, but the informal scaling method is often quicker and easier to use.

Exercise commentary and misconceptions

It may help the students to suggest methods for individual questions or to suggest they try different methods themselves.

The unitary method can be combined with the informal scaling method in questions 5–8, so that students do not have to find the cost of 1 g then 100 g, but find 100 g directly.

Question 9 is difficult to calculate with the unitary method.

Plenary

An Olympic athlete runs 100m in 10 sec. **What is this speed in km/h?** (Students may need reminding 1000 m = 1 km.) Students often begin by converting to km. Point out that km/h means how many km in 1 hour, so convert to 1 hour.

100 m : 10 sec ⟶ 600 m : 1 min ⟶
36 000 m : 1 hour ⟶ 36 km : 1 hour ⟶
36 km/h.

N4.2 Direct proportion

This spread will show you how to:
- Understand direct proportion and ratio
- Solve problems involving proportion and proportional change

Keywords
Direct proportion
Ratio
Variable

Quantities which can change are called **variables**.

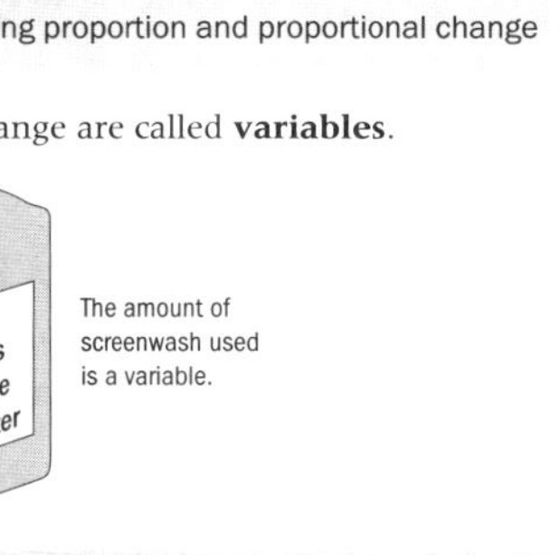

The amount of screenwash used is a variable.

- Two variables are in **direct proportion** if the **ratio** between them stays the same as the actual values vary.

When you multiply (or divide) one of the variables by a certain number, you have to multiply (or divide) the other variable by the same number.

Water (litres)	Capfuls
1 (×4)	4
5 (×4)	20

5 litres of water needs 20 capfuls of screenwash

Example

A pipe 2.5 metres long weighs 35 kilograms. How much would 5.5 metres of the same pipe weigh?

Here are three different ways to solve this problem:

2.5 m weighs 35
⇓ ÷5 ⇓ ÷5
0.5 m weighs 7
⇓ ×11 ⇓ ×11
5.5 m weighs 77 kg

This is an informal scaling method.

2.5 m weighs 35
⇓ ÷2.5 ⇓ ÷2.5
1 m weighs 14 kg
⇓ ×5.5 ⇓ ×5.5
5.5 m weighs 77 kg

This is called the unitary method.

The formula $w = kx$ tells you the weight, w kg, of a pipe x metres long.
The scale factor k is the weight of 1 metre of pipe, which is $35 \div 2.5 = 14$ kg,
so the formula is: $w = 14x$.
Substituting $x = 5.5$ gives $w = 14 \times 5.5 = 77$ kg

This is an algebraic method.

5.5 metres of pipe weighs 77 kg.

Example

The cost of 12 pencils is £2.16. Work out the cost of 9 pencils.

12 pencils cost £2.16, so 3 pencils cost 54p (dividing by 4), and 9 pencils cost £1.62 (multiplying by 3).

Exercise N4.2

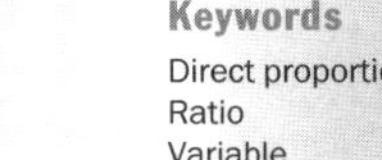

1 Ribbon costs £2.75 per metre. Find the cost of these lengths of ribbon.
a 3 m **b** 4.5 m **c** 6.85 m **d** 27.55 m

2 A shop sells shelving at £3.45 per metre. Find the cost of these lengths of shelving.
a 5 m **b** 3.45 m **c** 2.25 m **d** 4.85 m

3 A 2 m length of pipe weighs 8 kg. How much does a 3 m length of the same pipe weigh?

4 Four buckets of water weigh 60 kg. How much would 5 buckets of water weigh?

5 400 g of powder paint costs £2.40.
a Find the cost of 100 g of the paint.
b Use your answer to part **a** to find the cost of 300 g of the paint.

6 300 g of sherbet drops cost £1.20.
a How much do 100 g of sherbet drops cost?
b How much do 700 g of sherbet drops cost?

7 A pack of 250 tea bags contains 130 g of tea, and costs £7.50. Calculate
a the cost of one tea bag **b** the weight of tea in one bag.

8 A shop sells five different types of luxury tea. Calculate the cost of 100 g of each brand, given that
a 200 g of brand A costs £3.75 **b** 500 g of brand B costs £7.40
c 300 g of brand C costs £5.20 **d** 250 g of brand D costs £5.10
e 350 g of brand E costs £6.50.

DID YOU KNOW?
Each day Britons drink 165 million cups of tea.

9 Alan and Barry buy sand from a builders' merchant. Alan buys 35 kg of sand for £4.55. Barry buys 28 kg of the same sand. How much does he pay? Show your working.

10 A shop sells drawing pins in two different packs.
Pack A contains 120 drawing pins and costs £1.45.
Pack B contains 200 of the same drawing pins, and costs £2.30.
Calculate the cost of one drawing pin from each pack, and explain which pack is better value.

11 A store sells packs of paper in two sizes.

Regular	Super
150 sheets	500 sheets
Cost £1.05	Cost £3.85

Which of these two packs gives better value for money? You must show all of your working.

N4.3 Exchange rates

Objectives
- Solve problems involving exchange rates

Useful resources
- Calculators

Mental starter
You can donate blood from the age of 17 years. You can donate every 16 weeks. **How many times could you donate blood if you live to your 60th birthday?** (Based on 3 times a year gives 129 donations: based on 52 weeks per year gives 139 donations: based on 365 (or 365.25) days per year gives 140 donations.)

Introductory activity
Exchange rates give the amount of a currency you can buy when you have 1 unit of another currency. When dealing with exchange rates, the easiest way to convert is to multiply or divide by the exchange rate, depending on which way you are converting.

For example, in 2005 £1 was worth 1887 Korean Won (KRW). (Smallest unit is 1 Won.)
Convert £7.35 into KRW
(7.35 × 1887 = 13 869 KRW.)
Convert 1 000 000 KRW to £.
(1 000 000 ÷ 1887 = £529.94.)

How do you know whether to multiply or divide? (Recall that multiplying by a number larger than 1 will increase the amount, but multiplying by a number smaller than 1 will reduce it. Dividing by a number larger than 1 gives a decrease, but dividing by a number smaller than 1 gives an increase.) Students need to think whether the conversion needs to give a bigger or smaller answer.

For example, $1 is worth £0.54. **If you convert £12 into dollars, will the amount go up or down from 12?** (Up because £1 is worth more than $1.) (12 ÷ 0.54 = $22.22.) Remind students that answers should be rounded to the smallest unit of currency.

Discuss the examples in the student book.

Suppose you change $462.50 for £250.00 on a particular day. How can you find the exchange rate? What is the exchange rate for £1? (Divide by 250 to find £1 = $1.85.)

Exercise commentary and misconceptions
The questions focus on converting currency by multiplying and dividing with the exchange rate.

Some students may like to write their rates as ratio and use a unitary method when necessary. This is more complex but is also a good method.

If students have problems with the multiply/divide decision, it may help to recall the conversion of £s to pence.

Students are usually willing to estimate answers when dealing with exchange rates as they may have had experience on holiday and a discussion on banks and accuracy may help them with the topic of significant figures.

Plenary
£1 will buy you 4000 Venezuela Bolivares (VEB). How rich in pounds would a Venezuelan billionaire be if he brought all his money to the UK? (£250 000). If £1 = $0.60 (US), what does this mean about American millionaires? (They might not be millionaires if they lived in the UK because $1 million is worth less than £1 million.)

N4.3 Exchange rates

This spread will show you how to:

- Solve problems involving exchange rates

Keywords
Convert
Exchange rate

You can **convert** from one currency to another using an **exchange rate**.

Example

If €1 (1 euro) is worth 63p, find the value of

a €185 in pounds **b** £260 in euros.

a The exchange rate is €1 = £0.63, so €185 = £0.63 × 185 = £116.55

b As an estimate, £250 ÷ 0.5 = 250 × 2 = €500,
£260 = 260 ÷ 0.63 = €412.70 to the nearest cent.

To convert from one currency to another, simply multiply by the appropriate exchange rate.

If you need to do the 'reverse' conversion, divide by the given exchange rate.

Example

Samantha travels from Ottawa to Buenos Aires, where she changes 375 Canadian dollars into Argentine pesos.

a How many pesos does she receive, if the exchange rate is CAN$1 = 2.4027 pesos?

b After her trip, Samantha changes 58 pesos back into Canadian dollars, at a rate of CAN$1 = 2.5458 pesos. How many dollars does she receive?

a Estimate 400 × 2 = 800
$375 = 375 × 2.4027 = 901.01 pesos

b Estimate 60 ÷ 3 = 20
58 ÷ 2.5458 = CAN$22.78

Example

a Mary changed £900 into US dollars ($), when the rate of exchange was £1 = $1.75.
How many dollars did she receive?

b After her holiday Mary had $110 left. She changed them into pounds at an exchange rate of £1 = $1.79. How many pounds did she get?

a £900 = $1.75 × 900 = $1575.

b Divide by the conversion rate to convert dollars back to pounds.
So, $110 ÷ 1.79 = £61.45.

Example

Use the fact that 1 inch = 2.54 centimetres to convert

a 7.25 inches to centimetres **b** 15 cm to inches.

a 7.25 inches = 7.25 × 2.54 cm = 18.4 cm

b 15 cm = 15 ÷ 2.54 inches = 5.91 inches

Do these conversions in the same way as currency conversions.

Exercise N4.3

1 The exchange rate between pounds and dollars is £1 = $1.55. Convert these amounts from pounds to dollars.

a £20 **b** £35 **c** £10.50 **d** £38.55

2 The exchange rate between pounds and euros is £1 = €1.49. Convert these amounts from pounds to euros.

a £5 **b** £30 **c** £59 **d** £264

3 One Australian dollar (A$1) is worth 42.7p. Convert these amounts to pounds.

a A$2.50 **b** A$45 **c** A$299 **d** A$715

4 £1 is worth 205 Japanese yen (£1 = ¥205). Convert these amounts into pounds.

See example 1b

a ¥410 **b** ¥2050 **c** ¥300 **d** ¥750
e ¥6500 **f** ¥595

5 One US dollar is worth 0.8182 Canadian dollars (US$1 = CAN$0.8182).

a Convert US$50 into Canadian dollars.

b Convert CAN$50 into US dollars.

6 The table shows the exchange rates between two currencies.

£1 (pound)	is worth €1.67
$1 (dollar)	is worth €1.15

a Alan changes £300 into euros. How many euros does he receive?

b Barbara changes €875 into dollars. How many dollars does she receive?

7 One pint is exactly 0.568261 litres. Find the number of pints in one litre, giving your answer to 3 decimal places.

8 One pound weight (1 lb) is exactly 0.45359237 kg. Use this information to convert the following weights to kilograms, giving your answers to 3 decimal places.

a 28 lb **b** 2240 lb **c** 375.5 lb **d** 38.125 lb

9 One mile is exactly 1609.344 metres. Use this information to convert

a 25 miles into kilometres **b** 10 km into miles
c 40 000 km into miles **d** 4.5 miles into kilometres

10 Use the approximate conversion rate of 1 gallon = 4.5 litres to convert

a 35 gallons into litres **b** 50 litres into gallons
c 12.5 gallons into litres **d** 38.8 litres into gallons

Give your answers to a suitable degree of accuracy.

N4.4 Compound measures

Objectives
- Round answers to an appropriate degree of accuracy
- Solve problems involving compound measures, including speed and density

Useful resources
- OHTs of formulae triangles for density and speed

Mental starter
Pressure is measured in Pascals (Pa) and is calculated by dividing the force by the area of the object pushing down. A young elephant has an overall downward force of 30 000 Newtons (N). The area of each foot is 0.1 m^2.
If this elephant stepped on you, how much pressure would you feel?
(30 000 ÷ 4 = 7500 N for each foot → 7500 ÷ 0.1 = 75 000 Pa.)

Introductory activity
Discuss density.
The density of an object is a measure of how tightly the atoms are packed together inside.

The formula is $density = \frac{mass}{volume}$. It is easier to use this formula if you put it into a triangle.
You need to remember the order in which the letters D, M and V are written in the triangle.
To use the triangle, cover up the quantity you want to find. If the mass(M) of an object is 10 kg and the volume(V) is 2 cm^3, then the density(D) is $D = \frac{10}{2} = 5$ kg/cm^3.

There is a triangle which gives the formulae for questions of speed(S), distance(D) and time(T).
How can you remember the order in which S, D and T are written?

How far can you travel in 20 min if you travel at 5 m/s?
($D = S \times T = 5 \times 20 \times 60 = 6000$ m. Need to watch for units.)

You travel at 45 km in 4 hours and 15 min.
Find speed in km/h to 1 dp. (Estimate 40 ÷ 4 = 10 km/h 45 ÷ 4.25 = 10.6 km/h.)

Exercise commentary and misconceptions
Students may prefer to remember the formulae instead of using the triangles. The triangles should be used with caution as they are only intended as an aid to memory and are not intended to take the place of learning.

Question 2 is similar to questions in Ex N4.2.

Questions 1, 3–5 are on speed, questions 6–9 are on density.

Question 3–5 involve hours and minutes, so students may need reminding that 45 min is 0.75 hr, not 0.45.

Question 9 has cm and m mixed together, so students should proceed with care.

Plenary
I travel 30 km to work and 30 km back. I travel to work at 30 km/h, but travel back at 10 km/h. **What is the average speed over the two journeys?** (Students are likely to suggest 20 km/h as the middle or mean of the two speeds.) **What is the total time and total distance?** (4 hours, 60 km.) **What speed is this on average?** (60 ÷ 4 = 15 km/h. Surprising!) **What is the average speed if you travelled 60 km each way, 30 km/h there and 10 km/h back?** (15 km/h as before, the distance does not matter.)

N4.4 Compound measures

This spread will show you how to:

- Round answers to an appropriate degree of accuracy
- Solve problems involving compound measures, including speed and density

Keywords
Compound
Density
Speed

Compound measures involve a combination of measurements and units.

- The **density** of a material is its mass divided by its volume.
- **Speed** is the distance travelled divided by the time taken.

- **A formula triangle can be a useful way of remembering the relationships between the different parts of a compound measure.**
  For example:
  speed = distance ÷ time
  distance = speed × time
  time = distance ÷ speed

D
Distance
S Speed | T Time

The units for density can be grams per cubic centimetre (g/cm^3), or kilograms per cubic metre (kg/m^3).

Speed can be measured in miles per hour (mph), kilometres per hour (kph) or metres per second (m/s).

Example

a A cube of side 3.5 cm has a mass of 600 g. Find the density of the cube in g/cm^3, correct to 3 significant figures.
b A car travels 240 miles in 3 hours 45 minutes. Find the average speed of the car in miles per hour.
c Lubricating oil has a density of 0.58 g/cm^3. Find
i the mass of 2.5 litres of this oil
ii the volume of 10 grams of the oil.

a Volume = 3.5^3 cm^3 = 42.875 cm^3
Density = 600 g ÷ 42.875 cm^3 = 13.994 … g/cm^3 = 14.0 g/cm^3 to 3 sf

Volume of a cube = length^3.

Density = $\frac{\text{mass}}{\text{volume}}$

b Average speed = $\frac{\text{total distance}}{\text{total time}}$
$= \frac{240}{3.75}$
= 64 mph

Put the time in hours.

c i Mass = density × volume = 0.58 g/cm^3 × 2500 cm^3 = 1450 g
ii Volume = mass ÷ density = 10g ÷ 0.58g/cm^3 = 17.2 cm^3

1 litre ≡ 1000 cm^3

Example

A metal cuboid is 95 cm long and has a length of 2 cm, width 4 cm and height 6 cm.
a Find the density of the cuboid.
b If the mass of the cuboid is 7.8 kg, find its density in g/cm^3.

a Volume of cuboid = 2 × 4 × 6 = 48 cm^3

b Density of cuboid = $\frac{\text{mass}}{\text{volume}} = \frac{7800}{48}$ = 163 g/cm^3 to 3 sf

Volume of a cuboid = length × width × height.

Change the mass into grams.

Exercise N4.4

1 Rod cycles 18 miles in 2 hours. Find his average speed, in miles per hour (mph).

2 If 4 metres of fabric costs £8.40, find the price of the fabric in pounds per metre.

3 A car travels 24 miles in 45 minutes. Find the average speed of the car in miles per hour (mph).

4 A train leaves Euston at 8:57 a.m. and arrives at Preston at 11:37 a.m. If the distance is 238 miles find the average speed of the train.

5 Copy and complete the table to show speeds, distances and times for five different journeys.

Speed (kph)	Distance (km)	Time
105		5 hours
48	106	
	84	2 hours 15 minutes
86		2 hours 30 minutes
	65	1 hour 45 minutes

6 A cube of side 2 cm weighs 40 grams.
a Find the density of the material from which the cube is made, giving your answer in g/cm^3.
b A cube of side length 2.6 cm is made from the same material. Find the mass of this cube, in grams.

Volume of cube = length^3.

7 A box has a length and width of 22.50 mm, and a height of 3.15 mm. It has a mass of 9.50 g.
a Find the density of the metal from which the box is made, giving your answer in g/cm^3.
b How many boxes can be made from 1 kg of the material?

Volume of cuboid = length × width × height.

8 Emulsion paint has a density of 1.95 kg/litre. Find
a the mass of 4.85 litres of the paint.
b the number of litres of the paint that would have a mass of 12 kg.

9 A steel cable weighs 2450 kg.
The cable has a uniform circular cross-section of radius 0.85 cm.
The steel from which the cable is made has a density of 7950 kg/m^3.
Find the length of the cable.

Volume of a cylinder = πr^2 × length.

N4.5 Proportional change

Objectives

- Solve problems involving proportion and proportional change

Useful resources

- Calculators

Mental starter

Write down some products of decimals and integers for students to evaluate mentally.
For example:

0.3×15
0.8×20
1.2×36
2.3×12

Introductory activity

Increasing or decreasing by a certain proportion, (fraction or percentage) can be done in different ways:

Increase £50 by 20%.

£50 × 20% = £10, so answer is £50 + £10 = £60
or Amount will now be 120%,
£50 × 120% = £60.
Emphasise that the operation of '+' has been done on 100%, instead of £50.

Decrease £90 by 30%.

£90 × 30% = £27, so answer is £90 – £27 = £63
or Amount will now be 70%,
£90 × 70% = £63.

Again the '–' has been done on 100%.

Increase £45 by $\frac{2}{5}$.
(£45 × $\frac{2}{5}$ = £18, so answer is £45 + £18 = £63.)

Now reduce your answer by 40%.
(40% of £63 = £25.20, so answer is
£63.00 – £25.20 = £37.80.)

$\frac{2}{5}$ = **40% so why didn't you get back to £63?** (The fraction and percentage are the same but you are finding them for different amounts.)

A collectable vase is worth £5 000 in 2003. It gains 12% in value in the first year, then loses 8% the next year. **What is its value in 2005?** (£5 152.)

Discuss the examples in the student book.

Exercise commentary and misconceptions

Encourage students to use the method of adding to or subtracting from 100% as well as adding or subtracting amounts.

Question 4 can be done using methods learned in N4.2.

Plenary

Compound and simple interest is not explicitly mentioned but could be discussed at the end of the lesson.
Is there a quick way to work out VAT ($17\frac{1}{2}$%)? You may need to explain what VAT is for. **Find the VAT to be added onto the price of a PC costing £300.**
(10% = £30 ⟶ 5% = £15 ⟶ $2\frac{1}{2}$% = £7.50
⟶ $17\frac{1}{2}$% = £52.50).

Some students may suggest that it would have been easier to multiply 17.5 by 3 as £300 is 3 × 100. Ask them if this would be easy to use for other starting values.

N4.5 Proportional change

This spread will show you how to:

- Solve problems involving proportion and proportional change

Keyword
Proportional change

Proportional change means increasing or decreasing a quantity by a certain percentage or fraction.

Daily Tabloid

Town population down a quarter

Chairman's Report

'Revenue increased by 5%'

CEO Mr James Jameson's announcement in his Review of the company's Annual Accounts for 2005-2006.

Example

Find the new amount when £350 is increased by a quarter.

£350 ÷ 4 = £87.50 — Work out the increase.

£350 + £87.50 = £437.50 — Add it to original amount.

When a quantity is increased by 20% its new value is 120% of the original value, for example

£50 increased by 20% = 120% of £50

When a quantity is decreased by 20% its new value is 80% of the original value, for example

£50 decreased by 20% = 80% of £50

You can work out the new amounts in one step as in this example.

Example

When a new motorway was built between Utopia and Edenlandia the 6-hour journey time was decreased by 30%.

a Find the new journey time.

The motorway from Utopia to Edenlandia is being resurfaced resulting in a 15% increase in journey time.

b Find the new journey time due to resurfacing.

UTOPIA EDENLANDIA

a New time is 70% of old time — 100% − 30% = 70%

New time = 70% × 6 hours
= 0.7 × 6 — 70% = 0.7
= 4.2 hours or 4 h 12 min

b New time due to resurfacing is 115% of old time. — 100% + 15% = 115%

New time = 115% × 4.2 hours
= 1.15 × 4.2 — 115% = 1.15
= 4.83 hours or 4 h 50 min

Exercise N4.5

1 Find the result when these amounts are increased by one quarter.

a 20 g **b** 160 cm **c** 6 hours **d** 500 kg

You should be able to do these mentally.

2 Decrease these amounts by one third.

a 600g **b** 30 sec **c** 45° **d** 72 hours

Again, you should be able to do these mentally.

3 Copy and complete the table to show the result of some proportional increases and decreases.

Original number	Proportional change	Result
42	Decrease by $\frac{1}{4}$	
110	Increase by $\frac{1}{5}$	
250	Increase by $\frac{1}{10}$	
450	Decrease by $\frac{2}{5}$	
965	Increase by $\frac{1}{10}$	

4 A recipe for making 16 scones includes these ingredients

60 g butter 3 teaspoons caster sugar 200 ml milk

Find the quantity of each ingredient needed to make 24 scones.

5 Increase these amounts by $\frac{1}{6}$.

a 240 g **b** 300 g **c** 200 g **d** 750 g

6 Increase these amounts by 5%.

a £120 **b** £240 **c** £500 **d** £72

7 Calculate these percentage changes.

a £450 increased by 10% **b** £600 decreased by 15%

c £900 increased by 6% **d** £740 decreased by 11%

8 Andrew earns £240 per week. He is awarded a pay rise of 4.5%. Bella earns £260 per week. She is awarded a pay rise of 4%. Whose weekly pay increases by the larger amount? Show all your working.

9 Mr and Mrs Jones receive their electricity bill. The details are

Present meter reading	23 087
Previous meter reading	20 893
Charge per unit	8.2 pence
Service charge	£13.75
VAT	5%

Find the total cost of the electricity including VAT. Show all your working.

N4 Exam review

Key objectives

- Use knowledge of operations and inverse operations, and of methods of simplification, in order to select and use suitable strategies and techniques to solve problems and word problems, including those involving ratio and proportion, fractions, percentages and measures and conversion between measures, and compound measures defined within a particular situation

Worked solution	Commentary and misconceptions
1	Encourage students to write down all their calculations and the justification behind them in order to achieve maximum marks.
a 1 cake needs 150g of flour. 5 = 1 x 5. So, 5 cakes need (150 x 5)g of flour. 150 x 5 = 750g of flour.	Ensure that students understand that the amount of flour is in proportion to the number of cakes baked.
b $\frac{150}{100}$ = 1.5 1.5 x 25 = 37.5 The bigger cake requires (150 + 37.5)g 150 + 37.5 = 187.5g of flour.	Students need to calculate 25% of 150 g and then add this to 150 g. Ensure that they use the correct order of operations.
2	Ensure that students understand what they are being asked to work out and the calculations involved. Encourage students to split the problem into two sets of calculations.
10% of 12 000 = $(\frac{12\,000}{100})$ x 10 = £1200. So, after 1 year, Value of car = 12 000 − 1200 = £10 800. 10% of 10 800 = $(\frac{10\,800}{100})$ x 10 = £1080. So, after 2 years, Value of car = 10 800 − 1080 = £9720	Remind students that their answer should be in £s. Encourage them to check if their answer seems feasible.

Transformations and congruence

Objectives

H Use congruence to show that translations, rotations and reflections preserve length and angle, so that any figure is congruent to its image under any of these transformations

F/H Understand that reflections are specified by a mirror line at first seeing a line parallel to an axis

F/H Use any point as the centre of rotation; measure the angle of rotation, using fractions of a turn or degrees

F/H Understand that translations are specified by giving a distance and direction (or a vector)

F/H Recognise and visualise rotations, reflections and translations including reflection symmetry of 3-D shapes, transform triangles and other 2-D shapes by combinations of these transformations

Unit overview

This unit consolidates Key Stage 3 knowledge of the 2-D coordinate plane and transformations. It includes reflection, rotation and translation and then extends to combining these transformations. The effect of transformations on 2-D shapes and congruence is focused on providing a basis for further work on transformations and enlargements in unit S7.

Prior knowledge

Before your students start this unit they should be able to:

- Recognise and plot points in all four quadrants of the coordinate plane
- Understand and use equations of straight lines
- Draw straight lines given their equations
- Understand and use angle measure

Differentiation

- **Foundation** focuses on simple descriptions of reflections and rotations, considering the corresponding symmetries.
- **Foundation Plus** extends to reflections, rotations, translations and a combination of these transformations in the coordinate plane, covering congruence and symmetry
- **Higher Plus** extends the Higher book to enlargement, proving congruence and similar shapes

S3.1 Reflection

Objectives

- Identify properties preserved under reflection
- Understand congruence
- Describe reflections, using mirror lines

Useful resources

- Graph paper
- Geometry tool

Mental starter

Arrange 12 dots in a rectangle. You can move between the dots either vertically or horizontally. **What is the minimum number of moves to move between dots at opposite corners?** Experiment for different sized rectangles all with 12 dots. (Best arrangement is 4 by 3, with 5 moves.)

Introductory activity

Ask students to draw a set of axes from –5 to 5. Plot a few points along the line $x = 4$. **What pattern do the points have?** Conclude that their x coordinate is always 4, so the equation of the line is $x = 4$. Label the line $x = 4$.

Repeat for points on the line $y = -3$ on the same axes.

Draw a second set of axes from –5 to 5. Plot a few points along the lines $y = x$. **What pattern do the points have?** Conclude that their x coordinate is always equal to their y coordinate. This means the name or equation is $y = x$.

Repeat for points on the line $y = -x$ on the same axes.

Comment on the fact that the lines are 45° to the axes, but that this is only true if the axes have the same scale.

Reflect triangle ABC (1, 2) (2, 4) (1, 5) in the line $y = x$. Very carefully count the squares diagonally from corner to corner from the shape to the mirror line then out the other side. Image has coordinates (2, 1) (4, 2) (5, 1). Label original triangle 'object' and reflection 'image'.

Exercise commentary and misconceptions

You may need to discuss the fact that $x = 4$ is parallel to the y axis and $y = -3$ is parallel to the x axis, which may be counter-intuitive. **What does that suggest about the equation (name) of the x axis?** ($y = 0$) **And the y axis?** ($x = 0$).

In questions 1, 2 and 4 ensure students draw the mirror line in the correct direction. Remind students that the $x =$ lines are parallel to the y axis. When reflecting, watch for students confusing their counting of squares by counting from the axes.

In question 3 students will need to count diagonally when reflecting in the line $y = x$.

Most students find it quite a challenge to reflect one point then draw the rest of the shape correctly, especially with a diagonal reflection, so encourage them to reflect all the points unless they are completely proficient.

Plenary

How do the co-ordinates change for a reflection in $y = x$? (x and y change over.) **And for a reflection in $y = -x$?** (x and y change over and change sign.)

S3.1 Reflection

This spread will show you how to:

- Identify properties preserved under reflection
- Understand congruence
- Describe reflections, using mirror lines

Keywords
Congruent
Mirror line
Perpendicular
Reflection

Just as you can see your reflection in a mirror, you can reflect a shape in a mirror line.

Corresponding points on the object and image are the same distance from the **mirror line**.

The line joining a point and its image is **perpendicular** to the mirror line.

mirror line $y = x$

object

image

Corresponding angles and lengths are the same in the image and the object.

The object and the image are **congruent**.

Reflection flips the shape over.

Example

Reflect the triangle with vertices at (1, 2), (2, 7) and (4, 7) in the line $x = 1$.

$x = 1$

The image of a point that is on the mirror line is the same point.

Example

Reflect the pink triangle in the line $y = 3$.

$y = 3$

When an object crosses the mirror line, so does its image.

Exercise S3.1

1 Copy this diagram.

a Reflect the triangle T in the line $x = 2$. Label the image U.

b Reflect the triangle T in the line $y = 1$. Label the image V.

2 Copy this diagram.

a Reflect the kite K in the line $x = -1$. Label the image L.

b Reflect the kite K in the line $y = 1$. Label the image M.

3 Copy this diagram and extend the y-axis to –8.

a Reflect the quadrilateral Q in the x-axis. Label the image R.

b Reflect the quadrilateral Q in the line $y = x$. Label the image S.

4 Copy this diagram.

a Reflect triangle A in the y-axis. Label the image B.

b Reflect triangle B in the y-axis. Label the image C.

c What do you notice? Does this always happen? Check with some other reflections.

S3.2 Rotation

Objectives
- Identify properties preserved under rotation
- Understand that rotations are specified by a centre and an (anticlockwise) angle

Useful resources
- Tracing paper
- Graph paper
- Geometry tool

Mental starter
Using vertical or horizontal reflections, which capital letters cannot be reflected to make the same letter? (F, G, J, L, N, P, Q, R, S, Y, Z.) Some differ according to the font used.

Introductory activity
Ask students to draw axes –5 to 5 and draw the triangle ABC (–2, 3) (–2, 5) (–5, 3). Rotate triangle through 90° clockwise with a centre of rotation at (–1, 1) using tracing paper. Encourage students to line the paper up square with the lines on the page. Be as accurate as possible. Discuss the fact that tracing the shape gives a good approximation, but is not exact due to sharpness of the pencil and approximate angle. New coordinates are (–1, 2) (3, 2) (1, 5).

Discuss anticlockwise being positive angles, and clockwise being negative angles. **Does 180° give a different result from –180°?** (No.)

Exercise commentary and misconceptions
Ensure students draw their image with vertices on exact coordinates. Tracing paper may help for this exercise.

In question 1 students can rotate the triangle in either direction.

In question 2, remind students that angles are measured anticlockwise for their positive direction, therefore a negative angle is in the clockwise direction.

Question 3 uses the terms 'clockwise' and 'anticlockwise' rather than positive and negative angles.

Plenary
What would happen if you rotated an equilateral triangle about its centre by 120°? (Would produce the same image in either direction.)
What other regular shapes could this be done for and how many degrees do you need to rotate? (Square 90°, pentagon 72°, hexagon 60°, heptagon 360/7°, octagon 45° etc.)
How does this relate to the exterior angles of polygons?

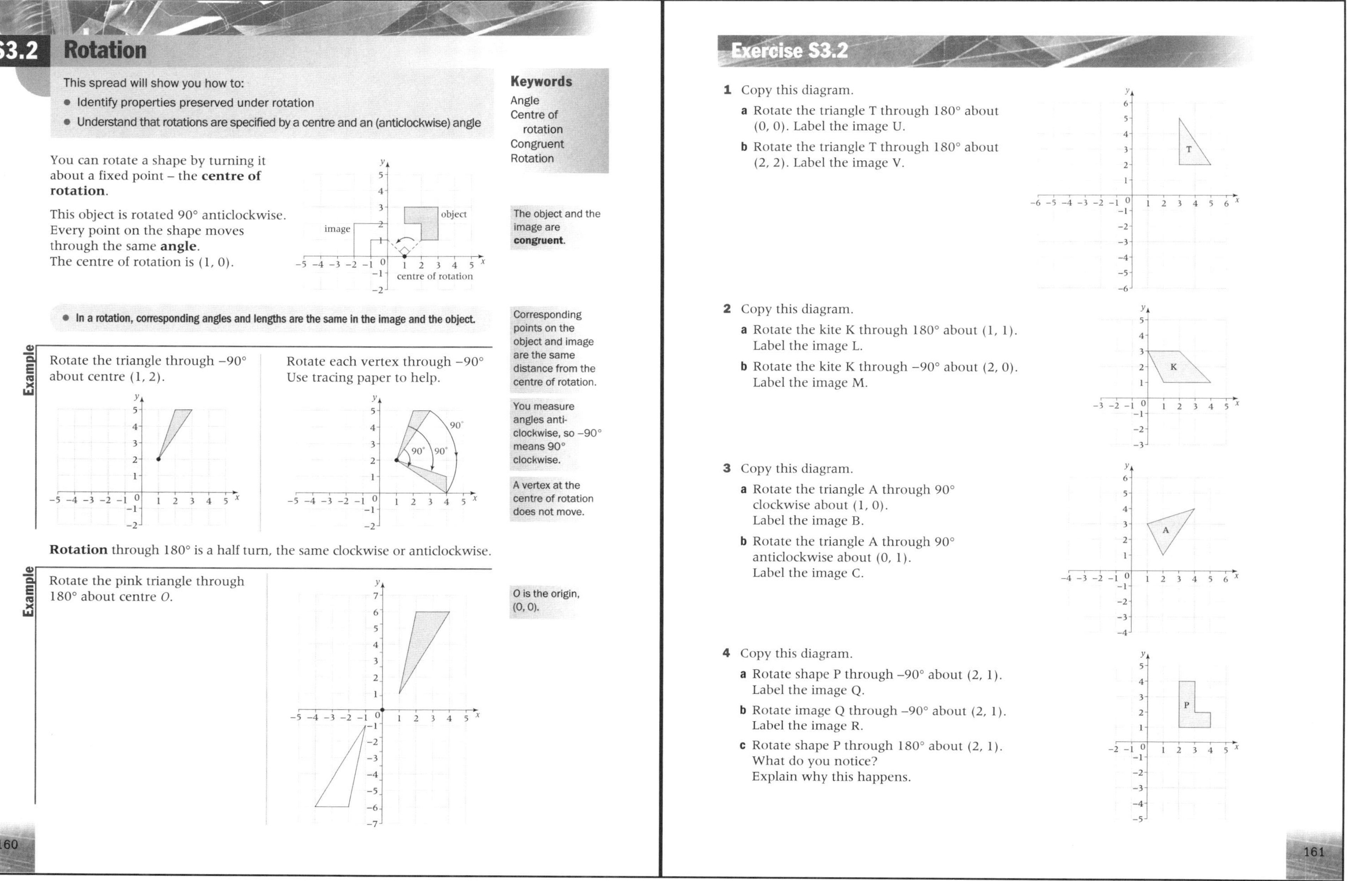

S3.2 Rotation

This spread will show you how to:
- Identify properties preserved under rotation
- Understand that rotations are specified by a centre and an (anticlockwise) angle

Keywords
Angle
Centre of rotation
Congruent
Rotation

You can rotate a shape by turning it about a fixed point – the **centre of rotation**.

This object is rotated 90° anticlockwise. Every point on the shape moves through the same **angle**.
The centre of rotation is (1, 0).

The object and the image are **congruent**.

- **In a rotation, corresponding angles and lengths are the same in the image and the object.**

Corresponding points on the object and image are the same distance from the centre of rotation.

Example

Rotate the triangle through −90° about centre (1, 2).

Rotate each vertex through −90°. Use tracing paper to help.

You measure angles anti-clockwise, so −90° means 90° clockwise.

A vertex at the centre of rotation does not move.

Rotation through 180° is a half turn, the same clockwise or anticlockwise.

Example

Rotate the pink triangle through 180° about centre *O*.

O is the origin, (0, 0).

Exercise S3.2

1 Copy this diagram.

a Rotate the triangle T through 180° about (0, 0). Label the image U.

b Rotate the triangle T through 180° about (2, 2). Label the image V.

2 Copy this diagram.

a Rotate the kite K through 180° about (1, 1). Label the image L.

b Rotate the kite K through −90° about (2, 0). Label the image M.

3 Copy this diagram.

a Rotate the triangle A through 90° clockwise about (1, 0). Label the image B.

b Rotate the triangle A through 90° anticlockwise about (0, 1). Label the image C.

4 Copy this diagram.

a Rotate shape P through −90° about (2, 1). Label the image Q.

b Rotate image Q through −90° about (2, 1). Label the image R.

c Rotate shape P through 180° about (2, 1). What do you notice? Explain why this happens.

S3.3 Translation

Objectives

- Identify properties preserved under translation
- Understand and use vector notation
- Describe translations by giving a distance and direction (or vector)

Useful resources

- Geometry tool
- Graph paper

Mental starter

A parallelogram is made from 6 congruent rhombuses. The perimeter of each rhombus is 15 cm. **What is the perimeter of the parallelogram?**

($10 \times \frac{15}{4} = 37.5$ cm for a 2 by 3 arrangement **or** $14 \times \frac{15}{4} = 52.5$ cm for a 6 by 1 arrangement.)

Introductory activity

There are four ways of changing (transforming) a shape: rotation and reflection have already been covered, the next one is translation. If you want to move a shape to another place, but don't want to change what it looks like or which way round it is drawn, then you can slide it right/left, up/down. This is a translation (not to be confused with transformation).

To move a shape use a special bracket called a **vector**. Moving a shape by vector $\begin{pmatrix} 4 \\ 2 \end{pmatrix}$ means moving the original shape 4 units right and 2 units up. You use negatives for left and down. $\begin{pmatrix} -1 \\ -5 \end{pmatrix}$ means 1 left and 5 down. The first number (top) is always right/left, the second (bottom) is always up/down, just like coordinates.

Ask students to draw a 10 by 10 grid. Shape A is a triangle with vertices at (1, 2) (2, 7) (4, 7).

Translate by $\begin{pmatrix} 5 \\ -2 \end{pmatrix}$ and $\begin{pmatrix} -3 \\ 0 \end{pmatrix}$ and $\begin{pmatrix} 0 \\ 4 \end{pmatrix}$ separately.

Label images B, C, D.

Exercise commentary and misconceptions

The most common mistake made by students is to count the squares in the gap between the original shape and its image: the space between the leading edge of the original and trailing edge of the image. Encourage students to translate one point and then draw the rest of the shape around the translated point.

Remind students that a vector is not an fraction, there is no dividing line.

Plenary

Discuss question 3 in the exercise. Encourage students to suggest other inverse translations.

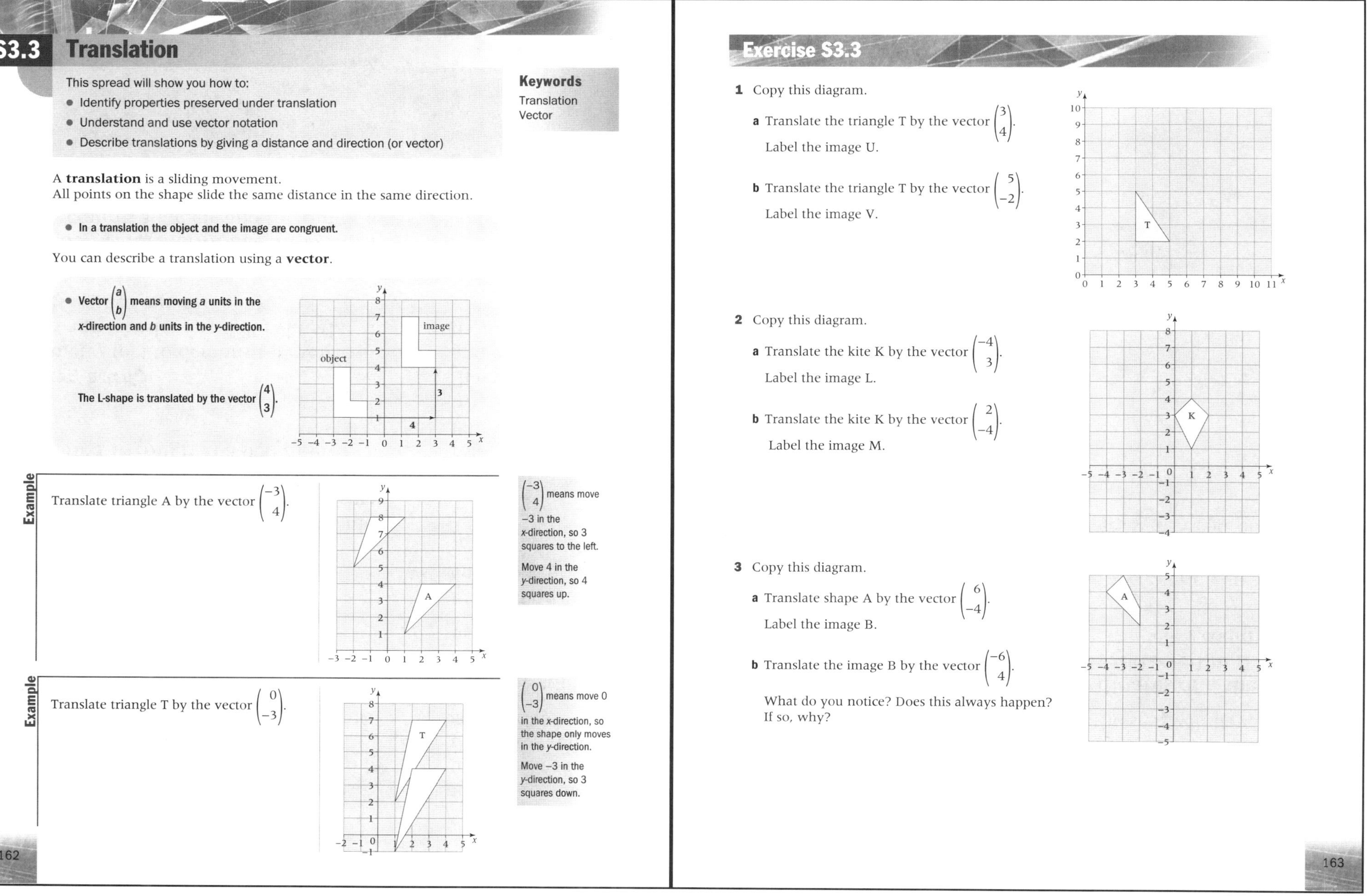

S3.3 Translation

This spread will show you how to:

- Identify properties preserved under translation
- Understand and use vector notation
- Describe translations by giving a distance and direction (or vector)

Keywords
Translation
Vector

A **translation** is a sliding movement.
All points on the shape slide the same distance in the same direction.

- **In a translation the object and the image are congruent.**

You can describe a translation using a **vector**.

- **Vector $\begin{pmatrix} a \\ b \end{pmatrix}$ means moving a units in the x-direction and b units in the y-direction.**

The L-shape is translated by the vector $\begin{pmatrix} 4 \\ 3 \end{pmatrix}$.

Example

Translate triangle A by the vector $\begin{pmatrix} -3 \\ 4 \end{pmatrix}$.

$\begin{pmatrix} -3 \\ 4 \end{pmatrix}$ means move −3 in the x-direction, so 3 squares to the left.

Move 4 in the y-direction, so 4 squares up.

Example

Translate triangle T by the vector $\begin{pmatrix} 0 \\ -3 \end{pmatrix}$.

$\begin{pmatrix} 0 \\ -3 \end{pmatrix}$ means move 0 in the x-direction, so the shape only moves in the y-direction.

Move −3 in the y-direction, so 3 squares down.

Exercise S3.3

1 Copy this diagram.

a Translate the triangle T by the vector $\begin{pmatrix} 3 \\ 4 \end{pmatrix}$.

Label the image U.

b Translate the triangle T by the vector $\begin{pmatrix} 5 \\ -2 \end{pmatrix}$.

Label the image V.

2 Copy this diagram.

a Translate the kite K by the vector $\begin{pmatrix} -4 \\ 3 \end{pmatrix}$.

Label the image L.

b Translate the kite K by the vector $\begin{pmatrix} 2 \\ -4 \end{pmatrix}$.

Label the image M.

3 Copy this diagram.

a Translate shape A by the vector $\begin{pmatrix} 6 \\ -4 \end{pmatrix}$.

Label the image B.

b Translate the image B by the vector $\begin{pmatrix} -6 \\ 4 \end{pmatrix}$.

What do you notice? Does this always happen?
If so, why?

S3.4 Describing transformations

Objectives

- Describe reflections, using mirror lines
- Understand that rotations are specified by a centre and an (anticlockwise) angle
- Describe translations by giving a distance and direction (or vector)

Useful resources

- Tracing paper
- Geometry tool

Mental starter

A trampolinist can perform 2 full revolutions on each bounce. **If he/she bounces every 2 seconds, how long will it take him/her to have turned through 108 000 degrees?** (5 min.)

Introductory activity

Use geometry tool to demonstrate reflections, rotations and translations, following the examples in the students' book.

Emphasise that to specify a reflection requires the equation of the mirror line; to specify a rotation requires an angle and the centre of rotation and to specify a translation requires two directions.

Exercise commentary and misconceptions

Encourage students to write proper sentences to describe the transformation and to include the words reflection, rotation or translation, as appropriate.

Reassure students that tracing paper is always available in an exam and encourage them to use it in this exercise.

In question 1 students should give the equations of the mirror lines.

In question 2 students should give the answers as vectors.

In question 3 students should give the angle of rotation and the centre of rotation. Students who are struggling to find the centre of rotation may be helped by the information that it will lie on an exact point (coordinate) and encouraged to use a systematic approach to finding it.

Plenary

How do the coordinates of a shape change when it is rotated by 180° about the origin? (The x and y coordinate will have the same number but the sign will have changed.)

How do the coordinates of a shape change when it is reflected in the x-axis? (x coordinate is unchanged, y coordinate changes sign.)

How do the coordinates of a shape change when it is reflected in the y-axis? (x coordinate changes sign, y coordinate is unchanged.)

S3.4 Describing transformations

This spread will show you how to:
- Describe reflections, using mirror lines
- Understand that rotations are specified by a centre and an (anticlockwise) angle
- Describe translations by giving a distance and direction (or vector)

Keywords
Maps
Reflection
Rotation
Transformation
Translation

- To describe a **reflection**, you give the equation of the line.
- To describe a **rotation**, you give the centre and the angle of rotation.
- To describe a **translation**, you give the distance and direction or you specify the vector.

Reflections, rotations and translations are all **transformations**.

Example

Describe the transformation that maps shape A on to
a shape B **b** shape C **c** shape D.

Maps means changes.

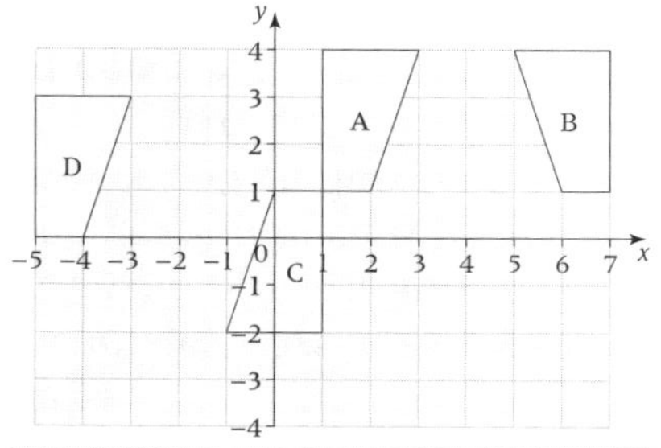

a In a reflection the mirror line bisects the line joining corresponding points on the object and image.

Shape B is a reflection of shape A in the line $x = 4$.

b The vertex (1, 1) does not move during the rotation, so it must be the centre of rotation.

Shape C is a rotation of shape A through 180°.

c Shape D is a translation of shape A by the vector $\begin{pmatrix} -5 \\ -1 \end{pmatrix}$.

Exercise S3.4

1 Describe fully the transformation that maps
- **a** shape A onto shape B
- **b** shape A onto shape C
- **c** shape A onto shape D
- **d** shape B onto shape D
- **e** shape C onto shape D
- **f** shape D onto shape C.

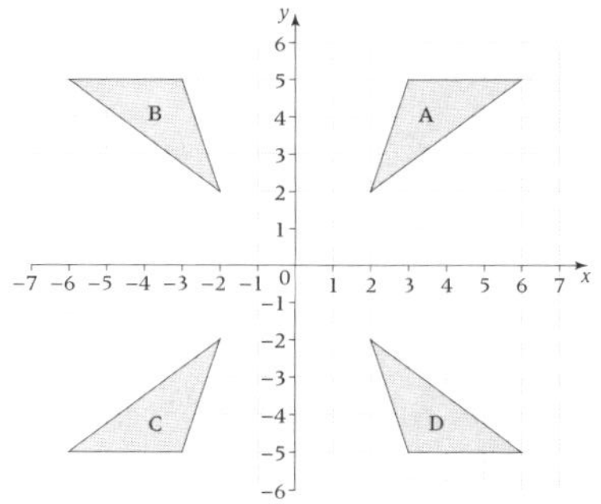

2 Describe fully the transformation that maps
- **a** shape J onto shape K
- **b** shape L onto shape K
- **c** shape M onto shape K
- **d** shape L onto shape M
- **e** shape J onto shape M
- **f** shape M onto shape J.

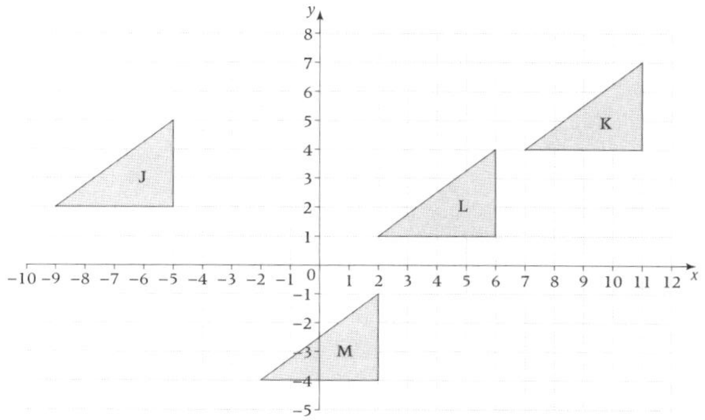

3 Describe fully the transformation that maps
- **a** shape W onto shape X
- **b** shape W onto shape Y
- **c** shape W onto shape Z
- **d** shape Z onto shape W
- **e** shape X onto shape Y
- **f** shape Z onto shape X.

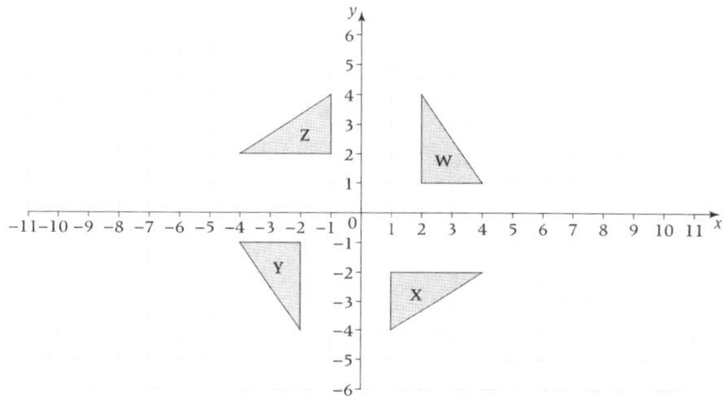

S3.5 Combining transformations

Objectives

- Transform 2-D shapes by translation, rotation and reflection and combinations of these transformations

Useful resources

- Tracing paper
- Graph paper
- Geometry tool

Mental starter

One of the longest cars in the world is 30.5 m long. If you want to turn it right round in the road in one go, you need the road to be 40 m wide. For every 50 cm less than 40 m you need to do an extra turn, for example a 39 m wide road needs a 2 point turn. **How wide is the road for a 7 point turn?** (36.5 m) **Why will you never need to do a 20 point turn?** (Road would be 30 m wide, car would not fit across the road.)

Introductory activity

Remind students of the list of transformation terms from S3.4. Discuss the concept that transformations can be combined.

Draw a shape and reflect it in the line $y = x$, then translate it by the vector $\begin{pmatrix} 5 \\ 1 \end{pmatrix}$ to produce an image. **What transformation would be needed to take the image back to the original shape?** (Work the transformations backwards. Translate by vector $\begin{pmatrix} -5 \\ -1 \end{pmatrix}$ followed by a reflection in the line $y = x$.)

The student book uses combined transformations applied to triangles to illustrate three key points.
Discuss these in turn.

Exercise commentary and misconceptions

Remind students that they need to use the words reflection, rotation or translation in their answers, as well as giving the details.

It is difficult to remember which combinations produce which single transformations. Students are better advised to use their own knowledge and previously practiced methods to find the single transformation.

The most common problems in any of these questions result from a mistake in one of the initial transformations, so advise students to check that they have carried out parts **a** and **b** of a question correctly before attempting **c**.

Plenary

Is it possible to transform 3-D objects?
Discuss how this might be achieved for each type of transformation.

For reflection you need a plane of symmetry instead of a line of symmetry. For rotation you need a 3-D coordinate as the centre and you need to rotate in a particular plane. For a translation you need a 3-D vector. Discuss the 3-D coordinate axes with x, y, z axes. Most commonly the z axis is vertical and the y axis is into the page.

S3.5 Combining transformations

This spread will show you how to:

- Transform 2-D shapes by translation, rotation and reflection and combinations of these transformations

Keywords
Reflection
Rotation
Transformations
Translation

You can combine **transformations** by doing one after the other.
You can describe a combination of transformation as a single transformation.

Example

In this diagram, triangle A undergoes three pairs of transformations.

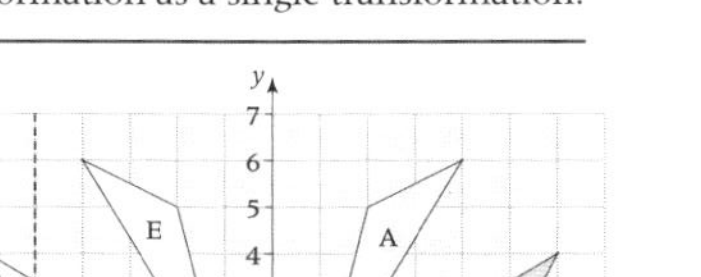
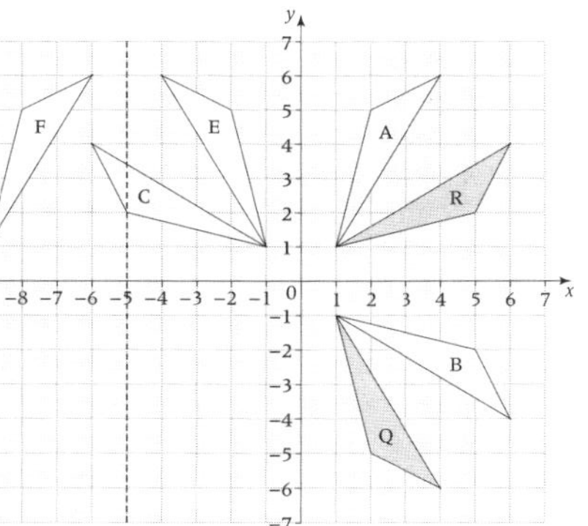

1. Triangle A is rotated 90° clockwise about (0, 0) to triangle B. Then triangle B is rotated 180° about (0, 0) to triangle C.
   What single transformation maps triangle A onto triangle C?
2. Triangle A is reflected in the line $y = 0$ (the x-axis) to triangle Q. Then triangle Q is rotated through 90° anticlockwise about (0, 0) to triangle R.
   What single transformation maps triangle A onto triangle R.
3. Triangle A is reflected in the line $x = 0$ (the y-axis) to triangle E. Then triangle E is reflected in the line $x = -5$ to triangle F.
   What single transformation maps triangle A onto triangle F?

1. A rotation of 90° anticlockwise about (0, 0) maps A onto C.

- **A combination of rotations that have the same centre is equivalent to a single rotation.**

2. A reflection in the line $y = x$ maps A onto R.

- **A combination of a reflection and a rotation is equivalent to a single reflection.**

3. A translation by the vector $\begin{pmatrix} -10 \\ 0 \end{pmatrix}$ maps A onto F.

- **A combination of reflections is a translation when the mirror lines are parallel.**

Exercise S3.5

1 Copy this diagram.

a Rotate triangle A 90° anticlockwise about centre (0, 0). Label the image B.

b Rotate triangle B 90° anticlockwise about centre (0, 0). Label the image C.

c Describe fully the single transformation that takes triangle A to triangle C.

d Rotate triangle B 90° clockwise about centre (2, 1). Label the image D.

e Describe fully the single transformation that takes triangle A to triangle D.

2 Copy this diagram.

a Reflect triangle E in the y-axis. Label the image F.

b Reflect triangle F in the x-axis. Label the image G.

c Describe fully the single transformation that takes triangle G to triangle E.

d Reflect triangle F in the line $x = 2$. Label the image H.

e Describe fully the single transformation that takes triangle E to triangle H.

3 Copy this diagram.

a Reflect triangle J in the x-axis. Label it K.

b Rotate triangle K 180° about centre (0, 0). Label it L.

c Describe fully the single transformation that takes triangle L to triangle J.

4 Copy this diagram.

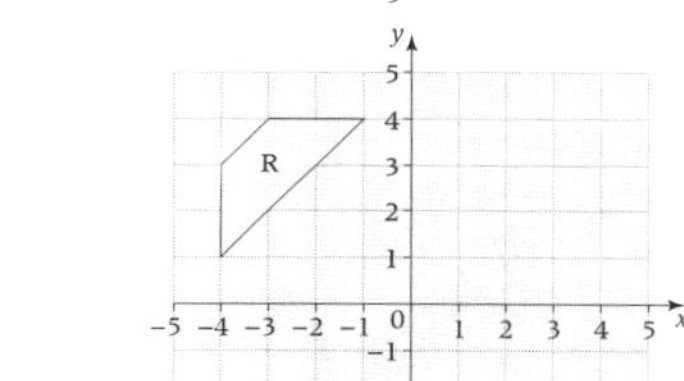

a Translate trapezium R by the vector $\begin{pmatrix} 5 \\ -3 \end{pmatrix}$. Label the image S.

b Translate trapezium S by the vector $\begin{pmatrix} -1 \\ 2 \end{pmatrix}$. Label the image T.

c Describe fully the single transformation that takes trapezium T to trapezium R.

S3 Exam review

Key objectives

- Recognise and visualise rotations, reflections and translations, transform 2-D shapes by combinations of these transformations
- Distinguish properties that are preserved under particular transformations

Worked solution | **Commentary and and misconceptions**

1 a, c

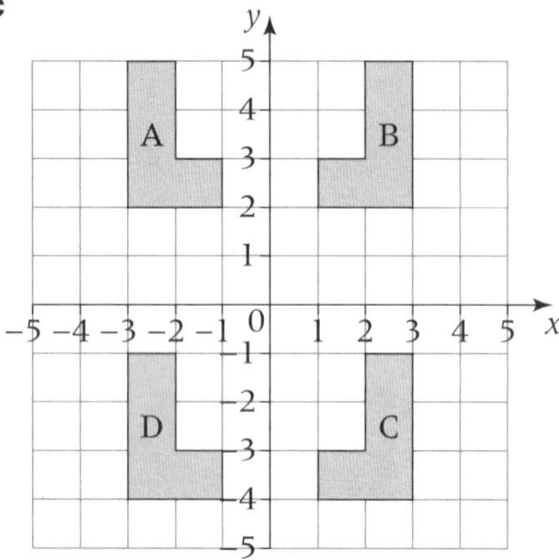

Students should know how to transform shapes.

Ensure students use the correct mirror line.

Ensure students understand the term 'symmetrical'.

b A translation with vector $\begin{pmatrix} 0 \\ -6 \end{pmatrix}$.

Ensure students use appropriate vocabulary.

The transformation from shape C to shape D is a reflection in the y-axis

Ensure students use appropriate vocabulary and describe the transformation from C to D.

2 a

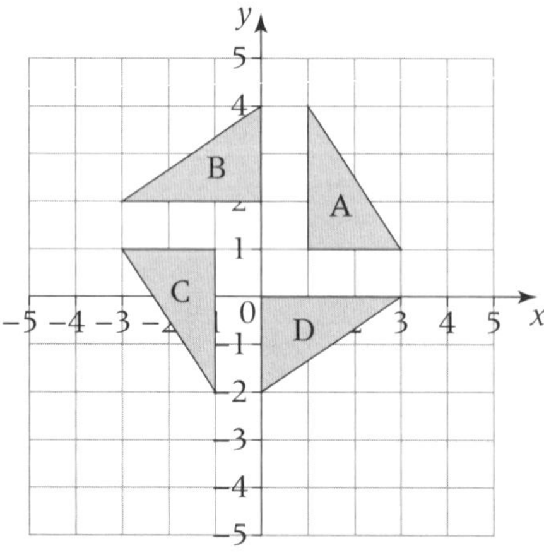

Ensure that students use the correct direction and centre of rotation throughout this question.

b The single transformation that takes shape C to shape A is an anticlockwise rotation through 180°, centre (0, 1).

Students should recognise that the combined effect of 90° rotations is a 180° rotation with the same centre. Ensure that students are able to answer this question without using tracing paper or any other methods.

Integers, powers and roots

Objectives

H Use index laws to simplify and calculate the value of numerical expressions involving multiplication and division of fractional and negative powers

H Use standard index form expressed in conventional notation and on a calculator display

Unit overview

This unit consolidates Key Stage 3 knowledge of calculations with whole numbers. It introduces index notation and index laws and then extends to using standard index form for large and small numbers.

Prior knowledge

Before your students start this unit they should be able to:

- Add, subtract, multiply and divide with integers and decimals
- Understand and use index notation for small positive powers
- Calculate the HCF and LCM of two integers
- Understand and use fraction and decimal notation
- Use a scientific calculator effectively

Differentiation

- **Foundation** focuses on squares, cubes, square roots and cube roots and standard index form for small powers of 10, including using a calculator and estimating. Prime numbers, factors and multiples are then introduced
- **Foundation Plus** extends to trial and improvement and powers and reciprocals and then proceeds to prime factors and prime factor decomposition
- **Higher Plus** extends the Higher book to irrational numbers and rationalising the denominator, calculations with surds, including using brackets, and algebraic laws

N5.1 Powers and indices

Objectives

- Understand and use powers and index notation
- Find the highest common factor and lowest common multiple of two numbers

Useful resources

- Calculators

Mental starter

Ask students to work out strings of multiplications:

$2 \times 2 \times 2 \times 2 \times 3$ $\quad$ $2 \times 2 \times 3 \times 3 \times 7$

$3 \times 3 \times 3 \times 5$ $\quad$ $4 \times 4 \times 4 \times 4 \times 5$

Introductory activity

Indices (also called powers) provide a way of abbreviating products. For example, $6^2 = 6 \times 6 = 36$.

How could 6^2 be said?

('6 squared', 'the square of 6', '6 to the power of 2'.)

5^3 means what? $5 \times 5 \times 5 = 125$. **How could 5^3 be said?** ('5 cubed', 'the cube of 5', '5 to the power of 3'.)

How can you write 64 as a power of 4? (4^3.)

How can you simplify $5^3 \times 5^4$? Demonstrate the key point that you add the powers. To calculate 5^7 you can use the power key on a calculator, or repeated multiplication. Discuss the third example in the student book.

Numbers can be written as the product of prime factors in index form. For example $48 = 2 \times 2 \times 2 \times 2 \times 3 = 2^4 \times 3$. The powers of 2 cannot be combined with the powers of 3. Remind students of the factor tree method.

Writing numbers as the product of prime factors in index form gives an easy method of finding the HCF and LCM of two or more numbers.

For example, $84 = 2^2 \times 3 \times 7$ and $280 = 2^3 \times 5 \times 7$.

The HCF of two numbers contains the highest powers of each prime factor which are common to **both** numbers.

So the HCF of 84 and 280 is $2^2 \times 7 = 28$ because 2^2 and 7 are in the lists for both 84 and 280.

The LCM of two numbers contains the highest powers of each prime factor which are in **either** number.

So the LCM of 84 and 280 is $2^3 \times 3 \times 5 \times 7 = 840$ because $2^2 \times 3 \times 7$ and $2^3 \times 5 \times 7$ are both contained in the LCM.

Exercise commentary and misconceptions

Questions 1 and 2 do not require students to find the value of the numbers.

Questions 3, 4 and 5 do require values, but encourage students to do question 4 without using calculators. Watch for weak students confusing raising to a power with multiplying (e.g. squaring with multiplying by 2).

For questions 6 and 8 remind students that they need to divide repeatedly by the number required. Check that students use prime numbers in question 8; question 6 uses some non-primes.

In question 7 some students may attempt to combine the base numbers.

As a hint in question 10, suggest that students write the numbers in product form.

Plenary

Discuss answers to question 10. Practice more questions of this type, for example the LCM and HCF of 32 and 48 (96 and 16).

N5.1 Powers and indices

This spread will show you how to:

- Understand and use powers and index notation
- Find the highest common factor and lowest common multiple of two numbers

Keywords
HCF
Index
LCM
Power

- **Repeated multiplications such as $2 \times 2 \times 2 \times 2$ can be written in index notation as 2^4.**

You read 2^4 as 'two to the **power** 4'.

- You use powers when factorising, for example,

$$24 = 2 \times 2 \times 2 \times 3 = 2^3 \times 3$$

But note that 2^2 is 'two squared', and 2^3 is 'two cubed'.

- **When powers of the same number are multiplied together, you can find the answer by adding the indices.**

For example,

$$2^2 \times 2^3 = (2 \times 2) \times (2 \times 2 \times 2) = 2^5$$

Similarly, $3^5 \times 3^6 = 3^{(5+6)} = 3^{11}$.

In 3^{11}, 11 is the index. The plural of index is indices.

Example

Find the value of **a** 2^3 **b** 3^2 **c** 5^3 **d** 10^4 **e** 2^8

a $2^3 = 2 \times 2 \times 2 = 8$ **b** $3^2 = 3 \times 3 = 9$ **c** $5^3 = 5 \times 5 \times 5 = 125$
d $10^4 = 10 \times 10 \times 10 \times 10 = 100 \times 100 = 10\,000$
e $2^8 = 2 \times 2 \times 2 \times 2 \times 2 \times 2 \times 2 \times 2 = 256$

Example

Write **a** 625 as a power of 5 **b** 100 000 as a power of 10
c 48 as a product of prime factors in index form

a $625 = 5 \times 125 = 5 \times 5 \times 25 = 5 \times 5 \times 5 \times 5 = 5^4$
b $100\,000 = 10 \times 10 \times 10 \times 10 \times 10 = 10^5$
c $48 = 2 \times 24 = 2 \times 2 \times 12 = 2 \times 2 \times 2 \times 6 = 2 \times 2 \times 2 \times 2 \times 3 = 2^4 \times 3$

Example

Find, in index form, the values of
a $4^3 \times 4^7$ **b** $5^2 \times 5$ **c** $3^2 \times 3^4 \times 3^3$ **d** $2^2 \times 3^4 \times 2^5 \times 3^3$

a $4^3 \times 4^7 = 4^{(3+7)} = 4^{10}$ **b** $5^2 \times 5 = (5 \times 5) \times 5 = 5^3$
c $3^2 \times 3^4 \times 3^3 = 3^{(2+4+3)} = 3^9$
d $2^2 \times 3^4 \times 2^5 \times 3^3 = 2^{(2+5)} \times 3^{(4+3)} = 2^7 \times 3^7$

Example

Find the **HCF** and **LCM** of 84 and 280.

$84 = 2 \times 2 \times 3 \times 7$
$= 2^2 \times 3 \times 7$

$280 = 2 \times 2 \times 2 \times 5 \times 7$
$= 2^3 \times 5 \times 7$

$HCF = 2^2 \times 7 = 28$
$LCM = 2^3 \times 3 \times 5 \times 7 = 840$

HCF is the highest number that is a factor of both 84 and 128.

LCM is the lowest number that both 84 and 280 will divide into exactly.

Exercise N5.1

1 Write these in index form.
a 3×3 **b** $2 \times 2 \times 2$ **c** $3 \times 3 \times 3$ **d** $5 \times 5 \times 5 \times 5$
e $7 \times 7 \times 7$ **f** $10 \times 10 \times 10$ **g** $6 \times 6 \times 6 \times 6$ **h** $5 \times 5 \times 5$

2 Write these numbers in product form. For example, $4^3 = 4 \times 4 \times 4$.
a 3^4 **b** 5^2 **c** 7^4 **d** 10^5 **e** 4^9 **f** 6^3 **g** 2^5 **h** 9^3

3 Find the value of each of these. For example, $5^3 = 5 \times 5 \times 5 = 125$.
a 4^2 **b** 4^3 **c** 2^5 **d** 10^2 **e** 10^3 **f** 3^3 **g** 2^3 **h** 3^2

4 Copy and complete the table to show the values of powers of 10.

Index form	Product	Value
10^6	$10 \times 10 \times 10 \times 10 \times 10 \times 10$	1 000 000
10^5		
10^4		
10^3		
10^2		
10^1		

5 Make a table, like the one in question **4**, to show the values of powers of 2 from 2^1 to 2^8.

6 Write
a 81 as a power of 9 **b** 125 as a power of 5
c 128 as a power of 2 **d** 100 000 as a power of 10
e 81 as a power of 3 **f** 343 as a power of 7

7 Write the answers to these multiplications in index form.
a $3^4 \times 3^2$ **b** $2^8 \times 2^1$ **c** $4^4 \times 4^4$
d $5^2 \times 5^3$ **e** $8^3 \times 8^5$ **f** $3^2 \times 3^5 \times 3^1$
g $2^3 \times 3^2 \times 2^4 \times 3^2$ **h** $5^2 \times 7^1 \times 5^2 \times 7^6$

8 Write each of these numbers as a prime number raised to a power. For example, $49 = 7^2$.
a 16 **b** 121 **c** 27 **d** 125 **e** 169 **f** 625 **g** 243 **h** 256

9 Write these numbers as products of their prime factors, using index notation. For example,
$72 = 2 \times 36 = 2 \times 2 \times 18 = 2 \times 2 \times 2 \times 9 = 2 \times 2 \times 2 \times 3 \times 3 = 2^3 \times 3^2$.
a 52 **b** 36 **c** 50 **d** 24 **e** 18 **f** 48 **g** 60 **h** 144

10 Find the HCF and LCM of
a 64 and 112 **b** 38 and 133

N5.2 Index laws

Objectives

- Use index laws, including fractional and negative indices

Useful resources

- Mini-whiteboards

Mental starter

Find the next two terms in each of these patterns:

(i) **12, 3, $\frac{3}{4}$** ($\frac{3}{16}$, divide by 4)
(ii) **18, 6, 2** ($\frac{2}{3}$, divide by 3)
(iii) **10, 2, $\frac{2}{5}$** ($\frac{2}{25}$, divide by 5)

Introductory activity

Remind students of the rule used in the previous spread, that you can add indices when multiplying powers of the same number.

This can be written algebraically as $x^a \times x^b = x^{a+b}$. The letters emphasise that x can be any number but must be the same number in both terms.

How can you simplify $6^5 \div 6^2$? With $6^5 \div 6^2$ you are cancelling out 2 of the 6's. So it leaves only 6^3. (It is worth showing the cancelling on the board.)

$7^6 \div 7^4 = 7^2$. (Again it is worth showing the cancelling on the board.)

What shortcut rule could you use?

Dividing means SUBTRACT the indices. This can be written as $x^a \div x^b = x^{a-b}$.
Discuss the power 0:

$4^3 \div 4^3 = 4^{3-3} = 4^0$ using the subtraction rule. But any number divided by itself gives the answer '1', so $4^0 = 1$.

More generally $x^0 = 1$, if $x \neq 0$.

Discuss the power 1.
$x^1 = x$ for all x. Ask for volunteers to demonstrate this.

Exercise commentary and misconceptions

Use the rules for multiplying and dividing with indices. Encourage students to write divisions out in fraction form to help them see the cancelling process more clearly.

In questions 4, 5 and 6 where there are more than 2 indices in a question make sure students work from left to right, but give precedence to brackets and that they simplify the numerators and denominators in question 6 separately.

In questions 7 and 8 some students may try to combine terms with different base numbers.

Plenary

Discuss answers to question 8. This involves both multiplication and division, with a mixture of base numbers.

N5.2 Index laws

This spread will show you how to:

- Use index laws, including fractional and negative indices

Keywords
Index
Indices
Power

There are rules that you can use when calculating with indices.

In these **index laws**, letters are used to represent numbers.

- **Add indices when multiplying powers of the same number.**
  $x^a \times x^b = x^{a+b}$
- **Subtract indices when dividing powers of the same number.**
  $x^a \div x^b = x^{a-b}$

$5^4 \times 5^3 = 5^{4+3} = 5^7$

$6^5 \div 6^2 = 6^3$

Using the rule for multiplication, $7^3 \times 7^0 = 7^3$. Since multiplying by 7^0 leaves the 7^3 unchanged, 7^0 must be equal to 1.

- **Any number (except 0) to the power 0 = 1: $x^0 = 1$ for any value of x, if $x \neq 0$.**

0^0 is *not defined* – it doesn't actually mean *anything*!

You know that $3^2 \times 3 = (3 \times 3) \times 3 = 3^3$.
You can write this as $3^2 \times 3^1 = 3^{(2+1)} = 3^3$, so $3 = 3^1$.

- **Any number to the power 1 is just the number itself: $x^1 = x$ for any value of x.**

Example

Solve these, giving your answers in index form.

a $2^3 \times 2^2$ **b** $5^7 \div 5^3$ **c** $6^4 \times 6^2 \div 6^3$
d $7^2 \times 5^3 \times 7^3 \times 5^4$ **e** $(2^5 \times 3^4) \div (2^3 \times 3^2)$

a $2^3 \times 2^2 = 2^{(3+2)} = 2^5$

b $5^7 \div 5^3 = 5^{(7-3)} = 5^4$

c $6^4 \times 6^2 \div 6^3 = 6^{(4+2-3)} = 6^3$

d $7^2 \times 5^3 \times 7^3 \times 5^4 = 7^{(2+3)} \times 5^{(3+4)} = 7^5 \times 5^7$

e $(2^5 \times 3^4) \div (2^3 \times 3^2) = 2^{(5-3)} \times 3^{(4-2)} = 2^2 \times 3^2$

Example

Write the value of **a** 19^0 **b** 8^1 **c** $(16^3 - 81 \times 17)^0$ **d** $(4.8)^1$

a $x^0 = 1$ for any value of x, so $19^0 = 1$.
b $x^1 = x$ for any value of x, so $8^1 = 8$.
c There is no need to evaluate the expression in the bracket (except to note that it is bigger than zero.)
The power of 0 means that $(16^3 - 81 \times 17)^0 = 1$.
d $x^1 = x$ for any value of x (including decimal numbers), so $(4.8)^1 = 4.8$.

Exercise N5.2

1 Write the answers to these multiplications in index form.

a $7 \times 7 \times 7$ **b** 3×3^2 **c** 5×5^2 **d** $6 \times 6 \times 6^2$
e $5^3 \div 5$ **f** $8^4 \times 8$ **g** $9^3 \times 9^2 \times 9$ **h** $8^7 \times 8$

2 Work out these, giving your answers in index form.

a $6^2 \times 6^3$ **b** $4^5 \times 4^4$ **c** $2^6 \times 2^7$ **d** $11^5 \times 11^2$
e $1^{17} \times 1^{13}$ **f** $7^8 \times 7^4$ **g** $3^6 \times 3^6$ **h** $9^9 \times 9^1$

3 Work out these, giving your answers in index form where appropriate.

a $7^8 \div 7^6$ **b** $8^6 \div 8^2$ **c** $3^3 \div 3^2$ **d** $9^{11} \div 9^8$
e $4^7 \div 4^1$ **f** $2^9 \div 2^9$ **g** $12^8 \div 12^6$ **h** $6^{13} \div 6^{13}$

4 Work out these, giving your answers in index form.

a $8^6 \times 8^2 \div 8^3$ **b** $5^7 \times 5^2 \div 5^4$ **c** $2^8 \times 2^3 \div 2^5$ **d** $9^6 \times 9^3 \div 9^7$
e $8^5 \times 8^5 \div 8^2$ **f** $7^6 \times 7^5 \div 7^4$ **g** $4^6 \times 4^8 \div 4^4$ **h** $11^2 \times 11^2 \div 11^3$

5 Write the answers to these in index form.

a $3^4 \times 3^2 \div (3^3 \times 3^2)$ **b** $(5^6 \div 5^2) \times 5^4 \times 5^2$
c $(4^5 \div 4^2) \div (4^6 \div 4^5)$ **d** $(7^9 \div 7^2) \div (7^2 \times 7^3)$
e $(8^7 \div 8^4) \times 8^5 \times 8^3$ **f** $9^3 \times (9^5 \div 9^2) \times 9^4$

6 Simplify these expressions, giving your answers in index form.

a $\frac{4^2 \times 4^2}{4^2}$ **b** $\frac{6^3 \times 6^4}{6^5}$ **c** $\frac{9^8}{9^2 \times 9^4}$ **d** $\frac{8^6 \div 8^3}{8^2}$
e $\frac{5^9 \times 5^4}{5^3 \times 5^7}$ **f** $\frac{6^3 \times 6^4}{6^5 \div 6^3}$ **g** $\frac{8^9 \div 8^2}{8^7 \div 8^2}$ **h** $\frac{10^6 \div 10^2}{10^2 \times 10^2}$

7 Simplify these expressions as far as possible, giving your answers in index form.

a $4^2 \times 3^3 \times 4^2$ **b** $8^5 \times 7^2 \div 8^2$ **c** $6^2 \times 5^3 \times 6^2 \times 5^3$
d $4^5 \times 3^3 \times 3^3 \times 4^4$ **e** $5^4 \times 2^3 \div 5^2$ **f** $9^5 \times 7^2 \times 7^2 \times 9^2$
g $8^2 \times 5^6 \times 8^3 \div 5^3$ **h** $3^4 \times 8^5 \times 3^4 \times 8^2$ **i** $9^3 \times 2^5 \div 2^3 \times 9^2$

8 Simplify these expressions, giving your answers in index form.

a $\frac{5^2 \times 8^5}{8^2}$ **b** $\frac{6^5 \times 7^2}{6^3}$ **c** $\frac{6^4 \times 5^4}{6^2 \times 5^2}$ **d** $\frac{7^8 \times 5^6}{5^3 \times 7^2}$
e $\frac{8^7 \times 3^5}{3^2 \times 8^5}$ **f** $4^3 \times \frac{4^5 \times 5^9}{4^3 \times 5^7}$ **g** $\frac{6^9 \times 7^5}{6^7 \times 7^3} \times 6^2$ **h** $4^3 \times \frac{7^6 \times 4^5}{4^4 \times 7^3} \times 7^2$

N5.3 More index laws

Objectives

- Use index laws, including fractional and negative indices

Useful resources

- Calculators

Mental starter

If you want to divide an amount by 100 you can do so in two identical smaller steps by dividing by 10 twice. **What number can you divide by twice so that overall you divide by 9** (3) **4** (2) **25** (5) **1**? (1)
(Note, could also take the negative values −3, −2, −5, and −1.) **How are these answers connected?**
(Square roots.)

Introductory activity

Recap integral indices and ensure students understand that

$5^3 = 5 \times 5 \times 5$

Progress to fractional indices, and argue that $5^{\frac{1}{2}} = \sqrt{5}$ (use the justification in the student book). Show this is true for any number.
Progress to other fractional indices and negative indices, using the student book for guidance. You may wish to use different numbers other than 5.
Extend to the power of a power. In each case, identify the general rule by using algebra.
Keep the general rules displayed on the board, with particular numerical examples next to them.

Exercise commentary and misconceptions

In question 10 it may help students to write expressions in factor form, so
$(2^3)^2 = 2^3 \times 2^3 = 2^6$.

In question 11 some students may try to combine terms with different bases.

Plenary

What is the value of x in each equation?
$x^{1/2} = 1$, $x^3 = 1$, $x^{-2} = 1$ (x always equals 1).
What does this mean about 1 to any power? (Answer is always 1.)

N5.3 More index laws

This spread will show you how to:

- Use index laws, including fractional and negative indices

Keywords
Fractional
Index laws
Negative
Power
Reciprocal

Indices can be fractions as well as whole numbers.

$$5^{\frac{1}{2}} \times 5^{\frac{1}{2}} = 5^{(\frac{1}{2}+\frac{1}{2})} = 5^1 = 5$$

Since $\sqrt{5} \times \sqrt{5} = 5$, $5^{\frac{1}{2}}$ must represent the square root of 5.

- **In general, $x^{\frac{1}{2}} = \sqrt{x}$ for any value of x Similarly, $x^{\frac{1}{3}} = \sqrt[3]{x}$ (the cube root of x), and so on.**

A **fractional** index means a root.

$$\tfrac{1}{5} = 1 \div 5 = 5^0 \div 5^1 = 5^{0-1} = 5^{-1}, \text{ so, } 5^{-1} \text{ means } \tfrac{1}{5}$$

- **In general, x^{-1} is equal to $\frac{1}{x}$, for any value of x. This is the reciprocal of x.**

A **negative** index means a reciprocal.

$$\tfrac{1}{5^2} = 1 \div 5^2 = 5^0 \div 5^2 = 5^{0-2} = 5^{-2}, \text{ so, } 5^{-2} \text{ means } \tfrac{1}{5^2}$$

- **In general, x^{-n} means $\frac{1}{x^n}$. This is the reciprocal of x^n.**

Note that zero has no reciprocal as 0^{-1} is not defined.

$(5^2)^3 = 5^2 \times 5^2 \times 5^2 = 5^{(2+2+2)} = 5^6$
The indices are multiplied: $(5^2)^3 = 5^{2\times3} = 5^6$

- **In general, when finding a 'power of a power', multiply the indices: $(x^m)^n = x^{mn}$.**

You can use these index laws to calculate quantities including powers.

Notice that $8 \times 8^{-1} = 1$. What about 6×6^{-1}? Or 50×50^{-1}?

Example

Evaluate these expressions.

a $16^{\frac{1}{2}}$ **b** 8^{-1} **c** $(3^3)^2$

a $16^{\frac{1}{2}} = \sqrt{16} = 4$ **b** $8^{-1} = \frac{1}{8}$ **c** $(3^3)^2 = 27^2 = 729$

Example

Write these expressions as powers of the numbers indicated.

a 2 as a power of 4 **b** 0.125 as a power of 8 **c** $\frac{1}{16}$ as a power of 2

a $2 = \sqrt{4} = 4^{\frac{1}{2}}$ **b** $0.125 = \frac{125}{1000} = \frac{1}{8} = 8^{-1}$ **c** $\frac{1}{16} = \frac{1}{2^4} = 2^{-4}$

Example

Evaluate these expressions.

a 10^{-4} **b** $2^{-\frac{1}{2}}$ **c** $100^{-\frac{1}{2}}$

a $10^{-4} = \frac{1}{10\,000} = 0.0001$

b $2^{-\frac{1}{2}} = \frac{1}{2^{\frac{1}{2}}} = \frac{1}{\sqrt{2}} = 0.707$ (by calculator)

c $100^{-\frac{1}{2}} = \frac{1}{\sqrt{100}} = \frac{1}{10} = 0.1$

Exercise N5.3

1 Find the value of each expression.

a 5^1 **b** 6^1 **c** 6^0 **d** 7^0
e $(4 + 88^2)^0$ **f** $(4^2 + 5^2)^1$ **g** $(92.5)^0$ **h** 0^1

2 Find the value of each expression.

a $100^{\frac{1}{2}}$ **b** $16^{0.5}$ **c** $49^{\frac{1}{2}}$ **d** $4^{0.5}$
e $64^{\frac{1}{2}}$ **f** $9^{0.5}$ **g** $121^{\frac{1}{2}}$ **h** $64^{0.5}$
i $144^{\frac{1}{2}}$ **j** $8^{\frac{1}{3}}$ **k** $27^{\frac{1}{3}}$ **l** $100^{0.5}$

3 Evaluate.

a $36^{\frac{1}{2}}$ **b** 36^1 **c** $81^{0.5}$ **d** 81^1

4 Write these in index form. For example, $\frac{1}{2} = 2^{-1}$

a $\frac{1}{3}$ **b** $\frac{1}{5}$ **c** $\frac{1}{7}$ **d** $\frac{1}{11}$
e 0.5 **f** 0.2 **g** 0.1 **h** $0.\dot{3}$

Remember that x^{-1} is the reciprocal of x, and vice versa.

5 Evaluate.

a 9^{-1} **b** 9^0 **c** $9^{0.5}$ **d** 9^1
e 9^2 **f** 9^3 **g** 9^5 **h** 9^{-3}

6 Write these in index form. For example, $\frac{1}{5^2} = 5^{-2}$

a $\frac{1}{7^2}$ **b** $\frac{1}{9^2}$ **c** $\frac{1}{2^2}$ **d** $\frac{1}{2^3}$
e $\frac{1}{2^5}$ **f** $\frac{1}{3^4}$ **g** $\frac{1}{5^3}$ **h** $\frac{1}{6^4}$

7 Write these in fraction form. For example, $7^{-3} = \frac{1}{7^3}$

a 8^{-2} **b** 7^{-3} **c** 5^{-2} **d** 9^{-4}
e 3^{-2} **f** 9^{-3} **g** 4^{-5} **h** 6^{-6}

8 Evaluate.

a 4^{-2} **b** 4^{-1} **c** 4^0 **d** $4^{0.5}$
e 4^1 **f** 4^2 **g** 4^3 **h** $4^{-0.5}$

9 Write these in index form. For example, $\frac{1}{\sqrt{2}} = \frac{1}{2^{\frac{1}{2}}} = 2^{-\frac{1}{2}}$

a $\frac{1}{\sqrt{3}}$ **b** $\frac{1}{\sqrt{5}}$ **c** $\frac{1}{\sqrt{7}}$ **d** $\frac{1}{\sqrt{11}}$

10 Simplify these expressions. For example, $(5^2)^3 = 5^{2\times3} = 5^6$

a $(2^2)^2$ **b** $(2^3)^2$ **c** $(3^2)^3$ **d** $(4^{0.5})^2$
e $(5^2)^4$ **f** $(4^{-2})^3$ **g** $(7^2)^6$ **h** $(5^2)^{-2}$

11 Calculate these, giving your answers in index form.

a $2^2 \div 2^4$ **b** $3^5 \div 3^6$ **c** $4^3 \div 4^9$
d $(3^4 \times 5^5) \div (3^5 \times 5^6)$ **e** $[(5^4 \times 7^3) \div (5^2 \times 7^5)]^2$

N5.4 Standard index form for large numbers

Objectives

- Understand and use standard form in calculations with large and small numbers
- Use calculators to calculate in standard form

Useful resources

- Place value table

Mental starter

Every morning I go for a 4 km run. The first time I ran it took me 25 minutes. I found that each time I ran I improved by 10%. **How fast did I run on the third occasion?**
(20 min 15 sec = 1215 sec. You might give the hint to convert 25 min into seconds.) **Why will this improvement not continue?**
(Eventually you will get run 4 km in an impossibly fast time.) Actually after 70 runs and improving by 10% each time you would run 4 km in under 1 sec!

Introductory activity

We use standard form as a convenient way of writing large numbers. Standard form numbers are always written by using 2 numbers multiplied together. The first is between 1 and 10 not including 10. The second is a power of 10. e.g. $450\ 000 = 4.5 \times 10^5$ (don't count the zeros, always count the places you move the decimal point).

3.24×10^4 means multiply 3.24 by 10, four times, move the decimal point 4 places (32 400).

Write this standard form number as an ordinary number 6.05×10^7. (60 500 000.)

Write 500 000 and 73 000 in standard form. Remind students that they must start with a number between 1 and 10, and multiply by 10 to a power. To find the power count the places from where you put the decimal point.
(5×10^5 and 7.3×10^4.)

If a number is not in standard form, for example 45×10^7, you can change it by making 45 into standard form and then using the laws of indices.
$45 \times 10^7 = 4.5 \times 10 \times 10^7 = 4.5 \times 10^8$.

Simplify $(2 \times 10^3) \times (4 \times 10^6)$. This is two number multiplied together, but you can think of it as four numbers. Work out $2 \times 4 = 8$ first. Then $10^3 \times 10^6 = 10^9$. Final answer is then 8×10^9.

Division is similar. Simplify
$(3.6 \times 10^9) \div (9 \times 10^6) = (3.6 \div 9) \times (10^9 \div 10^6)$
$= 0.4 \times 10^3$
$= 0.4 \times 10 \times 10^2$
$= 4 \times 10^2$.

Discuss the example in the student book, particularly the last example which applies standard form to astronomical data.

Exercise commentary and misconceptions

In question 7 allow students to use calculators to multiply or divide the numbers between 1 and 10, not the entire standard form numbers, so question 7**a** is $(2.5 \times 10^5) \times (3.9 \times 10^4)$
$= (2.5 \times 3.9) \times (10^5 \times 10^4)$
$= 9.75 \times 10^9$.
(**Note**: Using calculators to calculate with numbers in standard form is covered in the next spread.)

Plenary

There are approximately 1×10^{79} atoms in the universe. The number of atoms in the earth is approximately 1×10^{50}. **How many earths make up the universe?**
$1 \times 10^{79} \div 1 \times 10^{50} = 1 \times 10^{29} =$ 100 000 000 00 0 000 000 000 000 000 000 = hundred thousand trillion trillion.

N5.4 Standard index form for large numbers

This spread will show you how to:
- Understand and use standard form in calculations with large and small numbers
- Use calculators to calculate in standard form

Keywords
Standard form

You can use **standard form** to represent large numbers.

- **In standard form, a number is written as $A \times 10^n$.**
  - **A is a number between 1 and 10 (but not including 10). Using algebra, $1 \leq A < 10$.**
  - **The value of n is an integer.**

  For example, $856 = 8.56 \times 10^2$ and $43\,994 = 4.3994 \times 10^4$.

13×10^5 is *not* in standard form, because 13 is larger than 10.

The correct version is 1.3×10^6

0.75×10^4 is *not* in standard form, because 0.75 is less than 1.

The correct version is 7.5×10^3

You can calculate with numbers in standard form.

- Multiplication works like this:
  $(3 \times 10^5) \times (4 \times 10^3) = (3 \times 4) \times 10^{(5+3)} = 12 \times 10^8 = 1.2 \times 10^9$

Multiplication – add the indices

- Division works like this:
  $(1.4 \times 10^8) \div (7 \times 10^5) = (1.4 \div 7) \times 10^{(8-5)} = 0.2 \times 10^3 = 2 \times 10^2$

Division – subtract the indices

Example

Write these numbers in standard form.

a 235 **b** 12 492 **c** 15×10^4 **d** 0.23×10^6

a $235 = 2.35 \times 10^2$ **b** $12\,492 = 1.2492 \times 10^4$
c $15 \times 10^4 = 1.5 \times 10^5$ **d** $0.23 \times 10^6 = 2.3 \times 10^5$

Example

Calculate

a $(4.2 \times 10^3) \times (2 \times 10^2)$ **b** $(3.6 \times 10^5) \div (1.2 \times 10^3)$ **c** $(5.4 \times 10^4) \times (2 \times 10^3)$

a $(4.2 \times 10^3) \times (2 \times 10^2) = (4.2 \times 2) \times (10^3 \times 10^2)$
$= 8.4 \times 10^{(3+2)} = 8.4 \times 10^5$

b $(3.6 \times 10^5) \div (1.2 \times 10^3) = (3.6 \div 1.2) \times (10^5 \div 10^3)$
$= 3 \times 10^{(5-3)} = 3 \times 10^2$

c $(5.4 \times 10^4) \times (2 \times 10^3) = (5.4 \times 2) \times (10^4 \times 10^3)$
$= 10.8 \times 10^7 = 1.08 \times 10^8$

Example

The Andromeda Galaxy has a radius of about 1 040 700 000 000 000 000 km. Write this in standard form.

1 040 700 000 000 000 000 = 1.0407×10^{18}

Exercise N5.4

1 Write these numbers as powers of 10.

a 100 **b** 10 **c** 100 000 **d** 1

2 Write these numbers in standard form.

a 200 **b** 800 **c** 9000 **d** 650
e 6500 **f** 952 **g** 23.58 **h** 255.85

3 These numbers are in standard form. Write each of them as an 'ordinary' number.

a 5×10^2 **b** 3×10^3 **c** 1×10^5 **d** 2.5×10^2
e 4.9×10^3 **f** 3.8×10^6 **g** 7.5×10^{11} **h** 8.1×10^{18}

4 Although they are written as multiples of powers of 10, these numbers are not in standard form. Rewrite each of them correctly in standard form.

a 60×10^1 **b** 45×10^3 **c** 0.65×10^1 **d** 0.05×10^8

5 Work out these calculations, giving your answers in standard form. Do not use a calculator.

a $(2 \times 10^2) \times (2 \times 10^3)$ **b** $(3 \times 10^4) \times (3 \times 10^3)$
c $(5 \times 10^3) \times (5 \times 10^4)$ **d** $(8 \times 10^7) \times (3 \times 10^5)$

6 Evaluate these, showing your working.
Do not use a calculator; give your answers in standard form.

a $(4 \times 10^4) \div (2 \times 10^2)$ **b** $(8.4 \times 10^9) \div (4.2 \times 10^5)$
c $(2 \times 10^6) \div (4 \times 10^4)$ **d** $(3 \times 10^5) \div (4 \times 10^2)$

7 Use a calculator to find these.
Give your answers in standard form, to 3 significant figures.

a $(2.5 \times 10^5) \times (3.9 \times 10^4)$ **b** $(4.1 \times 10^6) \div (3 \times 10^2)$
c $(4.95 \times 10^3) \times (8.11 \times 10^7)$ **d** $(3.7 \times 10^{11}) \div (1.8 \times 10^3)$

8 The speed of light is approximately 3×10^8 metres per second. Copy and complete the table to show the time taken for light from the Sun to reach the various planets.

Planet	Mean distance from Sun (m)	Light travel time
Mercury	5.79×10^{10}	
Earth	1.50×10^{11}	
Mars	2.28×10^{11}	
Jupiter	7.78×10^{11}	
Pluto	5.90×10^{12}	

N5.5 Standard form for small numbers

Objectives

- Understand and use standard form in calculations with large and small numbers
- Use calculators to calculate in standard form

Useful resources

- Place value table
- Calculators

Mental starter

Challenge students to use their knowledge of place value to multiply and divide with powers of 10, for example:

8.243×1000	82.1×0.01
0.43×0.1	$0.0487 \times 10\,000$

Replace the × signs with ÷ . Discuss responses.

Introductory activity

You can also use standard form as a convenient way of writing very small numbers. For this, standard form numbers are still written by using two numbers multiplied together. The first is between 1 and 10 not including 10. The second is a power of 10.

For example: $0.045 = 4.5 \times 10^{-2}$.

3.24×10^{-4} means divide 3.24 by 10, four times, move the decimal point four places (0.000 324).

If a number is not in standard form, for example:
$45 \times 10^{-5} = 4.5 \times 10 \times 10^{-5} = 4.5 \times 10^{-4}$.

Write 0.0008 in standard form.
(To find the power count the places you move the decimal point: 8×10^{-4}.)

Simplify $2 \times 10^{-6} \times 3 \times 10^{4}$.
Work out $2 \times 3 = 6$ first.
Then $10^{-6} \times 10^{4} = 10^{-2}$.
Final answer is then 6×10^{-2}.

You can use a calculator to deal with standard form.

Simplify $\frac{4.08 \times 10^{-12}}{1.7 \times 10^{7}}$. There is a key for standard form, either [EXP] or [EE] .

([EXP] stands for 'exponent', meaning the power you want to use. Key in
4.08 [EXP] (–) 12 ÷ 1.7
[EXP] 7 = [EXP]. Answer should be 2.4×10^{-19}.
There are many ways of displaying standard form on calculator screens, so check that students are familiar with their own calculators. Some students will have been taught to use the power key, try to persuade them to use the designated key!

Exercise commentary and misconceptions

Most scientific calculators distinguish between the '–' key used for operations and the (–) or +/– key, so make sure students use the correct one.

In question 3 encourage students to write 'one hundredth' as $1 \times \frac{1}{100} = 1 \times \frac{1}{10^2} = 1 \times 10^{-2}$.

Question 6 is on addition and subtraction of numbers in standard form, and all but the strongest students will prefer to change the numbers into normal form before adding or subtracting.

Plenary

The world population is approximately 6×10^{9}.
The diameter of 1 atom is approximately 1×10^{-8} cm.
If every person in the world took 1 atom each and laid them in a line, how long would the line be? (Answer 6×10^{1} cm or 60 cm.)

N5.5 Standard form for small numbers

This spread will show you how to:

- Understand and use standard form in calculations with large and small numbers
- Use calculators to calculate in standard form

Keywords
Standard form

It is often useful to write small numbers, such as 0.000415, in standard form.

- Negative powers of 10, such as 10^{-4}, represent small numbers.

  For example, $10^{-4} = \frac{1}{10^4} = \frac{1}{10\,000} = 0.0001$

- You can write any small number in standard form.

  For example, $0.00312 = 3.12 \times 0.001 = 3.12 \times 10^{-3}$

- You can calculate with small numbers expressed in standard form.

  For example, $(4.25 \times 10^{-3}) \div (3.75 \times 10^{4}) = (4.25 \div 3.75) \times 10^{(-3-4)} = 1.13 \times 10^{-7}$

- You can obtain a small number as a result of a calculation involving large numbers.

  For example, $(3 \times 10^{5}) \div (4 \times 10^{8}) = 0.75 \times 10^{-3} = 7.5 \times 10^{-4}$

You can also work with numbers in standard form on a scientific calculator. The button for entering the power of 10 is often marked EXP or EE; so for 4.5×10^{-3}, you enter

[4] [·] [5] [EXP] [(−)] [3]

Example

Write these numbers in standard form.

a 0.003 **b** 0.000 000 416 **c** 0.45

a $0.003 = 3 \times 10^{-3}$
b $0.000\,000\,416 = 4.16 \times 10^{-7}$
c $0.45 = 4.5 \times 10^{-1}$

Example

Calculate. **a** $(4.8 \times 10^{4}) \times (3.6 \times 10^{-5})$ **b** $(4.5 \times 10^{3}) \div (9.7 \times 10^{8})$

These can be done directly, using a scientific calculator.
The answers, to 3 significant figures, are **a** 1.73 **b** 4.64×10^{-6}

Example

What is wrong with this calculation? $10^{4} \div 10^{-5} = 10^{-1}$

Be careful with the signs of the indices in examples like this. Because you are dividing, you need to subtract the indices. The correct power of 10 for the answer is $4 - (^{-}5) = 4 + 5 = 9$, so $10^{4} \div 10^{-5} = 10^{9}$

Exercise N5.5

1 Write these numbers in standard form.

a 0.3 **b** 0.0047 **c** 0.000 078 **d** 0.4485

2 Although these numbers are written as multiples of powers of 10, they are not in standard form. Write each of the numbers correctly in standard form.

a 28×10^{-2} **b** 0.4×10^{-1} **c** 13.5×10^{-4} **d** 12×10^{-8}

3 Write these measurements using standard form.

a One hundredth of a kilometre **b** Two thousandths of a gram
c Five millionths of a metre **d** 11 thousandths of a litre

4 Work out these calculations without a calculator.
Give your answers in standard form.

a $(2.5 \times 10^{-3}) \times (2 \times 10^{2})$ **b** $(4.6 \times 10^{-6}) \times (2 \times 10^{-2})$
c $(4 \times 10^{4}) \div (2 \times 10^{6})$ **d** $(8.4 \times 10^{-2}) \div (2 \times 10^{6})$

5 Use a scientific calculator to work out these. Give your answers in standard form, and to three significant figures.

a $(3.7 \times 10^{-4}) \times (3.1 \times 10^{-4})$ **b** $(5.3 \times 10^{5}) \div (2.9 \times 10^{8})$
c $(3.18 \times 10^{2}) \div (6.55 \times 10^{7})$ **d** $(1.79 \times 10^{5}) \times (2.8 \times 10^{-6})$

6 Work out these calculations without using a calculator, giving your answers in standard form.

a $(5 \times 10^{-1}) + (2 \times 10^{-2})$ **b** $(4 \times 10^{-2}) + (6 \times 10^{-3})$
c $(2 \times 10^{-2}) + (9 \times 10^{-4})$ **d** $(1.5 \times 10^{-2}) - (2 \times 10^{-3})$

You may find it easier to convert the numbers to 'ordinary' numbers first, and then convert the answers back to standard form.

7 Use a scientific calculator to check your answers to question 6.

8 Use a scientific calculator to find the volume of a cube of side length 4.5×10^{-3} metres. Give your answer in m^3, to 3 sf, in standard form.

9 A pack of 500 sheets of A4 paper weighs 2.65 kg. Find the mass in kg of a single sheet of paper, giving your answer in standard form.

10 A sheet of gold leaf is one ten thousandth of a millimetre thick.
The diameter of an atom of gold is about 0.26 nanometres.
(One nanometre is 10^{-9} metres.)
Approximately how many atoms thick is the sheet of gold leaf?
Show your working.

N5 Exam review

Key objectives

- Use index laws to simplify and calculate the value of numerical expressions involving multiplication and division of fractional and negative powers
- Use standard index form expressed in conventional notation and on a calculator display

Worked solution	Commentary and misconceptions
1	Ensure that students understand index notation and index laws before attempting this question. Some students may need reminding of the rules of manipulating powers.
a $3^{-1} = \frac{1}{3}$	
b $(\sqrt{3})^2 = (3^{\frac{1}{2}})^2 = (3^{\frac{1}{2} \times 2}) = 3$	Encourage students to manipulate powers in such calculations instead of automatically using their calculator. Ensure that students know when to add, subtract or multiply powers.
c $3^{-1} \times (\sqrt{3})^2 = \frac{1}{3} \times 3 = 1$	Ensure that students use their results from parts **a** and **b** to answer this question. Highlight the wording of this and similar questions; 'hence...', 'use your answers to...'.
2	Ensure that students understand standard index form before attempting this question.
a $0.000\ 000\ 001 = 1 \times 10^{-9}$	Some students may get confused by the sign of the power. Remind them that a negative power gives a small number (less than 1) and a positive power gives a large number (greater than 1).
b $5 \times 1 \times 10^{-9} = 5 \times 10^{-9}$ $1 \div (5 \times 10^{-9}) = 200\ 000\ 000$ $= 2 \times 10^{8}$	Ensure that students understand what they are required to work out. Encourage them to break such calculations into steps. Ensure that students give their solutions in standard form.

A5

Formulae

Objectives

F/H Know the meaning of and use the words 'equation', 'formula', 'identity' and 'expression'

H Use formulae from mathematics and other subjects

F/H Substitute numbers into a formula

H Use π in exact calculations

F/H Present answers to sensible levels of accuracy

H Generate a formula

H Change the subject of a formula including cases where the subject occurs twice and where a power of the subject appears

H Understand the difference between a practical demonstration and a proof

H Recognise the significance of stating constraints and assumptions when deducing results

Unit overview

This unit consolidates the knowledge of algebra from units A1 and A2, extending to writing and rearranging formulae and then an introduction to proof.

Prior knowledge

Before your students start this unit they should be able to:

- Use formulae from mathematics and other subjects
- Expand and factorise single and double brackets
- Solve simple linear and quadratic equations
- Understand and use the language of algebra
- Manipulate algebraic expressions using the correct order of operations
- Round answers to a suitable degree of accuracy
- Understand and use fractions, powers and surds in algebraic expressions
- Add, subtract, multiply and divide with integers and fractions

Differentiation

- **Foundation** begins with writing formulae in words and then proceeds to using letters and symbols and substituting numbers into simple formulae
- **Foundation Plus** extends to deriving and rearranging formulae using symbols and then gives a basic introduction to proof
- **Higher Plus** extends the Higher book to looking at more complex formulae and proof

A5.1 Identities, formulae and equations

Objectives

- Understand the difference between identities, formulae and equations
- Use formulae from mathematics and other subjects

Useful resources

- OHT of grid in question 1

Mental starter

If five people take six days to build one kit car, how many complete kit cars can be built in one day if you have 100 people available? (3) Could be set out as a ratio and days reduced to 1, then the people can be changed to 100; or calculate how many 'people days' are needed for one car.

Introductory activity

Can you solve $4x + 2 = 2(2x + 1)$? What values work for x? (All values.) **Why is this?** (Because each side is identical, the right-hand side becomes $4x + 2$ if you remove the brackets.) This is called an identity and you use 3 lines '≡', so you write $4x + 2 \equiv 2(2x + 1)$. '≡' is read as 'is identical to'.

Can you solve $A = \pi r^2$? (No, because you need a value for the radius r to find the area A of the circle.) This is not an equation where you can find an answer, it is a formula that works out something when you know the value of the other things in the formula.

Can you solve $2x + 1 = 5$? (Yes, $x = 2$.) So you can have identities, formulae and equations.

When you put values into a formula you must make sure you work it out in the correct order. Remember what BIDMAS stands for: 'Brackets' 'Indices' 'Divide and Multiply' 'Add and Subtract'. Emphasise that divide and multiply have the same degree of priority as each other, as do add and subtract.

For example, $P = 4(x - 2y^3)$.
Find the value of P if $x = 100$ and $y = 5$.

$P = 4(100 - 2 \times 5^2)$
$= 4(100 - 50)$
$= 4 \times 50 = 200$.

Discuss the examples in the student book.

Exercise commentary and misconceptions

Remind students to multiply out to remove brackets in question 1.

In question 3 encourage students to write the formula, put in the values and work the solution out in stages using the rules of BIDMAS. Always work down the page. Try to line up the working from the line above.

In question 4b, non-integer values are permissible, but not negative values as they are lengths of sides.

Plenary

Sketch a large cube with a smaller cube in the exact middle of it. Join the vertices of the larger cube to the matching (corresponding) vertices of the smaller cube. You will not be able to sketch all the edges. If the sides of the large cube are 4 m and the sides of the small cube are 2 m then the lines joining the vertices are $\sqrt{3}$ m.

Write an expression for the total length of all the lines.
($4 \times 12 + 2 \times 12 + 8 \times \sqrt{3} = 72 + 8\sqrt{3}$ m.)

This is a very odd shape, in fact it is a shadow of a 4-D cube! 4-D cubes cannot be physically made or seen in our universe, but they exist in the mind of a computer (or a human) who can think in higher dimensions than 3.

A5.1 Identities, formulae and equations

This spread will show you how to:
- Understand the difference between identities, formulae and equations
- Use formulae from mathematics and other subjects

Keywords
Equation
Formula(e)
Identity
Substitute

- An **identity** is true for all values of x. For example

  $x(x+1) \equiv x^2 + x$ Whatever value of x you try, this statement is always true.

$\equiv$ means 'is identical to'

- An **equation** is only true for a limited number of values of x. For example

  $2x + 1 = 5$ is only true when $x = 2$.

- A **formula** describes the relationship between two or more variables. For example the formula for the area of a triangle is

  $A = \frac{1}{2}bh$

- You can **substitute** numbers into a formula to work out the value of a variable. For example

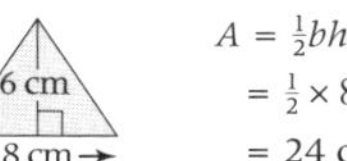

$A = \frac{1}{2}bh$
$= \frac{1}{2} \times 8 \times 6$
$= 24\text{ cm}^2$

Example

Decide if each of these statements is an identity, an equation or a formula.

a $x^3 - 2x = x(x^2 - 2)$ **b** $5x - 1 = 2x + 3$ **c** $A = \pi r^2$

a Expand the right-hand side: $x(x^2 - 2) = x^3 - 2x$
$x^3 - 2x =$ left-hand side so the statement is an identity.

b $5x - 3 = 2x + 3$
$3x - 3 = 3$ This statement is only true for one value of x.
$3x = 6$
$x = 2$
$5x - 3 = 2x + 3$ is an equation.

c $A = \pi r^2$ is a formula showing the relationship between the radius and the area of a circle.

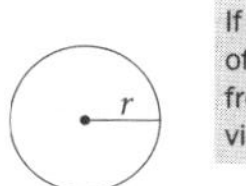

If you know the value of r, you can find A from the formula and vice versa.

Example

Use the formula $V = \pi r^2 h$ to find V, the volume of a can of beans, when $h = 11$ cm and $r = 7.5$ cm.

$V = \pi r^2 h$ — Write the formula.
$= \pi \times 7.5^2 \times 11$ — Substitute the values of r and h.
$= 1943.860\text{ cm}^3$
$= 1944\text{ cm}^3$ to 2 sf — Give the answer to 2 sf to match the accuracy of the data in the question.

Exercise A5.1

1 Copy these statements and say whether they are identities, equations or formulae.

a $c = 2\pi r$	**b** $3x(x+1) = 3x^2 + 3x$	**c** $3x + 1 = 10$
d $y \times y = y^2$	**e** $2x + 5 = 3 - 7x$	**f** $A = \frac{1}{2}(a+b)h$
g $a^2 + b^2 = c^2$	**h** $20 - x = -(x - 20)$	**i** $2x^2 = 50$

2 A campsite charges for the pitching of a tent and the number of people, p, that stay in it. If C is the total cost in pounds, the formula used by the campsite is $C = 2p + 5$

a Work out the cost of pitching a tent for 6 people.

b If the cost is £23, how many people are sleeping in the tent?

c Explain why the cost could never be £26.

3 Use these formulae to work out the required information.

a $F = \frac{9}{5}C + 32$. Find 12° Celsius in degrees Fahrenheit

b $P = \frac{1}{4}t - 8$. Find P when $t = 32$

c $A = \frac{bh}{2}$. Find h when $A = 10$ and $b = 5$

d $W = 3d^2$. Find W when d is 6 and d when W is 75

e $C = 2a - b$. Find C when a is -8 and b is -2

f $P + 2r = K$. Find K when P is $\frac{3}{4}$ and r is $\frac{1}{8}$

4 **a** The formula for the area of a trapezium is $A = \frac{1}{2}(a+b)h$, where a and b represent the lengths of the parallel sides and h is the perpendicular height.
Find the area of this trapezium.

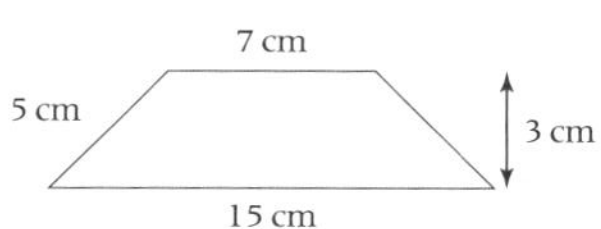

b If another trapezium with the same perpendicular height has area 15 cm^2, suggest as many possibilities for the lengths of the parallel sides as you can.

A5.2 Writing formulae

Objectives
- Write a formula from given information

Useful resources
- OHT of the shapes in the introductory activity

Mental starter
Arrange 16 dots in a square. You can move between the dots either vertically or horizontally. **What is the minimum number of moves to move between dots at opposite corners?** (6) Do the same for 25 and 36 dots. (8 and 10.) **Can you spot a pattern?** (Going up by 2 each time **or** (number of dots on the side of the square − 1) × 2 **or** $2(\sqrt{n} - 1)$ for n dots.)

Introductory activity
Sketch these shapes, label the sides and ask for the formula in each case.
What is the formula for the perimeter, p, of an equilateral triangle of side x? ($p = 3x$.)

What is the formula for the area, A, of a right angled triangle with sides a, b and c if a is the hypotenuse? ($A = \frac{bc}{2}$.)

What is the formula for the area, A, of a circle with a radius of r? ($A = \pi r^2$.)

What is the formula for the circumference, c, of a circle with a diameter of d? ($c = \pi d$.)

What is the formula for the volume, V, of a cuboid with sides x, $2x$ and $3x$? ($V = 6x^3$.)

What is the formula for the volume, P of a prism with a cross sectional area of A and a length of L? ($P = AL$.)

Exercise commentary and misconceptions
The formulae in question 1 are all derived from standard formulae. Students may need convincing that the 'L' shape is a hexagon!

The idea of **unwrapping** a cylinder to produce the rectangle that forms the curved surface should be looked at again for question 3**b**, previously covered in S1.5.

Plenary
A rugby ball roughly forms a shape called a prolate ellipsoid. The formula for its volume V when you measure its length 'a' and its width 'b' is given by $V = \frac{\pi ab^2}{6}$.

Estimate the volume of a rugby ball.
Students will have to use an estimate of what they think the length and width are. Note that the height of the ball when lying flat is roughly the same as the width. Length = 37 cm, width = 18.1 cm (size 5 adult size).
Volume = 6346.83921 = 6350 cm^3 (3 sf).
Estimates between 25 000 and 3000 should be reached by anybody, even if they do not have first hand experience of handling a rugby ball.

A5.2 Writing formulae

This spread will show you how to:

- Write a formula from given information

Keywords
Derive
Formula

- **You can derive a formula from information you are given.**

Example

Derive a formula for the perimeter of this pentagon.

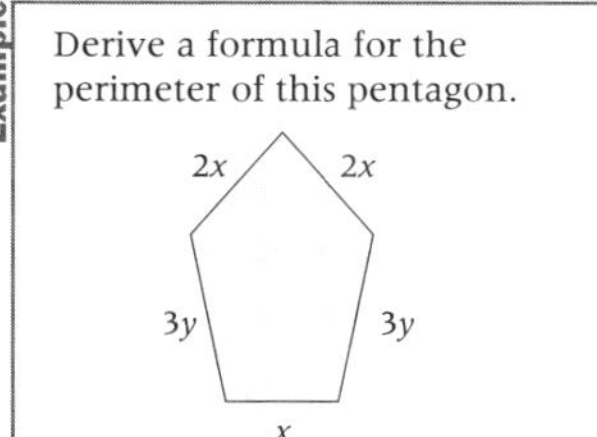

Let P represent the perimeter.
$P = 2x + 2x + 3y + x + 3y$
$P = 2x + 2x + 3y + 3y + x$
$P = 5x + 6y$

Perimeter is the distance all of the way around a shape.

Write formulae as simply as possible, using the rules of algebra.
For example

Write $S = \frac{D}{T}$, not $S = D \div T$ and $A = lw$, not $A = l \times w$

Example

a Write a formula to show your total amount of pocket money P (£s), if you receive £3 per month with an extra £2 for every job (j) you do at home.

b Write a formula for the total area of this shape. Let the total area be A.

2y
z
6y

a If I don't do any jobs, I get £3.
If I do 4 jobs I get £3 plus 4 × £2, which is £11.
If I do 10 jobs I get £3 plus 10 × £2, which is £23.
So, if I do j jobs I get £3 plus j × £2.
Hence
$P = 3 + 2j$

Try the situation with various numbers, before putting it into algebra.

Write $2j$ not $j \times 2$

b A = area of a rectangle plus the area of a triangle
$= (z \times 2y) + \frac{1}{2}((6y - 2y) \times z)$
$= 2yz + \frac{1}{2}(4y \times z)$
$= 2yz + \frac{1}{2}(4yz)$
$= 2yz + 2yz$
$= 4yz$

The area of a rectangle is $A = lw$ and the area of a triangle is $A = \frac{1}{2}bh$

Exercise A5.2

1 Write your own formula to represent these quantities.

a The perimeter, P, of a square.

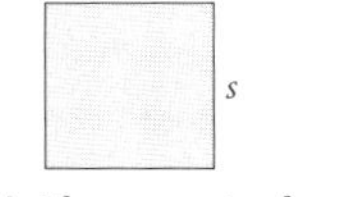

b The area, A, of a semicircle.

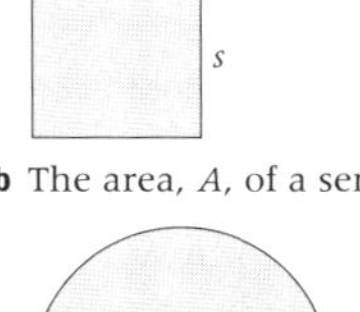

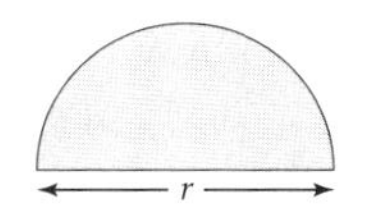

c The perimeter, P, of this hexagon.

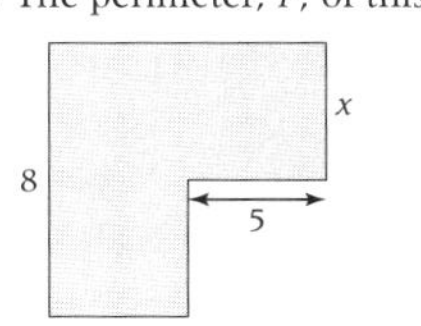

d The time in minutes, T, to complete my homework if I take 10 minutes to get my books organised and then 35 minutes to do each piece, p.

2 Megan is writing a formula to work out the volume of this prism.
Her formula is

$$V = \frac{abc}{2}$$

Explain what a, b and c stand for.
Explain why Megan's formula is correct.

3 This cylinder has base radius r and height h.

a Write a formula for V, the volume of the cylinder.

b Nicholas has written this formula for the surface area, A, of the cylinder.

$$A = 2\pi r^2 + 2\pi rh$$

Explain what the $2\pi r^2$ part represents. Repeat for the $2\pi rh$ term.

Remember a cylinder is a prism. Volume = area of cross section × height.

A5.3 Rearranging formulae

Objectives

- Rearrange a formula in order to change its subject

Useful resources

- Mini-whiteboards for mental starter

Mental starter

All regular hexagons are made from 6 congruent equilateral triangles. If the perimeter of each triangle is x cm, write an expression for the perimeter of the regular hexagon in its simplest form. ($2x$ cm.)

Introductory activity

To rearrange a formulae you need to isolate the new subject on one side of the equals sign. You can do this in the same way as solving equations, that is by using inverse operations.

In these examples, make x the subject of the formula. Each example has something different in it. Use the same methods as solving equations used in sections 2.2 to 2.4.

(i) $3x + y = a$.
Subtract y from a then divide by 3.
$x = \frac{a-y}{3}$.

(ii) $a(2 - x) = 5$.
Divide 5 by a then add x to both sides.
Finally subtract $\frac{5}{a}$.
$x = 2 - \frac{5}{a}$.

(iii) $\frac{x^2}{4} = ab$. Multiply ab by 4 then take the square root.
$x = \sqrt{4ab}$.

(iv) $\sqrt{x} - b = p$. Add b to p then square.
$x = (p + b)^2$.

Exercise commentary and misconceptions

In question 1**g**, AB can be treated as a single symbol.

For questions 3 and 4 remind students that cube and cube root will undo each other as will square and square root.

Question 6 is an unusual formula for changing degrees Celsius to degrees Fahrenheit – students may be interested to compare it with a more familiar version ($F = \frac{9}{5}C + 32$).

Plenary

T is the time it takes a child on a swing to go forward then back once. L is the length of the swing's chain. T and L are linked together by the formula $T = 2\pi\sqrt{\frac{L}{10}}$.

If $T = 3$ sec, how long is the chain?

Rearrange to
$(\frac{T}{2\pi})^2 \times 10 = L$
$\rightarrow L = 2.28$ m.

A5.3 Rearranging formulae

This spread will show you how to:

- Rearrange a formula in order to change its subject

Keywords
Rearrange
Subject

- The **subject** of a formula is the variable before the equals sign.
  For example, in $A = \pi r^2$, A is the subject of the formula.
- You can **rearrange** a formula in order to change its subject.

Example

Rearrange the formula $A = \pi r^2$ to make r the subject.

$A = \pi r^2$ Divide both sides by π.

$\frac{A}{\pi} = r^2$ Square root both sides.

$\sqrt{\frac{A}{\pi}} = r$ r is now the subject.

$r = \sqrt{\frac{A}{\pi}}$ Put r on the left-hand side.

Start by writing the formula. Use inverse operations to get r on its own.

Example

Make h the subject of the formula $A = \frac{1}{2}bh$.

$A = \frac{1}{2}bh$ Clear this fraction by multiplying both sides by 2.

$2A = bh$ Divide both sides by b.

$\frac{2A}{b} = h \rightarrow h = \frac{2A}{b}$

Example

Make x the subject of these formulae

a $V = u + bx$ **b** $M = axy - c^2$

a $V = u + bx$ Subtract u from both sides.

$V - u = bx$ Divide both sides by b.

$\frac{V-u}{b} = x$

$x = \frac{V-u}{b}$

b $M = axy - c^2$ Add c^2 to both sides.

$M + c^2 = axy$

$\frac{M+c^2}{ay} = x$ Divide both sides by ay.

$x = \frac{M+c^2}{ay}$

The function $x = \frac{k}{y}$ is called the **reciprocal** function.

Exercise A5.3

1 Make x the subject of each formula.

a $C = ax + b$ **b** $M = x - b - c$ **c** $K = \frac{x}{t} - q$ **d** $W = t + xy$

e $H = \frac{x+z}{p}$ **f** $D = p(x - q)$ **g** $AB = x - ct$ **h** $Y = mx + c$

2 James and Sebastian are rearranging the formula $C = a(x - b)$ in order to make x the subject. They both come up with solutions that look different but are, in fact, correct. Can you explain why?

James' solution: $\frac{C}{a} + b = x$

Sebastian's solution: $\frac{C + ab}{a} = x$

3 Make y the subject of each formula.

a $c = y^2$ **b** $k = \frac{1}{4}y - 2$ **c** $M = xyz + t$ **d** $2x = \sqrt{y}$

e $p = y^3 + 2$ **f** $T = ky^2$ **g** $R = \frac{1}{3}ayz$ **h** $^3\sqrt{y} = p$

4 Richard has made a mistake with his rearranging whilst trying to make p the subject of this formula. Copy his working and explain where he has gone wrong.

$K = mp^3$

$\sqrt[3]{k} = mp$

$\frac{\sqrt[3]{k}}{m} = p$

5 These are the stages in changing the subject of the formula $c = \frac{8(D+k)}{ab}$.

Put them in order.

$D + k = \frac{1}{8}abc$	$\frac{8(D+k)}{ab} = c$	$D = \frac{1}{8}abc - k$	$8(D + k) = abc$

6 A formula to change from degrees Celsius to degrees Fahrenheit is

$$F = \frac{9(C + 40)}{5} - 40$$

a Use this to change 30°C into °F.

b Rearrange to make C the subject of the formula.

c Use your new formula to find the Celsius equivalent of −32°F.

A5.4 Rearranging harder formulae

Objectives

- Rearrange a formula in order to change its subject

Useful resources

- Digit cards for mental starter

Mental starter

Practice substituting values into a formula. For example,

$$y = \frac{\sqrt{x} - 2}{5}$$

Find y when $x = 1, 2, 4, 9, 19$.
Find x when $y = 2, 3, -1, -3$.

Introductory activity

Challenge students to rearrange more complex formulae including ones that have the subject at the bottom of the fraction (denominator).

Write x in terms of the other letters.

(i) $\frac{a}{x} = 4$

Multiply by x then divide by 4

$x = \frac{a}{4}$

(ii) $v^2 = u^2 + 2xs$

Subtract u^2 then divide by $2s$

$x = \frac{v^2 - u^2}{2s}$

(iii) $\frac{y}{x} + ab = c^2$

Subtract ab then multiply by x then divide by $(c^2 - ab)$

$x = \frac{y}{c^2 - ab}$

(iv) $2x + ax = y$

Take x as a common factor outside $(2 + a)$ then divide y by $(2 + a)$

$x = \frac{y}{2 + a}$

(v) $4y + p = \sqrt[3]{2x}$

Cube $(4y + p)$ then divide by 2

$x = \frac{(4y + p)^2}{2}$

Go through each of these as a whole class.

Exercise commentary and misconceptions

In question 1**e** remind students of the rule for dividing by a fraction, that dividing by $\frac{1}{2}$ is the same as multiplying by 2.

In question 2**a**–**d**, stronger students could be introduced to the concept of multiplying by −1 which makes the question much quicker, but weaker students should continue to 'add the term with x to both sides'.

Observant students may notice that the two parts of question 3 use different units, but this makes no difference to the formula.

Question 4 is the hardest question and may prove too difficult for the weaker students.

In question 6, the strongest students could be encouraged to think about the units of measurement of g especially as they may have encountered the concept in physics.

Plenary

Make x the subject in each of these formulae:
$\sqrt[a]{x} = y$ and $x^a = y$ ($x = y^a$ and $x = \sqrt[a]{y}$).
Use this as a link to discuss the power and power root as inverses. Use as an example $\sqrt[3]{125} = 5 \leftrightarrow 125 = 5^3$.

A5.4 Rearranging harder formulae

This spread will show you how to:

- Rearrange a formula in order to change its subject

Keywords
Rearrange
Subject

Some formulae can be difficult to **rearrange**.
This is the case when

1. The new **subject** is subtracted in the original formula.
For example to make x the subject of the formula

$p - x = k$ — Start by adding x to both sides.
$p = k + x$ — Now subtract k from both sides.
$p - k = x$

This removes the 'subtracted from'.

2. The new subject is in the denominator in the original formula.
For example to make x the subject of the formula

$\frac{p}{x} = k$ — Start by multiplying both sides by x.
$p = kx$ — Now divide both sides by k.
$\frac{p}{k} = x$

Remember that x is the reciprocal of $\frac{1}{x}$.

You multiply by the reciprocal to remove the subject from the denominator.

Example

a Make x the subject of the formula $t(p - ax) = y$.
b Make y the subject of the formula $\frac{p}{y} + k = w$.

a $t(p - ax) = y$

$p - ax = \frac{y}{t}$ — Divide both sides by t.
$p = \frac{y}{t} + ax$ — Add ax to both sides.
$p - \frac{y}{t} = ax$ — Subtract $\frac{y}{t}$ from both sides.
$\frac{p - \frac{y}{t}}{a} = x$ — Divide both sides by a.

You can go one step further and tidy up the numerator.

$$\frac{p - \frac{y}{t}}{a} = x \rightarrow \frac{\frac{pt - y}{t}}{a} = x \rightarrow \frac{pt - y}{at} = x$$

b $\frac{p}{y} + k = w$

$\frac{p}{y} = w - k$ — Subtract k from both sides.
$p = y(w - k)$ — Multiply both sides by y.
$\frac{p}{w - k} = y$ — Divide both sides by $(w - k)$.

Exercise A5.4

1 Make p the subject of each formula.

a $m = px - q$ **b** $p^2 - r = w$ **c** $\sqrt[3]{p} + h = m$ **d** $\frac{p}{t} - g = h$
e $\frac{1}{2}p + r = q$ **f** $bp^2 = k$ **g** $apw = z$ **h** $2x + y = \sqrt{p}$

2 Make x the subject of each formula.

a $k - x = w$ **b** $t - ax = p$ **c** $y = b - tx$ **d** $m = n(a - x)$
e $\frac{k}{x} = w$ **f** $m = \frac{t}{x}$ **g** $\frac{h}{x} + p = g$ **h** $\frac{p}{x^2} = k$

3 The formula $S = \frac{d}{t}$ connects speed, distance and time.

a Use the formula to find S when a distance of 27 miles is travelled in $\frac{3}{4}$ hour.
b Rearrange the formula to make t the subject.
c Use the formula to find the time taken to travel 60 km at 42 km/h.

4 This formula contains two difficult operations. Make k the subject.

$$p - \frac{t}{k} = q$$

5 Put these cards in order to give the steps in changing the subject of the formula $x = \frac{2(p - y)}{ab}$.

a $abx = 2(p - y)$
b $y + \frac{1}{2}abx = p$
c $y = p - \frac{1}{2}abx$
d $x = \frac{2(p - y)}{ab}$
e $\frac{1}{2}abx = p - y$

6 The formula $T = 2\pi\sqrt{\frac{p}{g}}$ is used to find the time, T, that a pendulum of length p takes to swing freely under gravity, g.

a Rearrange to make g the subject.
b Find a value of g (to 2 sf), given that a pendulum of length 0.4 m takes 1.27 seconds to complete its swing.

7 You can find the volume of a cylinder using the formula $V = \pi r^2 h$.
Rearrange the formula to make the subject

a h **b** r

8 You can find the area of a trapezium using the formula $A = \frac{1}{2}(a + b)h$.
Rearrange the formula to make b the subject.

A5.5 Introducing proof

Objectives

- Understand the difference between a practical demonstration and a proof
- Recognise the importance of assumptions when deducing results

Useful resources

- OHT of second example

Mental starter

Write down these statements:
$a = b^2$, $b = 2c$, $c = a$.
The values of a, b and c are all between 0 and 1; they are all fractions.
Can you find them?
($a = \frac{1}{4}$, $b = \frac{1}{2}$, $c = \frac{1}{4}$. Also $a = b = c = 0$ works.)

Introductory activity

If you cube a positive whole number (integer) the result is always even. Is this statement true? If you can find a counter example the statement will be disproved: $1^3 = 1$ odd.
Emphasise that to prove a statement is true, you need to generalise.

How can you create a general even number? Start with the integer n, which may be odd or even. If you double it, it will be $2n$ which must be even because it always has two as a factor, which is the definition of an even number. Try it with a few integers.

How can you create a general odd number? (Because $2n$ is always even, $2n + 1$ is always odd.)
This means that any even number can be written as $2n$ and any odd number can be written as $2n + 1$.

Prove that the sum of any even and any odd is always odd.
Use the 2 numbers $2n$ (even) and $2m + 1$ (odd). Their sum is $2n + 2m + 1$. This factorises partially to $2(n + m) + 1$.
$2(n + m)$ must be even, because it has two as a factor, so $2(n + m) + 1$ is always odd.

Prove that the product of an even number and an odd number is always even.
Use the even number $2n$ and the odd number $2m + 1$.

Product is $2n(2m + 1) = 4nm + 2n$. Factorise with the number $2 \rightarrow 2(2\ nm + n)$. This means the answer has two as a factor, so must be even.

These are generalised proofs, because they hold true for any values of n and m. Discuss the last example, which is a geometrical proof.

Exercise commentary and misconceptions

Suggest to students that in question 4 they do not need to restrict themselves to integer numbers.

In question 5 students may need help in generalising the numbers in the pattern. It is probably best done by relating the second, third and fourth numbers to the first number.

Plenary

Prove that the angles in a triangle add up to 180°.
First you have to accept that there are 360° in a full turn, this is the definition of a degree.
Now draw a triangle and extend the base line. Then draw an extra line parallel to one of the sides. The 3 angles inside the triangle could be anything, so label them a, b and c. The other 2 angles are corresponding angles and alternate angles. Label them with the correct letter or value.
Notice that at the point where you have 3 angles, they make a straight line so
$a + b + c = 180$ (angles on a straight line) and so the angles in the triangle must equal 180°.
Or:
Discuss question 6.

A5.5 Introducing proof

This spread will show you how to:
- Understand the difference between a practical demonstration and a proof
- Recognise the importance of assumptions when deducing results

Keywords
Counter-example
Demonstration
False
Proof

- **You can show that a statement is false, by finding a counter-example.**

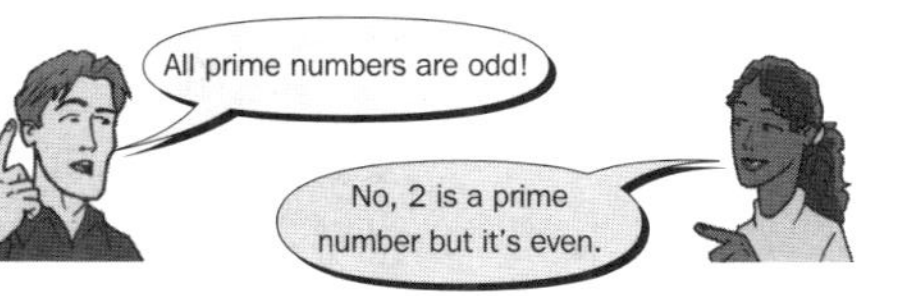

2 is a counter-example – it doesn't fit the statement.

To prove that a statement is true, you can't just find an example. There could still be an example that doesn't work! However, it is useful to demonstrate the statement to yourself first with numbers. For example

Show that the sum of two consecutive integers is always odd:
$3 + 4 = 7$ ... odd
$36 + 37 = 73$ ... odd

This is a **demonstration**.

- **To prove a statement is true, you need to generalise it to all possible examples.**

For example

Show that the sum of two consecutive integers is always odd:
Let the consecutive integers be n and $n + 1$: $n + n + 1 = 2n + 1$.
$2n + 1$ is always odd since it is one more $(+1)$ than a multiple of 2 $(2n)$.

Example

Prove that the sum of two odd numbers is even.

$2m + 1$ and $2n + 1$ are odd numbers. This is a **proof**. (They are each 1 more than a multiple of 2.)

$(2m + 1) + (2n + 1) = 2m + 2n + 2$ *or* $2(m + n + 1)$ by factorisation.

$2(m + n + 1)$ is even because it is a number multiplied by 2.

First, it may help to demonstrate it to yourself: $3 + 7 = 10$, $9 + 11 = 20$, $1 + 51 = 52$ and so on.

Example

Prove that, in this triangle, $d = a + b$.

$a + b + c = 180°$ (angles in a triangle)
$c + d = 180°$ (angles on a straight line)
So $a + b + c = c + d$
$d = a + b$

Exercise A5.5

1 Find a counter-example to show that each of these statements is untrue.
- **a** When you subtract 7 from a number, the answer is always odd.
- **b** When you square a number, the answer is always even.
- **c** When you treble a prime number, the answer is always odd.
- **d** When you find the product of two consecutive numbers, the answer is always odd.

2 The sum of any five consecutive integers is always a multiple of 5.
- **a** Demonstrate, with a few examples of your own, that this statement is true.
- **b** By letting the numbers be n, $n + 1$, $n + 2$, $n + 3$ and $n + 4$, prove the statement is true.

3 Repeat question **3** for these statements.
- **a** The sum of two even numbers is even.
- **b** Squaring an even number gives a number in the four times table.
- **c** The sum of an odd number and an even number is always odd.
- **d** If two consecutive numbers are multiplied, and the smaller number is subtracted from the result, you always get a square number.

4 **a** Can you find a counter-example to this statement?

> Squaring a number will always give you a value greater than the number you started with.

- **b** What range of values does not support this statement?

5 Prove that these statements are true.
- **a** When taking three consecutive integers, the square of the middle integer is always one more than the product of the other two.
- **b** The answer to each equation in this pattern will always be 4.

  $5 \times 8 - 4 \times 9 = 4$
  $6 \times 9 - 5 \times 10 = 4$
  $7 \times 10 - 6 \times 11 = 4$ ...
- **c** The difference between two square numbers will always be equal to the product of their sum and of their difference.

6 Prove that the angle in a semicircle is always a right angle.

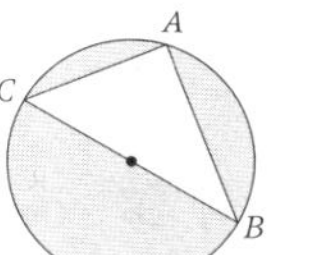

Join *A* to *O*.
Let angle *ACB* be *x* and angle *ABC* be *y*.

A5 Exam review

Key objectives

- Use formulae from mathematics and other subjects
- Substitute numbers into a formula
- Change the subject of a formula
- Generate a formula
- Understand the difference between a practical demonstration and a proof

Worked solution	Commentary and misconceptions
1	Ensure that students recall the formulae for the area and perimeter of polygons. Some students may need reminding of these formulae.
a $P = x + 1 + x - 2 + x$ $= 3x - 1.$	Ensure that students understand what they are required to work out and understand how to derive the relevant formulae. Encourage students to write the formula as they derive it before simplifying.
b i $P = 3x - 1$	Remind students that when rearranging formulae, the same operations must be done to both sides of the equation.
$P + 1 = 3x$ $\frac{P+1}{3}$ or $x = \frac{P+1}{3}$	Ensure that students divide the whole expression by 3.
ii $P = 8$ so $x = \frac{9}{3} = 3$	Emphasise the wording of the question to ensure that students use their results from part **i**.
2	Encourage students to approach such questions methodically. Here, suggest starting with the smallest prime and then looking at the prime numbers in ascending order.
When $n = 2$, $n^2 + 3 = 7$ 2 is a prime number and 7 is odd, so $n = 2$ is an example that shows that John is not correct.	Some students may find using a table to record their results useful. Ensure that students understand the question and what they are being asked to show. Some students may need to be reminded of the order of operations and how to substitute numbers into a formulae. Ensure that students understand square numbers and do not try to double *n*.

Properties of shapes

Objectives

F Use angle properties of equilateral, isosceles and right-angled triangles

F Understand congruence

H Understand similarity of triangles and of other plane figures, and use this to make geometric inferences

F/H Recall the definitions of special types of quadrilateral, including square, rectangle, parallelogram, trapezium and rhombus

F/H Classify quadrilaterals by their geometric properties

F/H Calculate perimeters and areas of shapes made from triangles and rectangles

F/H Understand, recall and use Pythagoras' theorem in 2-D problems

F/H Present answers to sensible levels of accuracy

F/H Given points A and B, calculate the length AB and find the coordinates of the midpoint of the line segment AB

F/H Understand that three coordinates identify a point in space, using the term '3-D'

Unit overview

This unit consolidates the students' prior knowledge of congruence and symmetry and the geometric properties of 2-D shapes. It provides an introduction to Pythagoras' theorem and using this to solve problems, including using Pythagoras' theorem with coordinates and the 3-D coordinate plane. This unit provides a basis for further work on Pythagoras' theorem in unit S8.

Prior knowledge

Before your students start this unit they should be able to:

- Recognise lines of symmetry in triangles and simple 2-D shapes
- Add, subtract, multiply and divide with integers
- Understand and use square numbers
- Understand and use angle measure
- Use the correct order of operations in calculations
- Understand and use perpendicular and parallel lines

Differentiation

- **Foundation focuses** on geometric properties of 2-D shapes, including congruence and coordinates, then proceeds to 3-D shapes
- **Foundation Plus** extends to further detail of 3-D shapes, including plans and elevations
- **Higher Plus** extends the Higher book to using trigonometry and solving multi-stage problems

S4.1 Congruence and symmetry

Objectives

- Understand and explain congruence and symmetry

Useful resources

- Geometry tool
- A triangular prism
- A cylinder

Mental starter

A triangular pyramid is made from baked bean tins. Top layer has 1 tin, next layer 2 tins, next 3 tins and so on. **How many tins altogether in 12 layers?** (78.) A second pyramid has 24 in the base layer. **How many layers will there be?** (24.) **How many tins in this pyramid?** (300) **How many tins in a pyramid with n layers?** ($\frac{1}{2} \times (n^2 + n)$.)

Introductory activity

Ask students what features they can identify as being exactly the same in the two triangles in the first example. Ask what features are different.
(The only difference is the orientation of the two triangles.) All corresponding sides equal, and all corresponding angles equal are the conditions for congruence. If the triangles were drawn to scale, a tracing of one triangle would fit exactly over the other.

Repeat with the second example. The lengths of the equal sides in Triangle D cannot be 6 cm because the triangle does not have angles of 60°. The triangles are not congruent.

Exercise commentary and misconceptions

In question 1 you cannot assume the angle in part **b** is a right angle.

Question 2 has an infinite set of answers as any line through the centre of the rectangle will do!

In questions 3 and 4, encourage students to find more than one answer.

In questions 4 and 5 solid shapes will help those students who find 3-D objects difficult to visualise.

Plenary

How many planes of symmetry are there in a cube? Draw sketches of a cube for each one you find and colour the plane of symmetry.

(9 planes of symmetry: 3 that cross the faces vertically or horizontally, 2 crossing the top diagonally, 2 crossing the front diagonally, 2 crossing the side diagonally.)

S4.1 Congruence and symmetry

This spread will show you how to:
- Understand and explain congruence and symmetry

Keywords
Congruent
Corresponding
Symmetry

- In **congruent** shapes, **corresponding** lengths are equal and corresponding angles are equal.

Example

Explain why these triangles are congruent.

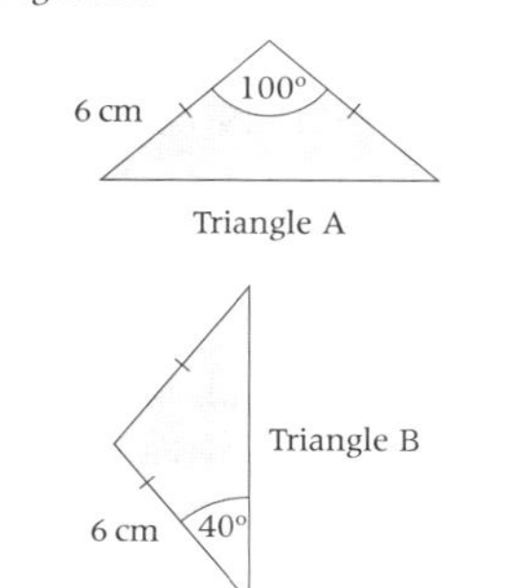

Both triangles are isosceles, with two equal sides of 6 cm.

Triangle A: angle 100° between equal sides, so other two angles must each be 40°.

Triangle B: base angles both 40°, so angle between two equal sides is 100°.

Side opposite 100° angle must be same length as side opposite 100° angle in Triangle A.

Three angles and three sides same in both triangles, so congruent.

Congruent shapes may be reflections, rotations or translations of each other.

Base angles of isosceles triangle are equal.
180° − 100° = 80°
80° ÷ 2 = 40°

180 − (2 × 40) = 100°

Example

Explain why these triangles are not congruent.

Triangle C — 100°, 6 cm

Triangle D — 40°, 6 cm

Both triangles are isosceles.

Both triangles have angles 40°, 40°, 100°.

Triangle D: lengths of the two equal sides not given. If they were 6 cm, as in Triangle C, then Triangle D would be equilateral.

Triangle D is not equilateral, so its equal sides are not 6 cm.

So the two triangles are not congruent.

Using base angles of isosceles triangle, as in the first example.

In an equilateral triangle, angles are 60°, 60°, 60°.

They have equal angles, but the side lengths are *not* equal.

A line of **symmetry** divides a shape into two congruent shapes.

A plane of symmetry divides a 3-D shape into two congruent 3-D shapes.

Exercise S4.1

1 Explain whether or not these pairs of triangles are congruent.

a

b

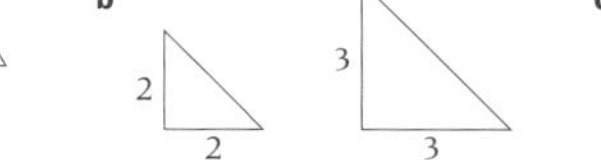

c

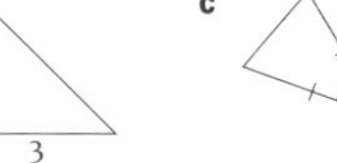

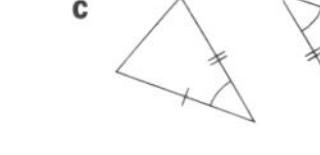

2 The dotted lines show two different ways of splitting a rectangle into two congruent shapes.

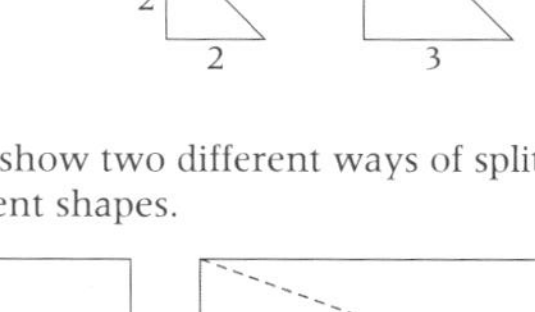

a Draw two copies of the rectangle.
Draw dotted lines to show two more ways of splitting the rectangle into congruent shapes.

b For each, state whether or not the dotted line is also a line of symmetry.

3 Copy these shapes.
Draw dotted lines to split them into congruent shapes.

You may need more than one copy of each shape.

4 Copy these 3-D shapes.
Show how each 3-D shape can be split into two congruent 3-D shapes.

You may need more than one copy of each shape.

5 How many planes of symmetry divide a cylinder into congruent shapes?

S4.2 Quadrilaterals

Objectives

- Recall the definitions of special types of quadrilaterals
- Classify quadrilaterals by their geometric properties

Useful resources

- Rulers
- Tracing paper

Mental starter

Imagine a square formed of rods linked together. If the square is distorted by moving the top rod sideways, the square will become a rhombus. The diagonals of the square intersect at 90°. The diagonals of the rhombus also intersect at 90°.

Now imagine a rectangle formed of similar rods. If the rectangle is distorted it will become a parallelogram. The diagonals of the rectangle intersect at different angles depending on the lengths of the sides.
Will the diagonals of the parallelogram intersect at the same angle as those of the rectangle?
(Yes. This can be shown by drawing a rectangle and several parallelograms with sides of the same length and tracing the angles.)

Introductory activity

Remind students of the properties of quadrilaterals. Students need to be aware that some classifications overlap. A square is also a rectangle because it has all the properties of a rectangle. A square is also a rhombus, parallelogram and kite, but not a trapezium, as a trapezium has exactly 1 pair of parallel sides. A rhombus is also a parallelogram and a kite. A rectangle is also a parallelogram.

Remind students how to find the area of a triangle, $\frac{base \times height}{2}$ is the easiest formula.
Give an example and remind students why you divide by 2. Emphasise that the base and height are at right angles (perpendicular).

Find the area of a rhombus with diagonals of 13 cm and 8 cm. (Use the fact that the diagonals bisect at right angles.)
(52 cm^2.)

Exercise commentary and misconceptions

Questions 1–4 use the fact that the diagonals are perpendicular, so the diagonals form the base and height of triangles.

Question 5 reminds students of the conditions for congruence.

Question 6 uses properties common to different quadrilaterals.

Plenary

The area of squares, rhombuses and kites can be found using a formula involving their diagonals. A square with diagonals 10 cm and 10 cm has an area of 50 cm^2. A rhombus with diagonals 5 cm and 9 cm has an area of 22.5 cm^2. A kite with diagonals 8 cm and 15 cm has an area of 60 cm^2.
What is the formula? (Half the product of the diagonals.) **Does this formula work for a rectangle?** (No.)
For a proof by counter example, draw a rectangle with sides 3 cm and 4 cm. The diagonals are 5 cm using Pythagoras. Area is 12 cm^2 by multiplying sides but 12.5 cm^2 by half the product of the diagonals.
Does it work for any other quadrilaterals?
(No.) Again you can use counter examples to disprove.

S4.2 Quadrilaterals

This spread will show you how to:
- Recall the definitions of special types of quadrilaterals
- Classify quadrilaterals by their geometric properties

Keywords
Kite
Parallelogram
Quadrilateral
Rectangle
Rhombus
Square
Trapezium

- A **quadrilateral** is a shape with four straight sides.
- Angles in a quadrilateral add up to 360°.

Properties of quadrilaterals

	Square	Rhombus	Rectangle	Parallelogram	Trapezium	Kite
1 pair opposite sides parallel	✓	✓	✓	✓	✓	
2 pairs opposite sides parallel	✓	✓	✓	✓		
Opposite sides equal	✓	✓	✓	✓		
All sides equal	✓	✓				
All angles equal	✓		✓			
Opposite angles equal	✓	✓	✓	✓		
Diagonals equal	✓		✓			
Diagonals perpendicular	✓	✓				✓
Diagonals bisect each other	✓	✓	✓	✓		
Diagonals bisect the angles	✓	✓		✓		
2 pairs of adjacent sides equal	✓	✓				✓

Example

Work out the area of this rhombus.

8 cm
13 cm

The diagonals bisect each other and are perpendicular.
Opposite angles are equal.
Diagonals bisect the angles.

A B C D

So the diagonal AC is a line of symmetry and the triangles ABC and ACD are congruent.

Each triangle has base 13 cm and height 4 cm.
Area of each triangle = $\frac{1}{2} \times 13 \times 4 = 26$ cm^2
Area of rhombus = $2 \times 26 = 52$ cm^2

Recap properties of a **rhombus**.

A line of symmetry divides a shape into two congruent shapes.

Area of triangle = $\frac{1}{2}bh$

Example

Are these statements true or false?
Give reasons for your answer.

a All squares are rhombuses

b Parallelograms are rectangles

a True, all properties of a rhombus are also properties of a square.
b False, a **parallelogram** does not have all its angles equal, nor are its diagonals perpendicular.

A **square** is a special type of rhombus with all angles equal.

A **rectangle** is a special type of parallelogram with all angles equal.

Exercise S4.2

1 The lengths of the diagonals of a rhombus are 5 cm and 9 cm.
Find the area of the rhombus.

2 A square has diagonals 10 cm long.
a Sketch the square.
b Find the area of the square.

3 Jenny draws these three kites each with diagonals 6 cm and 14 cm.

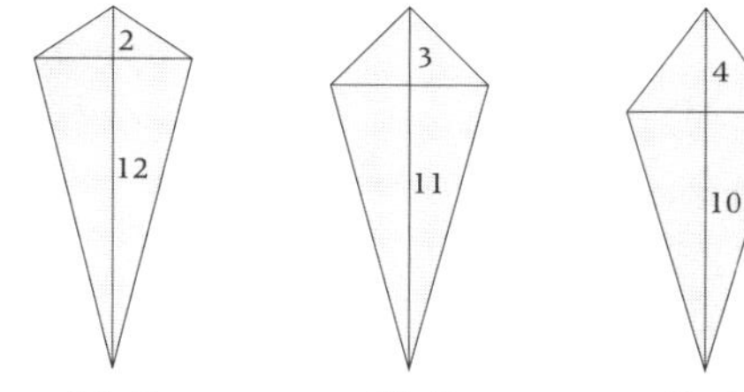

Kite X Kite Y Kite Z

a Find the areas of kites X, Y and Z.
b Describe how to work out the area of any kite.

DID YOU KNOW?
The kite is a rare bird of prey that hovers in the wind, which influenced the naming of the toy kite. This in turn influenced the naming of the kite shape.

4 The lengths of the diagonals of a kite are 8 cm and 15 cm.
Find the area of the kite.

8 cm
15 cm

Use your method from question **3**.

5 Here is a diagram of a parallelogram, P, and a rhombus, R.

P R

One diagonal been drawn inside each shape.
Has either shape has been split into two congruent triangles?
Give a reason for your answer.

6 Are these statements true or false? Give reasons for your answers.
a All squares are rectangles.
b All kites are rhombuses.
c All rhombuses are rectangles.

S4.3 Triangles and Pythagoras' theorem

Objectives

- Use angle properties of equilateral, isosceles and right-angled triangles
- Understand, recall and use Pythagoras' theorem in 2-D problems

Useful resources

- Calculators

Mental starter

The sum of the interior angles of a regular polygon with n sides is given by $180 \times (n - 2)°$. If a regular polygon with a sum of more than 5000° is shown on a computer screen, it will look like a circle from a distance. **What is the minimum number of sides the polygon will need?** (30 sides since
$180 \times (29 - 2) = 4860$ and
$180 \times (30 - 2) = 5040$).

Introductory activity

Remind students of the properties of equilateral, isosceles, scalene and right-angled triangles.

Draw 2 right-angled triangles with sides 6, 8, 10 and 5, 12, 13. Square each length and write it near the original length.
What is the relationship between the squared lengths of the sides? ($6^2 + 8^2 = 10^2$, $5^2 + 12^2 = 13^2$. Check that it is true for both triangles.)

Remind students of Pythagoras' Theorem. Students may be familiar with any of $c^2 = a^2 + b^2$ or $a^2 = b^2 + c^2$ or $h^2 = a^2 + b^2$, depending on the diagrams used in other textbooks. Emphasise the fact that it is the squares of the two shorter sides that add to make the square of the longest side (hypotenuse). The formula used in the students' book is $c^2 = a^2 + b^2$.

A right-angled triangle has short sides of 4.2 cm and 7.6 cm. **Find the third side (hypotenuse).**

(Answer $= \sqrt{(4.2^2 + 7.6^2)} = \sqrt{(17.64 + 57.76)}$
$= \sqrt{75.4} = 8.7$ cm.)

A right-angled triangle has a hypotenuse of 6.7 cm and a second side of 3.5 cm. **Find the third side (a short side).**

(Answer $= \sqrt{(6.7^2 - 3.5^2)} = \sqrt{(44.89 - 12.25)}$
$= \sqrt{32.64} = 5.7$ cm.)

Exercise commentary and misconceptions

Students often ask 'Do I add or subtract?' Encourage them to decide for each question whether they are finding the hypotenuse or one of the shorter sides. To find a hypotenuse, square, **add**, and square root. To find a shorter side, square, **subtract**, and square root. The key is to remember that the hypotenuse is the longest side, so must use addition.

Question 1 requires students to find the hypotenuse.

Question 2 requires students to find one of the shorter sides.

The most common mistakes are to forget to square at the beginning or to square root at the end, or to add instead of subtracting or vice versa. Encourage students to think about the problem they are solving rather than just going through a routine.

Remind students to give their answers to an appropriate level of accuracy.

Plenary

In question 2 the Pythagorean triples are 3, 4, 5 and 5, 12, 13; and in question 3 they are 6, 8, 10 and 10, 24, 26 and 9, 12, 15.
Is there any link between the numbers in question 2 and those in question 3? Can you find any other Pythagorean triples?

S4.3 Triangles and Pythagoras' theorem

This spread will show you how to:

- Use angle properties of equilateral, isosceles and right-angled triangles
- Understand, recall and use Pythagoras' theorem in 2-D problems

Keywords
Equilateral
Hypotenuse
Isosceles
Pythagoras' theorem
Right-angled triangle
Scalene

You need to know the properties of these triangles.

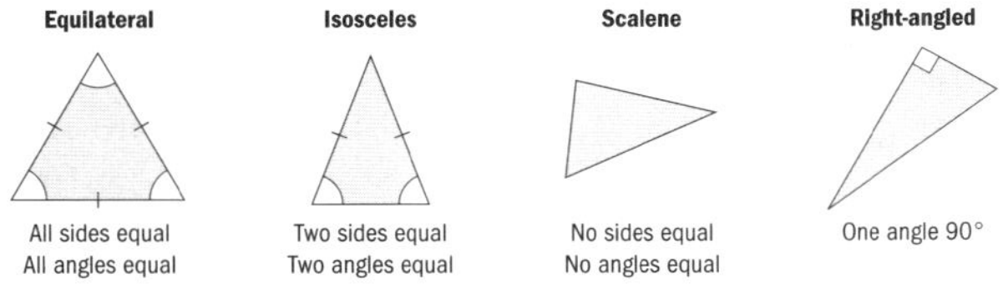

Equilateral
Isosceles
Scalene
Right-angled
All sides equal
All angles equal
Two sides equal
Two angles equal
No sides equal
No angles equal
One angle 90°

In a right-angled triangle the **hypotenuse** is the longest side.
It is opposite the right-angle.

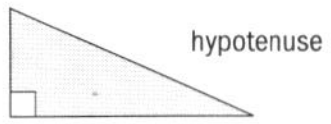

hypotenuse

In a right-angled triangle, the area of the square of the hypotenuse equals the sum of the areas of the squares of the other two sides.

- **This is Pythagoras' theorem.**

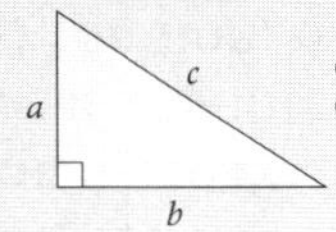

a
b
c

$c^2 = a^2 + b^2$
c is the hypotenuse

Example

Work out the missing sides in these triangles.

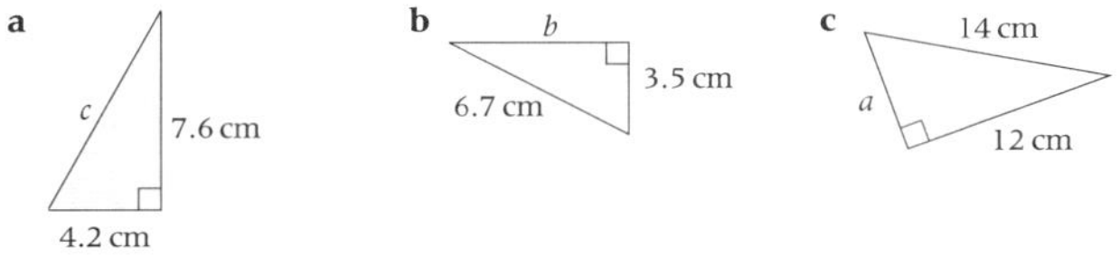

a
c
7.6 cm
4.2 cm
b
b
3.5 cm
6.7 cm
c
14 cm
a
12 cm

a $c^2 = a^2 + b^2$
$= 4.2^2 + 7.6^2$
$= 17.64 + 57.76$
$= 75.4$
$c = \sqrt{75.4}$
$= 8.7$ cm

b $c^2 = a^2 + b^2$
$b^2 = c^2 - a^2$
$= 6.7^2 - 3.5^2$
$= 44.89 - 12.25$
$= 32.64$
$b = \sqrt{32.64}$
$= 5.7$ cm

c $c^2 = a^2 + b^2$
$a^2 = c^2 - b^2$
$= 14^2 - 12^2$
$= 196 - 144$
$= 52$
$a = \sqrt{52}$
$= 7.2$ cm

To find the hypotenuse: square, add and square root.

To find a shorter side: square, subtract and square root.

Exercise S4.3

Give answers in this exercise to 1 dp where appropriate.

1 Find the hypotenuse in each of these right-angled triangles.

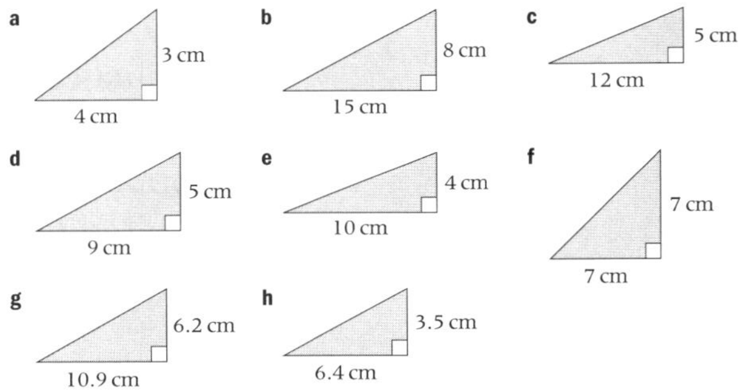

a
3 cm
4 cm
b
8 cm
15 cm
c
5 cm
12 cm
d
5 cm
9 cm
e
4 cm
10 cm
f
7 cm
7 cm
g
6.2 cm
10.9 cm
h
3.5 cm
6.4 cm

2 In some of the triangles in question **1**, all three sides have integer (whole number) values.
Such sets of three numbers are called Pythagorean triples.
Write the Pythagorean triples from question **1**.

3 Find the missing side in each of these right-angled triangles.

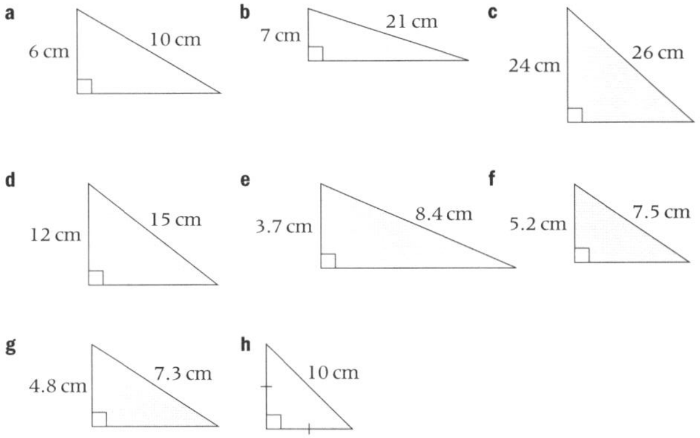

a
6 cm
10 cm
b
7 cm
21 cm
c
24 cm
26 cm
d
12 cm
15 cm
e
3.7 cm
8.4 cm
f
5.2 cm
7.5 cm
g
4.8 cm
7.3 cm
h
10 cm

4 Some of the triangles in question **3** are Pythagorean triples.

a Write down the Pythagorean triples in question **3**.

b Compare these with your answers to question **2**.

c Comment on anything you notice.

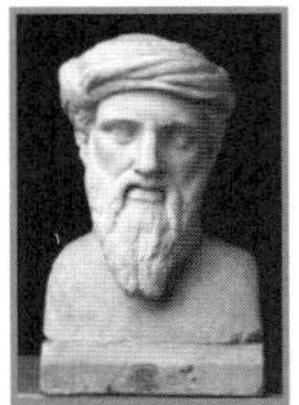

DID YOU KNOW?
Pythagoras was a Greek mathematician most famous for his theorem, who taught his students that absolutely everything was related to mathematics.

S4.4 Problem solving using Pythagoras' theorem

Objectives
- Understand, recall and use Pythagoras' theorem in 2-D problems

Useful resources
- Calculators

Mental starter
A right-angled triangle has an area of 6 cm^2 and a perimeter of 12 cm. **What are the lengths of the sides of the triangle?**

(3, 4, 5 cm.) **Can you find a right-angled triangle that has the same value for the perimeter and area?** You could give the hint that the sides are all integer, all even, all under 12. (Simplest solution, 6, 8, 10 cm.) If Pythagorean triples have been looked at in the previous lesson, they will be more familiar in this mental starter.

Introductory activity
Remind students of Pythagoras' theorem.

Some examples are more complex than solving a single triangle. A rectangle has sides 5 cm and 8 cm. Find the length of a diagonal. The diagonal forms a right-angled triangle with two of the sides. **Is the diagonal the hypotenuse?** (Yes.) **Do you need to add or subtract?** (Add $d = \sqrt{(5^2 + 8^2)} = \sqrt{89} = 9.43$ cm.) Discuss the accuracy that should be used in the answer.

Sometimes you use one side which has been found to find another side in the same diagram. (See the second example in the student book.)

In the triangle ABC, is AC the hypotenuse or one of the shorter sides? (hyp) **Do you need to add or subtract?** (add)

$AC = \sqrt{(24^2 + 7^2)} = \sqrt{625} = 25$ cm.

In the triangle ACD is AD the hypotenuse or one of the shorter sides? (hyp) **Do you need to add or subtract?** (add)

$AD = \sqrt{(25^2 + 9^2)} = \sqrt{706} = 26.57$ cm.

Exercise commentary and misconceptions
Remind students that Pythagoras will only work for right angled triangles. If the diagram does not have one, then you must find one and draw it in.

In questions 1–4 students need to mark the right-angles.

In questions 5 and 6 students need to use one result to find another side. In question 5, RP is the hypotenuse in one triangle but a shorter side in the second triangle.

In question 6 AC changes from hypotenuse to shorter side.

Plenary
I have a cylindrical pencil case. It is 18 cm long and has a diameter of 8 cm. **Can I fit a pencil that is exactly 19 cm long inside the pencil case?** (yes).
Students should draw a diagram and draw in the right-angled triangles (diagonal $= \sqrt{388} = 19.70$ cm). Some students may consider the approximate measures for the pencil case, in this case the minimum diagonal is $\sqrt{362.5} = 19.04$ cm, still fits. Note that the pencil length is considered as exact.

S4.4 Problem solving using Pythagoras' theorem

This spread will show you how to:

- Understand, recall and use Pythagoras' theorem in 2-D problems

Keywords
Hypotenuse
Pythagoras' theorem
Right-angled triangle

- **In a right-angled triangle, the square on the hypotenuse equals the sum of the squares of the other two sides.**

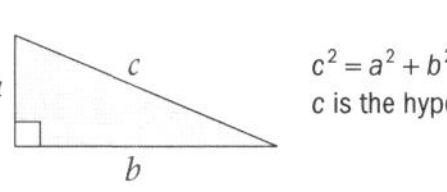

$c^2 = a^2 + b^2$
c is the hypotenuse

This is **Pythagoras' theorem.** The hypotenuse is always opposite the right-angle.

To solve problems using Pythagoras' theorem

- sketch a diagram and label the right-angle
- label the unknown side
- round your answer to a suitable degree of accuracy.

Unless the question tells you otherwise, round to 2 dp.

Example

A rectangle measures 5 cm by 8 cm.
Find the length of its diagonal.

Diagonal, d, is the hypotenuse of a right-angled triangle.
$d^2 = 5^2 + 8^2$
$d^2 = 89$
$d = 9.43$ cm

d   5 cm   8 cm

Mark all the facts you know on your diagram.

Example

Find the length AD.

A   24 cm   B   7 cm   C   9 cm   D

First find AC $AC^2 = 7^2 + 24^2$
$AC^2 = 625$
$AC = 25$ cm

A   24 cm   B   7 cm   C

AC is the hypotenuse of the right-angled triangle *ABC*.

In triangle ACD $AD^2 = 9^2 + 25^2$
$AD^2 = 706$
$AD = 26.57$ cm

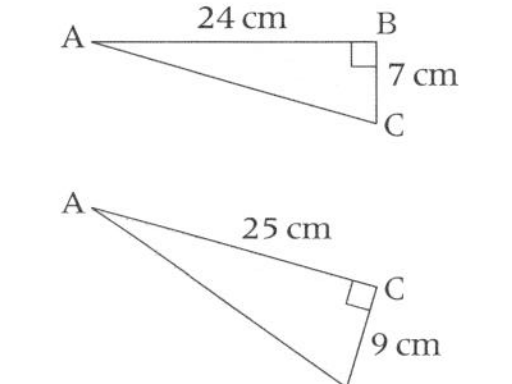

AD is the hypotenuse of the right-angled triangle *ACD*.

Exercise S4.4

Give answers in this exercise to 2 dp where appropriate.

1 Find the length of the diagonal of each rectangle.

a 8.3 cm, 2.4 cm   **b** 6 cm, 5.2 cm

2 A rectangle has one side 4 cm and diagonal 10.4 cm.
Find the length of the other side.

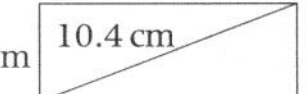

3 Find the length of the diagonal of a square with side length 8 cm.

4 Find the length of the side of a square with diagonal length 8 cm.

5 PQR and PRS are right-angled triangles.
Find the length PS.

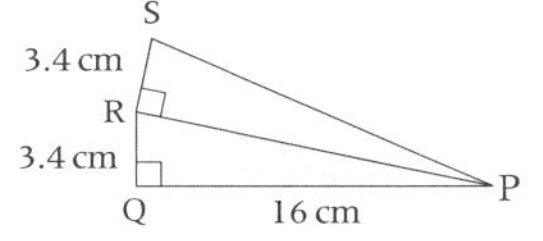

6 ABC and ACD are right-angled triangles.
Find the length AB.

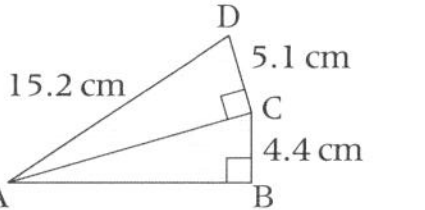

7 A ladder of length 5.5 m leans against a wall.
The foot of the ladder is 1 m from the wall.

How far up the wall does the ladder reach?

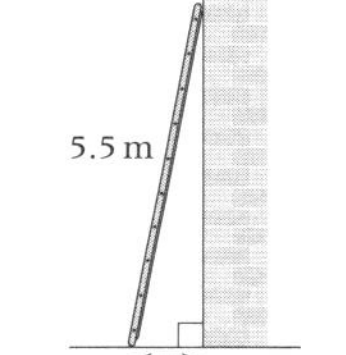

S4.5 Pythagoras' theorem and coordinates

Objectives

- Find the coordinates of the midpoint of the line segment *AB*, and its length, given the points *A* and *B*
- Understand and use coordinates to represent points in three dimensions

Useful resources

- Calculators
- Graph paper
- OHT of grid with A(2, 1), B(4, 4) and C(−2, 3) marked
- Metre rule and workman's tape measure
- Geometry tool
- OHT of 3-D grid

Mental starter

In how many ways can you arrange 4 dots to make a square or rectangle? (2)

How many ways for 8 dots? (2) **And 16 dots?** (3) **And 32 dots?** (3) **And 64 dots?** (4) **And 128 dots?** (4) **What is the rule?**

(For 2^n dots, one solution is that the number of ways $= \frac{1}{2} \times (n+1)$ if n is odd, and $\frac{1}{2}(n+2)$ if n is even.)

Introductory activity

Consider the coordinates A(2, 1), B(4, 4) and C(−2, 3). These are given in the student book. Using OHT or geometry tool, find the midpoint of AB(3, 2.5) and AC(0, 2) by drawing. **What mathematical method could be used instead of measuring?** (mean of coordinates)

Find length AB using Pythagoras.
$AB = \sqrt{((4-2)^2 + (4-1)^2)} = \sqrt{13} = 3.61$. **Is the line segment always the hypotenuse?** (Yes) Check answer by measurement and discuss what units should be used in the answer (none).

Finding the length of a line segment in 3-D can be shown using the classroom if it is approximately a cuboid. Find the 3 dimensions of the room to the nearest metre. Find the diagonal across the floor. Then find the diagonal across the middle of the room through the air to opposite corners. Emphasise that you use a previously found length in the second calculation and discuss the advantage in using the square as found, not the square root (re)squared.

Encourage students to expand on the method for 2-D to find a method for 3-D. They need to include the difference in the *z* direction and square and add to the sum of the differences squared in the *x* and *y* direction.

Use an OHT of a 3-D grid to find points with three coordinates. Plot the points P(3, 2, 5), S(3, 4, −2) and T(4, 0, 2). Find PS.

$PS = \sqrt{((3-3)^2 + (2-4)^2 + (5--2)^2)}$
$= \sqrt{(0+4+49)} = 7.28.$

Find PT and ST.
(PT = 3.74 ST = 5.74.)

Exercise commentary and misconceptions

Encourage students to sketch diagrams for the 2-D questions in question 2. They do not have to mark on every scale along the axes, just indicate the approximate place of each coordinate. This will help to prevent problems arising with negative coordinates.

Plenary

Some computers are capable of 'imagining' shapes in 4-D and 5-D etc. They have an advantage over humans because we like to be able to draw things before we can imagine them, but computers do not. **Using 4-D coordinates, what would be the distance between the two points A(3, 4, 5, 6) and B(7, 9, 3, 8)?** (7 units.)

S4.5 Pythagoras' theorem and coordinates

This spread will show you how to:

- Find the coordinates of the midpoint of the line segment *AB*, and its length, given the points *A* and *B*
- Understand and use coordinates to represent points in three dimensions

Keywords
Coordinates
Midpoint

Coordinates identify a point on a grid.

A (2, 1)   B (4, 4)

- **The midpoint of two points is the mean of their coordinates.**

- **Midpoint of (a, b) and (s, t) is $\left(\frac{a+s}{2}, \frac{b+t}{2}\right)$**

You can use Pythagoras' theorem to find the length of a line joining two points on a grid.

Example

Find the midpoint and length of the line joining A and B.

Midpoint
$\left(\frac{2+4}{2}, \frac{1+4}{2}\right) = (3, 2.5)$

Length
Draw a right-angled triangle and label the lengths of the two shorter sides.

Using Pythagoras' theorem

$AB^2 = 2^2 + 3^2$
$AB^2 = 13$
$AB = 3.61$ (2 dp)

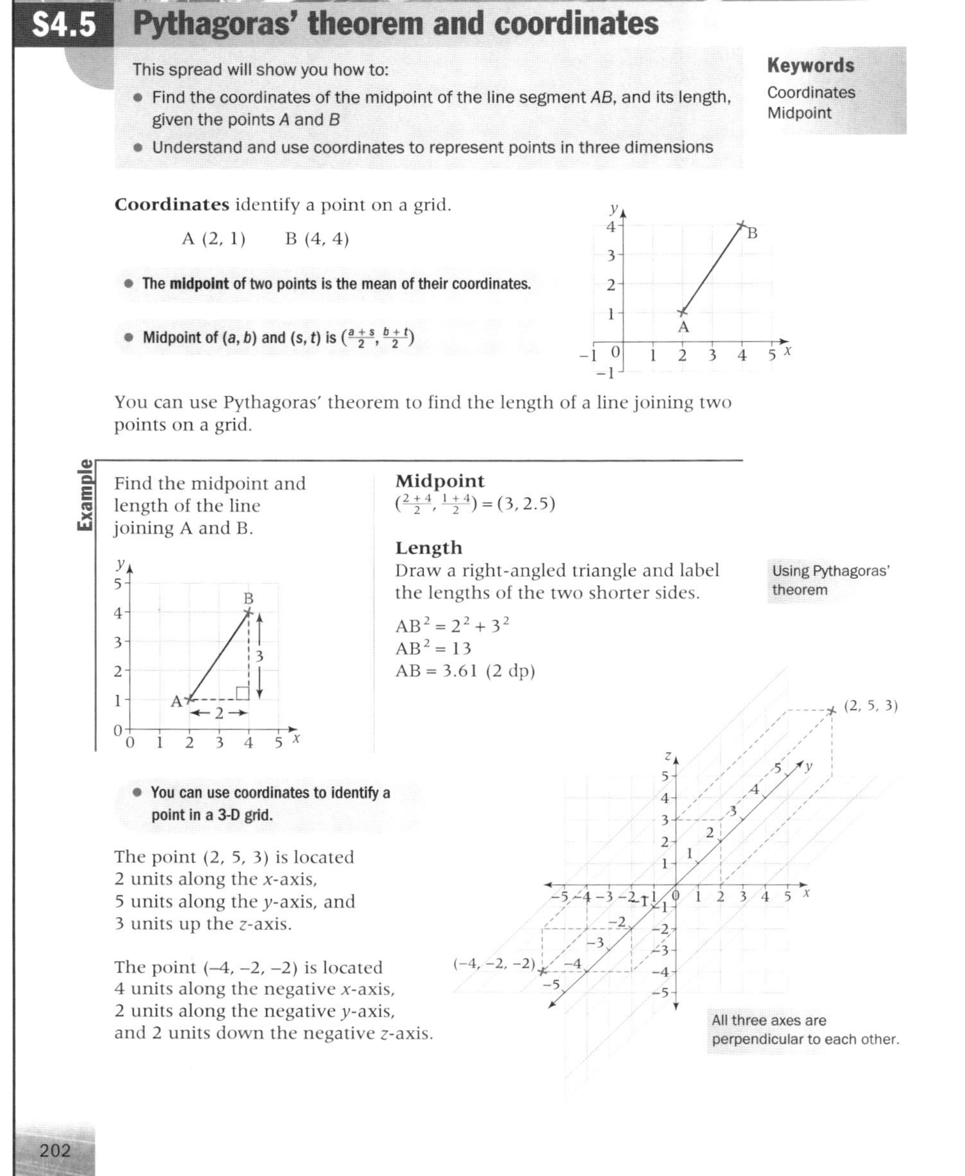

- **You can use coordinates to identify a point in a 3-D grid.**

The point (2, 5, 3) is located 2 units along the *x*-axis, 5 units along the *y*-axis, and 3 units up the *z*-axis.

The point (−4, −2, −2) is located 4 units along the negative *x*-axis, 2 units along the negative *y*-axis, and 2 units down the negative *z*-axis.

All three axes are perpendicular to each other.

Exercise S4.5

1 Find the midpoints of these line segments.

a A (6, 4) and B (2, 7)
b C (−2, 5) and D (3, −3)
c E (0, 6) and F (4, 0)
d G (7, −1) and H (−4, 5)

Use the diagram to check your answers.

Make a sketch to show the position of the points.

2 Use Pythagoras' theorem to find the length of the line segment joining each pair of points.

a (2, 5) and (6, 8)   **b** (7, 1) and (2, 8)
c (0, 7) and (1, −3)   **d** (−4, 5) and (4, −2)

3 Write the coordinates of these points.

4 Here is a grid in three dimensions.
The points on the grid have coordinates of

A(1, 2, 5)   B(1, 4, 2)   C(2, 3, 1)   D(5, 3, 4)

a Find the midpoints of these line segments.
i AB   **ii** CD   **iii** BC   **iv** AD

b Use Pythagoras' theorem to find the length of each line segment in part **ai** and **aii**.

c Can you extend the use of Pythagoras' theorem to find the length of BC? You will need to use it twice.

You need to identify the sides of the triangle.

S4 Exam review

Key objectives

- Understand congruence
- Use angle properties of equilateral, isosceles and right-angled triangles
- Recall the essential properties of special types of quadrilateral
- Classify quadrilaterals by their geometric properties
- Understand, recall and use Pythagoras' theorem in 2-D problems
- Find the coordinates of the midpoint of the line segment AB, given points A and B
- Understand that three coordinates identify a point in space, using the term '3-D'

Worked solution	Commentary and misconceptions
1	Ensure that students are able to read points from the coordinate plane.
a $AB^2 = 1^2 + 4^2 = 17$ $AB = \sqrt{17} = 4.1$ (2sf)	Students should recognise that the use of Pythagoras' theorem is required. Ensure that students recall the formulae and use squares and surds correctly. Encourage students to give their answer to an appropriate level of accuracy.
b $\left(\frac{(2+3)}{2}, \frac{(3+7)}{2}\right) = \left(\frac{5}{2}, 5\right)$	Ensure that students know how to calculate the midpoint and use the correct order of operations, including brackets. Encourage students to look at the diagram to check if their solution is feasible.
2 a i **D** Kite **ii** **B** Parallelogram **iii** **C** Rhombus. **b** **D** Kite.	Students should understand reflection and rotational symmetry.

N6

Estimating and calculating

Objectives

F/H Make mental estimates of the answers to calculations

F Use checking procedures, including use of inverse operations

F/H Present answers to sensible levels of accuracy

H Use surds and π in exact calculations

F Add, subtract, multiply and divide integers and then any number

F/H Understand where to position the decimal point

F/H Understand and use fractions as multiplicative inverses

H Use calculators effectively and efficiently, knowing how to enter complex calculations

F/H Understand the calculator display

H Use calculators to calculate the upper and lower bounds of calculations

Unit overview

This unit consolidates prior knowledge of the order of operations, including brackets and the use of surds and π in exact calculations, encouraging students to give answers to an appropriate degree of accuracy. A range of strategies for mental and written methods and estimating and checking solutions are included, as well as calculator methods.

Prior knowledge

Before your students start this unit they should be able to:

- Add, subtract, multiply and divide with integers, decimals and fractions
- Understand and use the correct order of operations in calculations, including brackets, powers and surds
- Round answers to any given number of significant figures
- Make mental estimates and use these to check answers
- Understand and use place value, decimals and fractions
- Use a calculator effectively
- Understand and use compound measures including speed

Differentiation

- **Foundation** focuses on basic calculating with decimals
- **Foundation Plus** extends to calculations involving powers and multi-step problems, using estimates to check answers
- **Higher Plus** extends the Higher book to include standard form and the limits of accuracy involved in calculations both with and without a calculator

N6.1 Order of operations

Objectives

- Perform the operations within a calculation in the correct order
- Estimate answers to calculations, using these to check the solution
- Round numbers to sensible degrees of accuracy

Useful resources

- Calculators

Mental starter

Make a sequence by square rooting all the integers from 0 to 16. Leave your answer as a square root if you do not know the exact answer, for example $\sqrt{3}$.
(0, 1, $\sqrt{2}$, $\sqrt{3}$, 2, $\sqrt{5}$, $\sqrt{6}$, $\sqrt{7}$, $\sqrt{8}$, 3, $\sqrt{10}$, $\sqrt{11}$, $\sqrt{12}$, $\sqrt{13}$, $\sqrt{14}$, $\sqrt{15}$, 4.)
What are the 169th and the 400th term in the sequence? (13 and 20.)

Introductory activity

What is $3 + 4 \times 2$? (11) Promote discussion about order of calculations. Establish that in maths, multiplying is always done before adding.

What is the value of $3x^2$ if $x = 4$? (48) Promote discussion about order of calculating it. Establish that powers (also known as indices) are always done before multiplying.

What is the order for the six operations +, −, (), indices, ×, ÷? (BIDMAS). Explain the meaning of each letter. Also emphasise that ÷ and × are grouped together, and so are + and −.

Find the value of $14 - 2 \times 3$
($14 - 2 \times 9 \rightarrow 14 - 18 = -4$.)
Put brackets in to make this statement true $5 \times 4 + 2^2 \div 2 = 20$. (Answer $5 \times (4 + 2^2) \div 2 = 20$.)
Estimate the value of $5 + 3 \times \sqrt{50}$
($5 + 3 \times 7 \rightarrow 5 + 21 = 26$.)
Explain clearly that $\sqrt{50} \approx \sqrt{49} = 7$ and also that roots are powers/indices and so come before ×, ÷, +, −.
Discuss the last example in the student book, which is a diagnostic example.

Exercise commentary and misconceptions

Students should do questions 1 and 2 without using a calculator, but may like to check their answers with a calculator.

In question 5 students may find it helpful to use inverse number machines.

In question 9 students may need reminding to write down the approximations they are using **before** doing the calculation as they sometimes try to do the calculation and then give an approximation for the answer. Although the rule of using 1 sf is useful, encourage students to look for numbers which cancel or which have integer square roots.

Throughout the exercise, encourage students to work down the page and to align their workings line by line. This is especially helpful if the result turns out to be negative.

Plenary

Use one set of brackets, each of the symbols ×, ÷, +, −, ^{2}, and the numbers 7, 6, 6, 3, 3 to make the largest possible number.

For example, $(7 + 6) \times 3^2 - 6 \div 3 = 115$. You may make this calculator or non-calculator. Possibly the largest answer is $(7 \times 6 + 6)^2 - 3 \div 3 = 2303$.

N6.1 Order of operations

This spread will show you how to:

- Perform the operations within a calculation in the correct order
- Estimate answers to calculations, using these to check the solution
- Round numbers to sensible degrees of accuracy

Keywords
BIDMAS
Operation
Order

When a calculation involves a number of steps, or **operations**, you need to do them in the right order.

The order in which operations are carried out is:

- **Brackets – start by working out the contents of any brackets**
- **Powers and indices – for example, squares, cubes or square roots – come next**
- **Multiplication and division are done next**
- **Addition and subtraction are done last.**

$(3+2)\times 4^2 - 6$
$= 5 \times 4^2 - 6$
$= 5 \times 16 - 6$
$= 80 - 6$
$= 74$

BIDMAS (**B**rackets, **I**ndices or powers, **D**ivision, **M**ultiplication, **A**ddition, **S**ubtraction) will help you to remember this.

Example

Evaluate **a** $4+3\times 2$ **b** $5+3^2$ **c** $\sqrt{(5+4\times 11)}$ **d** $\sqrt{(5^2-4^2)}$

a $4+3\times 2 = 4+6 = 10$
b $5+3^2 = 5+9 = 14$
c $\sqrt{(5+4\times 11)} = \sqrt{(5+44)} = \sqrt{49} = 7$
d $\sqrt{(5^2-4^2)} = \sqrt{(25-16)} = \sqrt{9} = 3$

Examiner's tip
Using brackets in parts **c** and **d** shows that **all** the values are contained in the square root.

Example

Estimate the answer to $\dfrac{6.3+\sqrt{9.7^2-17}}{149}$

Round all the numbers to a sensible amount.

$$\frac{6.3+\sqrt{9.7^2-17}}{149} \approx \frac{6+\sqrt{10^2-20}}{150} = \frac{6+\sqrt{80}}{150} \approx \frac{6+9}{150} = \frac{15}{150} = 0.1$$

Deal with the numerator and denominator separately.

Example

Adam explained how he would calculate $\dfrac{5+\sqrt{9}}{11}$

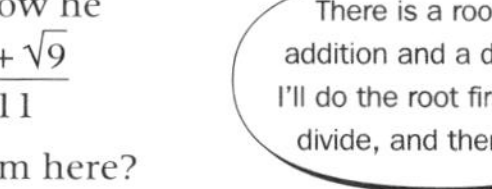

What is the problem here?

Even though there are no brackets in this expression, the whole of the 'top line' is divided by 11, so you need to find the square root, then add, then divide. The expression could be written as $(5+\sqrt{9}) \div 11$.

Exercise N6.1

1 Find the value each of these.
a $5+6\times 9$ **b** $4\times 9+1$ **c** $8-3-3$ **d** $4\times 3+7\times 2$

2 Find the value of each of these.
a $9-4\times 2$ **b** $(9-4)\times 2$ **c** $5\times 7+4\times 2$ **d** $5\times(7+4)\times 2$

3 Copy and complete these equations, replacing the □ with the correct operation.
a 5□3 × 7 = 26 **b** 4 × 6□2 = 22
c 4□7 + 1 = 29 **d** 17□2 + 2 = 6^2

4 Copy these calculations, inserting brackets to make the answers correct.
a $11-1\times 5=50$ **b** $12+3\div 3=5$
c $12-4-1=9$ **d** $8\div 4+4+1=2$

5 Copy and complete these equations, replacing the ● with the correct number.
a (● + 2) × 9 = 36 **b** 64 ÷ (● + 3) = 8
c √(● − 10) = 5 × 2 **d** √● − (5 × 2) = 2

6 Find the values of these expressions.
a $(5^2+3)\times 7$ **b** $(9-7)^2$
c $(5-3)\times(4^2-7)$ **d** $(5^2-8)^2$

7 Calculate the values of these expressions.
a $(4+7)^2$ **b** $(6+7)\times 9\div 3$
c $\dfrac{6\times(5^2-13)}{4}$ **d** $\sqrt{100-2\times 6^2}$

8 Find the values of these expressions.
a $\dfrac{28}{4}+\sqrt{100-(9^2+5\times 2)}$ **b** $\sqrt{28+4^2-(10-2)}+4\times 3$

9 Estimate the value of each of these expressions without using a calculator. Show all of your working.
a $\dfrac{5.2\times(4.8^2-12)}{3.9}$ **b** $73\times\dfrac{202-11}{38\times 5}+29$
c $18.4^2-\dfrac{592}{11.4}$ **d** $\sqrt{26.4\times\dfrac{12.5+7.4}{5.36}}$

N6.2 Exact calculations

Objectives

- Use surds and π in exact calculations

Useful resources

- Calculators

Mental starter

What is the circumference of a circle whose diameter is 5 m? Give the exact answer without using a calculator. (5π.)

The two shorter sides of a right-angled triangle are 2 cm and 3 cm. What is the exact length of the hypotenuse? ($\sqrt{13}$.)

Leaving the letter π or roots in your answer is often easier than writing out the answer with lots of decimal places.

Introductory activity

When you add and subtract multiples of π or roots, they behave just like algebra.

For example, $4\pi + 7 + 3\pi - 2 = 7\pi + 5$
and $3\sqrt{5} + \sqrt{5} + 6\sqrt{7} - 2\sqrt{7} = 4\sqrt{5} + 4\sqrt{7}$
and $\sqrt{9} \times \sqrt{9} = 9$.

It is often more accurate to simplify an expression first, and then find an approximation, so
$4\pi + 7 + 3\pi - 2 = 7\pi + 5$
$\approx 7 \times 3.14 + 5 = 26.98$
and $3\sqrt{5} + \sqrt{5} + 6\sqrt{7} - 2\sqrt{7} = 4\sqrt{5} + 4\sqrt{7}$
$\approx 4 \times 2.24 + 4 \times 2.65$
$= 8.96 + 10.6 = 19.56$

Discuss the first example in the student book, which illustrates how to deal with brackets.

Knowing the value of the square root of a prime number sometimes makes it possible to find the square root of another number.

For example, given that $\sqrt{3} = 1.73$ you can find
$\sqrt{243} = \sqrt{(3 \times 9 \times 9)} = \sqrt{3} \times \sqrt{9} \times \sqrt{9} = \sqrt{3} \times 3 \times 3$
$= 9\sqrt{3} = 9 \times 1.73 = 15.57$
$= 15.6$ to 3 sf

Discuss the second example in the student book.

Exercise commentary and misconceptions

Students should not use a calculator for questions 1–5.

In question 7 remind students to simplify as much as possible first, then use a calculator for the final answer. Make sure they write the simplification down.

In question 9, it is worth emphasising that $\sqrt{2} + \sqrt{3} \neq \sqrt{2+3}$, as this is a common mistake. Discuss why.

Throughout the exercise, remind students to calculate the value of roots if they can during simplifying.

Discuss why multiplying two square roots of the same number gives the number itself as the answer. You could explain that really you are squaring, the inverse of square rooting.

Plenary

Find an exact expression for the circumference of a circle whose area is 5 m^2.

$$(r = \sqrt{\frac{5}{\pi}} \rightarrow d = 2 \times \sqrt{\frac{5}{\pi}} \rightarrow c = 2 \times \pi \times \sqrt{\frac{5}{\pi}}$$
$$= 2 \times \sqrt{\pi} \times \sqrt{\pi} \times \frac{\sqrt{5}}{\sqrt{\pi}} = 2 \times \sqrt{\pi} \times \sqrt{5})$$

The last two steps are beyond a 'B' grade.

N6.2 Exact calculations

This spread will show you how to:
- Use surds and π in exact calculations

Keywords
Approximation
Surds
Surd form

- **Numbers like $\sqrt{2}$ and $\sqrt{5}$ are called surds.**

When calculating, first do the written calculation using surds.
Then use a calculator to find approximate values at the end, if required.

You can carry out written calculations with surds without converting them to decimals.

- **You can separate the square roots of a factorised number.**

For example:
- $36 = 4 \times 9$, so $\sqrt{36} = \sqrt{(4 \times 9)} = \sqrt{4} \times \sqrt{9} = 2 \times 3 = 6$
- $80 = 5 \times 16$, so $\sqrt{80} = \sqrt{(5 \times 16)} = \sqrt{5} \times \sqrt{16} = \sqrt{5} \times 4 = 4\sqrt{5}$

Decimal **approximations** are useful in practical contexts.

If you wanted to mark out a square of area 5 m^2, you would give the side length as 2.24 m, not $\sqrt{5}$ m.

Example

Evaluate **a** $\sqrt{2}(1+\sqrt{2})$ **b** $2\pi(4^2 + 4 \times 6)$

a Multiply each term in the bracket by $\sqrt{2}$.

$\sqrt{2}(1+\sqrt{2}) = \sqrt{2} \times 1 + \sqrt{2} \times \sqrt{2} = \sqrt{2} + 2$

You can then find a decimal approximation:

$\sqrt{2} = 1.414$ to 3 dp so $2 + \sqrt{2} \approx 3.414$

This is normally written as $2 + \sqrt{2}$.

b First simplify, leaving π in the expression.

$2\pi(4^2 + 4 \times 6) = 2\pi(16 + 24) = 2\pi \times 40 = 80\pi$

You can now find a decimal approximation:

$\pi = 3.14$ to 2 dp so $80\pi \approx 251$ to 3 sf.

Leaving an answer in **surd form** is more accurate than a decimal approximation.

Example

Given that $\sqrt{3} = 1.73$ and that $\sqrt{7} = 2.65$ (both to 3 sf), find approximate values for **a** $\sqrt{27}$ **b** $\sqrt{243}$ **c** $\sqrt{28}$ **d** $\sqrt{21}$, without using the square root function on your calculator. Show your working, and give your answers to a suitable degree of accuracy.

a $\sqrt{27} = \sqrt{(9 \times 3)} = \sqrt{9} \times \sqrt{3} = 3\sqrt{3} = 3 \times 1.73 = 5.19$ to 3 sf

b $243 = 3 \times 81 = 3 \times 9 \times 9$

$\sqrt{243} = \sqrt{(3 \times 9 \times 9)} = \sqrt{3} \times \sqrt{9} \times \sqrt{9} = \sqrt{3} \times 3 \times 3 = 9\sqrt{3} = 15.6$ to 3 sf

c $\sqrt{28} = \sqrt{(4 \times 7)} = \sqrt{4} \times \sqrt{7} = 2\sqrt{7} = 5.30$ to 3 sf

d $\sqrt{21} = \sqrt{(3 \times 7)} = \sqrt{3} \times \sqrt{7} = 4.58$ to 3 sf

Exercise N6.2

1 Simplify these expressions.

a $\sqrt{3} + \sqrt{3}$ **b** $\sqrt{5} + \sqrt{5}$ **c** $\sqrt{9} + \sqrt{4}$ **d** $\sqrt{7} + \sqrt{7} + \sqrt{7}$

2 Simplify these expressions.

a $\pi + \pi$ **b** $2\pi + \pi$ **c** $4 + \pi - 2 + \pi$
d $\pi(4^2 - 6)$ **e** $2\pi(4^2 - 7)$ **f** $4\pi(2 + \sqrt{4})$

3 Simplify these expressions.

a $\sqrt{16} + \sqrt{3}$ **b** $\sqrt{49} + \sqrt{2} - \sqrt{16}$
c $3\sqrt{7} + \sqrt{49} - \sqrt{7}$ **d** $17 + \sqrt{17} - \sqrt{9}$

4 Simplify these expressions.

a $\pi(6^2 - 4)$ **b** $\sqrt{49} + \pi(7 - 5)$ **c** $4(7 + \sqrt{2})$ **d** $\pi(8^2 - \sqrt{4})$

5 Work out these, giving your answers in surd form where necessary.

a $\sqrt{2} \times \sqrt{2}$ **b** $\sqrt{5} \times \sqrt{5}$ **c** $\sqrt{3}(\sqrt{3} + 3)$ **d** $\sqrt{4}(\sqrt{3} + 4)$

6 Use a calculator to find an approximate decimal value for each of these expressions. Give your answers to 2 decimal places.

a $4\sqrt{2}$ **b** $\sqrt{5} + 1$ **c** $2 + \sqrt{5}$ **d** $36\pi - 7$

7 Simplify these expressions, then use a calculator to find approximate decimal values for each one.
Give your answers correct to 3 significant figures.

a $4(3 + \sqrt{5})$ **b** $\sqrt{5}(5^2 - 5)$ **c** $\sqrt{7}(2 + \sqrt{7})$ **d** $\pi(2^3 + \sqrt{5})$

8 Simplify these expressions.

a $\sqrt{20}$ **b** $\sqrt{125}$ **c** $\sqrt{18}$ **d** $\sqrt{98}$

9 You are told that $\sqrt{2} \approx 1.414$, $\sqrt{3} \approx 1.732$ and $\sqrt{5} \approx 2.236$. Use this information to estimate the value of each of these, without using the square root button on your calculator. Show your working, and give your answers to 4 significant figures.

a $\sqrt{2} + \sqrt{3}$ **b** $\sqrt{10}$ **c** $\sqrt{125}$ **d** $\sqrt{24}$

10 Use the square root key on your calculator to work out the answers to question **9**. Compare your answers, and explain which answers are more accurate.

11 Simplify these expressions by multiplying out the brackets.

a $(3 + \sqrt{2})(4 + \sqrt{2})$ **b** $(4 + \sqrt{5})(3 + \sqrt{5})$
c $(6 + \sqrt{3})(3 - \sqrt{3})$ **d** $(5 + \sqrt{5})(5 - \sqrt{5})$

N6.3 Mental calculations

Objectives

- Use mental and written methods to multiply and divide with decimals
- Understand place value and where to place the decimal point

Useful resources

- Place value table

Mental starter

Which of these gives an answer smaller than 16?

16×0.94, $16 \div 1.45$, $16 \div 0.5$, 16^0, $\sqrt[3]{16}$, $16^{2.3}$.

(16×0.94, $16 \div 1.45$, 16^0, $\sqrt[3]{16}$.)

Discuss how you can tell without actually working out the answers.

Introductory activity

Remind students to always start decimal calculations with an estimate.

Start with the example of 12.3×0.2:
$12.3 \times 0.2 \approx 10 \times 0.2 = 2$
$123 \times 2 = 146$, so $12.3 \times 0.2 = 1.46$.
This follows the rule that multiplying a positive number by a number between 0 and 1 makes it smaller.

Now try $12.3 \div 0.2$.
$12.3 \div 0.2 \approx 10 \div \frac{1}{5} = 10 \times 5 = 50$
$123 \div 2 = 61.5$, so $12.3 \div 0.2 = 61.5$.
This follows the rule that dividing a positive number by a number between 0 and 1 makes it bigger.

What are 0.1, 0.2, 0.3, 0.4, 0.5, 0.6, 0.7, 0.8, 0.9, 0.25 and 0.75 when written as fractions?

($\frac{1}{10}$, $\frac{1}{5}$, $\frac{3}{10}$, $\frac{2}{5}$, $\frac{1}{2}$, $\frac{3}{5}$, $\frac{7}{10}$, $\frac{4}{5}$, $\frac{9}{10}$)

Exercise commentary and misconceptions

Questions 1, 2 and 3 are best done using the fractional equivalents of the decimal numbers given above.

Encourage students to cancel fractions used in multiplication.

Questions 9 and 11 are the only ones where calculators should be used.

The rules that multiplying a positive number by a number between 0 and 1 makes it smaller and that dividing a positive number by a number between 0 and 1 makes it bigger are difficult for some students to accept, so it may be useful to write them on the board and remind students of them throughout the exercise.

Plenary

Put these answers in order of size, smallest first: $20 \div 0.345$, $26 \div 0.057$, 17.58×0.9678, 4.789×1.6, $4.5 \div 1.487$

($4.5 \div 1.487$, 4.789×1.6, 17.58×0.9678, $20 \div 0.345$, $26 \div 0.057$.)
Discuss how you can achieve the answer without calculating any values.

N6.3 Mental calculations

This spread will show you how to:
- Use mental and written methods to multiply and divide with decimals
- Understand place value and where to place the decimal point

Keywords
Inverse
Place value

- **Multiplying a positive number by a number between 0 and 1 makes it smaller.**
  $6 \times 0.5 = 3$
- **Dividing a positive number by a number between 0 and 1 makes it bigger.**
  $6 \div 0.5 = 12$

Always start calculations with an estimate.

- Work out a mental calculation using the significant digits from the question; for example, for $12.5 \div 0.05$, work out $125 \div 5$.
- Finally, use your initial estimate to check your answer and adjust the **place value**.
- Use **inverses** where possible.

  $10 \div 0.2 = 10 \times 5$ and $10 \times 0.25 = 10 \div 4$

$12.5 \div 0.05 \approx 10 \div 0.05 = 200$

$125 \div 5 = 25$

$12.5 \div 0.05 = 250$

Example

Use a mental method to work out

a 320×0.4 **b** $320 \div 0.4$ **c** $3.2 \div 0.4$

a $320 \times 0.4 \approx 300 \times 0.4 = 120$
$32 \times 4 = 64 \times 2 = 128$
$320 \times 0.4 = 128$

b $320 \div 0.4 \approx 320 \div \frac{1}{2} = 640$
$32 \div 4 = 8$ so
$320 \div 0.4 = 800$

c $3.2 \div 0.4 \approx 3 \div \frac{1}{2} = 6$
$32 \div 4 = 8$ so
$3.2 \div 0.4 = 8$

Estimate first.

Compare with estimate to get final answer.

Example

Write a multiplication that is equivalent to $34.5 \div 0.25$.

Dividing by a quarter is the same as multiplying by 4.
$34.5 \div 0.25 = 34.5 \times 4$

$34.5 \div 0.25 = 138$
$34.5 \times 4 = 138$

Example

Calculate mentally. **a** $72.5 \div 0.05$ **b** 340×0.3 **c** $8.46 \div 0.2$

a $72.5 \div 0.05 \approx 70 \times 200 = 1400$
$725 \div 5 = 145$ so
$72.5 \div 0.05 = 1450$

b $340 \times 0.3 \approx 300 \div 3 = 100$
$340 \times 3 = 1020$ so
$340 \times 0.3 = 102$

c $8.46 \div 0.2 \approx 9 \times 5 = 45$
$8.46 \div 2 = 4.23$ so
$8.46 \div 0.2 = 42.3$

$0.05 = \frac{1}{200}$

$0.3 \approx \frac{1}{3}$

$0.2 = \frac{1}{5}$

Exercise N6.3

1 Use a mental method to find these.

a 5×0.2 **b** 4×0.3 **c** 0.5×3 **d** 16×0.5
e 7×0.2 **f** 30×0.25 **g** 40×0.4 **h** 0.6×25

2 Use a mental method to find these.

a $8 \div 0.2$ **b** $4 \div 0.4$ **c** $6 \div 0.3$ **d** $32 \div 0.4$
e $0.8 \div 4$ **f** $0.3 \div 0.03$ **g** $0.4 \div 0.04$ **h** $50 \div 0.01$

3 Use a mental method to find these. Start with an estimate, and show your working.

a 2×0.4 **b** 20×0.04 **c** 3×7 **d** 0.3×0.7
e 12×0.4 **f** $12 \div 0.3$ **g** $3.6 \div 4$ **h** $3.6 \div 0.9$

4 Use a calculator to check your answers to question **3**.

5 Write a division that is equivalent to each of these multiplications.

a 4×0.5 **b** 6×0.2 **c** 12×0.2 **d** 2×0.001

6 Write a multiplication that is equivalent to each of these divisions.

a $4 \div 0.5$ **b** $6 \div 0.25$ **c** $16 \div 0.01$ **d** $15 \div 0.05$

7 Use a mental method to find these. Show your method.

a $8 \div 0.25$ **b** $15 \div 0.5$ **c** $7 \div 0.2$ **d** 4×0.25
e 16×0.02 **f** $24 \div 0.12$ **g** 20×0.05 **h** $18 \div 0.025$
i 3×0.125 **j** $7 \div 0.25$ **k** 40×0.025 **l** 8×0.875

8 Use a mental method to find an estimate for each of these.

a $37 \div 0.47$ **b** $319 \div 0.3$ **c** 3.8×134 **d** $17 \div 0.031$

9 Use a calculator to find the exact answers to question **8**.

10 Use a mental method to work out these. Start each one with an estimate, and show your method.

a 31×0.3 **b** $49 \div 0.07$ **c** $3.66 \div 0.3$ **d** $4.24 \div 0.4$
e $13.9 \div 0.03$ **f** $3.9 \div 0.03$ **g** $171 \div 0.3$ **h** 5.2×0.125

11 Use a calculator to check your answers to question **10**.

N6.4 Written calculations

Objectives

- Use mental and written methods to multiply and divide with decimals
- Estimate answers to calculations, using these to check the solution

Useful resources

- 5 mm squared paper
- Calculators

Mental starter

Calculate

3.2 – 1.6 + 0.8 – 0.4 + 0.2 – 0.1 + 0.05 – 0.025 + ... How many terms can you use?

(From the list above the answer is 2.125.
Successive terms being added give the answers 2.1375, 2.13125, 2.134375, 2.1328125.
The actual answer if you carry on for ever is 2.13333333 ....)

Introductory activity

When adding and subtracting decimals line up the decimal point and add zeros in the decimals if necessary.

Multiplying and dividing decimals are done using a written method for the same numbers without decimal points, then using an estimate to give the correct place values.

Calculate 18.5 × 7.9
20 × 8 = 160
185 × 79 = 14 615,
so 18.5 × 7.9 = 146.15

Calculate 47.592 ÷ 1.8
50 ÷ 2 = 25
47 592 ÷ 18 = 2644,
so 47.592 ÷ 1.8 = 26.44

Exercise commentary and misconceptions

Questions 1–5 require the students to align their written work accurately; and those who find this difficult may be helped by using 5 mm squared paper, though this should only be used for a very few examples.

In questions 7–10 emphasise that students must show their working, including the numbers used to find the estimate.

Calculators are needed only for questions 6 and 11.

Plenary

A ball is dropped from a height of 8 m. Each time it bounces it only rises 0.2 times its previous height. **How many bounces before it rises less than 6 cm?**
(4 bounces. Height after each bounce is 1.6 m, 0.32 m, 0.064 m, less than 0.06 m.)

N6.4 Written calculations

This spread will show you how to:
- Use mental and written methods to multiply and divide with decimals
- Estimate answers to calculations, using these to check the solution

Keywords
Check
Estimate
Place value

For column methods of addition and subtraction, use the full decimal numbers as given in the question.

Use the decimal point to ensure that the columns are aligned correctly.

Example

Calculate

a 135.23 + 27.8 **b** 34.56 − 18.729

a Estimate 140 + 30 = 170

```
   1 3 5 . 2 3
 +   2 7 . 8
   1 6 3 . 0 3
     1 1
```

b Estimate 30 − 20 = 10

```
  2 3 13 4 . 15 5 6 10
 −  1  8 .  7  2  9
    1  5 .  8  3  1
```

Use your estimates to **check** your answers.

For standard methods of multiplication and division, work with the significant digits from the numbers in the question.

Use an estimate to adjust the place value correctly.

Example

Calculate

a 18.5 × 7.9 **b** 47.592 ÷ 1.8

a Estimate 20 × 8 = 160

```
      1 8 5
    ×   7 9
    1 6₇6₄5
  1 2₅9₃5 0
  1 4₁6₁1 5
```

Use your **estimate** to check your answer.

18.5 × 7.9 = 146.15

146.15 is close to 160, so this is about right.

b Estimate 50 ÷ 2 = 25

This could be done by long division (see unit N2.4) or by short division. Using short division you get:

```
      2  6  4  4
 18)4 7 ¹¹5 ⁷9 ⁷2
```

Use your estimate to adjust the **place values**.

So 47.592 ÷ 1.8 = 26.44

20 × 8 = 160

50 ÷ 2 = 25

Exercise N6.4

You should show your working for the questions in this exercise.

1 Use a written method to calculate these.

a 24.72 + 14.04 **b** 1.52 + 1.09 **c** 6.149 + 2.052 **d** 6.64 + 15.88

2 Use a written method to calculate these.

a 5.23 − 3.11 **b** 17.45 − 13.26 **c** 6.41 − 4.37 **d** 23.6 − 17.9

3 Use a written method to calculate these.

a 1.09 + 154 **b** 0.09 + 0.36 **c** 14.52 + 9.8 **d** 13.92 + 0.8

4 Use a written method to calculate these.

a 4.5 − 0.53 **b** 3.085 − 2.99 **c** 16.3 − 3.86 **d** 112.14 − 53.8

5 Use a written method to evaluate these.

a 11.1 − 8.29 **b** 2.09 − 1.333 **c** 102.8 − 14.79 **d** 978 + 148.72

6 Use a calculator to check your answers to questions **1** to **5**.

7 Use a written method to work out these multiplications.

a 15.9 × 4 **b** 17.9 × 0.3 **c** 16.9 × 0.8 **d** 0.048 × 0.07

8 Calculate these divisions, using a written method.

a 1.36 ÷ 0.8 **b** 3.01 ÷ 7 **c** 19.2 ÷ 0.4 **d** 13.45 ÷ 0.05

9 Use long multiplication (or an equivalent written method) to evaluate these.

a 8.8 × 1.9 **b** 190 × 0.054 **c** 189 × 4.2 **d** 214 × 0.037

10 Use long division (or an equivalent written method) to evaluate these.

a 211.68 ÷ 24 **b** 133.98 ÷ 0.66 **c** 292.38 ÷ 0.33 **d** 5.913 ÷ 0.27

11 Use a calculator to check your answers to questions **7** to **10**.

N6.5 Calculator methods

Objectives

- Use calculators effectively and efficiently, knowing when and when not to round the display
- Use calculators to calculate the upper and lower bounds of calculations

Useful resources

- Calculators

Mental starter

Ask students to calculate each of these products and quotients using a mental method:

12×34	$\frac{7}{20} \times 45$
123×14	$-0.1 \div 5$
$4.8 \div 0.3$	$3.2 \div 0.002$
$10.2 \div 4$	$19 \times -3\frac{2}{3}$

Introductory activity

Before starting the activity students will need reminding where the power button is on their calculator. Most scientific calculators also have a 'cube' button.

For example, calculate $(5.32^3 + \sqrt{3.41})^4$.
Calculator display is 539 653 809.5.
As 3 sf were used in the question, it is reasonable to use 3 sf in the answer, so answer is 540 000 000 to 3 sf. Some calculators can be set in scientific mode with a given number of significant figures.

Discuss the first example, which links order of operations to key sequences.
Progress to a discussion of upper and lower bounds in measurement.

If the diameter of a circle is given as 7.2 m what are the upper and lower bounds for the diameter? (7.15 m and 7.25 m.) Discuss why 7.25 m has to be used and not 7.24, 7.249, 7.24999999. Point out that 7.25 is not closer to 7.3 than 7.2, it is exactly in the middle.

What are the upper and lower bounds of the circumference of the circle? Give your answer to 2 sf.
$7.15 \times \pi = 22$ m (2 sf) $7.25 \times \pi = 23$ m (2 sf).
Students may need reminding of the formula.

Discuss the second example:
Speed = distance ÷ time. **If the distance is in the range 3.85 m to 3.95 m and the time is in the range 7.25 s to 7.35 s, what is the lowest value you can find for the speed?** ($3.85 \div 7.35 = 0.523809$) **And what is the highest value you can find for the speed?** ($3.95 \div 7.25 = 0.544827$) These are the lower and upper bounds for the calculation.

The maximum value from a division calculation is found by dividing the largest value by the smallest value, and the minimum value is vice versa. The maximum value from a multiplication is found by multiplying the two largest values together.

Exercise commentary and misconceptions

Questions 1 and 2 are straightforward, but some students will find question 3 difficult. It may be useful to use question 3 in the plenary session.

Question 4 gives the students detailed instructions for the method to be used, but they can be encouraged to experiment with the values in question 5.

Success in question 6 depends on the students' understanding of the upper and lower bounds of measurement, so it may be worthwhile reviewing this question in the plenary session.

Plenary

Discuss questions 5 and 6.
Focus on which numbers to select in the calculation in order to obtain the maximum and minimum values.

N6.5 Calculator methods

This spread will show you how to:

- Use calculators effectively and efficiently, knowing when and when not to round the display
- Use calculators to calculate the upper and lower bounds of calculations

Keywords
Display
Effective
Efficient
Round

You need to know how to enter calculations into a calculator, and how to interpret the calculator **display**.

Calculator answers often need to be **rounded** to a suitable degree of accuracy.

Measurements (like lengths and weights) are correct to a certain degree of accuracy. You can use a calculator to find the upper and lower bounds of calculations involving measurements.

The length of a pipe given as 3.95 m is rounded to the nearest cm. The actual length is between 3.945 m and 3.955 m.

3.945 3.955
3.94 3.95 3.96

Example

Use a calculator to find the value of (3.59 − 1.68) ÷ (2.4 × 6.9). Write all of the digits on the calculator display, and round your answer to a suitable degree of accuracy.

A possible key sequence is:

(3 . 5 9 − 1 . 6 8) ÷ (2 . 4 × 6 . 9) =

The answer on a 10-digit display is 0.115338164

Use 2 sf as a suitable degree of accuracy:

the answer is 0.12 (2 sf)

Do not round before the end of a calculation. You should not give an answer with more significant figures than there were in the question.

Example

A model boat travels 3.9 metres in 7.3 seconds. Both measurements are correct to 1 dp. Find the upper and lower bounds of the speed of the boat in metres per second.

Speed = distance ÷ time.

The distance 3.9 m is a rounded value in the range 3.85 m to 3.95 m, and 7.3 s is in the range 7.25 s to 7.35 s.

To get the maximum value of the speed, divide the largest possible distance by the shortest possible time:

The upper bound of the calculation is
3.95 ÷ 7.25 = 0.544827 ... metres per second

To get the minimum value of the speed, divide the smallest possible distance by the greatest possible time.

The lower bound of the calculation is
3.85 ÷ 7.35 = 0.523809 ... metres per second

$\text{Speed}_{\text{Max}} = \text{Distance}_{\text{Max}} \div \text{Time}_{\text{Min}}$

Similarly,
$\text{Speed}_{\text{Min}} = \text{Distance}_{\text{Min}} \div \text{Time}_{\text{Max}}$

Exercise N6.5

1 Use a calculator to evaluate these, giving each answer to 2 sf.

a $3.2 \times (2.8 - 1.05)$
b $2.8^2 \times (9.4 - 0.083)$
c $16 \div (5.1^2 - 7.2)$
d $(3.8 + 8.9) \times (2.2^2 - 7.6)$
e $1.8^3 + 4.7^3$
f $52 \div (4.6 - 1.8^2)$

2 Work out these, giving your answers to a suitable degree of accuracy.

a The total weight of two people weighing 68 kg and 73 kg.
b The area of a rectangle with sides 2.2 m and 3.8 m.
c The cost of 2.37 m of material at £5.75 per metre.
d The time taken to travel 1 mile at a speed of 80 mph.

3 Write the upper and lower bounds of these measurements, which are all given to 3 significant figures.

a 4.75 m
b 12.6 s
c 150 cm
d 24.5 kg
e 8.07 g
f 4.33 s

You can draw a number line to help you.

4 A model car was rolled down a track. The length of the track was measured as 2.55 m, to the nearest centimetre. The time for the journey was 1.7 seconds, measured to the nearest tenth of a second.

a Write the upper and lower bounds for the length of the track.
b Write the upper and lower bounds for the time of the journey.
c Use the relationship

$$\text{Speed}_{\text{Max}} = \text{Distance}_{\text{Max}} \div \text{Time}_{\text{Min}}$$

to find the maximum possible average speed of the car.

d Use the relationship

$$\text{Speed}_{\text{Min}} = \text{Distance}_{\text{Min}} \div \text{Time}_{\text{Max}}$$

to find the minimum possible average speed of the car.

5 Find the maximum and minimum area of squares with side lengths given as

a 8.00 m
b 6.40 cm
c 1.05 m
d 3.00 mm
e 3.75 m
f 9.99 cm

Give your answers to 3 sf.

6 The numbers in these calculations all relate to measurements. Find the upper and lower bound of each calculation.

a $6.5 \times (1.2 + 4.6)$
b $1.2 - 0.8$
c $(3.77 - 3.22) \times (2.43 - 1.75)$
d $\dfrac{(3.2 - 1.9)^2}{4.5 + 8.8}$

N6 Exam review

Key objectives

- Use the hierarchy of operations
- Use calculators effectively and efficiently
- Develop a range of strategies for mental calculation
- Use standard column procedures for multiplication of integers and decimals
- Use calculators, or written methods, to calculate the upper and lower bounds of calculations
- Check and estimate answers to problems

Worked solution	Commentary and misconceptions
1	Encourage students to estimate answers to calculations in order to check their solutions and to set out their working clearly.
a **i** $\begin{array}{r} 250 \\ \times\ 5 \\ \hline 1250 \end{array}$ **ii** $\begin{array}{r} 250 \\ \times\ 0.5 \\ \hline 125 \end{array}$	Ensure that students understand place value and its effect on calculations. They should recognise that these solutions differ by one place value although some students may need prompting.
b $250 \div 0.5 = 250 \div \frac{1}{2}$ $= 250 \times 2.$	Prompt students to convert 0.5 to $\frac{1}{2}$ if needed. This question tests students' understanding of multiplicative inverses. Students are not required to provide a solution to this calculation.
2	These questions test the students' understanding of the order of operations and using brackets. Some students may need reminding of the related rules.
a $(2.3 + 1.8)^2 \times 1.7 = 28.577$	Ensure that students use the correct order of operations, including squaring the whole bracket. Encourage them to write the intermediate steps of the calculation, $4.1^2 \times 1.7$, if this helps. Some students may still confuse squaring and doubling.
b $(1.6 + 3.8 \times 2.4) \times 4.2 = 45.024$	Ensure that students understand what the question is asking. Note that no brackets are needed around 3.8×2.4.

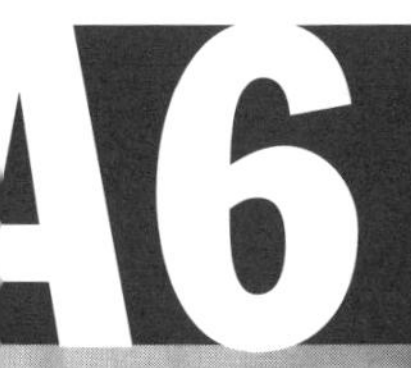

Simultaneous and quadratic equations

Objectives

F/H Solve linear equations that require prior simplification of brackets

H Solve quadratic equations by factorisation

F/H Use systematic trial and improvement to find approximate solutions of equations where there is no simple analytical method of solving them

F/H Substitute numbers into a formula

F/H Present answers to sensible levels of accuracy

H Find the exact solutions of simultaneous equations in two unknowns by eliminating a variable

H Distinguish the different roles played by letter symbols in algebra, using the correct notational conventions for multiplying or dividing by a given number

H Set up and use equations to solve word and other problems

Unit overview

This unit uses the knowledge gained in previous algebra units to solve more complex linear and quadratic equations, including using trial and improvement methods and extending to solving simultaneous equations. It provides a basis for work on graphical solutions covered in unit A7.

Prior knowledge

Before your students start this unit they should be able to:

- Solve simple linear equations
- Expand and factorise single and double brackets
- Understand and use index notation and index laws
- Understand and use the correct order of operations in algebra
- Understand and use fraction and decimal notation
- Round solutions to any number of significant figures
- Use formulae from mathematics and other subjects
- Derive formulae to solve problems
- Substitute numbers into formulae

Differentiation

- **Foundation** focuses on solving simple equations using inverse operations
- **Foundation Plus** extends to equations with the unknown on both sides, including those involving brackets and fractions. It also covers solving simple quadratic and cubic equations using trial and improvement
- **Higher Plus** extends the Higher book to solving simultaneous equations involving quadratics, the equation of a circle and solving inequalities, representing the solution set on a number line

A6.1 Solving harder equations

Objectives

- Solve linear equations after simplifying them

Useful resources

- Mini-whiteboards

Mental starter

Say in words 'What is 12 take away the sum of 3 and 5?'
Say in words 'What is 7 take away the difference between 4 and 10?
Say in words 'What is 20 take away 2 times the sum of x and 7?' Once students have tried each of these, write them out in the following ways to check the answers.

$12 - (3 + 5) = 12 - 3 - 5 = 4$.
$7 - (4 - 10) = 7 - 4 + 10 = 13$.
$20 - 2(x + 7) = 20 - 2x - 14 = 6 - 2x$.
In the first two examples students can check the answers are as they expected.

Using these ideas remove the brackets and simplify this expression
$2(3x + 5) - 3(x - 2) = 6x + 10 - 3x + 6 = 3x + 16$.

Each time ask the question 'What is the bracket being multiplied by?' Answers are −1, −1, −2, 2 and −3.

Introductory activity

Highlight these top tips for solving linear equations:

(i) Remove brackets in order to get a string of terms.
(ii) Simplify both sides of the equation if possible before you solve it.
(iii) If there are letters on both sides always remove them from one side.
(iv) Get rid of minus xs, they cause real problems in equations.

For example

(a) $2 + 3(x + 4) + 2x = 19 - 3x$
$\Rightarrow 2 + 3x + 12 + 2x = 19 - 3x$
$\Rightarrow 14 + 5x = 19 - 3x$
$\Rightarrow 14 + 8x = 19$
$\Rightarrow 8x = 5$
$\Rightarrow x = \frac{5}{8}$
Check that $x = \frac{5}{8}$ satisfies the original equation.

(b) $(x + 3)^2 = 3 - (4 - 3x) + x^2$
$\Rightarrow x^2 + 6x + 9 = 3 - 4 + 3x + x^2$
$\Rightarrow 6x + 9 = -1 + 3x$
$\Rightarrow 3x + 9 = -1$
$\Rightarrow 3x = -10$
$\Rightarrow x = -\frac{10}{3}$
Check that $x = -\frac{10}{3}$ satisfies the original equation.

Challenge students to form an equation for this problem then solve it.
I think of a number, add 7 then multiply my answer by 3. I get the answer 15.

If the start number is x, equation is
$3(x + 7) = 15 \Rightarrow 3x + 21 = 15$
$\Rightarrow 3x = -6$
$\Rightarrow x = -2$.

Discuss the examples in the student book.

Exercise commentary and misconceptions

This exercise contains equations which must be solved using a number of steps. Ensure that students write their working in on both sides. Some students might like to use a different colour for their workings. Make sure students always think of the bracket as being multiplied by some number, positive or negative.

Questions 2**g** and **h** look like quadratic equations but the terms in x^2 cancel.

Question 5 requires students to form and solve an equation from geometrical information.

Plenary

Dave is 20 cm taller than Fran. Their heights sum to 326 cm. **How tall are they both?**
Try to do this by forming an equation. Call Dave's height x so Fran's height is $x - 20$. Add their heights to get $2x - 20$, this equals 326.
So $2x - 20 = 326$
$\Rightarrow 2x = 346$
$\Rightarrow x = 173$ cm.

So Dave's height is 173 cm and Fran's height is 153 cm.

A6.1 Solving harder equations

This spread will show you how to:

- Solve linear equations after simplifying them

Keywords
Expand
Negative term

When solving an equation with brackets, **expand** the brackets first.

$4(x + 2) = 7(x - 1)$ Expand the brackets.
$4x + 8 = 7x - 7$ Subtract 4x from both sides.
$8 = 3x - 7$ Add 7 to both sides.
$15 = 3x$ Divide both sides by 3.
$x = 5$

An equation with a negative x-term is easier to solve if you get rid of the **negative term** first.

$5 - 8x = 45$ Add 8x to both sides to remove the –8x
$5 = 45 + 8x$ Subtract 45 from both sides.
$-40 = 8x$ Divide both sides by 8.
$x = -5$

Example

Solve this equation. $(x + 3)(x + 5) = (x + 2)^2$

$(x + 3)(x + 5) = (x + 2)^2$
$x^2 + 3x + 5x + 15 = (x + 2)(x + 2)$ Remove brackets first (FOIL): *multiply* the terms.
$x^2 + 8x + 15 = x^2 + 2x + 2x + 4$ Collect like terms.
$x^2 + 8x + 15 = x^2 + 4x + 4$ Subtract x^2 from each side.
$8x + 15 = 4x + 4$
$4x + 15 = 4$
$4x = -11$
$x = \frac{-11}{4}$ or $-2\frac{3}{4}$

Don't use a calculator – fractions are exact. Decimals that need rounding, make the **solution** less accurate.

Example

The square and the rectangle have the same area.
Find the value of x.

x
x + 5
x − 2

Area of square = Area of rectangle
$x^2 = (x + 5)(x - 2)$ Remember brackets first (FOIL)
$x^2 = x^2 + 5x - 2x - 10$ Collect like terms.
$x^2 = x^2 + 3x - 10$ Subtract x^2 from each side.
$0 = 3x - 10$
$-3x = -10$ Divide by –1.
$3x = 10$
$x = 3\frac{1}{3}$

Exercise A6.1

1 Solve these equations.

a $4x - 9 = 15$ b $\frac{y}{7} + 2 = 4$
c $3x + 9 = 2x + 15$ d $5p - 9 = 3p + 7$
e $10 - 3t = 8t - 12$ f $3 - 7b = 9 - 10b$

2 Solve these equations by first expanding brackets and/or collecting like terms.

a $3x + 7x - 3 + 9 = 17$ b $3(2y - 1) = 4(7y + 6)$
c $2(3 - 8z) = 4(5 - 6z)$ d $2a + 3(2a - 7) = 20$
e $3a - (9 - 7a) = 34$ f $6 - (3x - 9) = -2(5 - x)$
g $(x + 3)(x + 4) = (x + 7)(x - 2)$ h $(y - 7)^2 = (y + 5)^2$

3 Solve these equations containing negative terms.

a $10 - 4x = 2$ b $20 - 3y = 11$ c $15 - 8z = 12$
d $14 - 7a = 19$ e $18 = 17 - b$ f $36 = 2(3 - 6c)$

4 In each case, write an equation to represent the information given. Solve the equation to find the starting number.

a I think of a number, treble it and subtract this *from* 40. My answer is 22.

b I think of a number, add 6 and double the result. This gives me the same answer as when I subtract 5 from the number and multiply the result by 6.

c I think of a number, add 7 and square it. This gives me the same answer as when I take 5 from the number and square it.

5 In each case, form an equation to represent the information given. Solve it to find x.

a

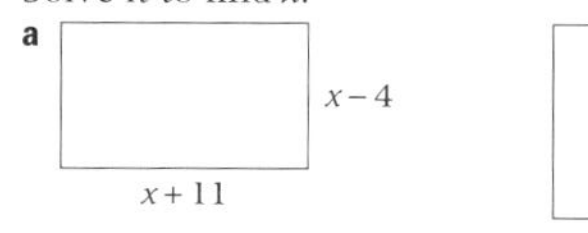

x + 3

The area of the square and rectangle are equal.

b

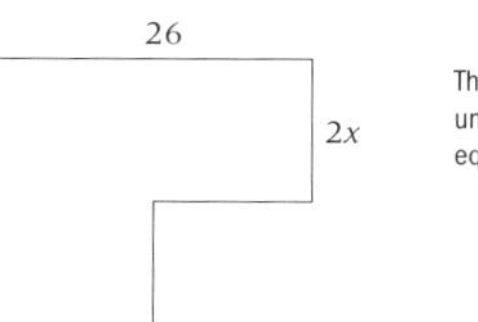

The lengths of the unmarked sides are equal.

A6.2 Introducing quadratic equations

Objectives

- Solve quadratic equations by factorisation

Useful resources

- OHT of examples in student book

Mental starter

What two numbers can you multiply to get the answer zero? Or phrase the question as if $a \times b = 0$ what could a and b equal?

Establish that if two numbers are multiplied together and the answer is zero then one or both of the numbers must be zero.

Introductory activity

First expand and simplify
$(x + 8)(x - 2) \Rightarrow x^2 + 8x - 2x - 16$
$\Rightarrow x^2 + 6x - 16$. Ask where the x^2 came from. Ask where the 16 came from. Ask where the $6x$ came from. Remind students that factorising involves putting terms back into brackets.

Students should have previously practiced factorising quadratics in section A1.5.
Go through the first two examples in the introduction to A1.5 again. Emphasise the order in which the three parts are created.

If a quadratic expression is equal to zero, it is sometimes possible to solve it by factorising the expression.

For example $x^2 + 4x + 3 = 0 \Rightarrow (x + 1)(x + 3) = 0$
$(x + 1)$ and $(x + 3)$ are both numbers, and one of them must equal zero because their product is zero.
So **either** $x + 1 = 0$, and $x = -1$,
or $x + 3 = 0$, and $x = -3$.

Students should try questions 1, 2 and 3 before the rest of the introduction.

Some quadratics factorise into a single bracket and another term.

For example $x^2 - 3x = 0 \Rightarrow x(x - 3) = 0$
x and $(x - 3)$ are both numbers, and one of them must equal zero because their product is zero.
So **either** $x = 0$
or $x - 3 = 0$ and $x = 3$.

Exercise commentary and misconceptions

Encourage students to keep checking their factorisation by multiplying out the brackets. When students have to tackle equations which do not equal zero, as in questions 4**e**, 6 and 7, ask them how they can make the RHS equal zero.

If students are finding factorisation difficult, encourage them to write down the pairs of numbers whose product is the constant term, and check whether any of these pairs have a difference which is the coefficient of the linear term.

In question 4**g** the expression is already factorised (some students who are working on auto-pilot may expand the brackets and then factorise.

Question 3**h** needs care. Refer students to the last example in the student book.

Plenary

Challenge students to attempt these three equations.

(i) $x^2 = -25$
(ii) $x^2 + 10 = 5$
(iii) $x^2 + x + 10 = 0$.

Suggest various approaches:

(a) Different values of x in your head
(b) Factorising
(c) Trial and improvement with a calculator. (There are in fact **no** answers. Not all quadratic equations have an answer.)

A6.2 Introducing quadratic equations

This spread will show you how to:

- Solve quadratic equations by factorisation

Keywords
Factorise
Product
Quadratic
Solution
Sum

- A **quadratic** equation contains an x^2 term as the highest power, for example
  $x^2 + 5x + 6 = 0$

Many quadratic equations can be solved by **factorising**.

For example, solve the equation $x^2 + 5x + 6 = 0$.

x^2 term, so equation is a quadratic.

$x^2 + 5x + 6 = 0$ — The two factors will each start with x.

$(x + \quad)(x + \quad) = 0$ — Now find two numbers with a **sum** of 5 (for $5x$) and a **product** of 6.

$(x + 3)(x + 2) = 0$ — Check; $(x + 3)(x + 2) = x^2 + 3x + 2x + 6 = x^2 + 5x + 6$

The two factors have a **product** of zero. This means that at least one of them is equal to zero.

Either $x + 3 = 0$ or $x + 2 = 0$
If $x + 3 = 0$, $x = -3$ and if $x + 2 = 0$, $x = -2$
The **solutions** are $x = -3$ and $x = -2$.

Check $(-3)^2 + 5(-3) + 6 = 9 - 15 + 6 = 0$
and $(-2)^2 + 5(-2) + 6 = 4 - 10 + 6 = 0$

- **The key to solving quadratic equations is that the product of the two factors is zero, so you can say that one or other (or both) of the factors is zero. This leads to the two solutions.**

Example

Solve these equations.

a $x^2 - 7x + 12 = 0$ **b** $x^2 + 3x = 0$

a $x^2 - 7x + 12 = 0$
$(x - 3)(x - 4) = 0$
Either $x - 3 = 0$ or $x - 4 = 0$
$x = 3$ $\quad$ $x = 4$

To factorise, you need two negatives to multiply to make +12 and add to make −7.

b $x^2 + 3x = 0$
$x(x + 3) = 0$
Either $x = 0$ or $x + 3 = 0$
$x = -3$

Don't forget to look out for common factors as well as double bracket style factorisation.

Example

Solve $x^2 = 2x + 15$

$x^2 - 2x - 15 = 0$ — Find two numbers with sum −2 and product −15.
$(x + 3)(x - 5) = 0$
Either $x + 3 = 0$ or $x - 5 = 0$
$x = -3$ $\quad$ $x = 5$

First make the left hand side of the equation equal to zero

Exercise A6.2

1 Here are six equations.

$5x + 3 = 3x$ $\quad$ $x^3 + x = 10$ $\quad$ $2x^2 = 50$ $\quad$ $x^2 = 100$ $\quad$ $x^2 = -49$ $\quad$ $x^2 = 169$

a Write the four quadratic equations.

b From the quadratic equations, find three that have two solutions. Write these three equations and their solutions.

2 Factorise these quadratic expressions.

a $x^2 + 8x + 12$ **b** $x^2 + 11x + 24$ **c** $x^2 + 13x + 36$
d $x^2 + 16x + 55$ **e** $x^2 - 7x + 12$ **f** $x^2 - 10x + 24$
g $x^2 + 3x - 28$ **h** $x^2 - 2x - 15$

3 Solve these quadratic equations by factorising them into double brackets.

a $x^2 + 7x + 12 = 0$ **b** $x^2 + 8x + 12 = 0$ **c** $x^2 + 10x + 25 = 0$
d $x^2 + 5x - 14 = 0$ **e** $x^2 - 4x - 5 = 0$ **f** $x^2 - 5x + 6 = 0$
g $x^2 - 10x + 21 = 0$ **h** $x^2 = 3x + 40$

4 Solve these quadratic equations by factorising them into a single bracket.

a $x^2 - 8x = 0$ **b** $x^2 + 4x = 0$ **c** $x^2 - 6x = 0$
d $y^2 + 5y = 0$ **e** $x^2 = 9x$ **f** $x^2 - 12x = 0$
g $2x^2 + 8x = 0$ **h** $6x - x^2 = 0$

5 Explain why $x^2 + 7x + 11 = 0$ cannot be solved by factorisation.

6 **a** Vicky is trying to solve $(x + 4)(x - 2) = 7$. Here is her attempt.

$(x + 4)(x - 2) = 7$
Either $x + 4 = 7$ or $x - 2 = 7$
$x = 3$ or $x = 9$

What is wrong with her method?

b Can you solve this equation correctly to show that the two solutions should really be $x = 3$ and $x = -5$?

7 Simplify and then solve this equation.

$x^2 + 4x - 5 = 2x(x - 1)$

A6.3 Solving equations using trial and improvement

Objectives

- Use trial and improvement to find approximate solutions of equations
- Substitute positive and negative numbers into expressions

Useful resources

- Calculators
- Blank tables for trial and improvement

Mental starter

I have a large 5 litre jug and I want to fill it as much as possible using 2 other jugs. The small one holds 300 ml, the other holds 1.5 litres. I can use these two smaller jugs as many times as I like but I must use all the liquid in them. **How can this be done?**

(One option is $3 \times 1.5 = 4.5$ litres plus $1 \times 300 = 300$ ml. Total = 4.8 litres. Other possibilities are 2 medium and 6 small (4.8 litres) or 1 medium and 11 small (4.8 litres) or 0 medium and 16 small (4.8 litres).)

Introductory activity

Some equations cannot be easily solved algebraically. The method of trial and improvement gets you closer and closer to the answer.

Start with the equation $x^2 + x = 80$.
This equation does not factorise, but can be solved by trial and improvement to any required degree of accuracy.

Write the first two rows of this table on the board.
Ask students to suggest further values of x – some are provided here.

x	x^2	$x^2 + x$	
9	81	90	Too large
8	64	72	Too small
8.5	72.25	80.75	Too large
8.4	70.56	78.96	Too small
8.47	71.7409	80.2109	Too large
8.46	71.5716	80.0316	Too large
8.455	71.487025	79.942025	Too small

The correct solution must lie between 8.455 and 8.46. Both these bounds round to 8.46, so the answer is $x = 8.46$ to 2 dp.

This method can be used for cubic equations, quartic equations and many other types of equations, so is a very important technique to master. Unlike the factorisation method, the RHS does not need to equal zero.

Exercise commentary and misconceptions

For exact solutions to questions 1–3, students need to be aware that it might take several trials, but that they should get an accurate answer in the end. Students will need to know where their power key on the calculator is (usually either x^y or y^x or ^).

Make sure students are writing down the x value as the answer to the question, not the value they are trying to make the expression equal. Students may try to make the right-hand side equal the value in the question up to a certain number of decimal places. Point out that it is the value of x that is to be found to a certain number of decimal places. Make sure students identify the bounds within which the solution lies.

Plenary

Can you solve $2x^3 = \sqrt{x}$ to 1 dp?
Note that $x = 0$ is one solution. Find the answer that is between 0 and 1.

It is easiest to rearrange equation to $2x^3 - \sqrt{x} = 0$.

x	0.5	0.6	0.7	0.8	0.75
$2x^3$	0.25	0.432	0.686	1.024	0.843
$\sqrt{x}$	0.707	0.774	0.836	0.894	0.866
$2x^3 - \sqrt{x}$	−0.457	−0.342	−0.150	0.130	−0.023

x must be between 0.75 and 0.8.
Both these bounds round to 0.8 (1 dp), so answer is $x = 0.8$ to 1 dp.

A6.3 Solving equations using trial and improvement

This spread will show you how to
- Use trial and improvement to find approximate solutions of equations
- Substitute positive and negative numbers into expressions

Keywords
Approximation
Cubic
Systematic
Trial and improvement

Some equations don't have exact solutions and can't be solved by an algebraic method. This includes most **cubic** equations.

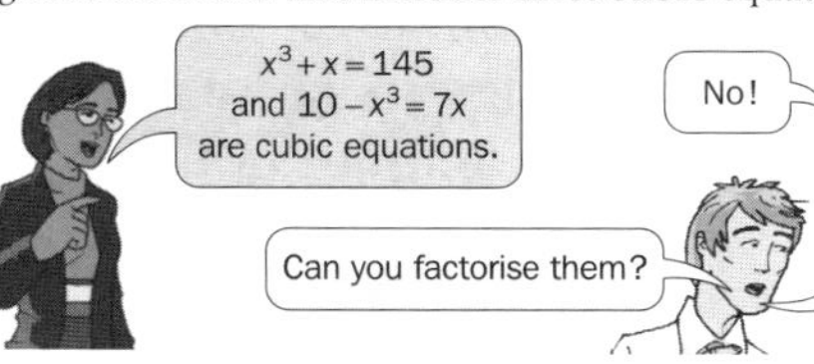

- You can find an approximate solution, correct to several decimal places if necessary, using **trial and improvement**.

Example

Solve the equation $x^3 + x = 145$.
Give your answer correct to 1 dp.

x	x^3	$x^3 + x$	
5	125	130	Too small
6	216	222	Too big
5.5	166.375	171.875	Too big
5.1	132.651	137.751	Too small
5.2	140.608	145.808	Too big
5.15	136.59088	141.74088	Too small

Put your trials in a table. Be **systematic**.

Start with whole numbers then find a better approximation.

The solution is between $x = 5.1$ and $x = 5.2$.

Test the x-value half way between 5.1 and 5.2.

The solution is between $x = 5.15$ and $x = 5.2$, so $x = 5.2$ correct to 1 dp.

You need to work to one more dp than is required in the answer.

Example

The volume of this cuboid is 20 cm³.
Find its dimensions, correct to 1 dp.

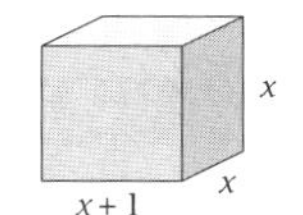

Volume $= x \times x \times (x + 1) = x^2(x + 1) = x^3 + x$
So $x^3 + x = 20$

x	x^3	$x^3 + x$	
2	8	10	Too small
3	27	30	Too big
2.5	15.625	18.125	Too small
2.6	17.576	20.176	Too big
2.55	16.581375	19.131375	Too small

The solution is between $x = 2.55$ and $x = 2.6$, so $x = 2.6$ to 1 dp.
The dimensions of the cuboid are 3.6 cm, 2.6 cm and 2.6 cm.

Exercise A6.3

1 Solve the equation, $x^3 - x^2 = 200$, by trial and improvement to find the solution correct to 1 dp.

2 Write an equation to solve each problem. Use trial and improvement to find the exact answer.

a The area of this rectangle is 77 cm².
What are the length and width?

x

$x + 4$

b The volume of this cuboid is 990 cm³.
What are the dimensions?

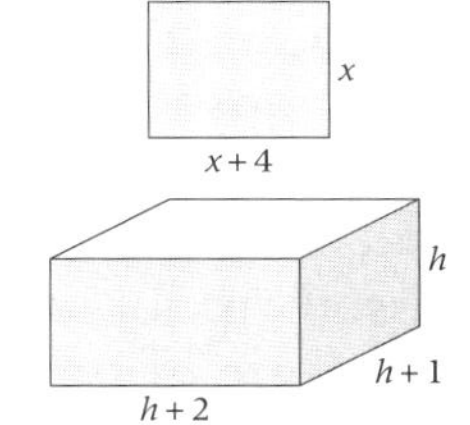

3 **a** Solve $x^2 + 2x = 15$ using trial and improvement in order to find the *exact* solution of this equation.

b Repeat **a** using an algebraic method and comment on which method you feel is most efficient.

4 Use trial and improvement to solve these equations, giving your answer to the stated degree of accuracy.

a $x^2 - x = 69$ (to 1 decimal place)

b $2x^3 + x = 197$ (to 1 decimal place)

c $p(p + 1) = 100$ (to 2 decimal places)

d $w^3 - 2w = 70$ (to 2 decimal places)

5 Write an equation to represent the given information. Use trial and improvement to find the value of the unknown in each case.

a The product of three consecutive numbers is 85 140.

b The area of a rectangle is 57 cm². Its length is 2 cm more than its width. Find the width to one decimal place.

c The surface area of this cuboid is 500 mm². Find x to 1 decimal place.

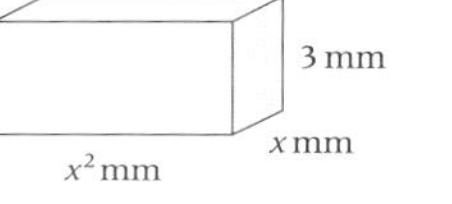

6 Find the number, to 2 dp, that satisfies this statement.

> The square of this number is
> 20 times its cube root.

A6.4 Simultaneous equations

Objectives

- Solve simultaneous equations by eliminating a variable

Useful resources

- Digit cards for mental starter

Mental starter

If I buy 3 adult tickets and 2 child tickets for the cinema and I pay £36 in total, how much might an adult and child ticket cost singly? Hint, the cost is an integer (whole number of pounds).
Solutions for adult and child are 0, 18; 2, 15; 4, 12; 6, 9; 8, 6; 10, 3; 12, 0.

If I also know that the total cost of an adult ticket and a child ticket is £14, the possible solutions are 14, 0; 13, 1; 12, 2; 11, 3; 10, 4; 9, 5; 8, 6; 7, 7; 6, 8; 5, 9; 4, 10; 3, 11; 2, 12; 1, 13; 0, 14. There is only one pair of numbers in both lists, so the correct prices must be £8 for an adult and £6 for a child.

Introductory activity

Discuss the solution to these equations.
Find possible values of x and y if $4x + 5y = 44$. (Integer solutions are $x = 1$, $y = 8$: $x = 6$, $y = 4$, allow other solution.)

Find possible values of x and y if $4x + 2y = 20$. (Integer solutions are $x = 0$, $y = 10$; $x = 1$, $y = 8$; $x = 2$, $y = 6$; $x = 3$, $y = 4$; $x = 4$, $y = 2$; $x = 5$, $y = 0$.)

What values of x and y solve both these equations? $x = 1$ and $y = 8$.
Discuss the need for a better way than just trial and error.

$4x + 5y = 44$
$4x + 2y = 20$ Subtract the terms in the second equation from those in the first equation.
↓ ↓ ↓
$0 + 3y = 24$ so $y = 8$. Substituting 8 for y in the first equation gives $4x + 5 \times 8 = 44$, $x = 1$.

The solution is the pair of values, $x = 1$ and $y = 8$. Check in the second equation, $4 \times 1 + 2 \times 8 = 20$.

Sometimes it is necessary to add the two equations to eliminate one of the variables.

$3x + 2y = 19$
$8x - 2y = 58$ Add the terms in the second equation to those in the first equation.
↓ ↓ ↓
$11x + 0 = 77$ so $x = 7$. Substituting 7 for x in the first equation gives $3 \times 7 + 2y = 19$, $y = -1$

The solution is the pair of values $x = 7$ and $y = -1$. Check in the second equation, $8 \times 7 - 2 \times (-1) = 58$.

Discuss the examples in the student book. The last example presents a problem in context, similar to the mental starter.

Exercise commentary and misconceptions

When solving these simultaneous equations students will need to decide which letter to eliminate. They should decide on the one which has the same numerical coefficient.
If these coefficients have the Same Sign Subtract (SSS).

Particular care is needed with negative numbers. Encourage students to line up the equations one on top of the other.

In question 5 students need to rearrange the equations so that the constant term is isolated on one side of the equation.

When forming simultaneous equations for the practical problem in question 6**a** students should make it clear what their letters represent.

In question 6**b** students may miss the fact that the triangle is isosceles!

Plenary

Ask students to draw axes from −10 to 10 on both axes. Plot the possible answers for $4x + 5y = 44$ from the introduction, then join the points to make a line, and label it. Do the same for the possible solutions of $4x + 2y = 20$.
Can you see how these graphs could help you to find how the answer works for both equations? Where they cross is the answer.
When might this not be a good method? When the numbers are very big or when they cross at a point which is hard to read accurately from the graph.

A6.4 Simultaneous equations

This spread will show you how to:

- Solve simultaneous equations by eliminating a variable

Keywords
Eliminate
Simultaneous
Solution

- **Two equations that have the same solution are called simultaneous equations.**
  For example,
  $x + y = 8$
  and $x - y = 2$
- **You can solve simultaneous equations using algebra.**

The solution is two numbers that add to 8 and with a difference of 2. They must be $x = 5$ and $y = 3$

Example

Solve the simultaneous equations: $3x + 2y = 12$ and $3x + 8y = 30$.

$3x + 2y = 12$ (1)
$3x + 8y = 30$ (2)

Label the equations (1) and (2).
The x terms are identical so you can **eliminate** them.

$(3x + 8y) - (3x + 2y) = 30 - 12$
$8y - 2y = 18$
$6y = 18$
$y = 3$

Subtract equation (1) from equation (2).

$3x + 6 = 12$
$3x = 6$
$x = 2$

Substitute 3 for y in equation (1).

Check: $3 \times 2 + 2 \times 3 = 12$ and $3 \times 2 + 8 \times 3 = 30$

Write the equations one under the other so that you can compare them.

Subtract because the x terms have the same sign. (SSS)

Example

Solve $4x - 3y = 5$ (1)
$8x + 3y = 1$ (2)

$(4x - 3y) + (8x + 3y) = 5 + 1$
$12x = 6$
$x = \frac{1}{2}$
$8\frac{1}{2} + 3y = 1$
$4 + 3y = 1$
$3y = -3$ so $y = -1$

In equation (1) $3y$ is positive, in equation (2) $3y$ is negative, so add the equations to eliminate the y-terms.

Substitute $\frac{1}{2}$ for x in equation (2).
Check in equation (1): $4 \times \frac{1}{2} - 3 \times (-1) = 5$

Example

In a sweet shop, I spend £3.20 on three cans of soft drink and four bars of chocolate. The next day, I buy a can of soft drink and four bars of chocolate for £2. How much does each item cost?

Choose letters for the variables.

Let c be the number of cans of drink I buy and b be the chocolate.
$3c + 4b = 320$ (1) $c + 4b = 200$ (2)
The differences between the two equations are 2 cans and 120p.
So, $2c = 120$ and $c = 60$.
$180 + 4b = 320$ Substitute 60 for c in (1).
$4b = 140$, so $b = 35$ Subtract (2) from (1).
A can of soft drink costs 60p and a bar of chocolate costs 35p.

Change from £ to pence to avoid working with decimals

Exercise A6.4

1 Which pairs of equations have the solution $x = 2$ and $y = 7$?

a $x + y = 9$, $x - y = -5$ **b** $2x + y = 11$, $3x - y = -1$ **c** $2x + 2y = 15$, $4x - y = 1$
d $x + 2y = 16$, $x - 2y = 8$ **e** $5x - y = 3$, $y - x = 5$

2 Solve these pairs of simultaneous equations by subtracting one equation from the other.

a $3x + y = 15$, $x + y = 7$ **b** $6x + 2y = 6$, $4x + 2y = 2$ **c** $x + 5y = 19$, $x + 7y = 27$
d $5x + 2y = 16$, $x + 2y = 4$ **e** $m + 3n = 11$, $m + 2n = 9$ **f** $4x + 3y = -5$, $7x + 3y = -11$

3 Solve these pairs of simultaneous equations by adding one equation to the other.

a $3x + 2y = 19$, $8x - 2y = 58$ **b** $5x + 2y = 16$, $3x - 2y = 8$ **c** $7a - 3b = 24$, $2a + 3b = 3$
d $2x + 3y = 19$, $-2x + y = 1$ **e** $4x - 7y = 15$, $2x + 7y = 4\frac{1}{2}$ **f** $6p - 2q = -2$, $6p + 2q = 26$

4 Solve these pairs of simultaneous equations by either adding or subtracting in order to eliminate one variable.

a $x + y = 3$, $3x - y = 17$ **b** $5x - 2y = 4$, $3x + 2y = 12$ **c** $5a + b = -7$, $5a - 2b = -16$
d $3v + w = 14$, $3v - w = 10$ **e** $20p - 4q = 32$, $7p + 4q = 22$ **f** $3x - 2y = 11$, $3x + 4y = 23$

5 Solve these simultaneous equations.

$a = 2b + 7$, $a + b - 1 = 10$ $6w = 38 - 2v$, $5w = 6 + 2v$

6 Solve these problems by using simultaneous equations.

a How much does a lemon cost?

3 lemons, 4 oranges, £1.27 *4 oranges, 5 lemons, £1.61*

b The perimeter of this triangle is 30 cm. How long is the base?

$2x - 1$ y x

A6.5 Further simultaneous equations

Objectives

- Solve simultaneous equations by eliminating a variable
- Use notation and symbols correctly and consistently within a given problem

Useful resources

- Digit cards for mental starter

Mental starter

Read out this problem:
The sum of two numbers is 12. The difference of the same two numbers is 4. What are the two numbers? (4 and 8.)

Repeat the problem with different values for the sum and difference, for example 28 and 16.

Introductory activity

In previous examples either the x term or the y term had the same numerical coefficient in both equations. In this lesson there will be an extra step to make the coefficients equal.

For example, to solve the equations

$2v + 3w = 12$
$5v + 15w = 60$

$2v + 3w = 12$
$10v + 15w = 60$ multiply each term by 5
$5v + 4w = 23$
$10v + 8w = 46$ multiply each term by 2

The coefficients of v are now equal and Same Signs Subtract,
$0 + 7w = 14$, so $w = 2$.
Substituting 2 for w in the first equation,
$2v + 3 \times 2 = 12$, so $v = 3$.
Check in second equation, $5 \times 3 + 4 \times 2 = 23$, so answer is $v = 3$ and $w = 2$.

In this example you could have chosen to eliminate the w terms by multiplying the first equation by 4 and the second by 3 to make the coefficients of w in both equations 12.

Sometimes it is only necessary to multiply one of the equations to make the coefficients equal.

Discuss the examples in the student book. The second example is a development of the mental starter.

Exercise commentary and misconceptions

In some of the parts in question 1 only one of the equations needs multiplying. Remind students to multiply every term and to use Same Signs Subtract. Subtracting with negatives in part **f** may cause problems.

To clear the fractions in question 3 students should multiply every term in the equation by the common denominator of the fractions.

In question 4 students should say what the variables they use mean.

Plenary

Discuss students' solutions to question 4. Students often find it hard to form equations from word problems, so addressing misconceptions as a whole class will help.

A6.5 Further simultaneous equations

This spread will show you how to:

- Solve simultaneous equations by eliminating a variable
- Use notation and symbols correctly and consistently within a given problem

Keywords
Coefficient
Eliminate
Simultaneous
Solution

Sometimes the **coefficients** of the x-terms (or the y-terms) in simultaneous equations are not the same.
You have to adjust the equations before you can **eliminate** the x-terms (or the y-terms).

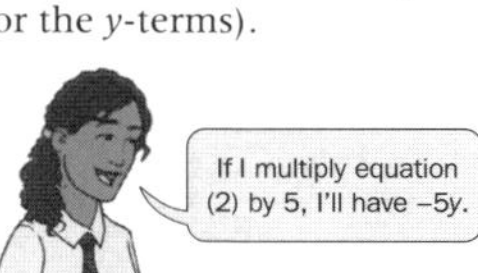
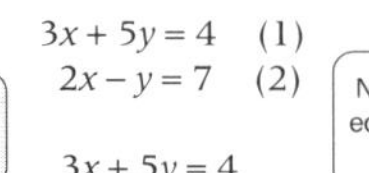

$3x + 5y = 4 \quad (1)$
$2x - y = 7 \quad (2)$

$3x + 5y = 4$
$10x - 5y = 35$

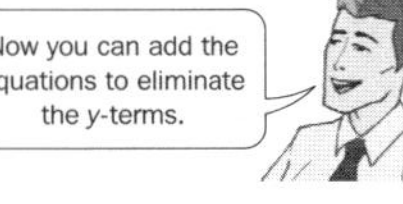

Example

Solve the simultaneous equations $3x + 5y = 4$
$2x - y = 7$

$3x + 5y = 4 \quad (1)$
$2x - y = 7 \quad (2)$ — Label the equations.

$10x - 5y = 35 \quad (3)$ — Multiply equation (2) by 5. *To get 5y in both equations*

$(3x + 5y) + (10x - 5y) = 4 + 35$ — Add equations (1) and (3) to eliminate the y-terms.
$13x = 39$
$x = 3$

$9 + 5y = 4$ — Substitute 3 for x in equation (1).
$y = -1$

*Check the **solution** in equation (2) $2 \times 3 - (-1) = 6 + 1 = 7$*

Example

The difference between two numbers is 4. Treble the smaller number, subtract double the larger number is 1. What are the numbers?

Let the larger number be x and the smaller be y.

$x - y = 4 \quad (1)$
$3y - 2x = 1 \quad (2)$
$-2x + 3y = 1 \quad (2)$ — Rearrange equation (2).

$3x - 3y = 12 \quad (3)$ — Multiply equation (1) by 3. *To get 3y in both equations*

$(-2x + 3y) + (3x - 3y) = 1 + 12$ — Add equations (2) and (3), to eliminate the y-terms.
$x = 13$

$13 - y = 4$ — Substitute 13 for x in equation (1).
$y = 9$

The numbers are 9 and 13.

Check in equation (2) $3 \times 9 - 2 \times 13 = 1$

Exercise A6.5

1 Solve these simultaneous equations.

a $2x + y = 8$, $5x + 3y = 12$ **b** $3x + 2y = 19$, $4x - y = 29$

c $8a - 3b = 30$, $3a + b = 7$ **d** $2v + 3w = 12$, $5v + 4w = 23$

e $9p + 5q = 15$, $3p - 2q = -6$ **f** $3x - 2y = 11$, $2x - y = 8$

2 Make as many pairs of simultaneous equations as you can using these three cards. Solve your pairs.

$2x + y = 12$ $y - x = 15$ $3x - 4y = 7$

3 Solve these simultaneous equations.
Remember to clear the fractions first.

a $\frac{x}{3} - \frac{y}{4} = \frac{3}{2}$, $2x + y = 14$ **b** $\frac{a}{2} + 3b = 1$, $5a - 7b = 47$ **c** $p - \frac{2q}{3} = \frac{26}{3}$, $\frac{p}{4} + 3q + 1 = 0$

4 Write a pair of simultaneous equations to solve each problem.

a Two numbers have a sum of 41 and a difference of 7.
What numbers are they?

b One number is 6 more than another. Their mean average is 20.
What numbers are they?

c 230 students and 29 staff are going on a school trip. They travel by large and small coaches. The large coaches seat 55 and the small coaches seat 39. If there are no spare seats and five coaches are to make the journey, how many of each coach are used?

d In an isosceles triangle, the largest angle is 30° more than double the equal angles. What are the angles in the triangle?

e Uncle Jack gave me a £25 book token at Christmas. At Firestone's Bookshop I can use the token to buy exactly 3 paperbacks and 1 hardback book or 1 paperback and 2 hardbacks.

Find the cost of a paperback and a hardback.

...am review

Key objectives

- Manipulate algebraic expressions; factorise quadratic expressions, including the difference of two squares, and cancel common factors in rational expressions
- Solve quadratic equations by factorisation
- Use systematic trial and improvement to find approximate solutions of equations
- Solve exactly, by elimination of an unknown, two simultaneous equations in two unknowns

Worked solution | **Commentary and misconceptions**

1

Encourage students to label equations when solving simultaneous equations. There are several possible methods for solving this problem. Students should be encouraged to examine the problem and to choose a method requiring the minimum level of calculation.

$5x - 2y = 16$ equation (1)
$3x + y = 14$ equation (2)
Equation (2) × 2:
$6x + 2y = 28$ equation (3)

Some students may need to be prompted to manipulate one of the equations before adding them.

Equation (1) + equation (3):
$11x = 44$
So $x = 4$.

Ensure that students understand how to add or subtract equations.

Substituting $x = 4$ into equation (2):
$y = 14 - 3 \times 4 = 14 - 12 = 2$.

Ensure that students substitute the value of x into the equation correctly, using the correct order of operations.

2

Ensure that students show all their working. Encourage them to use a table to record their results.
Remind students not to round in the intermediate steps of the calculation.

x	4	5	4.5	4.2	4.3	4.25
x^3	64	125	91.125	74.088	79.507	76.765625
$2x$	8	10	9	8.4	8.6	8.5
$x^3 - 2x$	56	115	82.125	65.688	70.907	68.265625
	Too low	Too high	Too high	Too low	Too high	Too high

The exact solution is between 4.2 and 4.25.
Therefore, $x = 4.2$ to 1 dp.

Ensure that students understand how their workings show where the solution lies and that they give the solution in the required rounded form.

Averages and box plots

Objectives

H Draw and produce, using paper and ICT, cumulative frequency tables and diagrams, box plots and histograms for grouped continuous data

F/H Identify the modal class for grouped data, calculate the mean, range and median of continuous data

F/H Calculate the mean for (and estimate of) large data sets with grouped data

H Compare distributions and make inferences, using shapes of distributions and measures of average and spread, including median and quartiles

H Identify seasonality and trends in time series

H Calculate an appropriate moving average

Unit overview

This unit consolidates the students' knowledge of data processing and using frequency tables from earlier data units. It covers finding the averages and range of large data sets and grouped data and using diagrams such as frequency polygons and time series graphs to represent, interpret and compare data, including calculating moving averages. The work covered in this unit is used and extended in unit D5.

Prior knowledge

Before your students start this unit they should be able to:

- Work out the midpoint of two numbers
- Understand and use fraction notation
- Add, subtract, multiply and divide with integers, decimals and fractions
- Use the correct order of operations in calculations, including brackets
- Calculate the mean, median, mode, quartiles, IQR and range of a data set
- Understand and use frequency tables
- Draw and interpret time series graphs

Differentiation

- **Foundation** focuses on calculating the mean, median and mode of discrete data and using charts and tables to represent, interpret and compare data
- **Foundation Plus** extends to grouped data
- **Higher Plus** extends the Higher book to comparing frequency polygons, predicting results from time series data and using scatter diagrams

D4.1 Large data sets – averages and range

Objectives

- Use frequency tables to find the averages and range of a data set
- Use estimates of averages and range to summarise large data sets

Useful resources

- Calculators
- Blank frequency tables with 3 columns

Mental starter

What is the median of the numbers 1 to 10? ((1 + 10) ÷ 2 = 5.5.) **What is the mean of the numbers 1 to 10?** (55 ÷ 10 = 5.5; same answer.) **What is the median of the numbers 1 to 100?** ((100 + 1) ÷ 2 = 50.5.) **What is the mean of the numbers 1 to 100?** (5050 ÷ 100 = 50.5; same answer.) **What is the median of the numbers 1 to 1000?** ((1000 + 1) ÷ 2 = 500.5.) **What is the mean of the numbers 1 to 1000?** (500 500 ÷ 100 = 500.5; same again!)

Introductory activity

Remind students how to find the mean, median and quartiles, and the mode of a small data set, using an example from D1.4.

What if the data is represented in a frequency table? Can the averages and IQR still be found?

Yes, because all the data results are still here, they are just in a table.

Present this data:

Ages of my friends	Number of people (frequency)
15	1
16	3
17	7
18	4

Mode = 17 yrs.

Median is the middle result. There are **15** results, not **4**! The 8th result is the middle one. Count through the frequencies until you find the 8th result, 17 yrs.

Mean – add up all the results:
$15 + 16 \times 3 + 17 \times 7 + 18 \times 4 = 254$,
then divide by the number of results,
$254 \div 15 = 16.9$ yrs (1 dp).

Range = 18 – 15 = 3 yrs.

LQ and UQ: There will be 7 results in each half. The 4th result from the bottom will be in the middle of the bottom half. The 4th result from the top will be in the middle of the top half. LQ = 16, UQ = 18, IQR = 2.

Discuss the example in the student book, which relates to the length of words in a crossword puzzle.

Exercise commentary and misconceptions

When finding medians and quartiles from frequency tables encourage students to think of counting through the frequencies until they reach the position they want.
With an odd total frequency it is the convention to exclude the median and then split the remaining data in 2 halves to locate LQ and UQ by finding the median of the lower half and upper half. With an even total frequency the data can just be split into a lower and upper half.

In question 1 'number of words' could be used instead of 'frequency', in question 2 'number of boxes' and in question 3 'number of matches'. A clue for exams is that the right-hand column is almost always the frequency.

In questions 2 and 3 an extra working column for calculating the mean is useful, especially for weaker students, and is a good habit to develop.

Plenary

I visit the doctor's surgery for a health check. The doctor informs me that I am in the top quartile in the population for my height, but in the bottom quartile for my weight. **What do I look like?** Tall and skinny!
You could extend to a brief discussion of deciles and percentiles, which are commonly used in medical data.

D4.1 Large data sets – averages and range

This spread will show you how to:
- Use frequency tables to find the averages and range of a data set
- Use estimates of averages and range to summarise large data sets

Keywords
Frequency table
IQR
Mean
Median
Mode
Range

- You can put large amounts of data in a **frequency table**.
- You use the averages and the **range** to summarise the data.

Example

The table shows the length of the words in the answers to a crossword puzzle.

For these data, work out the

a mode
b median
c mean
d IQR
e range.

Word length	Frequency
4	3
5	5
6	7
7	8
8	3
9	1

Word length	Frequency	Word length × frequency
4	3	4 × 3 = 12
5	5	5 × 5 = 25
6	7	6 × 7 = 42
7	8	7 × 8 = 56
8	3	8 × 3 = 24
9	1	9 × 1 = 9
Total	**27**	**168**

Total number of words.
Total number of letters.

a Mode = 7 Words with 7 letters have the highest frequency.

b $\frac{1}{2}(27 + 1) = 14$ so the 14th value is the median
The 14th value is in the 'Word length 6' group.
Median = 6

c Mean = $\dfrac{\text{Total number of letters}}{\text{Total number of words}}$
$= 168 \div 27 = 6.2$

d Lower quartile is $\frac{1}{4}(27 + 1)$th value = 7th value = 5
Upper quartile is $\frac{3}{4}(27 + 1)$th value = 21st value = 7
IQR = 7 − 5
= 2

e Range = 9 − 4 = 5 Longest − shortest word length

The techniques covered in this unit may be useful for your statistical coursework task.

IQR means interquartile range (see **D1.4**)

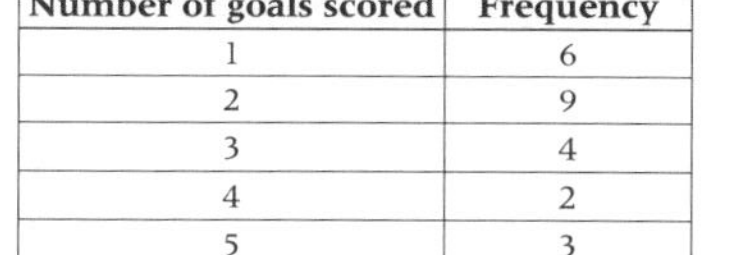

Exercise D4.1

1 The tables of data give information about the length of words in four different crosswords.
For each table, copy each table, add an extra working column and find the
i mode **ii** median **iii** mean **iv** range **v** interquartile range.

a

Word length	Frequency
4	2
5	5
6	4
7	2
8	2

b

Word length	Frequency
3	3
4	4
5	9
6	5
7	2

c

Word length	Frequency
4	6
5	3
6	5
7	4
8	2
9	5
10	2

d

Word length	Frequency
3	5
4	4
5	6
6	7
7	7
8	4
9	2

2 Jo had 16 boxes of matches.
She counted the number of matches in each box.
The table gives her results.

Number of matches	Frequency
41	2
42	7
43	4
44	3

Work out the mean number of matches in a box.

3 Brian played in 24 hockey matches one season.
The table gives information about the number of goals scored in these matches.

Number of goals scored	Frequency
1	6
2	9
3	4
4	2
5	3

Work out the mean number of goals scored.

D4.2 Averages of grouped data

Objectives

- Use a grouped frequency tables

Useful resources

- Calculators
- Blank frequency tables with 4 columns

Mental starter

Six numbers are all between 1 and 9, inclusive. There is no mode, the median is 4, the mean is 4, the range is 6, LQ is 2, UQ is 6.

What are the numbers? (1, 2, 3, 5, 6, 7 is the only solution.)

Introductory activity

Remind students of the work done in D4.1 on frequency tables.
Show this table:

Number of homeworks	Number of people (frequency)
0–6	2
7–9	3
10–20	10

(i) What is the modal class interval? (10–20.)

(ii) Which class interval contains the median result? (15 results, the middle one is the 8th result which is in the 10–20 group.)

(iii) Estimate the mean number of homeworks. Assume that two people have 3 homeworks, three people have 8 and ten people have 15. We are using the 'mid-point' as an estimate of the number of homeworks that people had, so the total number is
$2 \times 3 + 3 \times 8 + 10 \times 15 = 180$.
Divide by the total frequency (number of results, in this case, people)
$180 \div 15 = 12$ homeworks per person.

(iv) What is the range? ($20 - 0 = 20$.)

Exercise commentary and misconceptions

Because this exercise uses continuous data the midpoints of a group can be found by adding the boundaries of the group and dividing by 2. The mean calculated is an estimate, because the midpoints of a group are used to represent all the members of the group instead of using the actual values.

Weaker students often give the middle group as the median group, without reference to the frequencies. This is difficult to spot, especially as it often gives the correct answer!

Another common mistake is to quote the modal class as the frequency corresponding to the modal class.

Plenary

Taking question 3 as an example, discuss how you could find a more accurate estimate of the median.
Encourage thinking by asking questions such as: 'Will the median be closer to the lower end or the upper end?'

D4.2 Averages of grouped data

This spread will show you how to:
- Use grouped frequency tables

Keywords
Grouped frequency table
Mean
Median
Midpoint
Modal class

- You can put large amounts of continuous data into a **grouped frequency table**.

A grouped frequency table does not tell you the actual data values so you can only find estimates of the averages.

- You use estimates of averages to summarise the data.

Example

The table shows the time taken, to the nearest minute, by a group of students to solve a crossword puzzle.

Time, t, minutes	Frequency
$5 < t \leqslant 10$	2
$10 < t \leqslant 15$	14
$15 < t \leqslant 20$	13
$20 < t \leqslant 25$	6
$25 < t \leqslant 30$	1

For these data, work out an estimate for the

a modal class
b class containing the median
c mean

The **modal class** is the class with the greatest frequency.

a Modal class = $10 < t \leqslant 15$

b Class containing median = $15 < t \leqslant 20$

c

Time t minutes	Frequency	Midpoint	Word length × frequency
$5 < t \leqslant 10$	2	7.5	$7.5 \times 2 = 15$
$10 < t \leqslant 15$	14	12.5	$12.5 \times 14 = 175$
$15 < t \leqslant 20$	13	17.5	$17.5 \times 13 = 227.5$
$20 < t \leqslant 25$	6	22.5	$22.5 \times 6 = 135$
$25 < t \leqslant 30$	1	27.5	$27.5 \times 1 = 27.5$
Total	**36**		**580**

Total number of students   Total time

Put two extra columns in the table.

Find the totals of the Frequency and Word length × frequency columns.

$$\text{Mean} = \frac{\text{Estimated total time}}{\text{Total number of students}}$$

$$\text{Mean} = \frac{580}{36} = 16.1$$

Exercise D4.2

1 The grouped frequency tables give information about the time taken to solve four different crosswords.
For each table, copy the table, add extra working columns and find

i the modal class   **ii** the class containing the median
iii an estimate of the mean

a

Time, t, minutes	Frequency
$5 < t \leqslant 10$	2
$10 < t \leqslant 15$	14
$15 < t \leqslant 20$	13
$20 < t \leqslant 25$	6
$25 < t \leqslant 30$	1

b

Time, t, minutes	Frequency
$0 < t \leqslant 10$	3
$10 < t \leqslant 20$	6
$20 < t \leqslant 30$	4
$30 < t \leqslant 40$	5
$40 < t \leqslant 50$	2

c

Time, t, minutes	Frequency
$5 < t \leqslant 10$	8
$10 < t \leqslant 15$	5
$15 < t \leqslant 20$	7
$20 < t \leqslant 25$	4
$25 < t \leqslant 30$	0
$30 < t \leqslant 35$	1

d

Time, t, minutes	Frequency
$5 < t \leqslant 15$	3
$15 < t \leqslant 25$	9
$25 < t \leqslant 35$	7
$35 < t \leqslant 45$	8
$45 < t \leqslant 55$	2
$55 < t \leqslant 65$	1

2 Alfie kept a record of his monthly mobile phone bills for one year.

Phone bill, b, pounds	Frequency
$10 < b \leqslant 20$	6
$20 < b \leqslant 30$	2
$30 < b \leqslant 40$	3
$40 < b \leqslant 50$	1

a Write the class interval that contains the median.
b Calculate an estimate for the mean cost of Alfie's mobile phone bill.

3 The heights of 50 Year 10 students were measured. The results are shown in the table.

Height, h, cm	Number of students
$150 \leqslant h < 155$	3
$155 \leqslant h < 160$	5
$160 \leqslant h < 165$	15
$165 \leqslant h < 170$	25
$170 \leqslant h < 175$	2

a What is the modal group?
b Estimate the mean height.
c Which class interval contains the median?

D4.3 Frequency polygons

Objectives

- Draw frequency polygons
- Use frequency polygons to compare two data sets

Useful resources

- 2 mm graph paper

Mental starter

Ask students to work out calculations of this type:

$\frac{12}{18-12}, \frac{15}{48-18}, \frac{20}{40-30}, \frac{1000}{60-40}$ and so on.

Introductory activity

A postman recorded the number of letters he delivered to 33 different homes.

Number of letters	0–2	3–5	6–8	9–11	12–14
Number of houses (frequency)	2	13	9	7	2

Ask students to illustrate this data in a frequency diagram, showing the trend. Always plot frequency vertically and use an exact scale on both axes. Plot crosses at the mid-interval of the groups. Join the crosses with straight lines to show the trend. **Can you use the graph to estimate the number of houses that receive 6 letters?** (No, because the graph cannot be used to read off results except at the crosses you have plotted.)

Discuss the example in the student book, which involves a comparison between two different frequency polygons.

Exercise commentary and misconceptions

Check that students have drawn and labelled the axes correctly, and check also that the crosses are plotted in the right place.

Some students will be reluctant to write anything in part **c** of each question, others will make statements which are not based on the data. Stress that they must write something and that they must give reasons for their statements as these types of question are common in the exam.

Plenary

Discuss question 3, focusing on the comparison between the times taken by the two groups to travel to work.

D4.3 Frequency polygons

This spread will show you how to:
- Draw frequency polygons
- Use frequency polygons to compare two data sets

Keywords
Class interval
Frequency polygon
Midpoint
Modal

- **You can represent grouped data in a frequency polygon.**

To draw a frequency polygon for continuous data you plot the **midpoint** of each **class interval** against the frequency.

Example

The tables show the ages of people attending concerts to see the bands Badness and Cloudplay.

Badness

Age, a, years	Frequency
$20 < a \leq 30$	1600
$30 < a \leq 40$	4300
$40 < a \leq 50$	2100
$50 < a \leq 60$	1000

Cloudplay

Age, a, years	Frequency
$10 < a \leq 20$	2800
$20 < a \leq 30$	4600
$30 < a \leq 40$	3300
$40 < a \leq 50$	1200

a Draw frequency polygons for these data.
b Make comparisons, with reasons, between the ages of people attending these concerts.

a

Midpoint	25	35	45	55
Frequency	1600	4300	2100	1000

b

Midpoint	15	25	35	45
Frequency	2800	4600	3300	1200

b The **modal** age is greater at Badness concerts than Cloudplay concerts. The highest frequency for Badness is in the class interval 30–40 years old, whereas the highest frequency for Cloudplay is in the interval 20–30 years old.

Exercise D4.3

1 The tables show the ages of the first 100 people to visit a garden centre on a weekday and a Sunday.

Weekday

Age, a, years	Frequency
$0 < a \leq 20$	16
$20 < a \leq 40$	28
$40 < a \leq 60$	32
$60 < a \leq 80$	24

Sunday

Age, a, years	Frequency
$0 < a \leq 20$	22
$20 < a \leq 40$	45
$40 < a \leq 60$	18
$60 < a \leq 80$	15

a Draw frequency polygons for these data.
b Find, for each data set, the class which contains the modal age.
c Compare the ages of people at the garden centre on a weekday and a Sunday.

2 Jayne kept a daily record of the number of miles she travelled in her car during two months.

December

Miles travelled, m	Frequency
$0 < m \leq 20$	3
$20 < m \leq 40$	8
$40 < m \leq 60$	10
$60 < m \leq 80$	6
$80 < m \leq 100$	4

January

Miles travelled, m	Frequency
$0 < m \leq 20$	0
$20 < m \leq 40$	5
$40 < m \leq 60$	12
$60 < m \leq 80$	8
$80 < m \leq 100$	6

a Draw frequency polygons for these data.
b Find, for each month, the class which contains the modal number of miles
c Compare the number of miles Jayne travelled in December and January.

3 David carried out a survey to find the time taken by 120 teachers and 120 office workers to travel home from work.

Teachers

Time taken, t, minutes	Frequency
$0 < t \leq 10$	12
$10 < t \leq 20$	33
$20 < t \leq 30$	48
$30 < t \leq 40$	20
$40 < t \leq 50$	7

Office workers

Time taken, t, minutes	Frequency
$10 < t \leq 20$	2
$20 < t \leq 30$	21
$30 < t \leq 40$	51
$40 < t \leq 50$	28
$50 < t \leq 60$	18

a Draw frequency polygons for these data.
b Work out for each data set the class which contains the modal time taken.
c Make comparisons between the time taken by the teachers and office workers to travel home from work.

D4.4 Time series

Objectives

- Draw time series graphs

Useful resources

- Graph paper

Mental starter

Give students amounts to calculate as percentages, for example:
10 out of 25 (40%)
17 out of 68
90 out of 2000
3 out of 120.

Introductory activity

Discuss data that is collected regularly over a period of time. Can you suggest any situations in which this happens? (Retail sales of any kind, electricity, phone or other bills, SATS or GCSE results for schools.) It is useful to look at this type of data to identify trends.

A school recorded the percentage of students who stayed at the school's 6th form after they finished year 11.

Year	91–93	94–96	97–99	00–02	03–05
Percentage of students	0	60	65	60	95

Together with the students, draw the graph using the percentages for the vertical axis and the years for the horizontal axis. Plot the points with crosses in the centre of the year range, as with frequency polygons. **What does the graph show?** (The school probably opened its 6th form in 1994. The percentage of students staying on stayed roughly the same for about 8 years, but then grew rapidly.) Discuss the example in the student book.

Exercise commentary and misconceptions

If students use a regular scale for the time axes, they must plot the points in the centre of the class interval. Make sure students join the points with a ruler and are aware that they cannot read off results in between the plotted points. This would have no meaning.

In question 4 there are eight periods to plot along the horizontal axis as 2005 follows 2004. Some students will try to plot two graphs over four periods. Similarly for questions 5 and 6.

Plenary

A company's profits were 2.71, 2.74 and 2.78 million pounds in the years 2002, 2003, 2004. The company wants to draw a time series graph to show that their profits are rising rapidly. **How could they do this?** (Start the vertical axes at 2.7 instead of 0.)

D4.4 Time series

This spread will show you how to:
- Draw time series graphs

Keywords
Time series

- **Time series data are collected over a period of time at regular intervals.**

For example,
Electricity and gas bills are produced every quarter (three months).
Mobile phone bills are generated each month.
Unemployment rates are published each month.

- **You can plot time series data on a graph with time on the horizontal axis.**

Example

Jenny's quarterly gas bills over a period of two years are shown in the table.

	Jan–March	April–June	July–Sept	Oct–Dec
2003	£65	£38	£24	£60
2004	£68	£42	£30	£68

Draw a time series graph to represent these data.

Draw axes on graph paper with time on the horizontal axis.
Plot the coordinates as crosses on the grid.
Join them up with straight lines.

On the horizontal axis, J–M means Jan–March.

Exercise D4.4

1 The table shows Ken's monthly mobile phone bills.

Jan	Feb	Mar	April	May	June	July	Aug	Sept	Oct	Nov	Dec
£16	£12	£15	£18	£16	£18	£12	£10	£12	£15	£16	£20

Draw a time series graph for these data.

2 The table shows Mary's quarterly electricity bills over a two-year period.

	Jan–March	April–June	July–Sept	Oct–Dec
2004	£45	£20	£15	£48
2005	£54	£24	£18	£50

Draw a time series graph to represent these data.

You will need the graphs for questions **2**, **5** and **6** for Exercise D4.5.

3 The table shows monthly ice-cream sales at Angelo's shop during one year.

Jan	Feb	Mar	April	May	June	July	Aug	Sept	Oct	Nov	Dec
£16	£12	£15	£18	£38	£48	£52	£58	£18	£15	£16	£40

Draw a time series graph for these data.

4 A town council carried out a survey over a number of years to find the percentage of local teenagers who used the town's library. The table shows the results.

Year	1998	1999	2000	2001	2002	2003	2004	2005
%	14	18	24	28	25	20	18	22

Draw a time series graph for these data.

5 Christabel kept a record of how much money she had earned from babysitting during three years.

	January–April	May–August	Sept–Dec
2001	£12	£18	£30
2002	£21	£33	£60
2003	£39	£42	£72

Draw a time series graph for these data.

6 Steve kept a record of his quarterly expenses over a period of two years.

	Jan–March	April–June	July–Sept	Oct–Dec
2003	£35	£56	£27	£12
2004	£39	£68	£29	£18

Draw a time series graph to represent these data.

D4.5 Time series and moving averages

Objectives

- Calculate moving averages

Useful resources

- Graphs drawn in D4.4
- Calculators
- OHT of the example given in D4.5 of the students' book (Jenny's gas bills) with the table of results and the time series graph already drawn

Mental starter

A company's profits for the last 5 years, measured in millions of pounds are 12, 10, 9, 15, 18. **Are things getting better? Discuss**.

(Depends on how you look at it. Looking year to year, profits have gone up twice and down twice. But the overall trend is certainly good.)

Introductory activity

Some time series data can vary depending on the month or the season. For example, a retail shop will always sell more toys in the period before Christmas and less in January and February. The proportion of trains that run late increases every autumn because of leaves on the line. If a graph has many peaks and troughs it can be difficult to tell what the long term trend is. Calculating and plotting moving averages will have the effect of smoothing the graph.

We need to consider a complete cycle of a year each time, calculate the mean, and plot the value in the middle of the cycle.

Look at the example in the student book.
How many periods make up a year? (4)
When does the first complete year start? (Jan 2003) **When does it end?** (Dec 2003)
The mean of the bills for this year is £46.75. Plot this between June and July, the mid-point of the year.

When does the second complete year start and end? (April 2003 and March 2004, it overlaps the first year.)

When does the third complete year start and end? (July 2003 and June 2004.)

It is worth continuing this till all students have fully grasped the idea of complete, overlapping cycles.

Then calculate the averages and plot them.

Although moving averages are commonly used to compare cycles of a week, or a year, they can also be used to smooth out trends by calculating them over an arbitrary number of years. In this case an exam question would state how many points to use.

Exercise commentary and misconceptions

For 4 point moving averages (or any even point moving average) the points need to be plotted in between the 2nd and 3rd time values. Encourage students to use an exact scale for the time axes, although this is not essential.

Plenary

Are moving averages a way to cheat and mislead people about the true results?
They can be if you do not say that the graph is a moving average. The moving average graph has the advantage of smoothing out one-off bad results, but also, if you have a one-off good result, this will not show up, it gets smoothed over as well.

D4.5 Time series and moving averages

This spread will show you how to:

- Calculate moving averages

Keywords
Moving average
Time series

- **Time series data can vary depending on the month or season.**

For example,

Electricity and gas bills are likely to be higher in the winter months.
Unemployment rates can be higher in the summer when students leave school.

- **You can find a moving average of time series data to help you spot a trend.**
- **You calculate moving averages over a whole cycle.**

Example

Jenny drew this table for her quarterly gas bills over a period of two years.

	Jan–March	April–June	July–Sept	Oct–Dec
2003	£65	£38	£24	£60
2004	£68	£42	£30	£68

a Calculate 4-point moving averages for these data.
b Plot the moving averages on a graph.
c Describe the trend shown by the moving averages.

A year (4 quarters) is one cycle, so 4-point moving averages are calculated.

a

Jan–Mar 2003	Apr–June 2003	July–Sept 2003	Oct–Dec 2003	Jan–Mar 2004	Apr–June 2004	July–Sept 2004	Oct–Dec 2004
£65	£38	£24	£60	£68	£42	£30	£68

$(65 + 38 + 24 + 60) \div 4 = £46.75$

$(38 + 24 + 60 + 68) \div 4 = £47.50$

$(24 + 60 + 68 + 42) \div 4 = £48.50$

$(60 + 68 + 42 + 30) \div 4 = £50$

$(68 + 42 + 30 + 68) \div 4 = £52$

To calculate 4-point averages, calculate the average of each group of 4 consecutive values.

b

Plot each value at the midpoint of the group of 4.

c The trend shown by the moving averages is that, in general, gas bills are increasing.

Exercise D4.5

1 For the data given in Exercise D4.4 question **1**, calculate the three-point moving averages.

2 **a** For the data given in Exercise D4.4 question **2**, calculate four-point moving averages.

b Plot the moving averages on the graph you have already drawn for Exercise D4.4 question **2**.

c Describe the trend shown by the moving averages.

3 For the data given in Exercise D4.4 question **3**, calculate four-point moving averages.

4 For the data given in Exercise D4.4 question **4**, calculate

a two-point moving averages

b five-point moving averages.

5 **a** For the data given in Exercise D4.4 question **5**, calculate three-point moving averages.

b Plot the moving averages on the graph you have already drawn for Exercise D4.4 question **5**.

c Describe the trend shown by the moving averages.

6 **a** For the data given in Exercise D4.4 question **6**, calculate four-point moving averages.

b Plot the moving averages on the graph you have already drawn for Exercise D4.4 question **6**.

c Describe the trend shown by the moving averages.

7 Audrey recorded the number of cups of tea sold over three weeks to office staff at afternoon break.

	Monday	Tuesday	Wednesday	Thursday	Friday
Week 1	62	55	36	47	68
Week 2	60	54	35	47	70
Week 3	61	55	37	50	73

a Draw a time series graph to represent these data.

Audrey wants to find out about the trend in the number of cups of tea sold.

b Explain why it would be appropriate for Audrey to calculate five-point moving averages.

c Calculate the five-point moving averages.

d Plot the moving averages on the graph.

e Describe the trend shown by the moving averages.

D4 Exam review

Key objectives

- Draw and produce, using paper and ICT, pie charts for categorical data, and diagrams for continuous data, including line graphs for time series and frequency diagrams
- Identify the modal class for grouped data
- Find the median, quartiles and interquartile range for large data sets
- Calculate the mean (and estimate of) for large data sets with grouped data
- Calculate an appropriate moving average

Worked solution

1

Number of pets	Number of students	Number of pets x frequency
0	3	0
1	7	7
2	5	10
3	4	12
4	1	4
Total	20	33

$$\text{Mean} = \frac{\text{total number of pets}}{\text{total number of students}} = \frac{33}{20} = 1.65$$

2 a Total frequency = 40.
Median value = 20th value.
Median lies in $150 < C \leqslant 200$

b The new median would be the 20.5th value.
This will still lie in the class $150 < C \leqslant 200$

c Let p = price of car.

$$p - \frac{20p}{100} = 5200$$

$$\frac{80p}{100} = 5200$$

$$p = 5200 \times \frac{100}{80} = £6500$$

Commentary and misconceptions

1 Before attempting the question, ensure that students understand frequency tables and how they are used to calculate averages and range.
Encourage students to draw a table with a column showing the total number of pets.
Students should understand that the number of students is the frequency.

Ensure that students use the correct formula to calculate the mean.
Students should realise that, although you can not have a part of a pet, the mean is not always a whole number.
Encourage students to check their solution with the table of values to decide if it is feasible.

2 a Students should know how to interpret values from frequency tables.
Ensure students understand how to work out the median value and give the class interval as their answer.

b Ensure students explain their answer.

c Ensure students recognise that this is a reverse percentage question. Ensure they give their answer in £s.

S5 Constructions and loci

Objectives

F Understand angle measure using the associated language

F Use angle properties of equilateral, isosceles and right-angled triangles

F Use and interpret maps and scale drawings

H Use straight edge and compasses to do standard constructions including the perpendicular from a point to a line and the perpendicular from a point on a line

F/H Find loci, both by reasoning and by using ICT to produce shapes and paths

Unit overview

This unit consolidates the students' knowledge of angles and trigonometry, applying this to solving 2-D problems involving bearings. It then covers constructing triangles, bisectors and perpendiculars using a ruler and compasses before extending to simple loci.

Prior knowledge

Before your students start this unit they should be able to:

- Understand and use angle measure
- Understand and use angle properties of triangles
- Use compasses to draw circles and parts of circles
- Draw and measure lines and shapes accurately
- Understand and use trigonometry in 2 dimensions
- Understand and use units of measure for distances
- Understand and use maps and scale drawings

Differentiation

- **Foundation** focuses on measuring and constructing lines and 2-D shapes and using bearings, maps and scale drawings
- **Foundation Plus** extends to constructing perpendicular bisectors and angle bisectors and simple loci
- **Higher Plus** extends the Higher book to using Pythagoras' theorem and the sine and cosine rule to solve 2-D and 3-D problems

S5.1 Bearings and scale drawings

Objectives

- Use geometry to solve problems involving bearings
- Use and interpret maps and scale drawings

Useful resources

- Angle measurer for board
- Angle measurers for students
- Geometry tool

Mental starter

An alternative way of measuring angles is to use gradians instead of degrees. It is most often used in the armed forces. 90 degrees is equivalent to 100 gradians. **Change these quantities in degrees into gradians 45, 270, 135, 30, 360** (50, 300, 150, $\frac{100}{3}$, 400). **Change these quantities in gradians into degrees** $133\frac{1}{3}$, $66\frac{2}{3}$, 200, $166\frac{2}{3}$ (120, 60, 180, 150).

Introductory activity

What is a bearing?

- An angle.
- An angle always measured from North.
- An angle always measured from North clockwise round to the point to which you are measuring.
- Always uses 3 digits, so 20° becomes 020°. (James Bond is agent 007, not agent 7.)

Discuss the first example, which does not involve drawing.

Sketch the diagram from the second example on the board, so students can understand the stages in drawing it. The students need to draw accurately to get the correct answer.

If using a 180° measurer for a bearing above 180°, there are two methods. **Either** subtract the bearing from 360° then measure this angle anti-clockwise, **or** subtract 180° from the bearing, extend the 'North line' south of the point and measure the additional angle clockwise from the extended line. Most students tend to find the first method easier.

Encourage students to do rough sketches first, as the diagram may go off the paper if it is not correctly positioned!

Before students start the exercise, emphasise

(a) The importance or reading the question carefully and identifying the point you are measuring the bearing **from**.
(b) Always draw in a North line at the point you are measuring from and a line to the point you are going to.
(c) Always line up zero with North. (You could think of 'Naughty North!')
(d) Always count anticlockwise round from zero on the angle measurer.

Exercise commentary and misconceptions

Encourage students to read the question carefully. Some bearings questions require the use of angle facts, and other questions require students to make an accurate drawing.

Questions 3 and 4 are done by calculation, not accurate drawing, but a sketch helps.

Questions 5 and 6 use scale drawings, but a sketch will help to position the scale drawing on the paper.

Question 6 does not tell students what scale to use, but a scale of 1 cm to 1 km would be easy and appropriate. The lines pointing north at H and L can be aligned using a set square and ruler, or by calculating angles between parallel lines and using a protractor.

Plenary

Can you think of examples of people who use bearings in their work or leisure?
(Sailors, astronauts, orienteers, explorers, surveyors ... .)

S5.1 Bearings and scale drawings

This spread will show you how to:
- Use geometry to solve problems involving bearings
- Use and interpret maps and scale drawings

Keywords
Angle
Bearing

- **A bearing is an angle measured in a clockwise direction from north.**

You always write bearings with 3 digits, for example 070°, 190°, 230°

To find the bearing of A from B

- Imagine you are standing at B, facing north.
- Turn clockwise until you face A.

The angle you have turned through is the bearing of A from B.

The bearing of A from B is 256°.

Example

The bearing of G from B is 028°.
Find the bearing of B from G.

Bearing of B from G is the angle at G measured clockwise from north to B.

Bearing of B from G = 360° − 152° = 208°

Interior angles are supplementary.

Example

A church, C, is 10 km due west of a school, S.
Joe is 6 km from the school on a bearing of 320°.
He wants to walk directly to the church.

Draw a diagram to show the positions of Joe, the church and the school, and use it to find the bearing Joe should take.
Use a scale of 1 cm to 2 km.

Label point S.
Draw C 5 cm west of S.
Draw the north line at S.
Measure and draw the 320° bearing from S and, 3 cm from S, mark a point, J, to show Joe's position.

Draw the line JC.
Draw the north line at J.
Measure the clockwise angle between the north line and JC.
The bearing Joe needs is 233°.

Scale 1 cm to 2 km, so represent 10 km by a line 5 cm long.

Represent 6 km by a line 3 cm long.

Draw and measure lengths and angles carefully or your answer will be inaccurate.

Exercise S5.1

1 These diagrams are drawn accurately.
Measure the bearing of T from S in each.

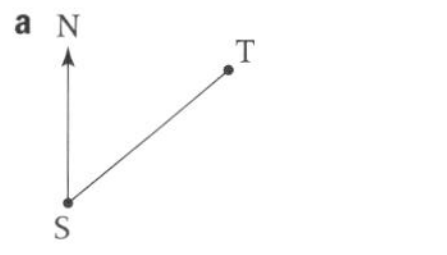

a **b**

2 P and Q are points 2 cm apart. Draw diagrams to show the position of points P and Q where the bearing of Q from P is

a 070° **b** 155° **c** 340° **d** 260°

3 These diagrams have *not* been drawn accurately.
Find the bearing of X from Y in each case.

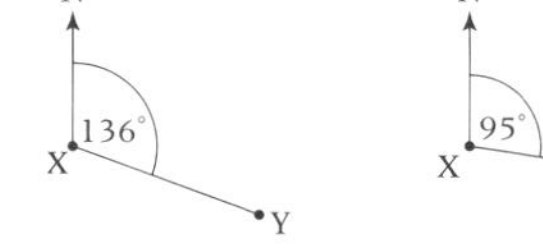

a **b** **c**

4 Find these bearings.

a The bearing of A from B is 104°. Work out the bearing of B from A.
b The bearing of E from F is 083°. Work out the bearing of F from E.
c The bearing of J from K is 297°. Work out the bearing of K from J.

Draw a sketch to help you.

5 A youth club (Y) is 4 km due east of a school (S).
Hazel leaves school and walks 5 km on a bearing of 042° to her house (H).

a Make a scale drawing to show the position of Y, S and H.
Use a scale of 1 cm to 1 km.
b Hazel walks directly from her house to the youth club.
What bearing does she take?

6 A lighthouse, L, is 6 km on a bearing of 160° from a point H at the harbour.
A boat, B, is 3 km from L on a bearing of 125°.

a **i** Make a scale drawing to show the positions of L, H and B.
ii What bearing should B travel to go directly to H?

The boat moves 4 km due west.

b **i** Mark on your drawing the new position of B.
ii What bearing should B now travel to go directly to H?

DID YOU KNOW?
At Fleetwood in Lancashire, boats navigate from the River Wyre to the sea by being positioned in relation to three lighthouses.

S5.2 Constructing triangles

Objectives

- Use a ruler and compasses to draw standard constructions

Useful resources

- Compasses for students
- Compasses for the board
- Rulers for students and a board ruler
- Angle measurers for students and a board angle measurer
- Geometry tool

Mental starter

What is wrong with each of these triangles?

(i) **Sides of 10 cm, 3 cm and 6 cm** (one side must be longer than the sum of the other two sides)

(ii) **Two obtuse angles and one acute angle** (angles will not sum to 180°)

(iii) **An isosceles triangle with two of the angles being 110° and 50°** (angles would sum to either 220° or 270° not 180°)

Introductory activity

Draw a triangle with sides 8 cm, 4 cm and 7 cm. Students should sketch the triangle first and then watch how it can be drawn accurately before drawing it themselves. Stress the importance of leaving construction lines on the page, as all the marks for a question will be lost if they are rubbed out.

Repeat for a triangle with sides 10 cm, 8 cm and 6 cm. **What is special about this triangle?** (Right angle.) **Could you have known that before you drew it?**
(Yes, Pythagoras' theorem.) This method of constructing a right angle was used before protractors were available.

Exercise commentary and misconceptions

In all construction questions, emphasise the benefits of drawing a sketch first.

Question 1**a** has already been done.

Question 3 should be done without trying to construct the triangles.

Questions 7 and 8 revise S4.2 on quadrilaterals.

Plenary

In the exercise you constructed some quadrilaterals from two triangles.
What does that mean about the sum of the angles inside a quadrilateral?
($2 \times 180 = 360°$.)
How might a hexagon be constructed from triangles, and what is the sum of the interior angles? (6 equilateral triangles.
$4 \times 180 = 54°$.)
How many triangles make a polygon with *n* sides and what is the sum of interior angles? ($n - 2$ triangles. Sum = $180(n - 2)°$.)

S5.2 Constructing triangles

This spread will show you how to:
- Use a ruler and compasses to draw standard constructions

Keywords
Arc
Compasses
Radius
Triangle

You can **construct** a unique triangle when you know

Two sides and the angle between them (SAS) or Two angles and a side (ASA) or Right angle, the hypotenuse and a side (RHS) or Three sides (SSS)

5 cm, 30°, 6 cm — 60°, 30°, 7 cm — 5 cm, 3 cm — R, P, Q, 6 cm, 8 cm, 10 cm

You will need a ruler and a protractor for SAS, ASA and RHS triangles.

You will need a ruler and compasses for SSS triangles.

Example

a Construct this equilateral triangle ABC with side length 4 cm.
b Construct another equilateral triangle with base AB and side length 4 cm.
c What special quadrilateral have you drawn?

A, C, B

a Draw BC 4 cm long.
Draw arcs of **radius** 4 cm from B and C to intersect at A.
Join AC and BC.

A, C, B, 4 cm, 4 cm, 4 cm

The construction **arcs** show your method, so do not erase them.

You can use this method to construct an angle of 60° (angles in an equilateral triangle = 60°).

b Draw arcs of radius 4 cm from A and B to intersect at D.
Join AD and DB.

A, B, C, 4 cm, 4 cm, 4 cm, 4 cm

c A rhombus

Four equal sides, two pairs of parallel sides, opposite angles equal.

Exercise S5.2

It helps to draw a rough sketch first.

1 Use a straight edge and compasses or a protractor to construct these triangles.

a Sides 8 cm, 4 cm, 7 cm (SSS) **b** 3 cm, 30°, 4 cm (SAS)
c Sides 10 cm, 7.5 cm, 6 cm (SSS) **d** 8 cm, 2 cm, 90° (RHS)
e Sides 6 cm, 9 cm, 5 cm (SSS) **f** 45°, 4 cm, 45° (ASA)

Try to construct the triangles to see what happens.

2 **a** Explain why you cannot construct a triangle with sides 9 cm, 4 cm, 3 cm.
b Explain what happens when you try and construct a triangle with sides 9 cm, 4 cm, 5 cm.

3 Without drawing the construction, write whether these sets of 3 sides will make a triangle.

a Sides 5 cm, 5 cm, 9 cm **b** Sides 2 cm, 2 cm, 2 cm
c Sides 29 cm, 26 cm, 4 cm **d** Sides 22 cm, 12 cm, 10 cm
e Sides 20 cm, 7 cm, 9 cm **f** Sides 14 cm, 8 cm, 6 cm
g Sides 15 cm, 60 mm, 100 mm **h** Sides 120 mm, 8 cm, 9 cm

4 Construct isosceles triangles with sides
a 7 cm, 7 cm, 5 cm **b** 5 cm, 5 cm, 7 cm

5 **a** Construct an equilateral triangle ABC with sides 5 cm.
b Construct a second equilateral triangle with base AB and sides 5 cm to get a rhombus.
c Follow the steps in **a** and **b** to construct a rhombus with sides 3.5 cm.

6 **a** Construct a triangle with sides 5 cm, 12 cm, 13 cm.
b What type of triangle is this?

7 **a** Construct a triangle ABC with sides AB = 3 cm, BC = 4 cm, CA = 5 cm.
b Construct triangle ADC with side CA from the triangle in part **a**, and side AD = 4 cm and side DC = 3 cm.
c What special type of quadrilateral is this?

8 **a** Construct a triangle ABC such that BC = 12 cm, AC = 7 cm and ∠B = 30°.
b Now try to draw a second, **different** triangle with the same measurements. (Move the position of A.)
c Are SSA triangles unique?

This triangle is called SSA because you have two sides and the non-included angle.

S5.3 Constructing bisectors

Objectives

- Use a ruler and compasses to draw standard constructions

Useful resources

- Compasses for students and a board compass
- Rulers for students and a board ruler
- Tracing paper
- Geometry tool

Mental starter

A formula states that if a solid shape has F number of faces, E number of edges and V number of vertices (corners) then

$$F - E + V = 2.$$

Test this formula for a cube, triangular based pyramid (tetrahedron), triangular prism and a cylinder ($6 - 12 + 8 = 2$; $4 - 6 + 4 = 2$; $5 - 9 + 6 = 2$; $3 - 2 + 0 = 1$.)
Why does this formula not work for a cylinder? (It has curved faces.) In fact the formula works only for polyhedra and was discovered by Euler.

Introductory activity

What is the difference between 'dissect' and 'bisect'? (Cut into pieces **or** cut into 2 equal pieces.) **What is the difference between 'parallel' and 'perpendicular'?** (Lines going the same direction **or** lines crossing at 90°.)

Demonstrate bisecting an angle as described in the students' book. Allow students time to draw their own, then to check that the two angles are indeed equal using tracing paper. Ask students to choose a point on the bisector and then measure the shortest distance between the point and each of the two lines which form the original angle. These should be equal.

Demonstrate the construction of the perpendicular bisector of a line as described in the students' book. Allow students time to draw their own, then to check that the two angles are indeed equal using tracing paper and to measures the two line segments.
Ask students to choose a point on the bisector and then measure the distance between this point and the ends of the original line.

Exercise commentary and misconceptions

Questions 2, 5 and 6 review S5.2, constructing triangles with three known sides.

Only rulers and compasses should be used for this exercise.

Plenary

How can you construct angles of 120°, 300°, 150° and 210° from an equilateral triangle using only a ruler and compasses?

S5.3 Constructing bisectors

This spread will show you how to:

- Use a ruler and compasses to draw standard constructions

Keywords
Bisect
Equidistant
Perpendicular

- **Bisect** means cut into two equal parts.

You can use a straight edge and compasses to construct an angle bisector.

- To bisect angle ABC

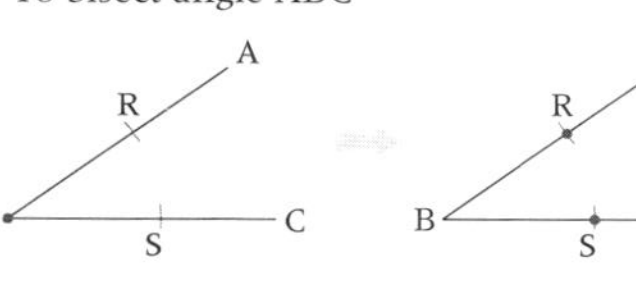

Use the same compass radius throughout the construction. Start at the red dots.

BRTS is a rhombus.

- All points on the angle bisector are equidistant from the arms of the angle.

Equidistant means equal distance from.

- The **perpendicular** bisector of a line bisects the line at right angles.

- To construct the perpendicular bisector of line AB

A —— B   A —— B   A —— B

Use the same compass radius throughout the construction.

Start at the red dots.

- All points on the perpendicular bisector of AB are equidistant from A and B.

Example

Construct an angle of 45°.

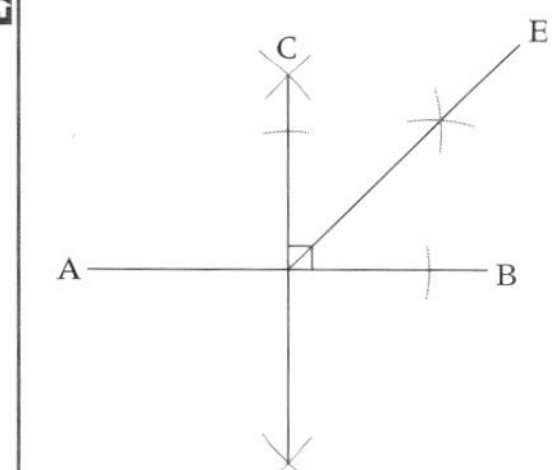

Draw a line AB
Construct the perpendicular bisector CD.
Construct the angle bisector of ∠BCD.
∠BCE = 45°

45° is $\frac{1}{2}$ of 90°. Construct a perpendicular bisector to AB (90°) and then bisect the angle.

Exercise S5.3

1 Trace these angles.
Construct the angle bisector of each angle, using ruler and compasses.

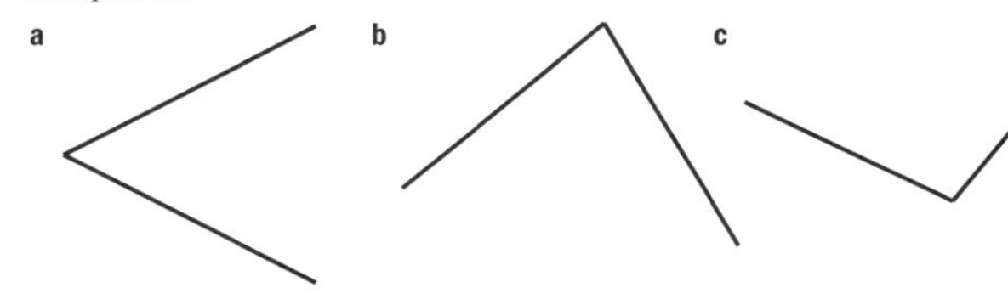

You will need to extend the lines.

2 **a** Construct an equilateral triangle with sides 5 cm.
b Construct the angle bisector of each angle of the triangle.
c What do you notice about the three angle bisectors?

See the example in S5.2 for help with constructing an equilateral triangle.

3 Follow these steps to construct an angle of 30°.
a Construct an equilateral triangle with sides 4 cm.
b Bisect one of the base angles.

Angles in an equilateral triangle = 60°.

4 Draw lines AB for these lengths and construct their perpendicular bisectors.
a 6 cm **b** 9 cm **c** 5.6 cm **d** 10 cm **e** 11.2 cm
Check by measuring that each bisector intersects the line AB at its midpoint.

5 **a** Construct an equilateral triangle with sides 5 cm.
b Construct the perpendicular bisectors of each side of the triangle.
c Compare your diagram with the one for question **2**. Write down what you notice.

6 **a** Construct two triangles with sides 8 cm, 5 cm, 7 cm. Label them triangle A and triangle L.
b On triangle A construct the angle bisector of each internal angle of the triangle.
c On triangle L construct the perpendicular bisectors of each side of the triangle.
d Compare and comment on your answers to **b** and **c**.

7 **a** Draw a line AB, 8 cm long, and construct its perpendicular bisector.
b Construct an angle of 45° where the perpendicular bisector intersects AB.
c What other angles have you created in this construction?

S5.4 Further constructions

Objectives
- Use a ruler and compasses to draw standard constructions

Useful resources
- Compasses for students and a board compass
- Rulers for students and a board rule
- Protractors for students and a board protractor
- Geometry tool

Mental starter
A snail crosses a railway track as fast as possible. It can move at a speed of 2 feet every 3 min. The track is 4 feet 8 inches wide.
(12 inches in a foot.)
How long will it take to cross the track?
(7 min)
What assumption must you make about the way the snail crosses the track?
(The rails are parallel and to cross at the shortest distance you must cross the rails at 90°.)

Introductory activity
There are two more types of bisector, and they are similar to the perpendicular bisector of a line.

Demonstrate constructing a perpendicular from a point **to** a line as described in the students' book. Allow students time to draw their own, then check that the two angles are equal using tracing paper.

Demonstrate constructing a perpendicular from a point **on** a line as described in the students' spread. Allow students time to draw their own, then check that the two angles are equal using tracing paper.

Exercise commentary and misconceptions
Emphasise that the setting of the compasses should remain the same throughout a construction.

Remind students to leave in the construction lines and arcs.

Protractors should only be used for checking, not in place of construction techniques.

Plenary
Suppose you have a plate in the shape of an equilateral triangle. **How can you find the point underneath it that it will balance on (centre of mass)?**
(The intersection of the perpendicular bisectors of each side. Only two are necessary, but a third side is a good way to check the accuracy of the construction.)

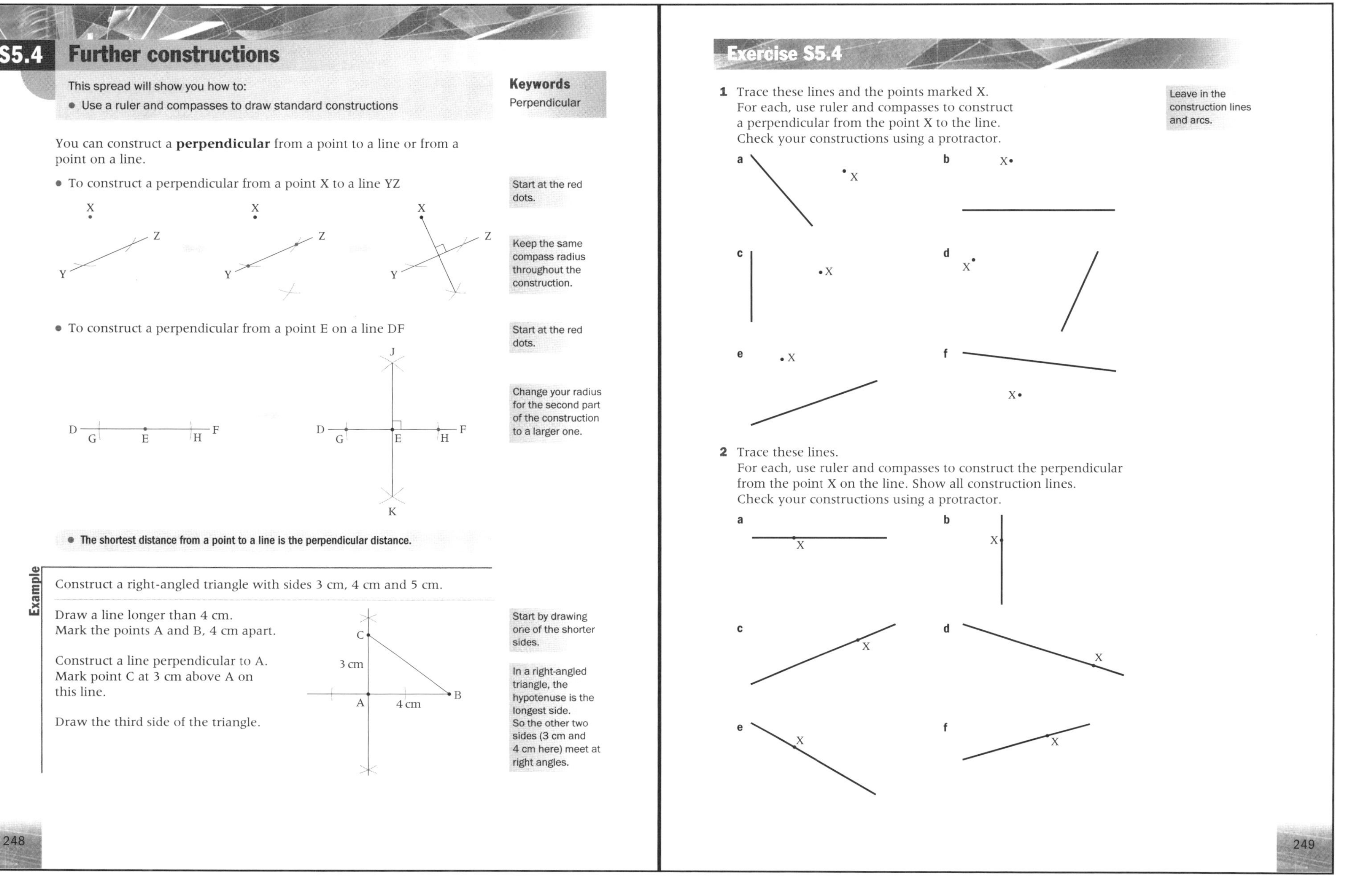

S5.4 Further constructions

This spread will show you how to:
- Use a ruler and compasses to draw standard constructions

Keywords
Perpendicular

You can construct a **perpendicular** from a point to a line or from a point on a line.

- To construct a perpendicular from a point X to a line YZ

Start at the red dots.

Keep the same compass radius throughout the construction.

- To construct a perpendicular from a point E on a line DF

Start at the red dots.

Change your radius for the second part of the construction to a larger one.

- **The shortest distance from a point to a line is the perpendicular distance.**

Example

Construct a right-angled triangle with sides 3 cm, 4 cm and 5 cm.

Draw a line longer than 4 cm.
Mark the points A and B, 4 cm apart.

Construct a line perpendicular to A.
Mark point C at 3 cm above A on this line.

Draw the third side of the triangle.

Start by drawing one of the shorter sides.

In a right-angled triangle, the hypotenuse is the longest side.
So the other two sides (3 cm and 4 cm here) meet at right angles.

Exercise S5.4

Leave in the construction lines and arcs.

1 Trace these lines and the points marked X.
For each, use ruler and compasses to construct a perpendicular from the point X to the line.
Check your constructions using a protractor.

a **b** **c** **d** **e** **f**

2 Trace these lines.
For each, use ruler and compasses to construct the perpendicular from the point X on the line. Show all construction lines.
Check your constructions using a protractor.

a **b** **c** **d** **e** **f**

S5.5 Loci

Objectives

- Find loci by reasoning and using diagrams

Useful resources

- Compasses and rulers for students
- Geometry tool

Mental starter

An empty cube has dimensions 5 m by 5 m by 5 m. You have two sizes of smaller cubes, some with sides of 2 m and some with sides of 3 m. You want to fit as many as possible of the smaller cubes into the empty one, but you have to use at least one of each size.
How can you do it? (Only solution is one with sides of 3 cm and seven with sides of 2 cm.)
What volume of space is not filled?
($125 - 27 - 7 \times 8 = 42$ m^3.)

Introductory activity

Discuss the main types of loci:

- construct distance from a point (circle)
- equidistant from two points (perpendicular bisector)
- construct distance from a line (parallel line)
- equidistant from two intersecting lines (angle bisector)

Discuss the first example, which relates to a perpendicular bisector.
The second example is contextual, and relates to an angle bisector.

Exercise commentary and misconceptions

Remind students that they will need to use the constructions they have learned to construct loci for practical problems.

The most common confusion is between points A, B, C and lines AB, BC etc. This could lead to the wrong type of bisector being drawn.

Accuracy is very important; time and care in construction will gain marks.

The terms locus and loci are not always used in a question, remind students that they can replace this word in their minds with 'position'.

Remind students that the 'perpendicular bisector of AB' and 'the locus of the points equidistant from A and B' mean the same thing.

Plenary

Challenge students to draw the locus of the points an equal distance round the outside of these shapes. Split the class into groups and ask one member of each group to put their solution on the board. Remind students that sharp corners get rounded off, just like making something in DT, where you sand off sharp corners. Complete the first example as a demonstration.

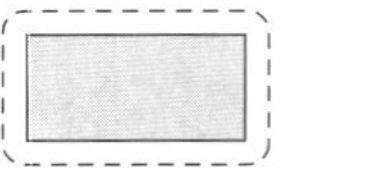

S5.5 Loci

This spread will show you how to:
- Find loci by reasoning and using diagrams

Keywords
Bisector
Equidistant
Locus
Perpendicular

Loci is the plural of locus.

A **locus** is the path traced out by a moving point.

- The locus of a point which is a constant distance from another point is a circle. 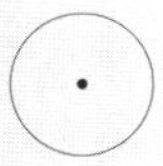
- The locus of a point at a constant distance from a fixed line is a parallel line. 
- The locus of a point that is **equidistant** from two other fixed points is the **perpendicular bisector** of the line joining the fixed points. 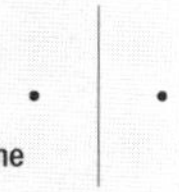
- The locus of a point equidistant from two intersecting lines is the angle bisector of the lines.

Example

P and Q are two points 2.5 cm apart.

P • Q •

Shade in the region that satisfies all these conditions:

- Right of the perpendicular to the line at point P.
- Closer to P than to Q.
- More than 1 cm from P.

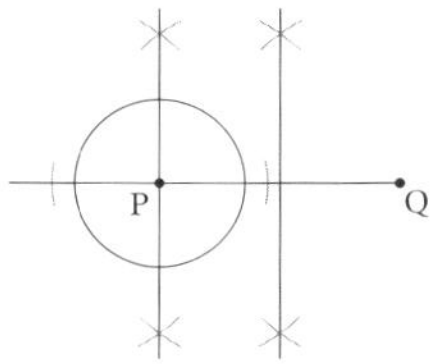

Construct the perpendicular to the line at point P.

Construct the perpendicular bisector of PQ. Points to the left are nearer to P than Q.

Draw a circle radius 1 cm, centre P. Points outside are more than 1 cm from P.

Example

ABCD is the plan of a garden.

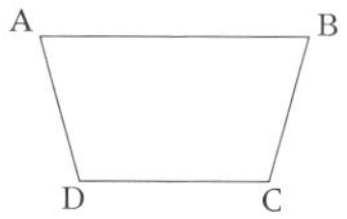

A tree is to be planted in the garden so that it is

- nearer to BC than to BA
- nearer to AD than AB.

Shade the region where the tree may be planted.

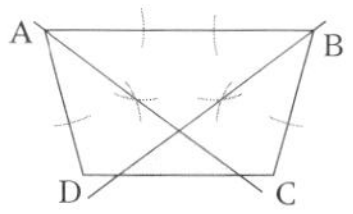

Construct the angle bisector of ∠ABC. Shade the region between this line and BC.

Construct the angle bisector of ∠BAD. Shade the region between this line and AD.

The tree can be planted in the region where the shadings overlap.

Exercise S5.5

1 **a** Draw points A and B, 6 cm apart.
b Shade in the region that satisfies both these conditions.
  i Closer to A than to B
  ii Less than 4 cm from B

2 **a** Draw points J and K, 5 cm apart.
b Shade the region that satisfies both these conditions.
  i More than 4 cm from J
  ii More than 3 cm from K

3 **a** Trace the points X, Y and Z.
b Shade the region that satisfies all three of these conditions.
  i Closer to X than to Y
  ii Closer to the line XZ than to the line XY
  iii More than 1 cm from X

X • Z • Y •

4 **a** Draw a rectangle PQRS where PQ = 5 cm and QR = 3 cm.
b Shade the region of the rectangle that is within 4 cm of P and within 2.5 cm of R.

5 **a** Construct a right-angled triangle ABC where angle ABC = 90°, AB = 6 cm, BC = 4.5 cm.
b Shade the region that satisfies all three of these conditions.
  i Closer to A than to B
  ii Less than 4 cm from A
  iii Less than 4 cm from C

6 The diagram shows the rectangular garden of a house.

There are two trees, *T*, in the garden.

A radio mast is to be placed in the garden.

It must be more than 5 m from the rear of the house.

It must be more than 3 m from a tree.

Using a scale of 1 cm : 2 m, draw a scale diagram and shade the possible site for the radio mast.

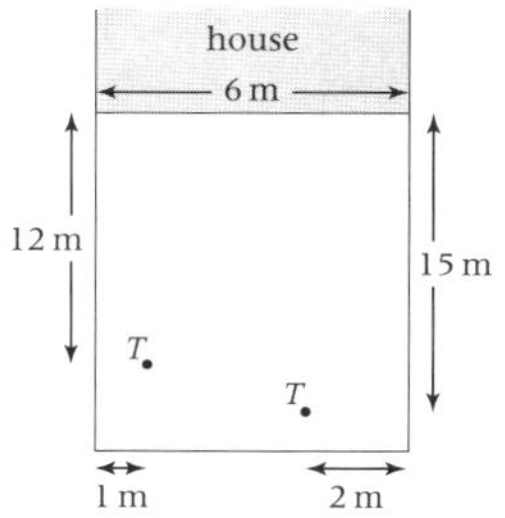

S5 Exam review

Key objectives

- Understand angle measure using the associated language
- Use and interpret maps and scale drawings
- Use a straight edge and compass to do standard constructions
- Find loci, both by reasoning and by using ICT to produce shapes and paths

Worked solution	Commentary and misconceptions
1	This question tests students' understanding of bearings and scale drawings.
a Bearing of school from John's house is 200°	
b **i** 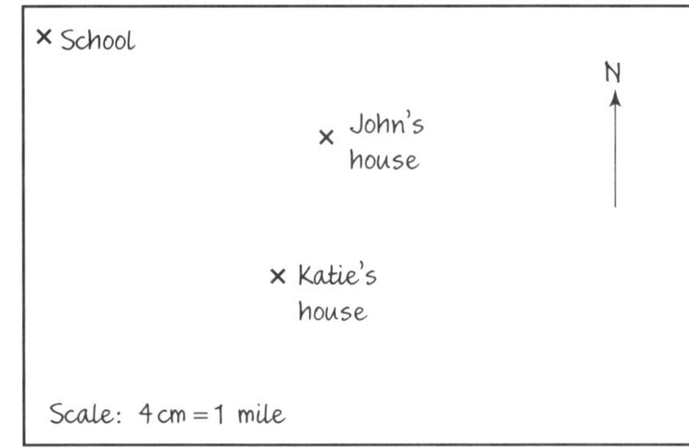	
ii 105°	Ensure that students give their answer correct to 3 sf.
2 **i** **ii** **iii** 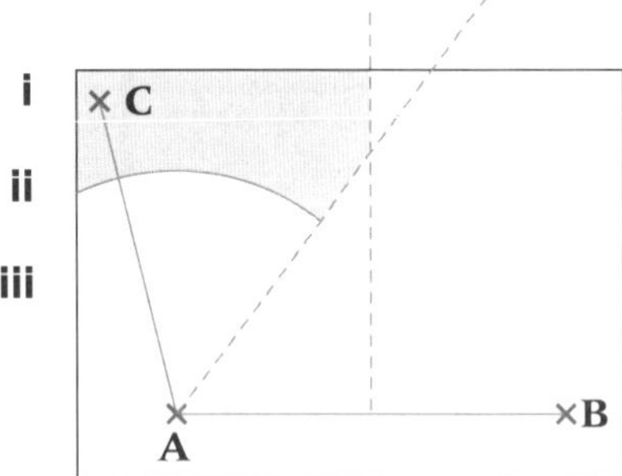	Ensure that students draw the correct lines and shade the correct regions. Some students may need to be prompted whether their loci should be a line or a circle and which side of the line to shade.

N7 Fraction and percentage calculations

Objectives

F/H Use efficient methods to calculate with fractions, including cancelling common factors before carrying out the calculation, recognising that, in many cases, only a fraction can express the exact answer

F/H Calculate a given fraction of a given quantity, expressing the answer as a fraction

F Convert simple fractions of a whole to percentages of the whole and vice versa

F/H Solve percentage problems, including percentage increase and decrease

F/H Use percentages to compare proportions

H Calculate an original amount when given the transformed amount after a percentage change

Unit overview

This unit consolidates the knowledge of calculating with fractions, decimals and percentages from unit N3. It begins by finding fractions and percentages of amounts and extends to percentage increase and decrease, simple and compound interest and using further percentage techniques to solve word problems. This unit prepares students for unit N8 which covers more complex problems including ratio, proportion and percentages.

Prior knowledge

Before your students start this unit they should be able to:

- Understand and use fraction, decimal and percentage notation
- Convert between fractions, decimals and percentages
- Calculate with integers, fractions, decimals and percentages
- Understand and use percentage as an operator
- Understand and use factors and multiples

Differentiation

- **Foundation** focuses on calculating a fraction and percentage of a given quantity using mental and written methods and then extends to calculating percentage increase and decrease and simple interest
- **Foundation Plus** extends to solving more general percentage problems and simple and compound interest. Students are also encouraged to use checking procedures and appropriate levels of accuracy
- **Higher Plus** contains an algebra unit (A7) on quadratic equations

N7.1 Finding fractions of quantities

Objectives
- Calculate a fraction of a quantity
- Calculate with fractions effectively following simplification
- Use fractions to express an exact answer

Useful resources
- Calculators for question 8

Mental starter
You have £53 made from: 3 × £10 notes, 2 × £5 notes, 10 × £1 coins, 6 × 50p pieces. **How would you divide this as fairly as possible between four people?** (Give £13 to each person and you have £1 left to keep or give 50p extra to two people.)

Introductory activity
What do we mean by $\frac{3}{5}$ of £45?
(Divide £45 into five equal parts and take three of the parts.)

We can find $\frac{1}{5}$ first by dividing by 5, then $\frac{3}{5}$ by multiplying by 3.

$45 \div 5 = 9 \qquad 9 \times 3 = £27$

It is often better to use fractions in the calculation, because cancelling makes the calculation easier and also avoids problems with awkward decimals.

For example, $\frac{2}{3}$ of 100 km:

Using the first method,
$100 \div 3 = 33.333333333333$
$33.3333... \times 2 = 66.666666... \approx 66.67$ km.
This is not an exact answer, because we have rounded it.

The second method is
$\frac{2}{3} \times 100 = \frac{200}{3} = 66\frac{2}{3}$ km which is an exact answer.

Discuss the examples in the student book. The second example is a problem in context.

Exercise commentary and misconceptions
Questions 1–7 all require fraction multiplication, usually with some cancelling.

Question 8 emphasises the approximation sometimes involved in using decimals.

In question 10 some students may want to change the initial number of hours into minutes before doing the calculation – this is **not** a good method!

In question 11 students need to give sensible answers rounded to the nearest penny.

Plenary
Discuss what happens when you repeatedly multiply a quantity by a proper fraction.
Discuss what happens with an improper fraction.
Encourage students to offer a justification.

N7.1 Finding fractions of quantities

This spread will show you how to:

- Calculate a fraction of a quantity
- Calculate with fractions effectively following simplification
- Use fractions to express an exact answer

Keywords
Cancel
Common factor
Exact

- **You find fractions of a quantity by multiplying.**

  For example, two thirds of $5 = \frac{2}{3} \times 5 = \frac{2 \times 5}{3} = \frac{10}{3} = 3\frac{1}{3}$

- **When the quantity and the denominator of the fraction have a common factor, cancel this factor before multiplying.**

  For example, $\frac{2}{_1\cancel{3}} \times \cancel{24}^8 = \frac{2}{1} \times 8 = 16$

Notice that you cannot give $3\frac{1}{3}$ as an **exact** answer in decimals because $\frac{1}{3}$ is a recurring decimal.

Example

Calculate **a** $\frac{3}{4}$ of 28 **b** $\frac{5}{8}$ of 6 **c** $\frac{4}{9}$ of 12 **d** $\frac{5}{9}$ of 25

a $\frac{3}{_1\cancel{4}} \times \cancel{28}^7 = 3 \times 7 = 21$

b $\frac{5}{_4\cancel{8}} \times \cancel{6}^3 = \frac{5 \times 3}{4} = \frac{15}{4} = 3\frac{3}{4}$

c $\frac{4}{_3\cancel{9}} \times \cancel{12}^4 = \frac{4 \times 4}{3} = \frac{16}{3} = 5\frac{1}{3}$

d $\frac{5}{9} \times 25 = \frac{125}{9}$

$= 13\frac{8}{9}$

Cancel by common factor 4 before multiplying.

You *can* also write $5\frac{1}{3}$ as $5.\dot{3}$, but it is simpler to leave it as a fraction.

In part **d**, you cannot cancel by 5. You can only cancel common factors where one is in a numerator (or a whole number) and the other is in a denominator.

You can extend this method to finding a fraction of a fraction of a quantity.

Example

Tom has £42. He spends $\frac{1}{3}$ of it on Monday. On Tuesday he spends $\frac{3}{4}$ of the remainder. How much does he spend on Tuesday?

Tom spends $\frac{1}{3} \times £42$ on Monday so he has $\frac{2}{3} \times £42$ on Tuesday.
On Tuesday he spends

$$\frac{\cancel{3}^1}{\cancel{4}_{\cancel{2}_1}} \times \frac{^1\cancel{2}}{\cancel{3}_1} \times \cancel{42}^{21} = £21$$

Exercise N7.1

1 Calculate these. Give your answers as fractions or mixed numbers.

a $\frac{1}{2}$ of 7 **b** $\frac{1}{5}$ of 8 **c** $\frac{1}{3}$ of 10 **d** $\frac{1}{3}$ of 2 **e** $\frac{2}{5}$ of 6

2 Work out these. Show how common factors can be cancelled in each.

a $\frac{1}{2}$ of 20 **b** $\frac{1}{4}$ of 84 **c** $\frac{1}{3}$ of 36 **d** $\frac{1}{5}$ of 65 **e** $\frac{2}{3}$ of 33

You should show all of your working for the questions **3**, **4** and **5**. In particular, show how the calculations can be simplified by cancelling common factors.

3 A reel holds 60 m of wire when new. $\frac{2}{5}$ of the wire has been used.

a What length of wire has been used?

b What length of wire is left on the reel?

4 Calculate the amount of liquid in these containers.

a A 40 litre barrel that is $\frac{3}{8}$ full. **b** A 240 cl jar that is $\frac{3}{4}$ full.

c A 120 cl glass that is $\frac{2}{5}$ full.

d A 750 ml litre bottle that is $\frac{2}{3}$ empty.

5 Calculate these.

a $\frac{5}{8}$ of 48 m **b** $\frac{2}{9}$ of 36 km **c** $\frac{4}{7}$ of 28 mm **d** $\frac{3}{4}$ of 120 m

The answers to questions **6** and **7** will not be whole numbers. You should show your working as before, and give your answers as fractions or mixed numbers.

6 Work out these.

a $\frac{1}{6}$ of 10 **b** $\frac{1}{4}$ of 22 **c** $\frac{3}{10}$ of 15 **d** $\frac{1}{12}$ of 8 **e** $\frac{4}{9}$ of 21

7 Calculate these lengths.

a $\frac{1}{9}$ of 24 miles **b** $\frac{5}{6}$ of 40 miles **c** $\frac{5}{18}$ of 45 miles **d** $\frac{3}{20}$ of 25 miles

8 Calculate these and convert your answers to (approximate) decimal numbers.

a $\frac{5}{12}$ of 16 m **b** $\frac{4}{9}$ of 12 mm **c** $\frac{3}{22}$ of 64 cm **d** $\frac{3}{14}$ of 104 km

9 An empty swimming pool is to be filled with water. It takes 12 hours to fill the pool, and the full pool contains 98 m^3 of water. How much water will the pool contain after 5 hours? Show your working.

10 Calculate these times using fractions, and then convert each of your answers into hours and minutes.

a $\frac{7}{12}$ of 9 hours **b** $\frac{5}{8}$ of 22 hours **c** $\frac{7}{10}$ of 24 hours **d** $\frac{7}{18}$ of 63 hours

11 Calculate these.

a $\frac{2}{3}$ of £7 **b** $\frac{5}{12}$ of £40 **c** $\frac{3}{8}$ of £34 **d** $\frac{4}{9}$ of £66

N7.2 Finding a percentage of a quantity

Objectives
- Calculate a percentage of a quantity

Useful resources
- Calculators

Mental starter
Ask the question:
80% of an amount is £5000.
What is the total amount?
Discuss solutions and methods.

Introductory activity
Find 30% of £45.

Can find 10% ($=\frac{1}{10}$) by ÷10.
10% is £4.50.

Can then multiply by 3 to find 30%.
30% is £13.50.

Find 32% of 500 kg.

Find 1% ($=\frac{1}{100}$) by ÷100.
1% is 5 kg.

Then multiply by 32, so 32% of 500 is 160 kg.

What is an easy way to find 25% of 80 using a mental method?
(Divide by 4, answer is 20.)
Which other percentages are easy to calculate using mental methods? (20%, $33\frac{1}{3}$%, 40%, 50%, 60%, $66\frac{2}{3}$%, 75%, 80%.)

Can also change a percentage into a decimal fraction, which is an easy calculator method.

Find 34% of 85 kg.

34% = 0.34.

34% of 85 kg = 0.34 × 85 = 29 kg to 2 sf.

You can also use multiplication by fractions.

16% of £25 = $\frac{16}{100}$ × 25 = £4

Discuss the examples in the student book. The second and third examples offer real-life contexts.

Exercise commentary and misconceptions
Students may be used to a variety of methods for finding the percentage of something.
Ensure students do not use a calculator for questions 1–5. The decimal fraction method is particularly useful when dealing with repeated percentage change.

Remind students that in 'real-life' questions 8–11 they need to round answers 'sensibly'.

Plenary
Will an increase of 10% in price followed by a decrease of 10% in price return you to the original price?
Encourage students to make their own examples up to demonstrate this, for example if you start with £200 ⟶ £220 ⟶ £198, it does not work.

What percentage of the original do you end up with?
Use a couple of examples to show you get the same answer whatever you start with. From the example above $\frac{198}{200} = \frac{99}{100}$ = 99%.
Discuss the reasons why.

N7.2 Finding a percentage of a quantity

This spread will show you how to:
- Calculate a percentage of a quantity

Keywords
Decimal equivalent

You often need to calculate a percentage of a quantity.

- **Use mental methods to find simple percentages.** — 75% of £38: work out one quarter of £38 (which is £9.50), and multiply by 3.
- **For more complicated examples, you can find 1% of the quantity, and then multiply by the required percentage.** — 27% of 48 m: first find 1% (which is 0.48 m), and multiply by 27.
- **A quick method, especially when using a calculator, is to multiply by the appropriate decimal number.** — To find 38% of a quantity, multiply by 0.38.

Example

Calculate **a** 45% of 60 cm **b** 34% of 85 kg **c** 16% of £25

a 50% of 60 = 30 (50% is a half.)
5% of 60 = 3
45% of 60 cm = 30 cm − 3 cm = 27 cm

b $34\% = \frac{34}{100} = 0.34$ (0.34 is the **decimal equivalent** of 34%.)
34% of 85 kg = 0.34 × 85
= 28.9 kg (By calculator)
= 29 kg to 2 sf.

c 16% of £25 $= \frac{^{4}\not{16}}{\not{100}_{4}} \times \not{25}^{1}$
= £4

Example

At Fitz High School, 95% of the students have never been absent.
There are 1180 students at the school.
How many of them have a perfect attendance record?

95% = 0.95
95% of 1180 = 0.95 × 1180
= 1121

So 1121 students have a perfect attendance record

Example

Tom is saving 6% of his salary in a pension fund.
His current salary is £25 000.
How much will he save this year?

10% of £25 000 = £2500
5% of £25 000 = £1250
1% of £2500 = £250
6% of £2500 = £1250+£250
= £1500

This calculation is easy to do mentally.

Exercise N7.2

1 Find the percentages of these numbers mentally.
a 25% of 42 **b** 90% of 140 **c** 20% of 1200
d 60% of 500 **e** 30% of 440 **f** 11% of 900

2 Calculate these, using a mental method wherever possible.
a 30% of £750 **b** 55% of 1800 m
c 90% of 2800 kg **d** 60% of €240

3 Use an appropriate method to work out these. Show all your working, and do not use a calculator.
a 9% of 1500 **b** 13% of 700 **c** 31% of 2400
d 36% of 50 **e** 43% of 900 **f** 6% of 3200

4 Calculate these, using an appropriate mental or written method. Show all your working, and do not use a calculator.
a 23% of 4800 mm **b** 61% of 3200 kg
c 39% of €3700 **d** 17% of £2900

5 Write a decimal number equivalent to each percentage.
a 50% **b** 60% **c** 25% **d** 51% **e** 64%
f 22% **g** 15% **h** 70% **i** 7% **j** 8.5%

6 Calculate these, using an appropriate method.
Use a calculator where necessary.
a 15% of 38 **b** 25% of 800 **c** 27% of 59
d 96% of 104 **e** 41% of 41 **f** 80% of 25

7 Calculate these, rounding your answers to the nearest penny.
a 16% of £24 **b** 63% of £85 **c** 93% of £15
d 42% of £405 **e** 88% of £32 **f** 6% of £265

8 Mrs Jones has a conservatory built which costs £12 000.
She pays an initial deposit of 15%. The remainder is to be paid in 24 equal monthly instalments. How much is each of these instalments?

9 A restaurant adds a 12% service charge to the bill. What will be the total cost for a meal that is £65.80 before the service charge is added?

10 A school has 1248 pupils, and 48% of them are girls. How many boys are there in the school? Show your working.

11 Julia earns a salary of £47 800 per year. She is awarded a 2.7% pay rise. Calculate her new salary.

N7.3 Percentage increase and decrease

Objectives
- Solve problems, involving percentage increase and decrease

Useful resources
- Calculators

Mental starter
A new car was bought for £32 000. **If it loses 50% of its value every year, how long before you would give it away?**

(Value after number of years 1, £16 000; 2, £8000; 3, £4000; 4, £2000; 5, £1000; 6, £500; 7, £250; 8, £125; 9, £62.50; 10, £31.25; 11, £15.62. Answers are likely to be around 9 years. It costs about £50 to have a car taken away for you if you can't sell it.)

Introductory activity
Explain that we are going to increase or decrease amounts by a certain percentage.

First without a calculator:

Increase £70 by 98%.

Easiest way to find 98% is to find 2% first.
$70 \div 100 = 0.70$ (1%) $\longrightarrow 0.70 \times 2 =$ £1.40 (2%).

So 98% of 70 is $70 - 1.40 =$ £68.60.

Increase by £68.60 $\longrightarrow$ £138.60.

Try the same question again with a calculator.

Instead of adding the amounts of money at the end of the calculation, we could add the percentages. We will have our original 100% plus a further 98%, so 198% or 0.198 in total.

$1.98 \times 70 =$ £138.60 as before.

Decrease £40 by 12%.

What percentage would we have left if we decreased by 12%? (88%) **What decimal number must we multiply by?** (0.88)
$0.88 \times 40 =$ £35.20.

Exercise commentary and misconceptions
Encourage students to use a variety of methods, especially multiplying by a decimal number when using a calculator.

Questions 3, 4 and 7 ask for the actual increases or decreases, while questions 1, 2, 8 and 9 ask for the increased or decreased amounts. Students often misread the questions, especially in exams!

Questions 10 and 11 may lead to some interesting discussions. Students will need to understand what is meant by a 100% increase and 100% decrease.

Plenary
Suppose an antique is worth £5000 and increases in value by 15% each year.
What can you multiply by to find its value in 4 years time? ($1.15 \times 1.15 \times 1.15 \times 1.15$ (or 1.15^4) $= 1.74900625$.)
How can you find the value after 100 years? ($5000 \times 1.15^{100} = 5\ 871\ 567\ 253$ which is nearly £6 billion.)

N7.3 Percentage increase and decrease

This spread will show you how to:
- Solve problems, involving percentage increase and decrease

Keywords
Decimal equivalent
Decrease
Increase

You often need to work out problems involving percentage **increase** and **decrease**.

- **To work out the new amount after a percentage increase or decrease,**
  - **first find the increase or reduction**

    Increase £25 000 by 5%
    5% of £25 000 = £1250

    then simply add or subtract

    £25 000 + £1250 = £26 250
  - **Using a calculator, you can find the new amount by multiplying by the decimal equivalent.**

    To find the result of a 12% decrease, multiply the original amount by 0.88.
    To find the result of a 23% increase, multiply the original by 1.23.

Example

Find the result when

a 45 cm is increased by 20% **b** 44 kg is decreased by 17%.

a 20% of 45 cm = 9 cm,
new length 45 cm + 9 cm = 54 cm

b 100% − 17% = 83%
44 kg × 0.83 = 36.52 kg

New mass is 83% of original.

Example

This is how Jackie worked out 22% of 85 cm.

What is wrong here?

Examiner's tip
The words 'of' and 'off' can cause confusion – especially when you are in a hurry in an exam!

This question simply asks her to work out 22% **of** 85 cm – she is not being asked to work out a percentage reduction.

22% of 85 cm = 0.22 × 85
= 18.7 cm

Example

Alan's car depreciates by 15% every year.
It is valued at £9750 now. What will be its value in one year's time?

Value after depreciation = 100% − 15% = 85% of original value
0.85 × 9750 = £8287.50 = £8300 to nearest £100

85% = 0.85

Exercise N7.3

1 Calculate the results when these amounts are increased by the percentages given.
a 72 by 50% **b** 60 by 20% **c** 45 by 10%
d 600 by 3% **e** 480 by 25% **f** 500 by 40%

You should be able to do all of these mentally.

2 Decrease each of these amounts by the percentages given.
a 28 by 10% **b** 45 by 20% **c** 60 by 15%
d 75 by 50% **e** 380 by 40% **f** 65 by 1%

Again, you should be able to do these mentally.

3 Use a written method to calculate each of these percentage increases. Show your working, and do not use a calculator.
a 300 by 14% **b** 200 by 32% **c** 800 by 26%
d 250 by 30% **e** 750 by 83% **f** 940 by 18%

4 Use a written method to find these percentage decreases.
a 800 by 16% **b** 700 by 24% **c** 400 by 32%
d 450 by 33% **e** 350 by 61% **f** 260 by 52%

5 Write the decimal number you must multiply by to find these percentage increases.
a 20% **b** 30% **c** 45% **d** 85% **e** 6.5%

6 Write the decimal number you must multiply by to find these percentage decreases.
a 40% **b** 60% **c** 35% **d** 72% **e** 18.5%

7 Use a calculator to find these percentage increases and decreases.
a Increase 53 by 7% **b** Decrease 42 by 4%
c Increase 620 by 16% **d** Decrease 300 by 18%

8 Calculate the new salaries after these pay increases.
a £32 000 increased by 5% **b** £18 450 increased by 4.7%
c £26 500 increased by 3.2% **d** £52 850 increased by 6%

9 Calculate the new prices after these price cuts.
a £450 decreased by 22% **b** £860 decreased by 35%
c £1250 decreased by 42% **d** £740 decreased by 3.5%

10 Alan says, 'The cost of computer chips increased by 120% last year'.
Billy says, 'That's not possible. They couldn't have gone up by more than 100%'.
Explain why Billy is wrong.

11 Carina says, 'The cost of computer memory fell by 120% last year.'
Deni says, 'That's not possible. Prices couldn't have fallen more than 100%'.
Explain why Deni is correct.

N7.4 Simple and compound interest

Objectives

- Calculate simple and compound interest

Useful resources

- Calculators
- Interest calculator, for example a spread sheet program

Mental starter

You have £1000 to invest for up to three years and have two options.

(A) An account offering you interest of 10% of £1000 every year.

(B) An account offering you interest of 10% of the amount you have at the start of each year and added every year. But you cannot take out any of your money until the end of the three years.

Which would you go for, A or B? (A gives £1300 in total, B gives 1100 ⟶ 1210 ⟶ £1331.)

Introductory activity

There are two main types of interest you can earn when investing money – simple interest and compound interest.

Suppose you deposit £400 for three years at 5% interest p.a. (per annum/per year).

How much money will you have at the end of the three-year period?

If this is 5% simple interest:
5% of £400 = £20. For three years, $3 \times 20 = 60$. Therefore $400 + 60 =$ £460 after three years.

If this is 5% compound interest:
5% of £400 = £20. After one year you have £420. For the next year you get 5% of £420. (£21). After two years you have £441. Then 5% of 441 (=22.05). After three years you have £463.05 and £63.05 interest has been earned. (Compounding something means building it up and up. If you 'compound a problem' you make it worse and worse. You are earning interest on the interest.)

How can you speed up the calculation with a calculator? ($1.05 \times 1.05 \times 1.05 \times 400$ or even quicker $1.05^3 \times 400$)

Suppose the population of a town was 84 000. The population fell steadily by 4% every year for ten years. **Estimate the population after this time.**

This is a fall of 4% every year for 10 years. $0.96 \times 84\,000$ will work out 96% (fall of 4%).

$0.96^{10} \times 84\,000 = 55\,845$ (round to 56 000, same accuracy as the question). Students will need to be reminded where the power button is. It may appear as x^y, y^x or ^.

Exercise commentary and misconceptions

The exercise focuses on calculating simple and compound interest and using decimal multiplication to a power.

Remind students to read the questions very carefully – are they being asked for the interest or the total amount after a certain number of years?

Plenary

When you borrow money you have to pay it back with interest. For example you might want to borrow £1000 and pay it all back at once two years later. If you are charged 3.4% interest p.a. you would pay back £1069.16. ($1.034 \times 1.034 \times 1000$) **Why do you think the bank has a savings rate of less than 3.4% p.a.?** (This would mean you could take out a loan and then put it into savings and make more money. This is never possible. That is why savings rates are always lower than loan rates.)

In practice the calculations are more complex because you normally repay a loan gradually over the loan period, not at the end, so interest is calculated on a decreasing debt, and is calculated more frequently than once a year.

N7.4 Simple and compound interest

This spread will show you how to:

- Calculate simple and compound interest

Keywords
Borrow
Compound interest
Interest
Invest
Principal
Rate
Simple interest

You earn **interest** when you **invest** in a savings account at a bank. However, you pay interest if you **borrow** money for a mortgage.

Interest is either

- **simple interest** – it is not added to the principal, or
- **compound interest** – added to the principal and will itself earn interest.

The original sum you invest is called the **principal**.

- **To calculate simple interest, use the interest rate to work out the amount earned.**
- **If simple interest is paid for several years, the amount paid each time stays the same, because the interest is paid elsewhere and the principal stays the same.**
- **To calculate compound interest, work out the interest in the same way, but add the interest earned to the principal.**
- **If compound interest is paid for several years, the amount of interest earned each year increases, because the principal increases.**

Example

Calculate the interest when £1000 is invested for 4 years at

a 5% simple interest (SI) **b** 5% compound interest (CI).

a SI for 1 year = 5% × £1000
= £50

SI for 4 years = 5% × £1000 × 4
= £200

Total SI = £200
Principal + SI = £1200

b 1st year's CI = 5% × £1000
= £50
new principal = £1000 + £50
= £1050
2nd year's CI = 5% × £1050
= £52.50
new principal = £1102.50
3rd year's CI = 5% × £1102.50
= £55.13
new principal = £1157.63
4th year's CI = 5% × £1157.63
= £57.88
new principal = £1215.51

Money invested at compound interest earns more than the same sum invested at simple interest.

- **When you have to calculate compound interest over a number of years you can do it quickly using a calculator.**

Example

£2000 is invested at 6.5% compound interest.
Find the principal after 15 years.

Principal at end of 1 year = £2000 × 1.065 = £2130

Principal at end of 2nd year = (£2000 × 1.065) × 1.065
= £2000 × 1.065^2 = £2268.45

After 15 years principal = £2000 × 1.065^{15} = £5143.68

To increase by 6.5% multiply by decimal equivalent of 6.5% = 1.065

Each year the principal increases by a factor of 1.065

Exercise N7.4

1 Find the total interest earned when these amounts of money are invested at these annual rates of **simple interest**.

a £100 at 5% for 1 year **b** £200 at 6% for 2 years
c £1400 at 7.5% for 3 years **d** £650 at 3.5% for 10 years

2 Find the final value of each of these amounts invested at **compound interest**.

a £250 at 5% for 1 year **b** £400 at 2% for 2 years
c £1200 at 6% for 2 years **d** £1000 at 2% for 3 years

3 A sum of money is invested at 5% compound interest.
- To find the amount after one year, multiply the principal by 1.05.
- To find the amount after two years, multiply the principal by $1.05^2 = 1.1025$

a What decimal number do you need to multiply the principal by to work out the amount after earning interest for one year at a rate of 6%?

b What decimal number do you need to multiply the principal by to work out the amount after earning compound interest for two years at 6%?

4 Use your answers to question **3** to work out the final amount when a principal of £500 is invested.

a for one year at an interest rate of 6%
b for two years at 6% compound interest.

5 Find the decimal number you should multiply the principal by to find the final amount after earning compound interest at these rates.

a 5% per year for 3 years **b** 6.5% per year for 5 years

6 Use your answers to question **5** to work out the final amount when

a £5000 is invested for 3 years at 5% compound interest.
b £800 is invested for 5 years at 6.5% compound interest.

7 Which of these options earns the most interest when £5000 is invested for

a 8 years at 6% simple interest
b 6 years at 8% compound interest?

Explain your answer.

8 **a** What decimal number do you multiply by, to find a 15% reduction?
b Explain how you would then find a second 15% reduction.

N7.5 More percentage techniques

Objectives

- Express a number as a percentage of another
- Calculate the original amount before a percentage increase or decrease

Useful resources

- Calculators

Mental starter

The cost of a CD is reduced by $\frac{1}{3}$ to £12. **How much did it cost originally?** (£18.)
Students should equate the £12 to $\frac{2}{3}$ of the price. So $\frac{1}{3}$ is £6.

Why might some people think the answer is £16? Finding a $\frac{1}{3}$ of £12 and adding it on is incorrect because you will be finding $\frac{1}{3}$ of the wrong number. The price went down by $\frac{1}{3}$ of the original, not $\frac{1}{3}$ of 12.

Introductory activity

Three people received test scores. A got 18%, B got 4 out of 25 and C got 6 out of 40.
How could you compare the marks?
(Change them to percentages.) Write them as fractions first, then multiply by 100 to make them percentages.

$\frac{4}{25} \times 100 = 16\%$ $\quad \frac{6}{40} \times 100 = 15\%$
so A was first, B second and C third.

The value of a house in 2005 was £62 000, in 2006 its value was £80 000.

Express the value in 2005 as a percentage of the value in 2006.

As before, write as a fraction $\frac{62\,000}{80\,000} = \frac{31}{40}$
then $\frac{31}{40} \times 100 = 77.5\%$.

Try to push the link between decimals and percentages rather than just learning to multiply by 100.

Suppose the number of people living a village has gone down by 40% and is now only 180 people. **How many people were there originally?**

Some people find 40% of 180 and add it onto 180. **Why is this wrong?** (Population went down by 40% of the original, not 40% of 180.)
What percentage do the 180 people stand for? (60% of the original population.)
So 180 people = 60%. But you want to find the original number or 100%. Use this answer like a ratio to find the original.

$180 \times \frac{100}{60} = 180 \div \frac{60}{100}$
$= 180 \div 0.6 = 300$

It may be useful to calculate 40% of 300 and subtract from 300 to show that the answer is correct.

It may also be helpful to use a number machine and its inverse.

If the population of another town rose by 3% to 5768, what was original population? (5768 ÷ 1.03 = 5600)

Exercise commentary and misconceptions

The exercise focuses on using calculator and non-calculator methods to express one number as a percentage of another number and find original amounts.

Question 5 is the first reverse percentage question. Ensure that the students get the first statement correct, so that they identify an amount with a certain percentage. Writing the percentages and amounts in a contingency table can make the ratios easier to see.

Plenary

Jamil scored 40% in his test. The teacher said he needed to improve by 15% in the next test. Jamil said 'I will definitely achieve this if I get over 50%'. **Is he correct?** (Improve by 15% of 40 means improving by 6%, so getting 46% overall or more.)

N7.5 More percentage techniques

This spread will show you how to:

- Express a number as a percentage of another
- Calculate the original amount before a percentage increase or decrease

Keywords
Original price

- **To write one number as a percentage of another, start by writing the first number as a fraction of the other.**
  - **Write '23 as a percentage of 25' as $\frac{23}{25}$.**
- **Now convert the fraction to a percentage.**
  - $\frac{23}{25} = \frac{23 \times 4}{25 \times 4} = \frac{92}{100} = 92\%$.
  - **Sometimes you will need to use a calculator. For example, to find 9 as a percentage of 17: $\frac{9}{17} = (9 \div 17) \times 100\% = 52.9\%$ (to 1 decimal place).**

- **To find the original amount before a percentage change, use the final amount to work out 1% of the original amount, and then find the original amount by multiplying.**

Example

Find

a 13 as a percentage of 20 b 39 as a percentage of 75
c 8 as a percentage of 23

a $\frac{13}{20} = \frac{13 \times 5}{20 \times 5} = \frac{65}{100} = 65\%$
b $\frac{39}{75} = \frac{13}{25} = \frac{52}{100} = 52\%$
c $\frac{8}{23} = (8 \div 23) \times 100\% = 34.8\%$

Example

Find the **original price** of a denim jacket reduced by 15% to £32.30.

85% of original price = £32.30
1% of original price = £32.30 ÷ 85
100% of original price = £32.30 ÷ 85 × 100
= £38

100% − 15% = 85%

Example

Following a 5% price increase, a car radio costs £168. How much did it cost before the increase?

105% of original price = £168
1% of original price = £168 ÷ 105 = £1.60
100% of original price = £1.60 × 100 = £160

Exercise N7.5

1 For these pairs of numbers, find the smaller number as a percentage of the larger one. You should be able to do all of these mentally.

a 16 and 20 b 7 and 25 c 6 and 50
d 20 and 40 e 17 and 34

2 Use a written method to find

a 12 as a percentage of 20 b 36 as a percentage of 75
c 24 as a percentage of 40

3 Use a calculator to find

a 19 as a percentage of 37 b 42 as a percentage of 147
c 8 as a percentage of 209

4 Jason scored 53 out of 60 in a science test, and 39 out of 45 in a maths test.

a Convert each of his marks to a percentage.
b Jason says, 'I did better in maths, because I only dropped 6 marks; I dropped 7 marks in science.' Explain why Jason is wrong.

5 A book costs £4 after a 20% price reduction. How much did it cost before the reduction? Show your working.

6 These numbers are the results when some amounts were increased by 10%. For each one, find the original number.

a 55 b 44 c 88 d 121

7 Find the original cost of the following items.

a A vase that costs £7.20 after a 20% price increase.
b A table that costs £64 after a 20% decrease in price.

8 a Francesca earns £350 per week. She is awarded a pay rise of 3.75%. Frank earns £320 per week. He is awarded a pay rise of 4%. Who gets the bigger pay increase? Show all your working.
b Bertha's pension was increased by 5.15% to £82.05. What was her pension before this increase?

9 Carys is trying to find the original price of an item that is on sale. Its sale price is £56 and it was reduced by 20%. Carys' working is shown to the right. What is wrong with her working?

20% of 56 = 11.2
56 + 11.2 = 67.2
The answer is £67.20

N7

Exam review

Key objectives

- Use efficient methods to calculate with fractions, including cancelling common factors before carrying out the calculation, recognising that, in many cases, only a fraction can express the exact answer
- Solve percentage problems, including percentage increase and decrease

Worked solution	Commentary and misconceptions
1	Students may find the combination of different values and forms within one question confusing. Encourage them to write down their working clearly. Remind students to read word problems carefully. Ensure that students understand what they are required to work out in each part of the question and give their solutions in the correct form.
a $\frac{2}{5} \times 80 = £32$	Ensure that students understand fraction as a multiplicative inverse.
b £80 − £32 = £48 spending money 25% of £48 = 0.25 × 48 = £12	Ensure that students calculate the amount of spending money correctly and use this value in the following calculation.
c **i** £48 − £12 = £36 **ii** $\frac{36}{80} = \frac{9}{20}$ **iii** $\frac{9}{20} = 0.45 = 45\%$	Ensure that students use the original amount of money, £80.
2	Remind students to read word problems carefully and ensure they know what they are required to work out.
a (269.30 − 56.80) ÷ 42.50 = 5 So he worked 5 + 1 hours = 6 hours	Students may need prompting to ensure they incorporate both of the hourly rates in their calculation. Ensure that students include the first hour in their final answer.
b 5% of £269.30 = 0.05 × 269.30 = 13.365 269.30 − 13.365 = £255.94 to 2dp	Students should give their solution to 2 dp.

Cumulative frequency

Objectives

H Draw and produce, using paper and ICT, cumulative frequency tables and diagrams, box plots and histograms for grouped continuous data

F/H Identify the modal class for grouped data

F/H Interpret a wide range of graphs and diagrams and draw conclusions

H Find the median, quartiles and interquartile range for large data sets

F/H Understand and use estimates or measures of probability from theoretical models, or from relative frequency

H Compare distributions and make inferences, using shapes of distributions and measures of average and spread, including median and quartiles

Unit overview

This unit consolidates the students' knowledge of frequency density and statistical diagrams from units D3 and D4, extending this to drawing cumulative frequency diagrams and using them to make statistical inferences and compare data sets. It then extends the work on box and whisker plots from unit D2, including using them to compare data sets.

Prior knowledge

Before your students start this unit they should be able to:

- Calculate a given fraction of a given quantity
- Use and interpret graphs modelling real life situations
- Read values and make inferences from a range of graphs and charts
- Calculate the averages and range of data sets
- Understand and use units of measure
- Understand and use frequency tables

Differentiation

- **Foundation** focuses on interpreting pictograms, pie charts and bar charts and using these to compare data sets. Using stem-and-leaf diagrams to find the median and range of a data set and identifying exceptions in data using time series graphs are also covered
- **Foundation Plus** extends to interpreting a wider range of charts and diagrams, including scatter diagrams and their correlation
- **Higher Plus** focuses on drawing and using histograms. It covers missing data and comparing data sets and provides an introduction to statistical reporting

D5.1 Cumulative frequency diagrams

Objectives
- Draw cumulative frequency polygons for grouped data
- Find the modal class of a data set

Useful resources
- 2 mm graph paper
- OHT of cumulative frequency graph of ages for plenary session

Mental starter
Write a string of numbers, such as:
7, 15, 21, 27, 12, 5, 2
Ask students to find the running total (7, 22, 43, ...)
Repeat for larger numbers.

Introductory activity
The ages of 40 people who were banned from driving were recorded.

Age years (y)	$16 \leqslant y < 20$	$20 \leqslant y < 24$	$24 \leqslant y < 28$	$28 \leqslant y < 32$	$32 \leqslant y < 36$
Number of people (frequency)	8	14	10	6	2
Cumulative frequency	(8)	(22)	(32)	(38)	(40)

How many people were banned up to the age of 20? (8.)

How many people were banned up to the age of 24? (22.) This is the accumulated number of people all together up to 24.

Ask students to draw axes on graph paper, and plot the cumulative frequencies against the upper class boundaries.
Sketch the result on the board for guidance.

Cumulative frequency diagrams are sometimes said to be S-shaped. They often have a shallow gradient at the start, then become much steeper and finally shallow again. They can have portions with a zero gradient, but never a negative gradient.

Exercise commentary and misconceptions
In spread D5.4 students will be asked to draw box plots under their cumulative frequency graphs, so advise them to leave a space at the bottom of each page.

There are a large number of common errors students make with cumulative frequency graphs.

- Points must be plotted at the top of each class interval, not the centre.
- The first point of the graph is the lowest value in the first class interval, this has a cumulative frequency of zero. This needs to be put onto the graph if the scale allows, otherwise the graph should be left hanging. A common error is to join the graph back to the origin.
- A regular scale **must** be used for the horizontal axis, some students will label the axis in groups matched to the class intervals.
- A regular scale must be used for the vertical axis. Some students will use the cumulative frequencies as the scale, so in the example above they would place the values 8, 22, 32, 38 and 40 at equal intervals.
- When joining the points a smooth curve or a straight line can be used. Some students will wrongly draw a line of best fit.
- Some students have the cumulative frequency axis horizontally, this is likely to lose the student marks.

Plenary
Discuss question 6. What can the graph tell you about the amount people spent?

D5.1 Cumulative frequency diagrams

This spread will show you how to:

- Draw cumulative frequency polygons for grouped data
- Find the modal class of a data set

Keywords
Cumulative frequency
Grouped frequency
Modal class
Upper bound

If you have a large amount of data, you can group the data in a **grouped frequency** table.

- **You can represent grouped data on a cumulative frequency diagram.**

Examiner's tip
The diagrams shown in this unit may be useful for your statistical coursework task.

Example

The heights of 120 boys are given in the table.

Height, h, cm	$145 \leqslant h < 150$	$150 \leqslant h < 155$	$155 \leqslant h < 160$	$160 \leqslant h < 165$	$165 \leqslant h < 170$
Frequency	8	27	48	31	6

a Draw a cumulative frequency table for these data.
b Use the table to draw a cumulative frequency diagram.
c Write the **modal class** interval.

a

Height, h, cm	< 150	< 155	< 160	< 165	< 170
Cumulative frequency	8	35	83	114	120

Upper bound of each class.

Add frequencies to get cumulative frequency
$8 + 27 = 35$;
$35 + 48 = 83$ and so on.

b

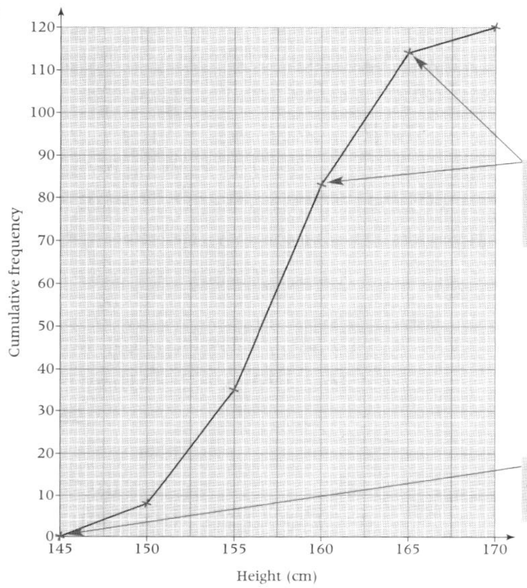

Plot the cumulative frequency against the upper bound of each class.

For a cumulative frequency polygon, join points with straight lines.

For a cumulative frequency curve, join points with a smooth curve.

Lower bound of first class is 145 so plot (145, 0).

c Modal class interval = $155 \leqslant h < 160$

The modal class is the class with the highest frequency.

Exercise D5.1

You will need the cumulative frequency diagrams from this exercise in Exercises D5.2 and D5.4.

For each of these data sets

a draw a cumulative frequency table
b draw a cumulative frequency diagram
c write the modal class interval.

1 The heights of 100 girls

Height, h, cm	$145 \leqslant h < 150$	$150 \leqslant h < 155$	$155 \leqslant h < 160$	$160 \leqslant h < 165$	$165 \leqslant h < 170$
Frequency	7	25	46	17	5

2 The ages of teachers in a school

Age, A, years	$20 \leqslant A < 30$	$30 \leqslant A < 40$	$40 \leqslant A < 50$	$50 \leqslant A < 60$	$60 \leqslant A < 70$
Frequency	18	37	51	28	16

3 The times taken to complete a crossword puzzle

Time, t, minutes	$0 \leqslant t < 10$	$10 \leqslant t < 20$	$20 \leqslant t < 30$	$30 \leqslant t < 40$	$40 \leqslant t < 50$	$50 \leqslant t < 60$
Frequency	4	11	29	37	27	12

4 The weights of a sample of cats and kittens

Weight, w, grams	$1500 \leqslant w < 2000$	$2000 \leqslant w < 2500$	$2500 \leqslant w < 3000$	$3000 \leqslant w < 3500$	$3500 \leqslant w < 4000$
Frequency	9	22	37	20	12

5 The heights of sunflowers growing in one field

Height, h, cm	$40 \leqslant h < 60$	$60 \leqslant h < 80$	$80 \leqslant h < 100$	$100 \leqslant h < 120$	$120 \leqslant h < 140$	$140 \leqslant h < 160$
Frequency	2	17	28	39	24	10

6 The total spent by 100 shoppers at Tesbury's superstore

Amount, p, £	$0 \leqslant p < 10$	$10 \leqslant p < 20$	$20 \leqslant p < 30$	$30 \leqslant p < 50$	$50 \leqslant p < 70$	$70 \leqslant p < 100$
Frequency	16	14	23	17	15	15

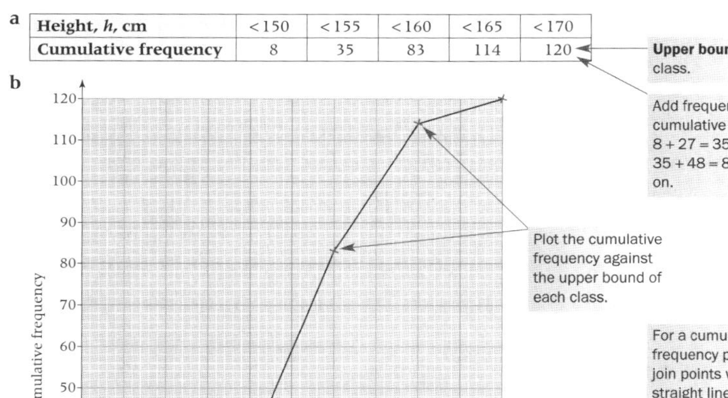

D5.2 More cumulative frequency diagrams

Objectives

- Estimate the median and the upper and lower quartiles from a cumulative frequency diagram

Useful resources

- OHT of cumulative frequency graph
- Graphs drawn by the students in D5.1

Mental starter

If you have 11 people lined up in height order which one is in the middle? (6th)
How can you work the number 6 from the number 11? ($(11 + 1) \div 2 = 6$.) So when you have a set of data with n items, you calculate $(n + 1) \div 2$ to find the middle position (**not** the median itself, but the position it is in).
If you have 20 results in order, where is the median? (10.5^{th}, so half way between the 10th and the 11th.)

A piece of wood is 20 cm long. **Where do you cut it so that it is two exactly equal pieces?** (At the 10 cm mark.) **Why don't you add 1 and divide by 2?** Discuss reasons.

Introductory activity

Look at the example used in the introduction for D5.1. The total frequency is 40 and when the data set is as large as this you use $\frac{n}{2}$ to find the median position, not $\frac{n+1}{2}$.
Where will you find the median result? (At the 20th value.) Read it off from the graph. (About 23.5 years, just below 24.)
Where do you find the LQ and UQ? (10th value, 20.5 and 30th value, 27 years.)
What is the IQR? ($27 - 20.5 = 6.5$ years.)
Note: all these values are approximate, whether the graph is joined with straight lines or a smooth curve.
How many people were over the age of 30? (Approximately 5, because you can estimate from the graph that 35 people were under 30.)

Exercise commentary and misconceptions

When finding results from cumulative frequency diagrams students should try to give the result as accurately as possible. The accuracy will depend on the size of the scale. Encourage students to draw in their lines to show their working. Some confusion can come from mixing the meanings of the two axes. Remind students that it is the horizontal axis that shows the values. Some students will record the median as the position of the middle data value rather than its actual value.

Plenary

If you had a set of raw data (just numbers) how could you put this into a cumulative frequency graph?
You would have to put it into a grouped frequency table first and then work out the cumulative frequencies.
Discuss question 6, and link to the plenary for D5.1.

D5.2 More cumulative frequency diagrams

This spread will show you how to:

- Estimate the median and the upper and lower quartiles from a cumulative frequency diagram

- **You can estimate the median and the upper and lower quartiles from grouped data.**
- **You use a cumulative frequency diagram to estimate measures.**

Keywords
Cumulative frequency
Interquartile range
Lower quartile
Median
Upper quartile

Example

The heights of 120 boys are summarised in the cumulative frequency graph.
Use the graph to estimate

a the median
b the **interquartile range**
c the number of boys with height
i less than 153 cm **ii** greater than 163 cm.

> In grouped data you do not know the individual values so you can only make estimates.

> To estimate the measures, draw a line from the known values across the graph, then down to the horizontal axis.

UQ → 90; Median → 60; LQ → 30
Cumulative frequency (0–120); Height (cm) (145–170)

> Read estimates on the horizontal axis.

a Median = 60th value
$= 157\frac{1}{2}$ cm — Total 120: $\frac{1}{2}$ of 120 = 60th value

b UQ = 90th value
= 161 cm — $\frac{3}{4}$ of 120 = 90th value

LQ = 30th value
= 154 cm — $\frac{1}{4}$ of 120 = 30th value

IQR = 161 − 154 = 7 cm — IQR = UQ − LQ

c i 24 boys are less than 153 cm — Read up from 153 to the graph and across to the vertical axis.
ii 19 boys are greater than 163 cm — Read up from 163 and across, then subtract 19 from the total 120.

> When the data set is large, use $\frac{1}{2}n$ and not $\frac{1}{2}(n+1)$ to find the median and quartiles.

Exercise D5.2

> You will need some of your answers from this exercise in Exercise D5.4.

1 Use the table and graph you have drawn in Exercise D5.1 question **1** to estimate
a the median
b the interquartile range
c the number of girls with height
i less than 152 cm **ii** greater than 163 cm.

2 Use the table and graph you have drawn in Exercise D5.1 question **2** to estimate
a the median
b the interquartile range
c the number of teachers who are aged
i less than 35 **ii** greater than 55.

3 Use the table and graph you have drawn in Exercise D5.1 question **3** to estimate
a the median
b the interquartile range
c the number of people who took
i less than 25 minutes
ii more than 45 minutes to complete the puzzle.

4 Use the table and graph you have drawn in Exercise D5.1 question **4** to estimate
a the median
b the interquartile range
c the number of cats and kittens that weighed
i less than 2200 g **ii** more than 3600 g.

5 Use the table and graph you have drawn in Exercise D5.1 question **5** to estimate
a the median
b the interquartile range
c the number of sunflowers that were
i less than 130 cm **ii** greater than 90 cm.

6 Use the table and graph you have drawn in Exercise D5.1 question **6** to estimate
a the median
b the interquartile range
c the number of shoppers who spent
i less than £20
ii more than £80.

D5.3 Comparing data sets

Objectives

- Use cumulative frequency diagrams to compare two data sets

Useful resources

- OHT of cumulative frequency graphs of men's and women's weights, given as example in the students' book

Mental starter

Ask students to find $\frac{1}{4}$, $\frac{1}{2}$ and $\frac{3}{4}$ of these amounts: 120, 80, 350, 400, 750.

Introductory activity

Show the OHT and identify the median value for each group, the range for each group and the IQR for each group.

Remind students that when describing the differences between two groups they should write proper sentences and that the comparison must be based on the data, not on their ideas of how these differences may arise.

Exercise commentary and misconceptions

When drawing only two comparisons students should comment on the median and IQR; when drawing 3 comparisons students can mention the range as well.

A more sophisticated description of the IQR is not really necessary. Students could say 'there is more variation in the middle half of the data' but this does not really add any more practical information than saying there is more variation (spread).

Plenary

What are the various ways data can be presented in order to make comparisons using the median and IQR?

Raw data, frequency table, cumulative frequency diagram, box plot, stem and leaf diagram, a histogram or line graph.

D5.3 Comparing data sets

This spread will show you how to:
- Use cumulative frequency diagrams to compare two data sets

Keywords
Cumulative frequency
Interquartile range
Median

- **You can compare two data sets using information from cumulative frequency diagrams.**
- **You can compare data using a measure of average, such as the median and a measure of spread, such as the interquartile range.**

Example

These cumulative frequency graphs summarise the weights of a sample of 100 men and 100 women.
Make three comparisons between the weights of the men and women.

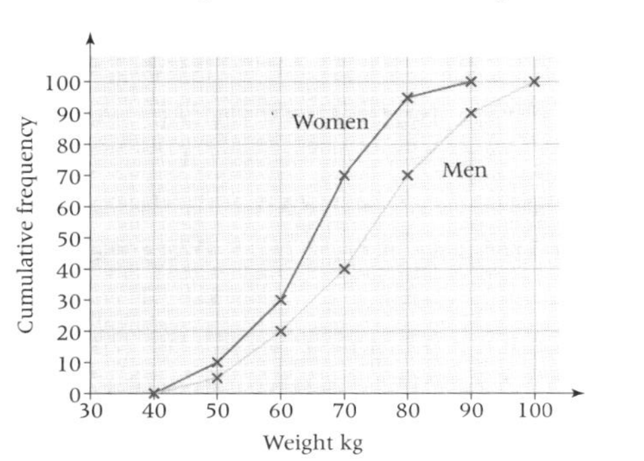

1 Median weight of women = 65 kg
Median weight of men = 74 kg
On average, the women are lighter than the men.

2 Range of women's weights = 90 – 40
= 50 kg

Range of men's weights = 100 – 40
= 60 kg

3 For the women, upper quartile = 71 kg
lower quartile = 56 kg
IQR = 15 kg

For the men, upper quartile = 82 kg
lower quartile = 62 kg
IQR = 20 kg

The middle half of the women's weights varies less than the middle half of the men's weights.

Exercise D5.3

1 Write three comparisons between the test results of a group of girls and a group of boys.

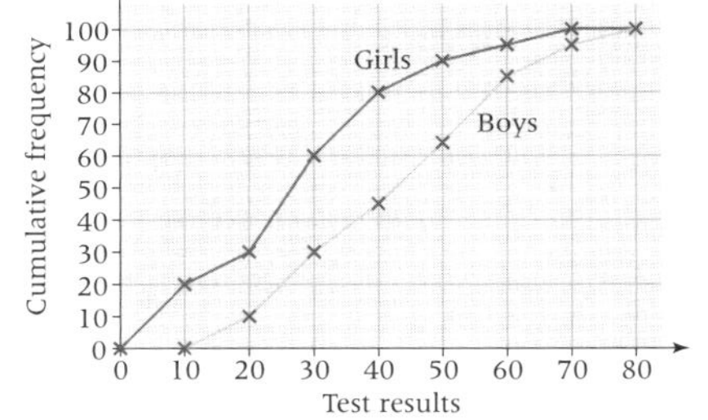

2 Write three comparisons between the heights of samples of sunflowers grown by two farmers.

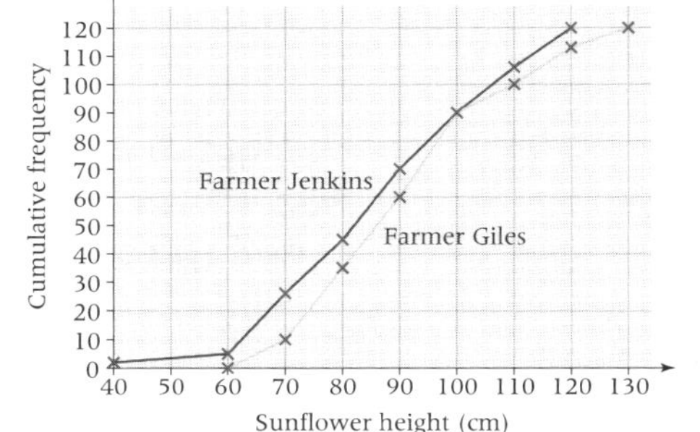

3 Write three comparisons between the mobile phone bills paid by samples of boys and girls.

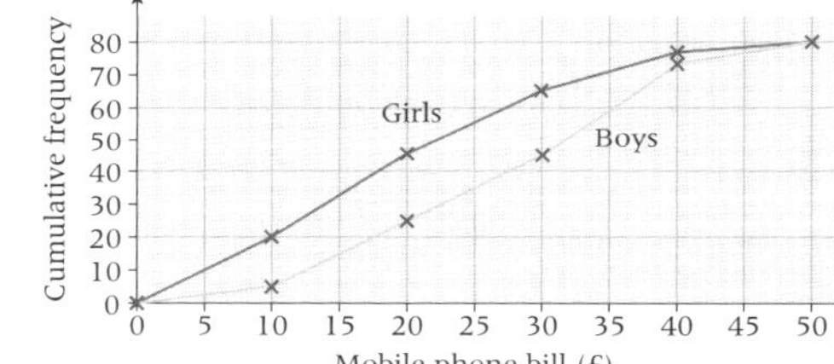

D5.4 Box plots – large data sets

Objectives
- Draw box plots

Useful resources
- OHT of cumulative frequency polygon
- Students' work from D5.1 and D5.2

Mental starter
Ask students to express 10 out of 120 as a fraction in its lowest terms ($\frac{1}{12}$).
Repeat for 20 out of 120 ($\frac{1}{6}$), 30 out of 120 ($\frac{1}{4}$) and so on.

Introductory activity
Discuss the example in the student book, which is based on the heights of 120 boys.
Ask these questions:
What are the five statistics shown by a box plot? (Minimum value, LQ, median, UQ, maximum value.)
How can you draw a boxplot from the cumulative frequency diagram? (Read off the median, LQ and UQ. Use the first value with zero cumulative frequency as the minimum value and the last value with the total cumulative frequency as the maximum value.)
Will you get an exactly correct box plot? (No, it will be an estimate, you need the original raw data to get the exact values needed for the box plot.)

Exercise commentary and misconceptions
Note that the terms 'box plot' and 'box and whisker diagram' are both acceptable, the whiskers refer to the extended lines to the left and the right (like cats' whiskers).

If students left enough space under their cumulative frequency graphs in D5.1 they can draw the box plots using the same horizontal scale. If not, they can replicate the scale on a separate piece of paper.

Plenary
What does a symmetrical box plot tell you about the spread of the data? The data is evenly spread, the bottom 25% of the results are in the bottom 25% of the range of values. The next 25% of results are within the next 25% of the range of results etc.
Could a country have a symmetrical box plot for its age distribution?
No, because there is not an even distribution of ages; for example the number of people in the 0–25 age group is much higher than the 75–100 age group.

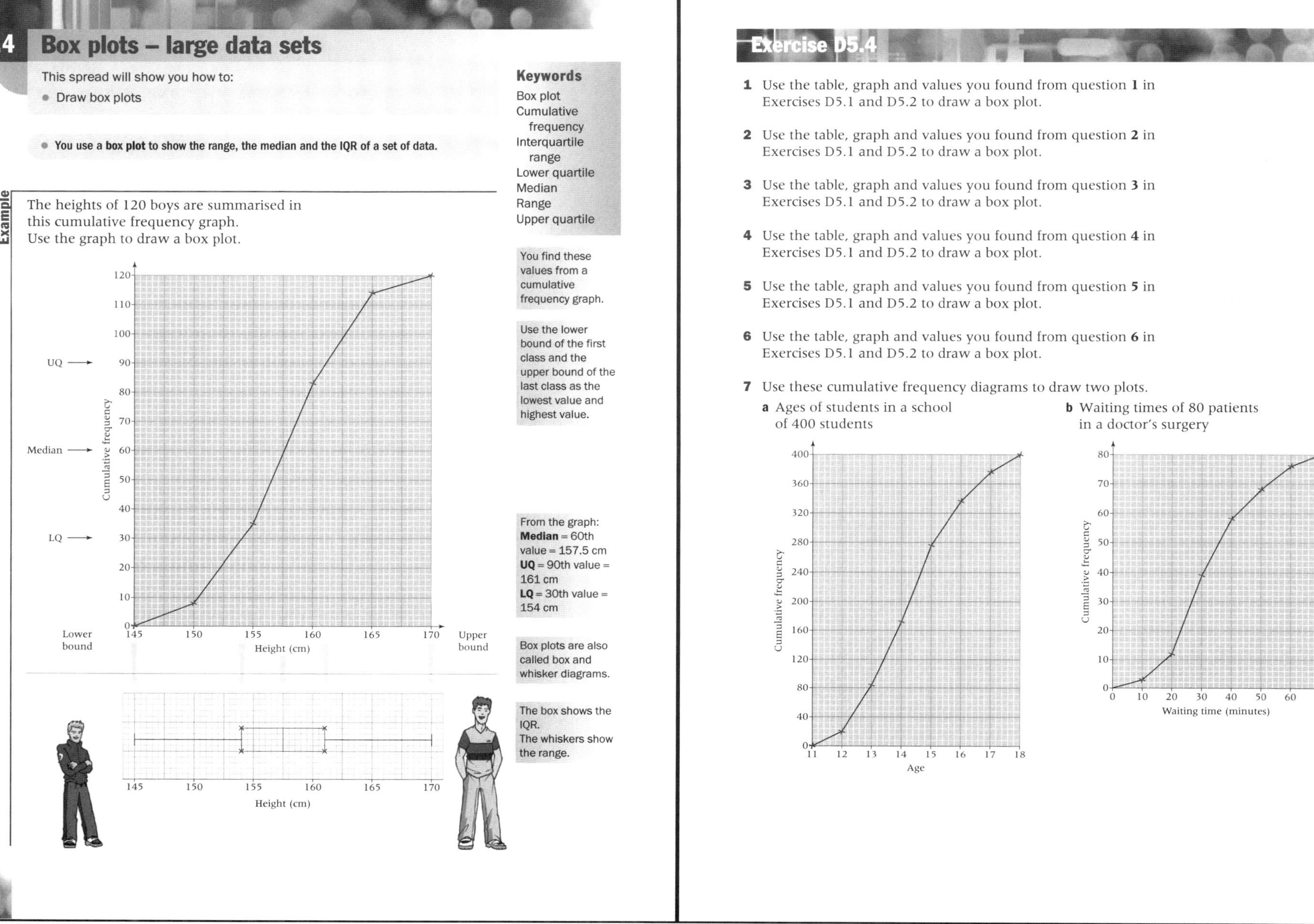

D5.4 Box plots – large data sets

This spread will show you how to:

- Draw box plots

- **You use a box plot to show the range, the median and the IQR of a set of data.**

Keywords
Box plot
Cumulative frequency
Interquartile range
Lower quartile
Median
Range
Upper quartile

Example

The heights of 120 boys are summarised in this cumulative frequency graph.
Use the graph to draw a box plot.

You find these values from a cumulative frequency graph.

Use the lower bound of the first class and the upper bound of the last class as the lowest value and highest value.

From the graph:
Median = 60th value = 157.5 cm
UQ = 90th value = 161 cm
LQ = 30th value = 154 cm

Box plots are also called box and whisker diagrams.

The box shows the IQR.
The whiskers show the range.

Exercise D5.4

1 Use the table, graph and values you found from question **1** in Exercises D5.1 and D5.2 to draw a box plot.

2 Use the table, graph and values you found from question **2** in Exercises D5.1 and D5.2 to draw a box plot.

3 Use the table, graph and values you found from question **3** in Exercises D5.1 and D5.2 to draw a box plot.

4 Use the table, graph and values you found from question **4** in Exercises D5.1 and D5.2 to draw a box plot.

5 Use the table, graph and values you found from question **5** in Exercises D5.1 and D5.2 to draw a box plot.

6 Use the table, graph and values you found from question **6** in Exercises D5.1 and D5.2 to draw a box plot.

7 Use these cumulative frequency diagrams to draw two plots.

a Ages of students in a school of 400 students

b Waiting times of 80 patients in a doctor's surgery

D5.5 Using box plots to compare data sets

Objectives

- Compare two sets of data using box plots

Useful resources

- OHT of box plots from heights example in students' book

Mental starter

A palindromic number is a number that has the same value if you reverse its digits. e.g. 424 or 1771.
Can you construct a palindromic number so that when you take each digit separately the mean, median and mode are all the same value? (Note, you cannot use all the same digit.) Possible solutions include 1955591 and 9155519. Mode = mean = median = 5.

Introductory activity

Show an OHT comparing the heights of a sample of boys and girls, as shown in the student book.

Point out the way that skewness is measured. If the difference between the median and the upper quartile is greater than the difference between the lower quartile and the median, then the distribution is said to be positively skewed. A way to remember this is that the wider interval is on the right, and when drawing coordinate axes, the positive numbers are to the right.

Exercise commentary and misconceptions

When comparing two sets of data encourage students to comment on the differences in the median and say what this tells you in the context of the question. The range gives a crude measure of spread, but the IQR gives a more sophisticated measure. If the median lies closer to the LQ than the UQ, then you have negative skewness, or the opposite way round, positive skewness. The closer the median is to the quartiles, the more skewed the data.

Plenary

Ask students to think of examples where analysing and graphing data is important to mankind. Some examples could include:

- Medical data to see whether a new treatment is effective.
- Age data to help the government plan for enough retirement homes and large enough pensions in the future.
- Measurements from the earth to help predict and understand earthquakes and volcanoes.
- Weather data to make transport by ship or plane safer by forecasting conditions.

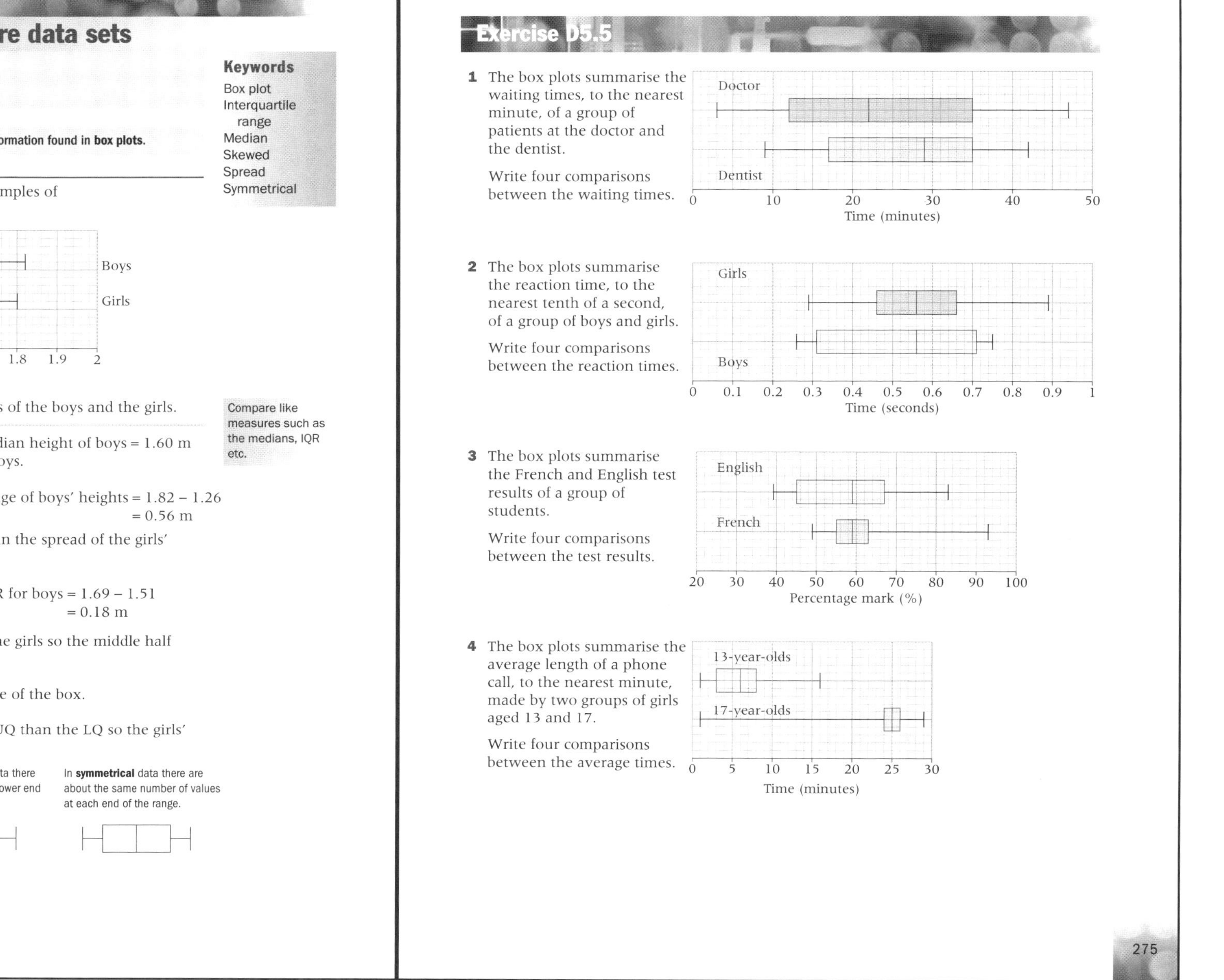

D5.5 Using box plots to compare data sets

This spread will show you how to:

- Compare two sets of data using box plots

Keywords
Box plot
Interquartile range
Median
Skewed
Spread
Symmetrical

- **You can compare two or more data sets by using information found in box plots.**

Example

These box plots summarise the heights of samples of 13- and 14-year-old boys and girls.

Write four comparisons between the heights of the boys and the girls.

Compare like measures such as the medians, IQR etc.

1 Median height of girls = 1.63 m   Median height of boys = 1.60 m
On average, the girls are taller than the boys.

2 Range of girls' heights = 1.8 – 1.42 = 0.38 m   Range of boys' heights = 1.82 – 1.26 = 0.56 m

The **spread** of boys' heights is greater than the spread of the girls' heights.

3 IQR for girls = 1.67 – 1.54 = 0.13 m   IQR for boys = 1.69 – 1.51 = 0.18 m

The IQR for the boys is greater than for the girls so the middle half of the heights is more varied for the boys.

4 The boys' median height is near the centre of the box.
The boys' heights are symmetrical.
The girls' median height is nearer to the UQ than the LQ so the girls' heights are negatively skewed.

In **negatively skewed** data, there are more values at the upper end of the range.

In **positively skewed** data there are more values at the lower end of the range.

In **symmetrical** data there are about the same number of values at each end of the range.

Exercise D5.5

1 The box plots summarise the waiting times, to the nearest minute, of a group of patients at the doctor and the dentist.

Write four comparisons between the waiting times.

2 The box plots summarise the reaction time, to the nearest tenth of a second, of a group of boys and girls.

Write four comparisons between the reaction times.

3 The box plots summarise the French and English test results of a group of students.

Write four comparisons between the test results.

4 The box plots summarise the average length of a phone call, to the nearest minute, made by two groups of girls aged 13 and 17.

Write four comparisons between the average times.

D5 Exam review

Key objectives

- Draw and produce, using paper and ICT, cumulative frequency tables and diagrams, box plots and histograms for grouped continuous data
- Interpret a wide range of graphs and diagrams and draw conclusions
- Compare distributions and make inferences using shapes of distributions and measures of average and spread, including median and quartiles

Worked solution | **Commentary and misconceptions**

1

Ensure that students understand how to use frequency tables and the difference between frequency and cumulative frequency.

a

Height	$140 \leqslant h < 145$	$145 \leqslant h < 150$	$150 \leqslant h < 155$	$155 \leqslant h < 160$	$160 \leqslant h < 165$
Frequency (girls)	4	5	3	2	1
C.F (girls)	4	9	12	14	15
Frequency (boys)	2	4	5	3	3
C.F (boys)	2	6	11	14	17

b

c

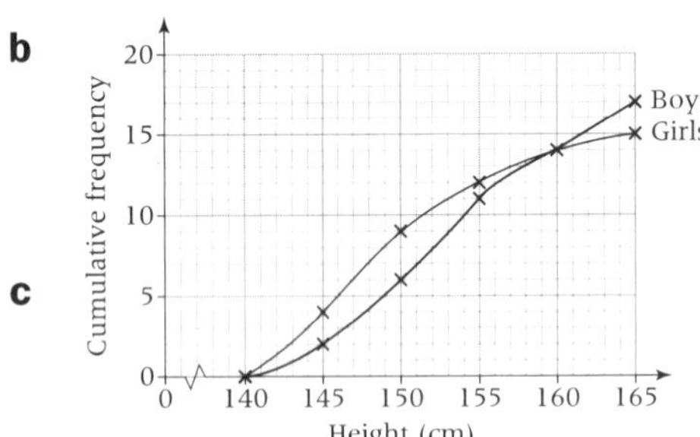

Ensure that students use the cumulative frequency values in their diagram and that they choose appropriate axis values and label their diagram correctly.

Encourage students to use the correct vocabulary when comparing data. They should be comfortable with words such as median, range and quartiles.

2 a Median = 20th value = 32 seconds

Ensure that students know how to read the median value from a CF diagram. Students should state the units of their answer; seconds in this case.

b

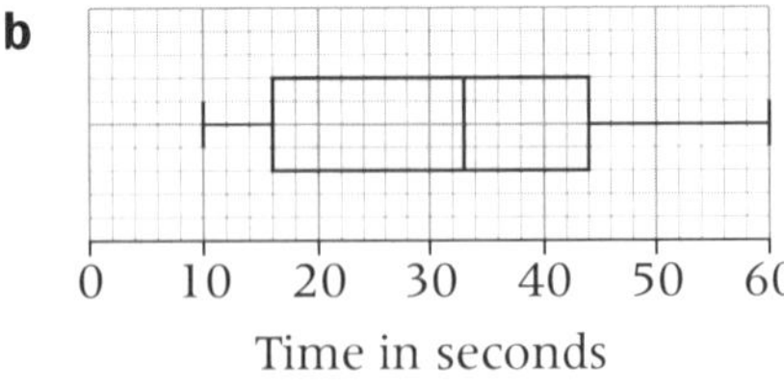

Encourage students to label their diagram clearly with appropriate axis values and units.

c The average times for boys and girls are similar.
The range of times is greater for boys.

Encourage students to use the correct vocabulary when comparing data. They should be comfortable with words such as median, range and quartiles.

S6 Perimeter, area and volume

Objectives

F/H Use 2-D representations of 3-D shapes and analyse 3-D shapes through 2-D projections and cross-sections

F/H Find the surface area of simple shapes by using the formulae for the areas of triangles and rectangles

F/H Calculate volumes of right prisms and of shapes made from cubes and cuboids

F/H Solve problems involving surface areas and volumes of prisms

F/H Convert between area measures and volume measures

F/H Understand and use compound measures

Unit overview

This unit provides an introduction to analysing 3-D shapes and uses knowledge of area and perimeter gained in unit S1 to calculate surface area and volume. It proceeds to solving problems involving units of measure and compound measures.

Prior knowledge

Before your students start this unit they should be able to:

- Recognise and use nets of simple 3-D shapes
- Calculate the area and perimeter of shapes made from triangles and rectangles, recalling the relevant formulae
- Distinguish between measurements of length, area and volume
- Add, subtract, multiply and divide with integers, fractions and decimals
- Understand the vocabulary associated with 2- and 3-D shapes
- Recall, understand and use Pythagoras' theorem
- Multiply and divide by powers of 10
- Understand and use place value
- Use formulae from mathematics and other subjects

Differentiation

- **Foundation** covers area and perimeter of 2-D shapes, including the circumference of a circle and then proceeds to surface area and volume of cuboids. Distinguishing between units of measurement for length, area and volume is also covered
- **Foundation Plus** extends to surface area and volume of prisms and measures, including compound measures, their bounds and conversions
- **Higher Plus** covers calculating the area of a general triangle and using the sine and cosine rules in 2- and 3-D problems. It extends the Higher book to the volume and surface area of cones, frustrums and spheres

S6.1 3-D solids: plans and elevations

Objectives

- Analyse 3-D shapes through 2-D projections and cross-sections, including plans and elevations

Useful resources

- Multi-link cubes, 8 per student
- 5 mm squared paper

Mental starter

At the front of a block of flats, the first floor has 16 windows and the number of windows halves each time you go up one floor. All floors have windows. There are equal numbers of windows on each of the four sides of the building. **How many windows altogether?** ($16 + 8 + 4 + 2 + 1 = 31$: $31 \times 4 = 124$.) A nervous window cleaner takes 5 min to clean windows on the first floor. Every time he goes up one floor it takes him twice as long to clean a window. **How long will it take him to clean all the windows?** (1600 min or 26 hr 40 min.) Some students will realise that since the number of windows is halving but the time is doubling the time for each floor stays the same. (80 min for 1 side of each floor.)

Introductory activity

What is a prism? (3-D solid with the same cross-section throughout its length.) Show students a 3-D model of the first diagram in the student book, made from multilink cubes, and discuss why it is a prism.

Show students the isometric drawing of the prism and explain it. Show students the plan and elevation of the prism and explain them.

Ask students to construct 3-D models of the shapes in the two examples.
Students should also draw the side elevation of each of these solids.

Discuss any problems that have arisen.

Exercise commentary and misconceptions

Questions 1**a**, **b**, **c**, **e** and **f** require students to draw the plan and elevation of prisms from sketches of the solids. Question 1**d** has the same requirement, though it is not a prism. Some students find visualisation very difficult, and may find it helpful to build models with the multi-link cubes. Remind students that an elevation is a 'head on' look at an object. Depth is shown by lines across the 2-D diagram. Encourage students to check the measurements on the diagrams.

Question 2 requires the students to draw 3-D sketches from the plans and elevations of solids. Students may find it helps to model the shapes with the cubes.

Plenary

Pair students together. Give each student 6 multi link cubes. Sit students back to back. One student has to link the cubes together and then describe to his/her partner how the cubes are arranged so that he/she can replicate the arrangement. **How successful were students? What methods of communication worked best?**

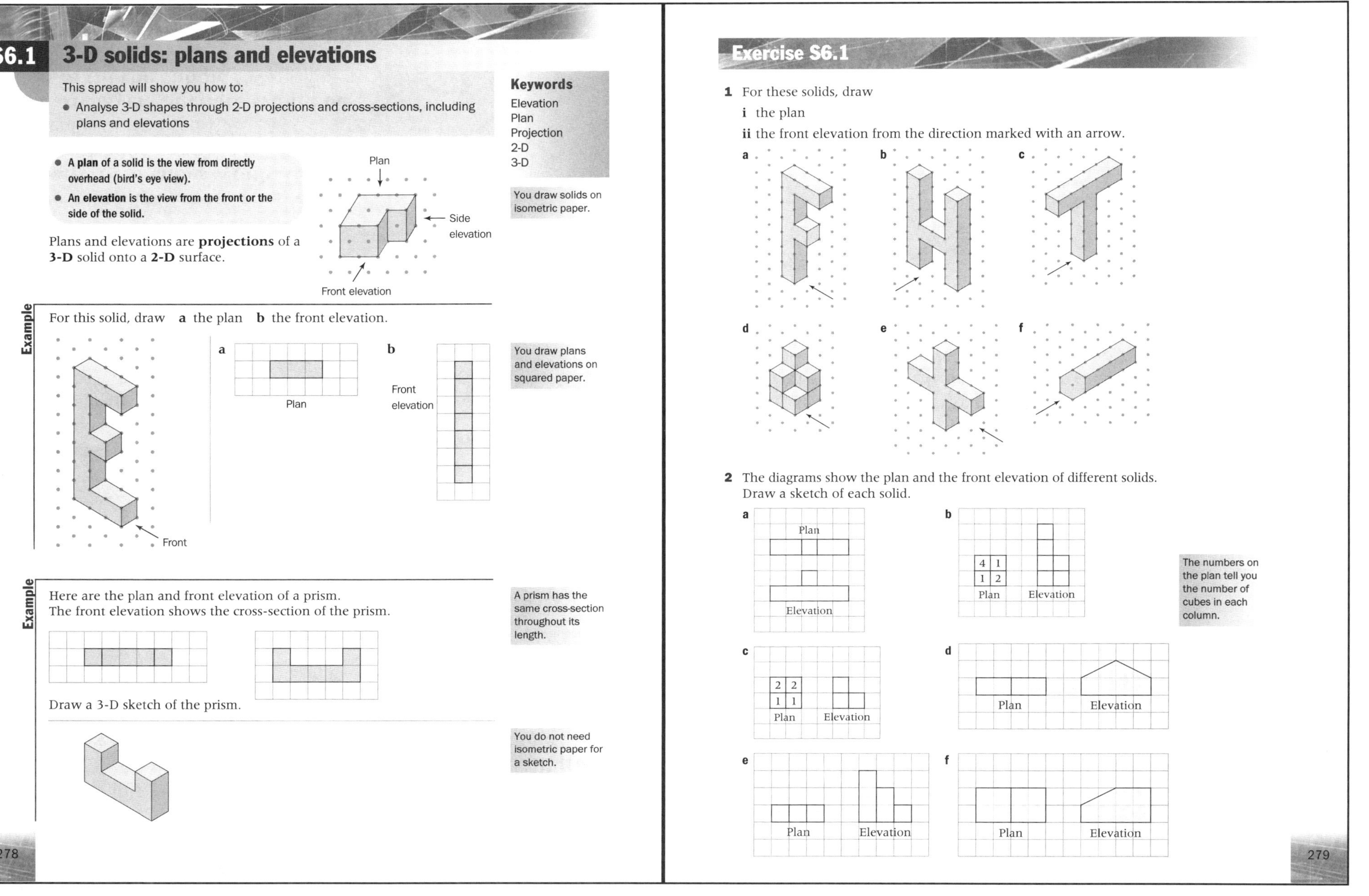

S6.1 3-D solids: plans and elevations

This spread will show you how to:

- Analyse 3-D shapes through 2-D projections and cross-sections, including plans and elevations

Keywords
Elevation
Plan
Projection
2-D
3-D

- A **plan** of a solid is the view from directly overhead (bird's eye view).
- An **elevation** is the view from the front or the side of the solid.

Plans and elevations are **projections** of a **3-D** solid onto a **2-D** surface.

You draw solids on isometric paper.

Example

For this solid, draw **a** the plan **b** the front elevation.

You draw plans and elevations on squared paper.

Example

Here are the plan and front elevation of a prism.
The front elevation shows the cross-section of the prism.

Draw a 3-D sketch of the prism.

A prism has the same cross-section throughout its length.

You do not need isometric paper for a sketch.

Exercise S6.1

1 For these solids, draw

i the plan

ii the front elevation from the direction marked with an arrow.

a **b** **c** **d** **e** **f**

2 The diagrams show the plan and the front elevation of different solids. Draw a sketch of each solid.

The numbers on the plan tell you the number of cubes in each column.

S6.2 Volume of prisms

Objectives

- Calculate volumes of right prisms

Useful resources

- OHTs of prisms in student book
- Box of prisms

Mental starter

There are different atmospheres above the earth. They are the troposphere, stratosphere, mesosphere and ionosphere. Their heights above the earth range from 0–18 km: 18 km–50 km: 50 km–90 km: 90 km–350 km. A new rocket can travel at 1 km/sec through the troposphere, but its speed doubles every time it enters the next thinner level of atmosphere.

How long before the rocket clears the ionosphere? (Distances to travel 18, 32, 40, 260 km: speeds 1, 2, 4, 8 km/sec: times 18, 16, 10, 32.5 sec: total = 76.5 sec.)

Introductory activity

What is a prism? (Object with constant cross-section.)

Referring to the first example part **a**:
How many 1 cm cubes would it take to cover the shaded area? ($4 \times 1.5 = 6$ allowing for half cubes to be used.)
How many layers of cubes would be needed to fill the cuboid? (7)

Remembering that each layer is the same size because it is a prism, **how many cubes will it take to fill the cuboid?** ($7 \times 6 = 42$.)
The volume of the cuboid is 42 cm^3.

Students will have encountered the formula of length × breadth × height for the volume of a cuboid and this is a way of deriving it.

Referring to part **b**:
What is the area of the shaded surface of the prism? ($\frac{1}{2} \times 3 \times 4 = 6$ cm^2.)
Volume of prism $= 6 \times 8 = 48$ cm^3.

Now look at the second example:
What is the area of the shaded surface of the cylinder? ($\pi \times 3^2 = 28.274 \ldots$ cm^2.)
Volume of cylinder $= 28.274 \times 7 = 198$ cm^3.

Cuboids and cylinders are two examples of prisms. The formula of **volume = area of cross-section × length** holds for both.

Remind students that volume is measured in cubic units, and that in the formulae for both cuboid and cylinder there are three linear measurements multiplied together.

Exercise commentary and misconceptions

For question 1 students need the formula for the volume of a cuboid.

For question 2 students need the formula for the volume of a cylinder. Weaker students may find it difficult to appreciate that 'length' is perpendicular to the constant cross-section, so can be length or height, depending on the orientation of the cylinder.

In question 3 ensure that students can identify the cross-section and the length. In question 3**c** they need to divide the cross-section into a rectangle and a triangle **or** a rectangle and a trapezium.

Plenary

What possible formula do you think could be used to find the volume of an oblique cylinder? Encourage students to be imaginative and consider which measurements they will need to make. (Answer is area of circle × perpendicular height.)

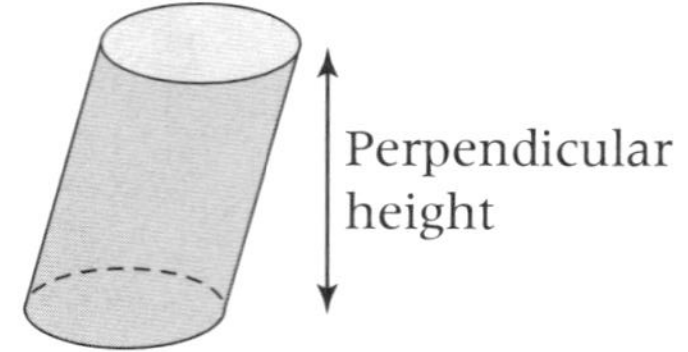

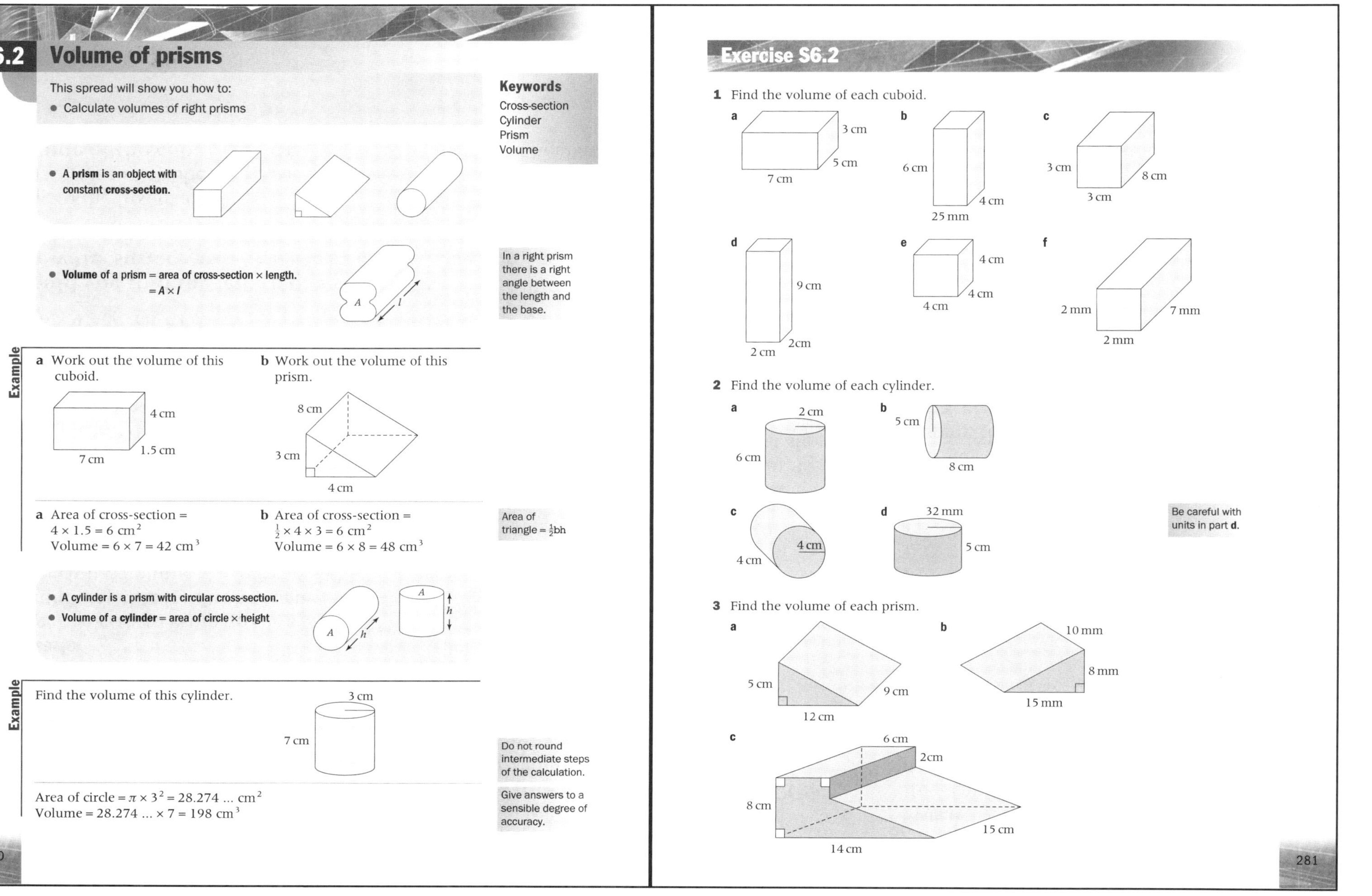

S6.2 Volume of prisms

This spread will show you how to:

- Calculate volumes of right prisms

Keywords
Cross-section
Cylinder
Prism
Volume

- A **prism** is an object with constant **cross-section**.

- **Volume** of a prism = area of cross-section × length.
  $= A \times l$

In a right prism there is a right angle between the length and the base.

Example

a Work out the volume of this cuboid.

b Work out the volume of this prism.

a Area of cross-section =
$4 \times 1.5 = 6 \text{ cm}^2$
Volume = $6 \times 7 = 42 \text{ cm}^3$

b Area of cross-section =
$\frac{1}{2} \times 4 \times 3 = 6 \text{ cm}^2$
Volume = $6 \times 8 = 48 \text{ cm}^3$

Area of triangle = $\frac{1}{2}bh$

- A cylinder is a prism with circular cross-section.
- Volume of a **cylinder** = area of circle × height

Example

Find the volume of this cylinder.

Area of circle = $\pi \times 3^2 = 28.274 \ldots \text{ cm}^2$
Volume = $28.274 \ldots \times 7 = 198 \text{ cm}^3$

Do not round intermediate steps of the calculation.

Give answers to a sensible degree of accuracy.

Exercise S6.2

1 Find the volume of each cuboid.

2 Find the volume of each cylinder.

Be careful with units in part **d**.

3 Find the volume of each prism.

S6.3 Volume and surface area

Objectives
- Solve problems involving surface areas and volumes of prisms

Useful resources
- Box of solids

Mental starter
Challenge the class to estimate the volume of a doughnut. Say that it is in the shape of a torus, which has volume $V = 2\pi^2 r^2(R + r)$, where r and R are the radii of the smaller and larger concentric circles respectively. Use $\pi = 3$ and ask for answers in cm^3 (around 1000 cm^3).

Introductory activity
What is **meant by the surface area of a cube?** (The sum of the area of all the faces.)

What is the surface area of a cube of side 4 cm? ($6 \times 4 = 96$ cm^2.)

How could you find the length of a side if you knew the surface area? (Divide by 6 and square root the answer.)

Referring to the first example:
The surface area of a cube is 150 cm^2. **What is the volume of the cube?** (125 cm^3.)

Referring to the second example:
What shape is the cross-section of this prism? (Equilateral triangle.) **What dimensions do we need the find the area of the triangle?** (Base and height of triangle.) **How can we find the height of the triangle**?
(Pythagoras, $\sqrt{(4^2 - 2^2)} = 3.464 \ldots$.)

$$\begin{aligned}\text{Area of triangle} &= \tfrac{1}{2} \times 4 \times 3.464\\ &= 6.928 \ldots \text{ Volume of prism}\\ &= 6.928 \ldots \times 9\\ &= 62.4\ \text{cm}^3.\end{aligned}$$

Exercise commentary and misconceptions
Remind students that they should not use rounded numbers in calculations, but should round off only for the final answer.

Question 1 involves finding the volume of cubes with known surface areas.

Question 2 involves finding volumes of prisms with equilateral triangle cross-sections.

Question 3 involves finding the surface areas of cubes with known volumes.

Questions 4 and 5 need a trial and improvement method and weaker students may need help in getting started on it.

Plenary
How can you find the surface area of this prism? What surfaces make it up? What measurement is missing? (Use Pythagoras.)

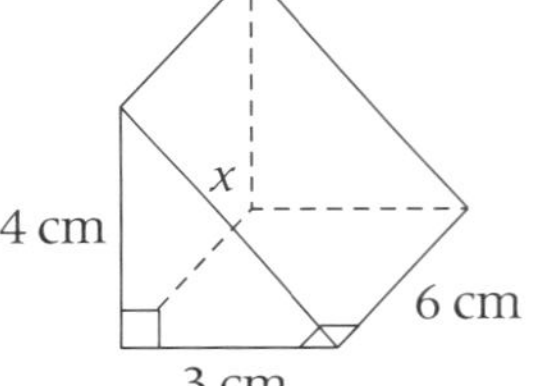

$x = 5$ m
Front = Back = 6 m^2
Base = 18 m^2
Left side = 24 m^2
Sloping right side = 30 m^2
Surface area = 84 m^2

What name could you give to this shape? (Right-angled triangular prism.)

S6.3 Volume and surface area

This spread will show you how to:
- Solve problems involving surface areas and volumes of prisms

Keywords
Pythagoras' theorem
Surface area
Volume

- **Surface area** is the total area of all the faces of a solid.

Example

This cube has surface area 150 cm^2.
Find the volume of the cube.

A cube has six faces.
Each face has the same area.

So area of one face is $150 \div 6 = 25$ cm^2

Each face is a square, so length of each side $\sqrt{25} = 5$ cm

Volume of cube = area of cross-section × length
$= 25 \times 5$
$= 125$ cm^3

Each face is a cross-section.

You can use **Pythagoras' theorem** to find the perpendicular height of a triangle.

- Pythagoras' theorem: $a^2 + b^2 = c^2$

Example

This triangular prism has length 9 cm and its end faces are equilateral triangles with side length 4 cm.
Work out its volume.

Using Pythagoras' theorem to find the height, h, of the triangle.

$h^2 = 4^2 - 2^2$

$h = \sqrt{12} = 3.464 \ldots$

Area of cross-section $= \frac{1}{2} \times 4 \times 3.464 \ldots$
$= 6.928 \ldots$

Volume of prism: $6.928 \ldots \times 9 = 62.4$ cm^3

To work out the area of the cross-section you need the perpendicular height of the triangle.

Exercise S6.3

1 Find the volumes of the cubes with these surface areas.

a Surface area 54 cm^2

b Surface area 294 cm^2

c Surface area 96 cm^2

d Surface area 1.5 m^2

2 Find the volumes of these triangular prisms which have equilateral triangles as cross-sections.

a

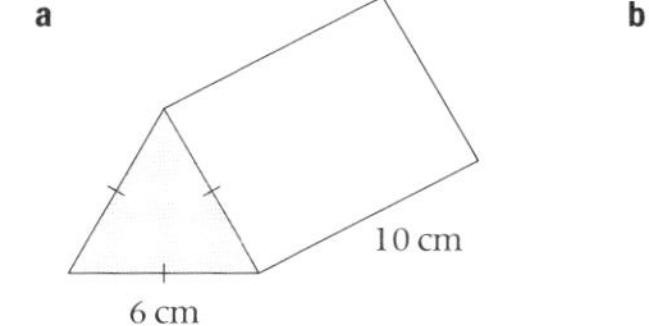

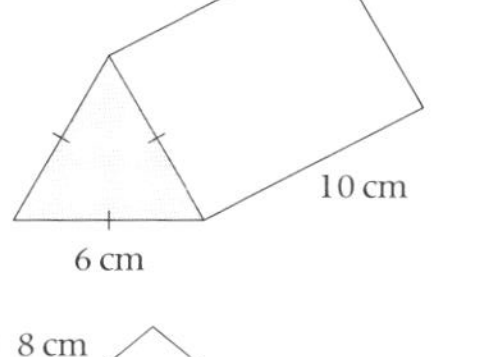

b

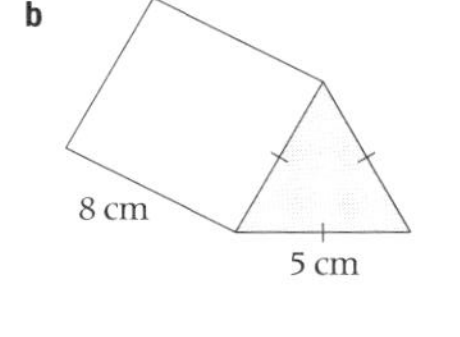

c

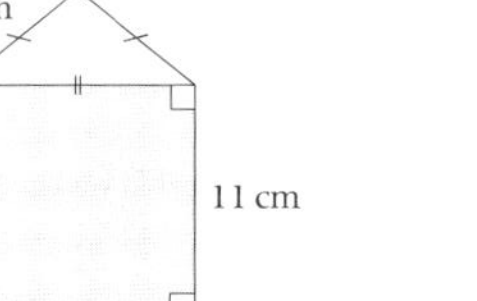

d

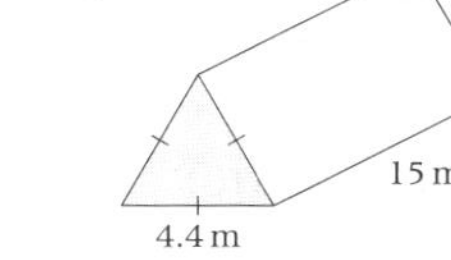

3 Find the surface areas of the cubes with these volumes.

a Volume 512 cm^3

b Volume 1000 cm^3

c Volume 216 cm^3

d Volume 1 m^3

4 The volume of a cuboid is 80 cm^3.
The cuboid has square ends.
The sides of the cuboid are whole numbers of centimetres.

a Find the possible dimensions of the cuboid.

b Find

i the smallest possible surface area

ii the largest possible surface area of this cuboid.

5 Repeat question **4** for cuboids with volume

a 24 cm^3

b 64 cm^3

S6.4 Measures

Objectives

- Convert between volume measures including cm^3 and m^3
- Understand the difference in dimensions of perimeter, area and volume

Useful resources

- OHT of the expressions in question 4

Mental starter

A normal dice used in board games has a volume of about 2 cm^3. A normal dose of medicine given on a plastic spoon or teaspoon is 5 ml. **If you melted some dice down to a liquid, how many would you need to fill the medicine spoon?** (2.5, since 1 cm^3 = 1 ml.)

Introductory activity

Area can be measured in cm^2 or m^2.

How many cm^2 are there in 1 m^2? Draw a metre square and label the sides 100 cm and 1 m. The area is either 10 000 cm^2 or 1 m^2.

How many mm^2 are there in 1 cm^2?
(10 × 10 = 100.)
How many mm^2 are there in 1 m^2?
(1 000 × 1 000 = 1 000 000.)

The conversion factor between cm^2 and m^2 is 10 000. **To convert 3 700 000 cm^2 to m^2 do you multiply or divide?** (Divide, 370 m^2.)

To convert 4 000 m^2 to cm^2 do you multiply or divide? (Multiply, 40 000 000 cm^2.) Remind students to think of the relative sizes of the units and divide if converting to a larger unit, and multiply if converting to a smaller unit.

How many cm^3 are there in 1 m^3?
(100 × 100 × 100 = 1 000 000.)
How many mm^3 are there in 1 cm^3?
(10 × 10 × 10 = 1 000.)
Discuss the first two examples.

Students will now be able to convert between units of measurement for area and volume. Emphasise that they should not try to **learn** all the conversions, but remember how to find them by considering a square or cube.

(Students should try questions 1–3 in Exercise 6.4 before an introduction to dimensions in formulae.)

You can tell from looking at an expression whether it represents the length, area or volume of a shape.

Show these three expressions: πr^2: $\frac{1}{2}$,bh, 1 bh. **Which of these expressions represents an area and which represents a volume?**

For each expression, ask students to decide how many lengths are being multiplied. Discuss the last example.

Exercise commentary and misconceptions

In questions 1–3 students will need to decide whether to multiply or divide. Students should be encouraged not to use calculators.

In question 4 suggest students use colour to underline the lengths in each of the expressions and then decide how many times a length has been multiplied by a length.

In questions 6 and 7 encourage students to find their own formulae for perimeters, areas and volumes.

Plenary

Question 4 can be used as the plenary as students usually find this a difficult topic.

S6.4 Measures

This spread will show you how to:

- Convert between volume measures including cm^3 and m^3
- Understand the difference in dimensions of perimeter, area and volume

Keywords
Area
Dimension
Perimeter
Volume

- **Perimeter** is the distance around a shape.
  It is a length measured in mm, cm and m.

- **Area** is the space covered by a two-dimensional shape.
  It is the product of length × length, measured in mm^2, cm^2 and m^2.

- **Volume** is the space inside a three-dimensional solid.
  It is the product of length × length × length, measured in cubic mm^3, cm^3 and m^3.

Example

Change

a 3 700 000 cm to m **b** 3 700 000 cm^2 to m^2 **c** 3 700 000 cm^3 to m^3

a 3 700 000 ÷ 100 = 37 000 m
b 3 700 000 ÷ (100 × 100) = 370 m^2
c 3 700 000 ÷ (100 × 100 × 100) = 3.7 m^3

Larger unit means smaller number , so divide.
Divide twice for squared units.
Divide three times for cubed units.

Example

Change

a 0.042 m to mm **b** 0.038 m^2 to cm^2 **c** 7 cm^3 to mm^3

a 0.042 × 1000 = 42 mm
b 0.038 × (100 × 100) = 380 cm^2
c 7 × (10 × 10 × 10) = 7000 mm^3

Smaller unit means larger number, so multiply.
Multiply twice for squared units.
Multiply three times for cubed units.

You can tell whether an expression represents length, area or volume by considering its dimensions.

Length: 1-D
Area = length^2: 2-D
Volume = length^3: 3-D

- Constants such as π and numbers have no **dimension**.

Example

In these expressions, r, s and t represent lengths.
For each, decide whether the expression represents length, area, volume or none of these.

$2r + s + 3t$	length
rst	volume
s^2t	volume
$rs - st$	area
$rs + t$	none

Length + length + length = length
Length × length × length = length^3
Length^2 × length = length^3

Length × length = length^2 = area
Area – area = area
Area + length = none

Exercise S6.4

1 Change these lengths to the units given.

a 40 000 cm to m **b** 63 000 cm to m **c** 42 m to cm
d 1200 cm to m **e** 80 000 m to km **f** 45 000 mm to m
g 0.05 m to mm **h** 0.0003 km to mm **i** 0.6 m to cm
j 0.007 km to mm **k** 0.00004 km to cm **l** 18.05 km to m

2 Change these areas to the units given.

a 2 600 000 mm^2 to m^2 **b** 700 000 000 cm^2 to m^2
c 0.00045 m^2 to cm^2 **d** 0.12 m^2 to cm^2
e 0.008 m^2 to cm^2 **f** 0.00045 m^2 to cm^2
g 840 000 000 mm^2 to cm^2 **h** 3 000 000 cm^2 to m^2
i 2 m^2 to cm^2 **j** 1 km^2 to m^2

3 Change these volumes to the units given.

a 3 cm^3 to mm^3 **b** 0.0002 m^3 to cm^3
c 0.0048 m^3 to cm^3 **d** 3 000 000 mm^3 to cm^3
e 10 000 000 cm^3 to m^3 **f** 50 000 000 cm^3 to m^3

4 In the following expressions the letters a, b, d, h, r each represent a length.
For each expression, decide whether it represents length, area, volume or none of these.

a ab **b** $\frac{1}{2}dh$ **c** $\frac{1}{3}\pi r^2 - \frac{1}{2}r^2$ **d** abh
e $2\pi r$ **f** $h + r$ **g** $a^2 + b^2$ **h** $\frac{1}{3}d + br$
i $\pi r^2 + ab$ **j** $a^2h + r^3$ **k** $a^2b + 2bh$ **l** $\frac{1}{2}h + d$

5 Jon said the volume of this shape is $\frac{5}{8}\pi rl$.

a Explain why the expression cannot be correct.
b What could the expression $\frac{5}{8}\pi rl$ represent?

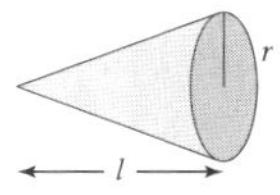
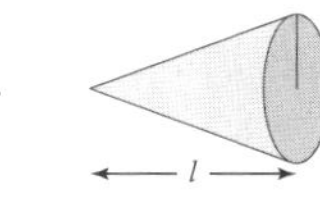
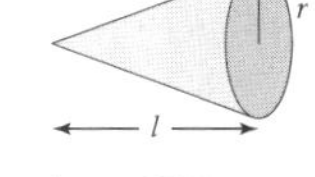

6 The photo shows a window that is rectangular with semi-circular top.
Write an expression for

a the perimeter
b the area.

7 The diagram shows a door wedge.
Write an expression for

a the surface area
b the volume.

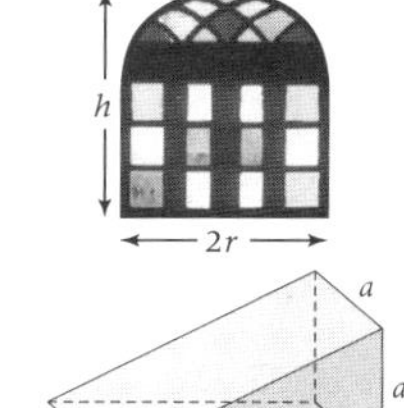

S6.5 Compound measures

Objectives

- Understand and use compound measures, including speed and density

Useful resources

- OHT of triangles for speed, density, population density and rate of flow (see students book)
- Calculators

Mental starter

Arrange these countries in ascending order of densities of population: Russia, USA, Monaco, India and China. (Population density is the number of people/km^2.) Populations are 145 million, 281 million, 32 000, 1045 million and 1284 million. Areas are 17 million km^2, 9 million km^2, 1.5 km^2, 3 million km^2 and 10 million km^2.
(People/km^2 are 9, 30, 21 000, 350 and 130. Order is Russia, USA, China, India and Monaco.)

Introductory activity

What are the units of measurement for speed? (m/s, km/h, miles per hour or mph.) '/' or 'per' means 'divided by', so speed = distance ÷ time. Remind students of the triangle of S, D and T.

Similarly, density is measured in g/cm^3 or kg/m^3, so density = mass ÷ volume. Remind students of the triangle of D, M and V.

Population density is population/km^2, so population density = population ÷ area. Remind students of the triangle of P, D and A.

In all these the units of measurement give a clue to the relationship between the quantities.

Show students how knowing one relationship helps in drawing the triangle. Remind them how to use the S, D and T triangle in this example (taken from the student book).

Kerry jogs at an average speed of 5 km/h for $1\frac{1}{2}$ hours. **What distance does she jog?**
(Distance = $5 \times 1\frac{1}{2} = 7.5$ km.)

Remind students how to use the D, M, and V triangle in this example:

Find the density of a piece of wood with a constant cross-section of area 42 cm^2, length 12 cm and mass 693 g.

(Volume = $42 \times 12 = 504$ cm^3,
density = $693 \div 504 = 1.375$ g/cm^3.)

For liquids, rate of flow = volume ÷ time;
for solids such as sand, rate of flow = mass ÷ time.

Exercise commentary and misconceptions

Questions 1–3 are on rate of flow of liquids.

Questions 4 and 5 use the S, D and T triangle, but remind students to take care with units.

Questions 6 and 7 are on density, and again care must be taken with the units used.

Plenary

An alien plant lands on the earth and begins to grow and cover the earth at a rate of 0.1 km^2/sec. (The area of a large secondary school is about 0.1 km^2.) The area of the earth is 40 000 000 km^2.
How long before the earth is covered by the alien plant? (Between 12 and 13 years.)

S6.5 Compound measures

This spread will show you how to:

- Understand and use compound measures, including speed and density

Keywords
Capacity
Density
Rate of flow
Speed

Compound measures describe one quantity in relation to another. These are examples of compound measures.

- **Speed** = $\frac{\text{total distance travelled}}{\text{total time taken}}$ **Units such as m/s; km/h** (D / S T)
- **Density** = $\frac{\text{mass}}{\text{volume}}$ **Units such as g/cm^3** (M / D V)
- **Population density** = $\frac{\text{population}}{\text{area}}$ **Units such as number of people/km^2** (P / D A)

Use the triangle to work out which calculation to use.

D / S T

Cover D (for distance)
You multiply S(speed) × T(time)

Example

Kerry jogs at an average speed of 5 km/h for $1\frac{1}{2}$ hours.
What distance does she jog?

Distance = $5 \times 1\frac{1}{2}$
= 7.5 km

Example

Find the density of a piece of wood with cross-section area 42 cm^2, length 12 cm and mass 693 g.

Volume = $42 \times 12 = 504$ cm^3

Density = $693 \div 504$
= 1.375 g/cm^3

M / D V

Density = $\frac{\text{mass}}{\text{volume}}$

Mass in grams
Volume in cm^3
So Density in g/cm^3.

- **Capacity is the volume of liquid that a container can hold.**
  Metric units of capacity are litre, centilitre, millilitre.

Rate of flow is a compound measure. It is the volume of liquid that passes through a container in a unit of time.

- **Rate of flow** = $\frac{\text{volume}}{\text{time}}$ **Units such as litres/s** (V / R T)

Example

a 12 litres of water flows from a hosepipe in 15 seconds.
What is the rate of flow in litres/s?

a Rate of flow = $\frac{\text{volume}}{\text{time}}$
= $12 \div 15$
= 0.8 litres/s

b Sand was falling from the back of a lorry at a rate of 0.4 kg/s.
It took 20 minutes for all the sand to fall from the lorry.
How much sand was the lorry carrying?

b 20 minutes = 20×60 s
= 1200 s

Rate = $\frac{m}{t}$
$0.4 = \frac{m}{1200}$
$m = 1200 \times 0.4 = 480$ kg

Here sand is flowing, not a liquid.

Rate of flow = $\frac{\text{mass}}{\text{time}}$

Units kg/s
You could use the triangle:

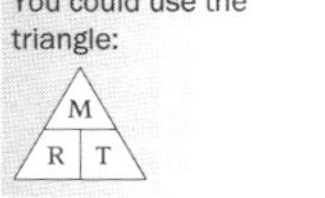

Exercise S6.5

1 Find the rate of flow for pipes A and B in litres/s.

a Pipe A: 20 litres of water in 8 seconds.

b Pipe B: 48 litres of water in 30 seconds.

2 Water empties from a tank at a rate of 2 litres/s.
It takes 10 minutes to empty the tank.
How much water was in the tank?

3 An engine uses oil at a rate of 0.3 ml/km.
How much oil will it use on a journey of

a 100 km **b** 80 km **c** 42 km?

4 A car travelled at an average speed of 48 km/h.

a How far did it travel in

i 2 hours **ii** 15 minutes **iii** 20 minutes?

b How long did it take to travel

i 144 km **ii** 72 km **iii** 8 km?

5 The table shows information about some journeys Shaun made in one week.
Copy and complete the table.
Remember to show the units.

Distance	Time taken	Average speed
120 km	$1\frac{1}{2}$ hours	
250 miles	4 hours	
4 km		16 km/hour
	24 seconds	5 m/s
300 m	15 seconds	
0.4 km	160 seconds	
3 km		24 m/s
	20 minutes	60 km/h

6 The table shows the densities of different metals.
Use the information in the table to find

Metal	Density
Zinc	7130 kg/m^3
Cast iron	6800 kg/m^3
Gold	19 320 kg/m^3
Tin	7280 kg/m^3
Nickel	88 kg/m^3
Brass	8500 kg/m^3

a the mass of 0.8 m^3 of zinc

b the mass of 0.5 m^3 of cast iron

c the mass of 3.2 m^3 of gold

d the volume of 910 g of tin

e the volume of 220 g of nickel

f the volume of a brass statue that has mass 17 kg.

7 In this question, give your answers in kg/m^3.

a The volume of 24 g of silver is 3 cm^3.
Work out the density of silver.

b The volume of 18 g of titanium is 4 cm^3.
Work out the density of titanium.

c A sheet of aluminium foil has volume 0.4 cm^3 and mass 1.08 g. Work out the density of aluminium foil.

S6 Exam review

Key objectives

- Use 2-D representations of 3-D shapes and analyse 3-D shapes through 2-D projections and cross-sections, including plan and elevation
- Solve problems involving surface areas and volumes
- Understand the difference between formulae for perimeter, area and volume by considering dimensions
- Convert measurements from one unit to another and understand and use compound measures including speed and density

Worked solution	Commentary and misconceptions
1 **a** i ii iii	This question test students' ability to analyse 3-D shapes represented in 2 dimensions. Ensure that students include all the relevant lines in their diagrams.
b volume = 8 cm^3	This question is a straightforward counting cubes exercise. Ensure that students use the correct units.
2 Volume = area of cross-section x thickness $= \pi r^2 \times 2.5$ $= \pi \times 3.8^2 \times 2.5$ $= 113.4114948\ cm^3$	Students should recall formulae for compound measures such as speed and density. Some students may need help with rearranging such formulae. Ensure that students understand what they are required to work out and how to use the formula for density to achieve their solution. Encourage students to write down all workings.
So, mass = 1.5×113.4114948 $= 170.1172422$ = 170 grams to 3 sf	Ensure that students state their answers with the correct units. Remind students not to round in the intermediate steps of a calculation. Ensure that students round correctly to the degree of accuracy required and that they state the correct units of their solution.

Graphical solutions

Objectives

F/H Generate points and plot graphs of simple quadratic functions, then more general quadratic functions

H Plot graphs of simple cubic functions

H Find the intersection points of the graphs of a linear and quadratic function, knowing that these are the approximate solutions of the corresponding simultaneous equations representing the linear and quadratic functions

Unit overview

This unit extends the work on simultaneous and quadratic equations in unit A6. It covers plotting quadratic and simple cubic functions and proceeds to solving linear simultaneous equations and other general equations graphically. It then extends to using graphs to find the approximate solutions of simultaneous equations when one is linear and one quadratic, providing a basis for more work on graphs in unit A8.

Prior knowledge

Before your students start this unit they should be able to:

- Substitute numbers in algebraic expressions
- Set up and solve equations for word problems
- Manipulate algebraic expressions
- Recognise, describe and plot the graphs of linear functions
- Add, subtract, multiply and divide with integers, fractions and decimals
- Use the correct order of operations in calculations
- Use calculators effectively, including brackets
- Use a range of methods to check solutions
- Solve simultaneous equations algebraically

Differentiation

- **Foundation** focuses on sequences derived from patterns and then more general sequences. It also covers square numbers and then extends to generating sequences and substituting into the general term
- **Foundation Plus** focuses on plotting linear and quadratic graphs from a table of values, extending to solving simultaneous and quadratic equations and related problems graphically
- **Higher Plus** extends the Higher book to graphs of exponential and reciprocal functions and circles and finding graphical solutions to more complex equations and problems

A7.1 Plotting curves

Objectives

- Generate points and plot graphs of simple quadratic functions

Useful resources

- Graph plotting tool
- 2 mm graph paper

Mental starter

What is 3×14^2? ($3 \times 196 = 588$.)
What is the value of x^2 if $x = -16$?
($-16 \times -16 = 256$.)
What is the value of -17^2?
($-(17 \times 17) = -289$.)

Introductory activity

Imagine this rule for changing a starting number:
Start → square it → divide by 2 → add 1 → end.
What would be the end number if the start number was 2? (3) **What would the end number be if the start number was x?**
$\left(\frac{x^2}{2} + 1.\right)$ Call this end number y. So $y = \frac{x^2}{2} + 1$.

Work out a list of possible x values in a table of values. Ask students to find the y values.

x	−4	−2	0	2	4
y	9	3	1	3	9

Plot these points on axes. Use −5 to 5 for x and 0 to 10 for y.

Can you tell what sort of shape the graph will be? The numbers are symmetrical about (0, 1). Put a few extra points in to help.

x	−3	−1	1	3
y	5.5	1.5	1.5	5.5

Now we can see that the graph is a curve. Join the points smoothly, no ruler. This 'U' shape is known as a quadratic curve or a 'parabola'. It is different from equations like $y = 2x + 1$ which makes a straight line. **What is it about the equation $y = \frac{x^2}{2} + 1$ that makes it a curve?**
(The x^2 term.)

Exercise commentary and misconceptions

When filling out a table of values, point out that a number of points are needed to get a good clear curve, whereas only 3 points were needed for a straight line.
Students should work out the more straightforward examples without a calculator. The confusion usually comes when squaring the negative. The calculator will appear to give the wrong answer for example the value of x^2 if x is −4. On the calculator pressing the three keys (−) (4) (x^2) will produce −16, not 16 because of the rules of BIDMAS. Students can use brackets around the −4 but it is better to make them aware of the problem and the reason for it.

Emphasise that students should draw a smooth curve freehand rather than using a ruler, and also that the curve will be symmetrical. This should help them recognise any errors they make, either in calculating values or in plotting points.

Plenary

Check that students have completed question 5. If not, do it with the class as a whole. **Why is the curve different from the one drawn in question 2?** (It has a maximum value, instead of a minimum value.) Ask students to suggest other equations which would have maximum values, and use a graph plotter to verify this. **How does the graph of $y = x^2 + x$ (drawn in question 3d) differ from the others?** Ask students to suggest other equations of graphs where the minimum would be to the left of the y axis, and use a graph plotter to verify this. **What would the graph of $y = x - x^2$ look like?**

A7.1 Plotting curves

This spread will show you how to:
- Generate points and plot graphs of simple quadratic functions

- **The graph of a quadratic equation is a parabola, a U-shaped curve.**

Keywords
Curve
Minimum
Parabola
Quadratic

You need at least 6 points to plot a parabola.

Example

Draw the graph of $y = x^2 + 1$.

First make a table of x and y values.

$(-4)^2 = 16$

x	−4	−3	−2	−1	0	1	2	3	4
x^2	16	9	4	1	0	1	4	9	16
$y = x^2 + 1$	17	10	5	2	1	2	5	10	17

Draw the x-axis from −4 to 4 and the y-axis from 0 to 18.
Plot the points, (−4, 17), (−3, 10) and so on, and join them in a smooth curve.

Example

Plot the curve $y = x^2 - x$ for $-2 \leqslant x \leqslant 2$ and find the coordinates of its minimum point.

Make a table of values.

x	−2	−1	0	1	2
x^2	4	1	0	1	4
$-x$	2	1	0	−1	−2
$y = x^2 - x$	6	2	0	0	2

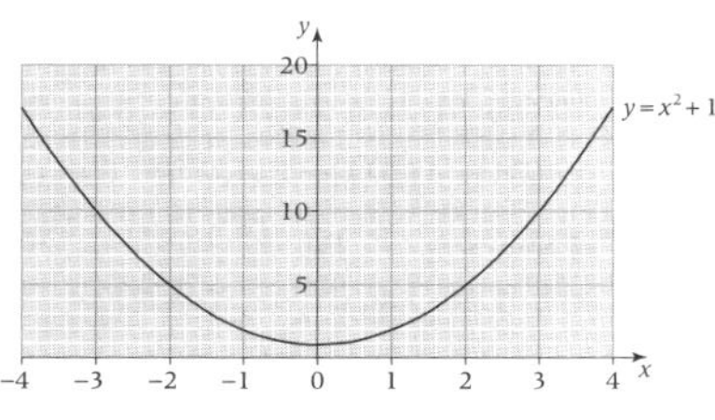

Draw the x-axis from −2 to 2 and the y-axis from −1 to 12.

Plot the points (−2, 6), (−1, 2), ..., (2, 2).

Join the points in a smooth curve.
The **minimum** point is $(\frac{1}{2}, -\frac{1}{4})$.

Exercise A7.1

1 a Copy and complete this table to show if each graph will be a straight line or a parabola.
b Add an equation of your own in each column.

Straight line	Parabola

$y = 3x - 2$ $y = x^2 - 2$ $3x + 2y = 8$ $y = 10 + x^2$ $y = x^2 + 2x + 1$ $y = x$

2 a Draw axes labelled from −4 to 4 on the x-axis and −5 to +15 on the y-axis.
b Copy and complete this table for $y = x^2 - 2$.

x	−4	−3	−2	−1	0	1	2	3	4
x^2	16							9	
$y = x^2 - 2$	14							7	

c Plot the points that you have found in part **b** on your axes from part **a**. Join them to form a smooth parabola.

3 For each equation
i Make a table with x-values from −4 to 4 and find the corresponding y-values.
ii Draw an x-axis from −4 to 4 and a suitable y-axis.
iii Plot the points and join them to form a parabola.
iv Write the coordinate of the minimum point of each parabola.
a $y = x^2 + 3$ **b** $y = 2x^2$
c $y = 3x^2 - 1$ **d** $y = x^2 + x$

In part **b**, square before you multiply by 2

4 True or false? The point (4, 10) lies on the graph $y = x^2 - 5$. Explain your answer.

5 a How do you think the graph $y = 10 - x^2$ will differ from those that you have plotted in questions 2 and 3?
b Draw and complete a table of coordinates with $-3 \leqslant x \leqslant 3$.
c Plot your points. Was your prediction correct?

6 a Plot the curve $y = x^2 + 5x + 6$ for $-4 \leqslant x \leqslant 4$.
b Write the coordinates of the points where the graph intersects the x-axis.
c Use your answer to part **b** to solve $x^2 + 5x + 6 = 0$.
d Compare your answers to **b** and **c**. What do you notice? Why?

A7.2 Further curve plotting

Objectives

- Plot graphs of simple cubic and reciprocal functions

Useful resources

- Graph plotting tool
- 2 mm graph paper

Mental starter

What happens if you square a negative? (Becomes positive.) **What happens if you cube a negative?** (It stays negative.) Give examples to back up these statements. **What value will $(-1)^9$ have?** (−1.) **What value will $(-1)^{20}$ have?** (1.) **What can you say about the powers of a negative number?** (If the power is even then the answer is positive, but if the power is odd then the answer is negative.)

Introductory activity

These equations produce straight lines: $x = 4$, $y = -3$, $y = x$, $y = -2x + 1$. Sketch them on the board.

Some equations produce a ∪ or ∩ shape. $y = x^2$, $y = -2x^2 + 1$. Sketch them on the board. Ask students how we can accurately draw them.

Why does this second set of equations produce a curve instead of a straight line? (The x^2 term produces a curve, the curve is actually symmetrical and smooth.)

Some equations produce more complex shapes, for example $y = x^3$. Ask students to complete this table.

x	−3	−2	−1	0	1	2	3
y	−27	−8	−1	0	1	8	27

Draw axes from 30 to −30 for y, use each square for two units; x axes from −3 to 3, use two squares for each unit.

Use a whole page to draw the graph. **Why do you get such a different sort of graph?** (Because of the x^3 term.)

Exercise commentary and misconceptions

Students need to recognise the link between the general shape of a function and the algebraic function. Students must take care when powering a negative value. Without a calculator, students need to be aware of what happens when negatives are multiplied together. When using a calculator students will need to be very wary of powering a negative (see the notes in the exercise commentary for A7.1). Some students are confused by the rows in the table of values that are designed to aid in the evaluation of the function. Students should be allowed to use 2 rows only, x and y if they feel more comfortable with that.

Question 4 may cause problems if students are unaware of the significance of the points of intersection of the two graphs.

Plenary

Draw the graph of $y = x^2 + 4$ on to the same axes as $y = x^3$ from the introduction. **Where do the curves intersect (cross)?** (2, 8) What is the value of x at this point? $x = 2$.

Do you know what a solution to the equation $x^2 + 4 = x^3$ is? Encourage students to use trial and error to see a solution.

Is there only one answer? (Yes, $x = 2$ only.)

Can any students see how the solution of this equation is linked to the 2 graphs? Where the graphs of $y = x^3$ and $y = x^2 + 4$ cross is the solution of the equation $x^3 = x^2 + 4$.

A7.2 Further curve plotting

This spread will show you how to:

- Plot graphs of simple cubic and reciprocal functions

Keywords
Cubic
Reciprocal
S-shaped

- **A cubic equation contains a term in x^3. It has a distinctive curved graph.**

Example

Draw the graph of $y = x^3 - 1$ and use it to estimate the value of y when $x = 2.5$.

First make a table of x and y values.

x	−3	−2	−1	0	1	2	3
x^3	−27	−8	−1	0	1	8	27
$y = x^3 - 1$	−28	−9	−2	−1	0	7	26

$(-3)^3 = -27$

Draw the x-axis from −3 to 3 and the y-axis from −30 to 30.
Plot the points, (−3, −28), (−2, −9) and so on, and join them in a smooth curve.

$y = x^3 - 1$

The value of $y = 14.5$ when $x = 2.5$

Cubic graphs have two bends.

The graph of a **cubic** equation is an **S-shaped** curve.

A reciprocal equation contains a term in $\frac{1}{x}$.

It has a different-shaped curve.

Example

Draw the graph of $y = \frac{1}{x}$.

Use x-axis from −3 to +3.

x	−3	−2	−1	1	2	3
y	$-\frac{1}{3}$	$-\frac{1}{2}$	−1	1	$\frac{1}{2}$	$\frac{1}{3}$

Draw the x-axis from −3 to 3 and the y-axis from −1 to 1.
Plot the points.

Note that you cannot plot $x = 0$ as $\frac{1}{0}$ is not a defined function.

Exercise A7.2

1 Match the graphs with their equations. For the equations that are left over, sketch the shape of their graphs.

$x = 4$ $y = x^2 - x - 6$ $y = 5$ $y = x^3$ $y = 3 - 2x$ $y = \frac{1}{x}$

2 **a** Draw an x-axis from −3 to 3 and a y-axis from −30 to +30.

b Copy and complete the table for $y = x^3 + 1$.

x	−3	−2	−1	0	1	2	3
y	−26						28

when $x = -3$, $(-3)^3 + 1$ which is $-27 + 1$

When using your calculator, remember brackets.

c Plot the coordinates that you have found in part **b** on your axes from part **a**. Join them to form a smooth, S-shaped curve.

d Use your graph to estimate the value of

i y when $x = 1.5$ **ii** $0.5^3 + 1$

3 For each equation, copy and complete the table of values. Plot these points on suitable axes and join them to form a smooth curve.

a $y = x^3 - 4$

x	−2	−1	0	1	2	3
x^3					8	
$x^3 - 4$					4	

b $y = \frac{2}{x}$

x	−2	−1	1	2	3
$\frac{1}{x}$				$\frac{1}{2}$	
y				1	

c $y = x^3 + x + 1$

x	−2	−1	0	1	2	3
x^3					8	
$x + 1$					3	
y					11	

d $y = \frac{3}{x} + 1$

x	−2	−1	1	2	3
$\frac{3}{x}$	$-\frac{3}{2}$				
y	$-\frac{1}{2}$				

4 **a** On a pair of axes for $-3 \leqslant x \leqslant 3$ and suitable y values, plot

i $y = 8 - x^3$ **ii** $y = 2x^2 - 3x$

b Use your graph to find the points where $8 - x^3 = 2x^2 - 3x$.

A7.3 Solving linear simultaneous equations graphically

Objectives

- Use graphs to find the solutions or approximate solutions of two simultaneous equations

Useful resources

- Graph plotting tool
- Mini-whiteboards

Mental starter

Give students the equations of different graphs, for example:

$y = 3x + 2$ $\quad y = x^2$ $\quad y = x^3$

$y = 3$ $\quad x = -4$ $\quad y = x^3 + 1$

Ask the students to sketch the shape of the graph on mini-whiteboards.

Introductory activity

Draw axes from −10 to 10 on both axes. **What values can you have for x and y if $y = 2x - 3$?** Obtain 3 possible answers from the students and plot these points, then draw the line, extending as far as possible. **What other values for x and y can you find from the line?** Get two more possible solutions. Discuss how the line actually gives you all the possible values, within the limits of the line.

Try the same for the line $x + y = 9$.

From the graph can you tell what values of x and y will work in both equations at the same time (simultaneously)? Lines cross at $x = 4$ and $y = 5$. What you have done is solve the simultaneous equations $y = 2x - 3$ and $x + y = 9$ by drawing both of them and seeing where they intersect (cross).

Discuss the examples in the student book. The second example involves forming equations from a description and is similar to question 4.

Exercise commentary and misconceptions

In question 2 drawing lines with equations $y = \ldots$ can usually best be achieved using x values of 0, 2, 4 to get a good spread and accurate line. Where the equation is given implicitly, for example, $2x + 3y = 12$, it is normally easiest to make $x = 0$, find y, and $y = 0$, find x. However on occasions when this is difficult, for example, $3x - 4y = 7$, students should try various values for x or y to find easy to plot values. In this example $x = 5$ gives $y = 2$, and $x = 1$ gives $y = -1$.

In question 4 where students need to form the simultaneous equations, remind them to state what the symbols or letters represent.

When forming simultaneous equations, students should re-read the words in the question to check they match the equation they have written.

Plenary

In the equation $y = x^2 - 2x$, what values of x make $y = 0$? (Use trial and improvement to establish that it is $x = 0$ and $x = 2$.) In reality you have solved the equation $x^2 - 2x = 0$.

How could we solve the equation graphically? (**Either** draw $y = x^2 - 2x$ and find the values of x where the curve crosses the x axis, that is where $y = 0$, **or** draw $y = x^2$ and $y = 2x$ and find the values of x at the points of intersection.) **Which would be the easier to do?**

A7.3 Solving linear simultaneous equations graphically

This spread will show you how to:

- Use graphs to find the solutions or approximate solutions of two simultaneous equations

Keywords
Intersection
Simultaneous
Solution

You can solve **simultaneous** equations on a graph.

Example

Solve $3x - y = 2$ (1) $2x + y = 8$ (2)

Plot their graphs. $3x - y = 2 \Rightarrow y = 3x - 2$

x	1	2	3
y	1	4	7

$2x + y = 8$

x	0	4	2
y	8	0	4

This line shows the solutions to $2x + y = 8$

At the point of **intersection** of the two lines, (2, 4) both equations have a **solution**. At the intersection $x = 2$ and $y = 4$

This line shows the solutions to $y = 3x - 2$

Check:
(1): $3 \times 2 - 4 = 2$
(2): $2 \times 2 + 4 = 8$

The solution is $x = 2$, $y = 4$.

- **The solution to simultaneous equations is where their graphs intersect.**

Example

Three times one number plus twice another is 9. Twice the first number subtract the second is 13. Use a graphical method to find the two numbers.

Let the numbers be x and y. Hence,
$3x + 2y = 9$
$2x - y = 13$ ...

$3x + 2y = 9$

x	1	3	2
y	3	0	1.5

$2x - y = 13$

x	0	$6\frac{1}{2}$	3
y	−13	0	−7

The lines intersect at (5, −3)

Don't just give the coordinates as the solution. Say what each number is.

Hence, the solution is $x = 5$, $y = -3$.
The two numbers are 5 and −3.

Exercise A7.3

1 Use the graph to solve these pairs of simultaneous equations.

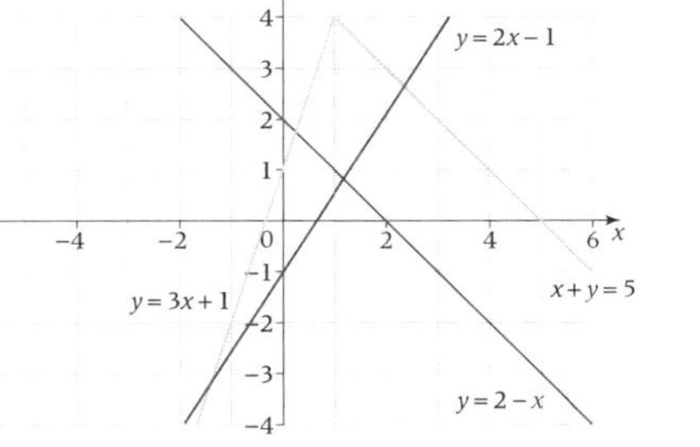

a i $y = 2x - 1$ and $x + y = 5$ ii $y = 3x + 1$ and $x + y = 5$
iii $y = 2 - x$ and $y = 2x - 1$ iv $y = 3x + 1$ and $y = 2 - x$

b Using the graph, explain why the simultaneous equations $x + y = 5$ and $y = 2 - x$ have no solution.

2 Plot graphs to solve these simultaneous equations.

a $y = 2x + 1$, $x + y = 10$ b $y = 3x - 2$, $x + y = 2$ c $2x + y = 5$, $x - y = 4$

3 a Solve $3x + 2y = 4$ and $x + 4y = 3$ graphically.

See A6.4.

b Solve $3x + 2y = 4$ and $x + 4y = 3$ algebraically.

c What should you notice about your answers to part **a** and part **b**?

4 Use a graphical method to solve these problems.

a Twice one number plus three times another is 4. Their difference is 2. What are the numbers?

b The sum of the ages of James and Isla is 4. The difference between twice Isla's age and treble James' age is 3. How old are they?

5 a By plotting graphs if necessary, explain why the simultaneous equations $y = 2x - 1$ and $y = 2x + 4$ have no solution.

b Is it possible to have a pair of simultaneous equations with more than one solution?

6 Two lines intersect. One has gradient 4 and y-axis intercept 3. The other has gradient 6 and cuts the y-axis at (0, 1).

a Write the simultaneous equations they represent.

b Solve the equations algebraically.

c What is the point of intersection of the lines?

A7.4 Solving quadratic and cubic equations graphically

Objectives

- Find approximate solutions of equations from their graphs, including one linear and one quadratic and simple cubic equations

Useful resources

- Graph plotting tool
- 2 mm graph paper

Mental starter

Sketch various straight-line graphs on blank coordinate axes, for example the graphs of $y = 2x + 1$, $y = 1 - x$, $y = 5$, $x = -2$.
Ask students to suggest the equation for each graph.

Introductory activity

Ask students to draw axes −4 to 4 for x and −4 to 10 for y. Draw the graph of $y = x^2 - 1$.
Suggest they compile a table first:

x	−3	−2	−1	0	1	2	3
y	8	3	0	−1	0	3	8

Now, as a whole class, solve these equations:

(i) $x^2 - 1 = 0$. **Where on the graph is $y = 0$?**
(At $x = -1$ and $x = 1$. These are the solutions to the equation.)

(ii) $x^2 - 1 = 5$. **Where on the graph is $y = 5$?**
(At $x = -2.4$ and 2.4.)

(iii) $x^2 - 1 = x + 5$. **What has y changed into?**
(y has changed into $x + 5$ or $y = x + 5$.)
Draw this line and see where it intersects (crosses) the curve $y = x^2 - 1$. (At $x = 3$ and −2.)

Exercise commentary and misconceptions

In question 1, if the intersection is at an integer value, the answer will be accurate, but if not it will be an approximation.

In question 2, students need to draw a smooth curve when copying the graph.

Students could try solving question **2f** by factorising so that they accept that the two methods agree.

Plenary

How could you solve the equation $x^2 = x^3 + 1$?

Draw the graphs of $y = x^2$ and $y = x^3 + 1$. Where they cross is the solution. ($x = -0.8$ approx.)

A7.4 Solving quadratic and cubic equations graphically

This spread will show you how to:

- Find approximate solutions of equations from their graphs, including one linear and one quadratic and simple cubic equations

Keywords
Intersect
Solution

Quadratic and cubic equations can be solved by drawing a graph.

x	−5	−4	−3	−2	−1	0
x^2	25	16	9	4	1	0
$5x$	−25	−20	−15	−10	−5	0
y	6	2	0	0	2	6

Since $y = x^2 + 5x + 6$,
then $x^2 + 5x + 6 = 2$ when $y = 2$.
Draw the line $y = 2$ on the graph.

The line $y = 2$ crosses the curve $y = x^2 + 5x + 6$ when $x = -1$ and $x = -4$.

$x^2 + 5x + 6 = 2$ when $x = -1$ and $x = -4$

The **solution**(s) to $x^2 + 5x + 6 = 2$ are where $y = 2$ and $y = x^2 + 5x + 6$. These are the points where the line and curve **intersect**.

Example

On the graph of $y = x^3 + x^2 - 6x$, where would you find the solutions to $x^3 + x^2 - 6x = 0$? Find them.

The solution of $x^3 + x^2 - 6x = 0$, is where the curve, $y = x^3 + x^2 - 6x$, and the line, $y = 0$, intersect.

The graph $y = x^3 + x^2 - 6x$ intersects the x-axis at $(-3, 0)$, $(0, 0)$ and $(2, 0)$, so the solutions of $x^3 + x^2 - 6x = 0$ are $x = -3$, $x = 2$ and $x = 0$.

$y = x^3 + x^2 - 6x$

x axis or $y = 0$

Solutions are where curve crosses the x-axis

You can check by substituting −3, 2 and 0 for x in the equation.

$y = 0$ is the same line as the x-axis.

Notice the characteristic 'S' shape of the cubic equation.

Exercise A7.4

1 Some graphs are drawn on these axes.

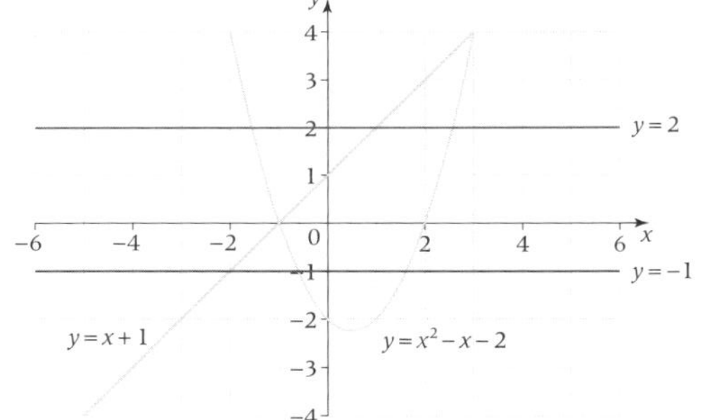

Use the graphs to find the approximate solutions of

a $x^2 - x - 2 = 2$ **b** $x^2 - x - 2 = -1$ **c** $x^2 - x - 2 = x + 1$

2 The graph shows $y = x^2 - 2x - 3$.

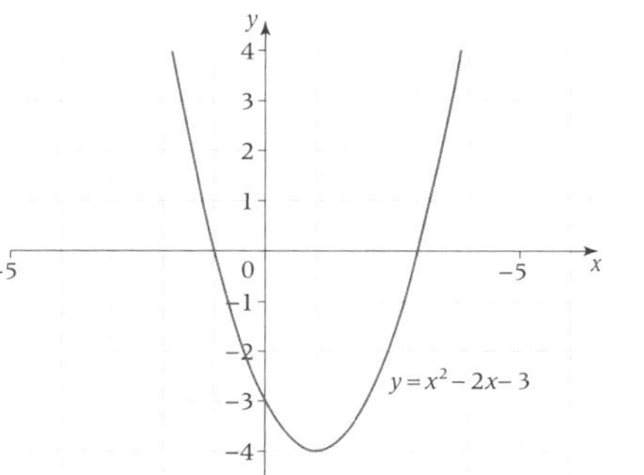

Copy the graph and, by adding lines, use it to find the approximate solutions of

a $x^2 - 2x - 3 = 1$ **b** $x^2 - 2x - 3 = -3$ **c** $x^2 - 2x - 3 = -4$

d $x^2 - 2x - 3 = x - 2$ **e** $x^2 - 2x - 3 = 1 - x$ **f** $x^2 - 2x - 3 = 0$

3 Draw appropriate graphs to find the approximate solutions of

a $x^2 - 2 = 5$ **b** $x^3 + x = 2x - 1$ **c** $2x^2 - x = 0$ **d** $x^3 - x^2 = 2$

A7.5 Solving quadratic and linear simultaneous equations

Objectives

- Find the approximate solutions of simultaneous equations from graphs, when one is linear and one quadratic

Useful resources

- 2 mm graph paper
- Graph plotting tool

Mental starter

When pumping up a rubber rowing boat I need to deliver 0.3 m^3 of air. This takes me 500 pumps on the air pump. **How much air do I deliver each time I pump?** (Could give the hint that there are 1 000 000 cm^3 in 1 m^3; 300 000 ÷ 500 = 600 cm^3 per pump, about 1 pint!)

Introductory activity

Draw axes −4 to 4 for x and −4 to 12 for y.
Draw the graph of $y = x^2 + x$.
Ask students to compile a table first:

x	−3	−2	−1	0	1	2	3
y	6	2	0	0	2	6	12

Solve these equations:

(i) $x^2 + x - 2 = 0$. **How can you change it into $x^2 + x$?** (Add 2 to each side $\longrightarrow x^2 + x = 2$.) Now you can see that the y has become 2, **or** $y = 2$. **Where is this on the graph?** ($x = -2$ and $x = 1$.)

(ii) $x^2 = 2$. **How can you change it into $x^2 + x$?** (Add x to each side $\longrightarrow x^2 + x = 2 + x$.) Now you can see that the y has become $2 + x$ **or** $y = x + 2$.
So draw $y = x + 2$ and see where they cross. ($x = -1.4$ and $x = 1.4$ approx.)

Exercise commentary and misconceptions

In question 2 students will need to adapt the equation to match the graph that has already been drawn. Use the normal skills of manipulating equations and write working in underneath on both sides.
Emphasise that really it is the y value that has changed into something new. Once you have found out what it is you could draw it, and the x values at the intersections would be the solutions to the equations.

Note that question 2 does not ask students to draw their suggested lines, but question 3 does!

Plenary

Will every possible straight line cross the cubic curve $y = x^3 + x^2 - 1$ (the curve from question 3)? (Yes.) **How many times will it cross?** (One, two or three times.)

A7.5 Solving quadratic and linear simultaneous equations

This spread will show you how to:

- Find the approximate solutions of simultaneous equations from graphs, when one is linear and one quadratic

Keywords
Linear
Quadratic
Solution

You can solve two simultaneous equations, one of which is **linear** and the other **quadratic**, by drawing a graph.
Look to the points of intersection.

Example

Solve the simultaneous equations $y = 11x - 2$ and $y = 5x^2$.

Make a table of values for each equation.

$y = 11x - 2$

x	-1	0	1
$11x$	-11	0	11
y	-13	-2	9

$y = 5x^2$

x	-3	-2	-1	0	1	2	3
$5x^2$	45	20	5	0	5	20	45

Plot the graphs.

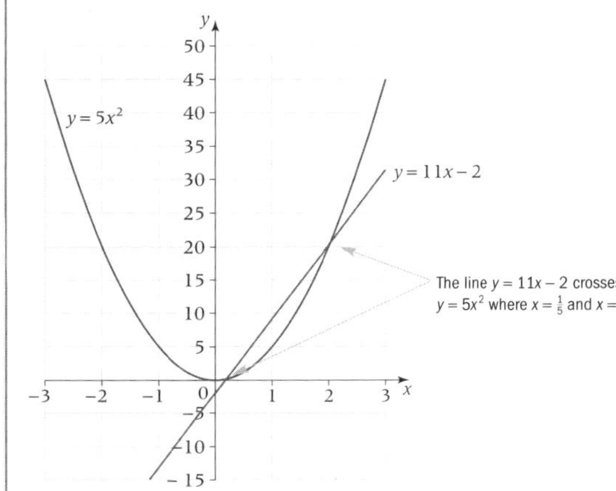

The **solutions** of the simultaneous equations
$y = 11x - 2$ and $y = 5x^2$ are
$x = \frac{1}{5}$, $y = \frac{1}{5}$ and $x = 2$, $y = 20$.

Exercise A7.5

1 Use the graphs to find approximate solutions of

a $x^2 + 2x - 3 = 0$ **b** $x^2 + 2x - 3 = x + 1$
c $x^2 + 2x - 5 = 0$ **d** $x^2 + x = 0$

2 Given that the graph of $y = x^2 + 4x - 2$ is already drawn, which one line would you need to draw in order to solve

You do not have to draw them.

a $x^2 + 4x - 2 = 3$ **b** $x^2 + 4x - 2 = 0$ **c** $x^2 + 4x - 2 = 2x + 1$
d $x^2 + 4x = 6$ **e** $x^2 + 5x = x + 4$ **f** $x^2 + 2 = 6x$?

3 The graph of $y = x^3 + x^2 - 1$ has been drawn below. Copy the graph and add on graphs of your choice in order to find the approximate solutions of

a $x^3 + x^2 - 1 = 2$ **b** $x^3 + x^2 = x + 2$ **c** $x^3 + x^2 - 1 = -2$
d $x^3 + x^2 - x = 0$ **e** $x^3 + x^2 - 2x + 4 = 0$ **f** $x^3 - 1 = 1 - 2x^2$

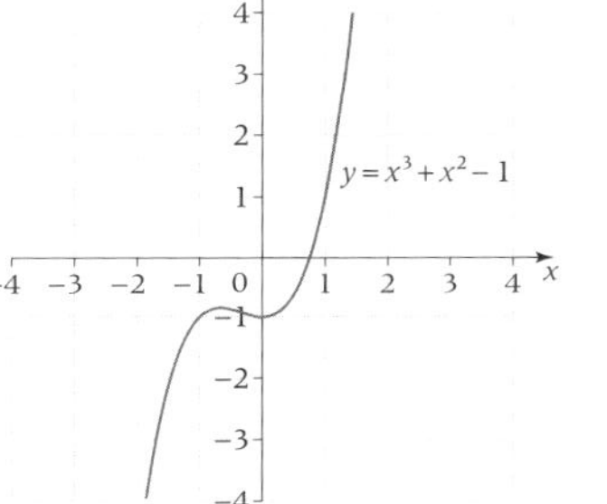

A7 Exam review

Key objectives

- Construct linear functions and plot the corresponding graphs
- Generate points and plot graphs of simple quadratic and cubic functions
- Find the intersection points of graphs of a linear and quadratic function, knowing that these are the approximate solutions of the corresponding simultaneous equations

Worked solution

1 a

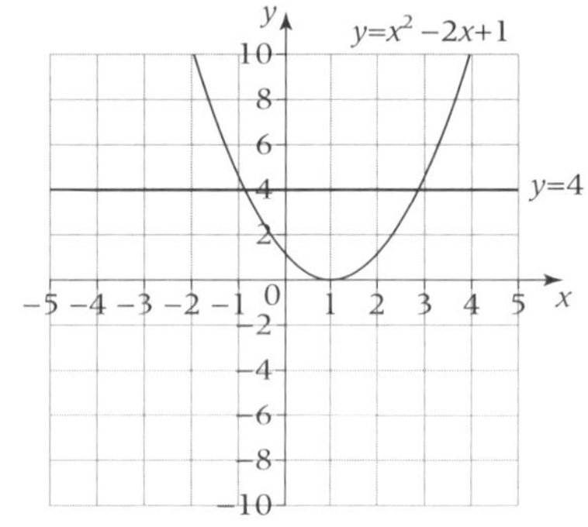

Commentary and misconceptions: Encourage students to plot graphs neatly and accurately. This will help when solving equations graphically.
Ensure that students plot the line $y = 4$ onto the same graph. This is needed for the following part of the question.

b The solutions to $4 + 2x = x^2 + 1$ are where their lines $y = x^2 - 2x + 1$ and $y = 4$ intersect.
The intersections are at (3, 4), and (−1, 4).
The solutions are $x = 3$, $y = 4$ and $x = -1$, $y = 4$.

Commentary and misconceptions: Ensure that students use the graph and do not find the solution algebraically.
Encourage them to check their solutions by substituting back into the equation.
Ensure that students explain their method.
Some students may not recognise the link between this equation and the simultaneous equations. They may find it easier in the form $4 = x^2 - 2x + 1$.
Prompt them to rearrange if necessary.
Ensure that students give both solutions for x and y.

2 a

x	−2	−1	0	1	2	3
y	−1	1	3	5	7	9

Commentary and misconceptions: Ensure that students substitute the numbers into the equation correctly. Drawing the graph accurately will help students to answer part **c** successfully.

b Students should draw a straight-line graph through coordinates: (−2, −1), (−1, 1), (0, 3), (1, 5), (2, 7), (3, 9).

Commentary and misconceptions: Ensure that students use their graph to answer this question. They should show their workings on the graph and explain their method.
Encourage students to check their solutions by substituting back into the equation.

c **i** When $x = -1.3$, $y = 0.4$
ii When $y = 5.4$, $x = 1.2$.

N8

Ratio and proportion

Objectives

F/H Use ratio notation, including reduction to its simplest form and its various links to fraction notation

F/H Divide a quantity in a given ratio

F Convert simple fractions of a whole to percentages of the whole and vice versa

H Convert a recurring decimal to a fraction and vice versa

H Solve problems and word problems, including those involving ratio and proportion repeated proportional change and reverse percentages, inverse proportion, surds, measures and conversion between measures, and compound measures defined within a particular situation

H Check and estimate answers to problems

Unit overview

This unit consolidates the knowledge of fractions, percentages and proportion from earlier number units and introduces students to ratio. It then develops to using their understanding of these forms and the other subjects met throughout the number units to solve word problems involving ratio, proportion and percentages, including reverse percentages.

Prior knowledge

Before your students start this unit they should be able to:

- Understand and use fractions, decimals and percentages and convert between the forms
- Solve problems involving proportion
- Add, subtract multiply and divide integers, decimals, fractions and percentages
- Use a range of methods to check answers to problems
- Round answers to any number of significant figures
- Find the factors of an integer and the HCF of two integers

Differentiation

- **Foundation** focuses on solving simple ratio and proportion problems and using efficient methods to calculate with fractions and percentages. It also covers checking procedures and calculator methods
- **Foundation Plus** extends to multi-step ratio and proportion problems, expressing ratios as a fraction and problems involving reverse percentages
- **Higher Plus** focuses on algebra (A9), in particular graphical modelling

N8.1 Introducing ratio

Objectives

- Use ratio notation
- Simplify a ratio by cancelling common factors
- Divide a quantity in a given ratio

Useful resources

- Mini-whiteboards for mental starter

Mental starter

Four people take three days to build five sheds. **How long would two people take to build 10 sheds?** ($4 \times 3 = 12$ 'people-days' for five sheds. 10 sheds require 24 'people-days', or 12 days with two people sharing the work.) Do not let students lay this out like a ratio since time and people are not in direct proportion.

Introductory activity

Discuss the meaning of a ratio.
With an example, discuss how ratios can be simplified.

Simplify the ratio 6 : 9 fully, then write it in the form n : 1.

Three is a factor common to six and nine, so $6 : 9 = 2 : 3 = \frac{2}{3} : 1$.

Share £300 between Ben and Laura in the ratio 2 : 3. **Who gets more?** (Laura.) **How many parts has the money been split into?** (Five.) One part is. $300 \div 5 = 60$. Ben gets two parts and Laura gets three 3 parts.

Ben gets $60 \times 2 =$ £120; Laura gets $60 \times 3 =$ £180.

Use a ratio when you don't know the overall total amount. For example, juice and water are mixed in the ratio 3 : 5. **If 96 ml of juice is used, how much water was added?** The 96 ml represents three parts of the drink. So $96 \div 3 = 32$ ml stands for 1 part. The five parts of water are $32 \times 5 = 160$ ml.

Use ratio to answer a proportion question. For example, you can buy 15 erasers for 70p. **What is the cost of 12 erasers?** It is tempting to find the cost of 1 eraser, but this is a recurring decimal. Instead use ratio, 15 : 70p $\rightarrow$ 3 : 14p $\rightarrow$ 12 : 56p (56p). Simplify the ratio first, then convert to the number you want.

Exercise commentary and misconceptions

The questions focus on simplifying ratios. Encourage students to work down the page and include all their working.

In questions 5–8, when dividing amounts in a given ratio, make sure students think of the total being split into many parts. They can find one part first.

Plenary

A bag contains coloured dice, either black or red. Ratio of black to red is 3 to 2. **If there are less than 11 black dice, what is the largest number of dice there can be?** (Ratios you could have are 3 : 2, 6 : 4, 9 : 6, 12 : 9, 15 : 10, but if there are less than 11 black dice, the ratio must be 9 : 6 = 15 dice in total.)

N8.1 Introducing ratio

This spread will show you how to:

- Use ratio notation
- Simplify a ratio by cancelling common factors
- Divide a quantity in a given ratio

Keywords
Ratio
Simplify

A **ratio** is used to compare quantities.

- **You simplify a ratio by cancelling common factors.**
  A class contains 16 girls and 12 boys. The ratio of boys to girls is 16 : 12, and you can simplify this ratio to 4 : 3.

- **To divide a quantity in a given ratio (for example, dividing 20 m in the ratio 3 : 1)**
  - **first find the total number of parts: 3 + 1 = 4**
  - **now find the size of each part: 20 ÷ 4 = 5**
  - **multiply to find each share: 5 × 3 = 15, and 5 × 1 = 5**

Example

In a group of 14 people, eight have brown eyes and six have green eyes. Write the ratio of eye colours in its simplest form.

brown : green = 8 : 6 = 4 : 3

Example

Simplify these ratios.

a 1 m to 40 cm **b** 6 hours to $2\frac{1}{2}$ days

a 1 m = 100 cm so the ratio of
1 m to 40 cm = 100 cm : 40
= 5 : 2

b $2\frac{1}{2}$ days = 60 hours so the ratio
6 hours to $2\frac{1}{2}$ days = 6 : 60
= 1 : 10

Both parts of the ratio must have the same units.

Example

Divide **a** 120 cm in the ratio 2 : 3 **b** £108 in the ratio 3 : 5

a 2 + 3 = 5 Total number of parts = 5
120 ÷ 5 = 24 One part = 24 cm
2 × 24 = 48
3 × 24 = 72
The two lengths are 48 cm and 72 cm.
To check your answer add the two parts:
48 + 72 = 120, correct

b 3 + 5 = 8
108 ÷ 8 = 13.5
3 × 13.5 = 40.5
5 × 13.5 = 67.5
The two amounts are £40.50 and £67.50.
Check your answer:
40.50 + 67.50 = 108, correct

Remember to add £ signs and zeros.

Exercise N8.1

1 Here are the numbers of boys and girls in some Year 11 classes. For each class, write down the ratio of boys to girls.

a Class 11A has 17 boys and 13 girls

b Class 11B has 11 boys and 19 girls

c Class 11C has 14 boys and 15 girls

2 The table shows the number of students in three different classes who own pets. Write the ratio of pet owners to non-pet owners in each class.

Class	Pet owners	Non-pet owners
11A	7	23
11B	13	17
11C	16	13

3 Write the ratio of the number of vowels to the number of consonants in these words. Give your answers in their simplest form.

a JAGUAR **b** PANTHER **c** TIGER **d** LEOPARD
e PAPERBACK **f** PERPETUATE **g** STAGGERS **h** MEIOSIS

4 Simplify these ratios.

a 6 : 4 **b** 12 : 3 **c** 5 : 10 **d** 2 : 8
e 6 : 9 **f** 12 : 8 **g** 10 : 15 **h** 21 : 14

5 Divide each of these numbers in the ratio 3 : 2. Show your working.

a 40 **b** 45 **c** 90 **d** 100
e 250 **f** 2000 **g** 7500 **h** 9250

6 Divide each of these quantities in the ratio 2 : 1. Show your working.

a 24 hours **b** 180° **c** 45 minutes **d** €360
e 246 cm **f** 120 kg **g** 81 km **h** 54 miles

7 Divide £360 in these ratios.

a 1 : 1 **b** 1 : 2 **c** 2 : 1 **d** 3 : 5

8 Karla and Wayne share the tips they receive for working in a café. One Saturday, Karla's share was £12.50, and Wayne's was £7.50.

a Write the ratio of Karla's tips to Wayne's tips in its simplest form.

b The next week Karla and Wayne shared tips of £22 in the same ratio. Find their shares. Show all your working.

9 Mrs Jones wins £500 in a competition. She keeps 40% of the money.

a How much money does Mrs Jones keep for herself?

b Mrs Jones shares the rest between her children, Annie (8 years old) and Ben (12), in the ratio of their ages.
Work out how much each child receives. Show your working.

N8.2 More ratio

Objectives

- Use ratio notation
- Simplify a ratio by cancelling common factors
- Divide a quantity in a given ratio

Useful resources

- Calculators
- Mini-whiteboards for mental starter

Mental starter

I can fit 6 people around a rectangular table. Two people sit on each long side, and one person at each short side.
Five such tables are placed together so that short sides are touching.
How many people can sit down?
How many people for *n* tables in a line?

Introductory activity

The previous lesson gave practice in ratio techniques, and triple ratios are introduced here.

Share £400 between A, B and C in the ratio 2 : 5 : 1. **How many parts is the £400 being split into?** (8 parts.)

Each part is 400 ÷ 8 = £50.
A gets 50 × 2 = £100, B gets 50 × 5 = £250, C gets 50 × 1 = £50.
Check that £100 + £250 + £50 = £400.

When I solve a problem, the times I spend reading it, solving it and checking it are in the ratio 5 : 6 : 1. **If I spent 24 min solving the problem, how long did I spend reading it?**
Make sure students realise that the 24 min refers to the six parts of the ratio, not the overall total. Six parts total 24 min, one part is 24 ÷ 6 = 4 min, five parts are 4 × 5 = 20 min.
Discuss the examples in the student book.

Exercise commentary and misconceptions

In this exercise students divide amounts in a given ratio and solve practical problems including those with triple ratios.
Some questions are best solved by changing the ratio into another equivalent ratio until you find the solution you need.
Ensure that students check their answers by adding up the individual parts at the end.

Plenary

£45 is shared in the ratio $x : x + 1$. The second person got £25. **What does x equal?**
(First person got £45 – £25 = £20, so ratio is 20 : 25, which cancels to 4 : 5 so $x = 4$.)

N8.2 More ratio

This spread will show you how to:

- Use ratio notation
- Simplify a ratio by cancelling common factors
- Divide a quantity in a given ratio

Keywords
Ratio
Simplify

You can use **ratio** to divide quantities into more than two amounts.

- **To divide 55 m in the ratio 2 : 3 : 5**
  - **first add: 2 + 3 + 5 = 10**
  - **then divide: 55 m ÷ 10 = 5.5 m.**
  - **then multiply: 2 × 5.5 m = 11 m, 3 × 5.5 m = 16.5 m, and 5 × 5.5 m = 27.5 m**

Check: 11 + 16.5 + 27.5 = 55

Example

Alan and Betty share a bingo prize of £48 in the ratio of their ages. Alan is 32, and Betty is 26. How much does each get?

Ratio of ages = 32 : 26 = 16 : 13 — **Simplify** the ratio

16 + 13 = 29, so one part = £48 ÷ 29 — Add the parts

Alan gets $\frac{£48}{29} \times 16 = £26.48$ — Round to nearest penny

Betty gets $\frac{£48}{29} \times 13 = £21.52$ — Check: £26.48 + £21.52 = £48

Example

Roni weighs 56 kg, Mike weighs 64 kg and Steffi weighs 72 kg. Write the ratio of their weights in its simplest form.

Roni's weight : Mike's weight : Steffi's weight = 56 : 64 : 72 = 7 : 8 : 9 — Cancel by 8

Example

The Smith, the Brown and the Jones families go on holiday to Cumbria. They stay in a farmhouse which they rent for £2000. They agree to share the rent according to the number of people in each family. There are 3 in the Smith family, 6 in the Brown family and 8 in the Jones family. How much does each family pay?

Ratio of family size = 3 : 6 : 8

3 + 6 + 8 = 17, so one part = £2000 ÷ 17

The Smiths pay $\frac{£2000}{17} \times 3 = £352.94$

The Browns pay $\frac{£2000}{17} \times 6 = £705.88$

The Joneses pay $\frac{£2000}{17} \times 8 = £941.18$

Check: £352.94 + £705.88 + £941.18 = £2000

Exercise N8.2

1 Write these ratios in their simplest form.

a 2 : 4 b 2 : 4 : 4 c 9 : 3 : 6
d 10 : 20 : 8 e 8 : 20 : 12 f 15 : 35 : 10

2 Split £240 in these ratios.

a 1 : 2 b 2 : 3 c 5 : 3
d 1 : 2 : 3 e 5 : 1 : 2 f 7 : 2 : 3

3 Divide 250 m in these ratios, giving your answers correct to the nearest centimetre.

a 1 : 4 b 2 : 3 c 2 : 5
d 6 : 1 e 6 : 7 f 9 : 5

4 Peter, Bob and Yasmin share prize money of £7500 between them in the ratio 9 : 5 : 11. How much do they each receive?

5 Ann, Charles and Edward divide prize money of £850 between them in the ratio 4 : 3 : 2. Find the amount that each receives, giving your answers to the nearest penny.

6 Copy and complete the table to show how each quantity can be divided in the ratio given. Give your answers to a suitable degree of accuracy.

Quantity	Ratio	Share 1	Share 2	Share 3
200 km	2:3:5			
38 kg	1:2:3			
450 cm	2:3:8			
£720	4:5:10			
95 litres	1:6:7			

7 A fruit drink contains mango juice, pineapple juice and passion fruit juice, in the ratio 4 : 3 : 2 by volume. Calculate the volume of each type of juice in a 1-litre pack of the fruit drink.

8 Mrs Williams won a £500 Premium Bond prize, and decided to share it among her three children in the ratio of their ages. The children are aged 5, 7 and 9.

a Calculate the amount that each child receives, giving your answers correct to the nearest penny.

b Exactly one year later, Mrs Williams wins another £500 prize. Again, she decides to divide it among her children in the ratio of their ages. Calculate the amount that each child receives this time.

9 Robert and Kathleen are business partners. At start up, Robert invested £150 000 and Kathleen invested £200 000. Each year they share the profits in the ratio of their investments. In 2004 the profit amounted to £29 540. Work out each partner's share.

N8.3 Ratio and proportion

Objectives

- Solve problems involving ratio and proportion

Useful resources

- Counting stick or number line to illustrate proportion

Mental starter

I had three pizzas. I ate $\frac{4}{5}$ of one pizza. **What is the ratio of the pizza left to the pizza eaten?** (Think in terms of fifths. 15 fifths overall. 11 left, 4 eaten. Ratio is 11 : 4.) **If I had *n* pizzas and I ate $\frac{1}{2}$ of one pizza, what would the ratio be?** $\{5 \times (n - 1) + 1\} : 4$ or $(5n - 4) : 1$. This algebraic example does not need to be constructed symbolically. A description along the lines of 'number of pizzas minus 1 gives number of whole pizzas. Then times 5 and add 1' gives one side of the ratio.

Introductory activity

The ratio of boys to girls in a full class is 3 : 5. If $\frac{1}{4}$ of the boys are away then there are only 9 boys. **How many students are there in total in the class?**
(9 boys = $\frac{3}{4}$ of the boys $\longrightarrow$ 3 boys = $\frac{1}{4}$ of the boys $\longrightarrow$ 12 boys. These 12 boys are 3 parts of the ratio. 1 part is 4 boys. So 5 parts is 20 girls. Total is 20 + 12 = 32 pupils.)

What proportion of the class are girls? ($\frac{20}{23} = \frac{5}{8}$.) **Can this be done without knowing the number of girls?** (Yes, you can use the ratio, five parts out of the total of eight parts $\longrightarrow \frac{5}{8}$.)

Discuss the examples in the student book. In each real-life problem, you need to work out the ratio given the proportion or vice versa.

Exercise commentary and misconceptions

The questions focus on expressing amounts in ratios and finding what proportions are being used. As always, encourage students to set their findings out clearly.

When dividing an amount in a ratio, encourage students to check that the portions add to the total.

Plenary

The Eiffel Tower in Paris is 1063 feet tall. Small and large models are available in sizes 2 inches and 12 inches.
What is the ratio of the model size to the real size for small and large in their simplest form? (Hint, you may need to remind students that there are 12 inches in 1 foot.)

Small model : real life = 2 : 1063 × 12
= 1 : 1063 × 6
= 1 : 6378 large model : real life = 1 : 1063.

What would you have if you built a model with a ratio of 1 : 1 with the real Eiffel tower?
(A model as high as the Eiffel Tower!)

N8.3 Ratio and proportion

This spread will show you how to:

- Solve problems involving ratio and proportion

Keywords
Fraction
Proportion
Ratio

- **Ratios compare one number with another.**
  The ratio of boys to girls is 3 : 2.
- **Proportions tell you what fraction of the whole amount something is.**
  $\frac{3}{5}$ of the students are boys.
- **A ratio can be used find a proportion.**
  A drink is made by mixing squash and water in the ratio 1 : 4, so one part out of every five is squash. The proportion of squash in the drink is $\frac{1}{5}$.

Example

A concrete mixer contains 5 kg of cement and 20 kg of sand. Find

a the ratio of sand to cement
b the proportion of sand in the mixture.

a Ratio sand : cement = 20 : 5 = 4 : 1
b Proportion of sand = $\frac{20}{25} = \frac{4}{5}$

Write the weight of sand as a fraction of the total.

Example

Pink paint is made from one-third red paint and two-thirds white paint.
Write the ratio of red paint to white paint.

Ratio red : white = $\frac{1}{3} : \frac{2}{3}$
= 1 : 2

Multiply by 3 to simplify.

Example

The ratio of hardback books to paperbacks on a bookshelf is 2 : 7.
What proportion of the books are paperbacks?

Proportion of paperbacks = $\frac{7}{2+7}$
= $\frac{7}{9}$

Write the fraction $\frac{\text{paperbacks}}{\text{total}}$

Example

The ratio of boys to girls in a class is 2 : 3.
Alex is working out the proportion of the class that are boys.
He says it's $\frac{2}{3}$.
What is wrong with Alex's reasoning?

The answer can't be right – there are more girls than boys in the class.
The proportion of boys = $\frac{2}{2+3} = \frac{2}{5}$

Remember to add the parts when you find a proportion.

Exercise N8.3

1 Orange paint is made from 2 parts red to 3 parts yellow paint.
a Write the ratio of red paint to yellow paint in the mixture.
b Find the proportion of red paint in the mixture, giving your answer as a fraction.

The answer is *not* $\frac{2}{3}$!

2 A bowl contains 200 grams of flour and 100 grams of sugar.
a Write the ratio of sugar to flour.
b Find the proportion of sugar in the mixture.

3 A string of decorative lights has 7 red bulbs, 21 green bulbs and 14 blue bulbs.
a Write the ratio of the number of bulbs of each colour in its simplest form.
b Calculate the proportion of
i red bulbs **ii** green bulbs **iii** blue bulbs.

4 The table shows the ratio of blue paint to yellow paint in mixtures to make different shades of green paint. For each one, calculate the proportion of the mixture that is blue paint.

a 1:1	**b** 1:2	**c** 2:3	**d** 3:5	**e** 4:3	**f** 5:2

5 Write the proportion of yellow paint in each of the mixtures given in question **4**.

6 A fruit drink contains $\frac{2}{5}$ water and $\frac{3}{5}$ fruit juice. Find the ratio of water to fruit juice in the drink.

7 A swimming club had a total income of £2000. It spent 30% of this income on pool hire, and the rest on instructors' fees.
a Calculate the amount spent on pool hire.
b Work out the ratio of the amount spent on pool hire to the amount spent on fees.

8 Tom has £2200. He gives $\frac{1}{4}$ to his son and $\frac{2}{5}$ to his daughter.
How much does Tom keep for himself? You must show all your working.

9 Andrew records how he spends his time over a 24-hour period.
He spends $\frac{1}{3}$ of the time sleeping and $\frac{1}{6}$ of the time travelling.
What proportion of his time does Andrew spend on activities other than sleeping and travelling? Show your working.

10 Mrs Smith inherits £16 000. She divides the money between her three children John, Sarah and Mark in the ratio 6 : 7 : 8, respectively. How much does Sarah receive? What proportion of the money does Mark receive?

N8.4 Ratio, proportion and percentages

Objectives

- Convert between ratios, proportions, percentages, fractions and decimals

Useful resources

- Calculators

Mental starter

Write these amounts on the board:
£42, £112, £329
Each amount is to be shared between Angelo, Bruno and Carlo in the ratio 2 : 3 : 5.
Ask students to work out how much each person receives for each amount.

Introductory activity

I throw a dice 20 times and score only one six. **What is the ratio of 'sixes' to 'not sixes'?** (1 : 19.) **What proportion of the throws are sixes?** ($\frac{1}{20}$.) **What percentage of the throws are sixes?** (5%.) You would expect the ratio of 'sixes' to 'not sixes' to be 1 : 5. **What proportion of throws would you expect to be sixes?** ($\frac{1}{6}$.) **What percentage of throws would you expect to be sixes?** (16.7%.) **Do you think the dice is fair?** The actual ratio of 1 : 19 is low, so the dice may not be fair. But 20 trials is not many. Repeat the test perhaps 100 more times. 120 trials should give a better idea and you would expect around 20 sixes.

To change ratio to proportion, you can think 'how many parts out of the total number'. In question 1 where yogurt and fruit are in ratio 9 : 1, out of the 10 parts one is fruit so the proportion of fruit is $\frac{1}{10}$. **What is the percentage of fruit?** (10%.)

The examples in the student book illustrate how ratios and proportions can be converted to percentages in context.

Exercise commentary and misconceptions

Question 1 has already been done.

Question 2 refers to the number of vowels (or consonants), not the number of **different** vowels (or consonants). Ensure that students know their vowels from their consonants.

Question 8 answers should be given to 2 dp because units are dollars.

Question 9 answers could be left as fractions or decimals.

Students need to be able to use non-calculator and calculator methods. Questions 1–7 can be treated as non-calculator, Questions 8 and 9 best as calculator.

Plenary

Three gears are arranged in a row with the largest gear in the middle. When this middle gear turns round twice, the gear to its left turns 18 times and the gear to its right turns three times.

The middle gear turns clockwise; in which direction will the gear on the left turn? (Anti-clockwise.) **In which direction will the gear on the right turn?** (Anti-clockwise.) **Which gear is bigger, the one on the left (18 turns) or the one on the right (three turns).** (Right.)

At the start the arrows point to the top of each gear. Draw 3 separate diagrams to show where the arrows are if (i) the middle gear goes round once (ii) the right gear goes round once (iii) the left gear goes around once.

N8.4 Ratio, proportion and percentages

This spread will show you how to:

- Convert between ratios, proportions, percentages, fractions and decimals

Keywords
Approximation
Fraction
Percentage
Proportion
Ratio

You can describe **proportions** using **percentages**.

- **Proportions expressed as percentages are easy to compare.**
  'The proportion of sugar is $\frac{1}{2}$ in recipe A and $\frac{2}{5}$ in recipe B'
  is the same as
  'Recipe A contains 50% sugar and recipe B contains 40% sugar'.

Sometimes it is impossible to express a proportion exactly as a percentage. For example, $\frac{1}{7}$ is an exact proportion, but 14% (or even 14.29%) is only an **approximation** to it.

Example

The metal alloy, constantan, is made from nickel and copper in a **ratio** of 2 : 3. What percentage of the alloy consists of nickel?

$2 + 3 = 5$

Proportion of nickel $= \frac{2}{5}$

Percentage of nickel $= \frac{2}{5} = \frac{2 \times 20}{5 \times 20} = \frac{40}{100} = 40\%$

Add the parts of the ratio.

Example

Here is Emma's fruitcake recipe.

Fruitcake (1kg)
125 g of butter
150 g of flour
100 g of sugar
450 g of dried fruit
170 g of eggs
5 g of spices

What percentage of the cake is butter?

Proportion of butter $= \frac{125}{1000} = \frac{1}{8}$

Percentage of butter $= \frac{125}{1000} \times 100\%$

$= 12.5\%$

Find the proportion of butter first.

Example

The ratio of the number of wins, draws and losses for a football team one season is 5 : 4 : 3. What percentage of the games are wins?

$5 + 4 + 3 = 12$

Proportion of wins $= \frac{5}{12}$

$= 42\%$ to nearest whole number

Exercise N8.4

1 A tub of fruit yogurt contains fresh yogurt and fruit in the ratio 9 : 1 by weight. Find the percentage of fruit in the contents of the tub.

2 **a** Write the ratio of the number of vowels to the number of consonants in the word CATERPILLAR.

b Find the percentage of the letters in the word CATERPILLAR that are vowels. Show your working.

3 The compositions of three different alloys of copper are shown in this table. Find the percentage of copper in each alloy.

Alloy	Composition
Nickeline	4 parts copper, 1 part nickel
US nickel coinage	3 parts copper, 1 part nickel
Medal bronze	93 parts copper, 7 parts tin

4 Scott and Maxine buy a present for their father. They share the cost in the ratio 3 : 2. What percentage of the cost does Scott pay?

5 The ratio of boys to girls in a class is 5 : 4. What percentage of the class is girls? Show your working.

6 The ABC mobile telephone company keeps records of calls made on its network. The ratio of the number of calls that are successfully connected to those that are missed for any reason is 4 : 1.

a Find the percentage of calls on the network that are successfully connected.

b The 123 mobile telephone company also keeps records of calls on their network. They say that 75% of calls are connected successfully.
Find the ratio of successful to missed calls on the 123 network.

7 Dan and Phil share £3800 between them in the ratio 3 : 7, respectively.

a Calculate the amount received by each person.

b Find the percentage of the total amount of money that was received by Dan.

8 Leon and Frieda divide \$500 in the ratio 61 : 33, respectively.

a Calculate the amount received by each person.

b Find the percentage of the total amount that Frieda receives.

9 German silver is made from copper, zinc and nickel in the ratio 16 : 5 : 3.

a Calculate the mass of each metal in 500 g of German silver.

b Work out the percentage of each metal in German silver.

N8.5 Reverse percentages

Objectives
- Solve problems using reverse percentages

Useful resources
- Calculators

Mental starter
Keira donates 5% of her salary to charity.
If she donates £130 per month, how much does she earn in total in the year? (£31 200.)

Introductory activity
This lesson builds on the work done in lesson N7.3, finding increased amounts by multiplying by decimal numbers. It may be helpful to remind students of this method before introducing 'reverse percentages'.

For example, to increase £40 by 23% there are two methods.

Either I can calculate 23% of £40, then add this amount to £40 (10% = £4, 1% = £0.40, 23% = £9.20, so answer is £49.20)
or $40 \times 1.23 =$ £49.20.

A car's cost is increased by 20% to £7200.
What did it cost originally?

Must equate £7200 to 120%, then 1% = £60 and 100% = £6000,
or the original was increased by 20%. This could have been done by multiplying by 1.2 to give the answer £7200. So you could work backwards by dividing by 1.2,
$7200 \div 1.2 =$ £6000 is the original.

Discuss the examples in the student book. They all involve reverse percentages in the context of money.
The second and fourth examples involve VAT.

Exercise commentary and misconceptions
In questions 1–4, students are multiplying to increase or decrease original amounts.
Questions 5 and 6 are on reverse percentages. It may be helpful to use some of the answers to questions 1–4 to reverse the calculations.

Firmly fix the concept that the original is 100%. The first line in a reverse percentage question is the most important. Students should begin by equating the amount in the question with the percentage it represents.

Plenary
I increase £400 by a certain percentage. Then I reduce my answer by half the percentage I increased by. The answer goes back to £400.
What percentage did I increase it by?
(Only possible answer is 100% increase then 50% decrease $400 \longrightarrow 800 \longrightarrow 400$). **Would this still work for an amount other than £400?**
(Yes.)

N8.5 Reverse percentages

This spread will show you how to:

- Solve problems using reverse percentages

Keywords
Decrease
Increase
Original
Percentage

In a reverse **percentage** problem, you are given an amount after a percentage change, and you have to find the **original** amount.

You have already solved some reverse percentage problems in N7.5.

Example

In a sale, a pair of shoes cost £38.25 after a 15% **decrease**.

Find the original price of the shoes.

Sale price is (100 – 15)% = 85% of original price
85% = 0.85
Original price × 0.85 = sale price

$$\text{Original price} = \frac{£38.25}{0.85} = £45$$

First work out the reverse percentage.

0.85 is the decimal equivalent of 85%.

Example

A table costs £88, including 17.5% VAT. Find the cost (to the nearest pound) before VAT was added.

£88 is 117.5% of cost before VAT.
117.5% = 1.175
Cost before VAT × 1.175 = £88

$$\text{Cost before VAT} = \frac{£88}{1.175} = £75 \text{ to nearest pound.}$$

Find the decimal equivalent.

Example

The price of unleaded petrol is increased by 4% to 93.7p per litre.
Find the price before the **increase**.

Increased price is 104% of old price.
100% = 1.04
Old price × 1.04 = increased price

$$\text{Old price} = \frac{93.7}{1.04} = 90.1\text{p per litre to nearest 0.1p}$$

Example

A car costs £15 000, including VAT at 17.5%. Veejay is working out the cost without VAT. He thinks: 'I'll just work out 17.5% of £15 000 and take that away, to get £12 375'.
What is wrong with Veejays's reasoning?

Original price + VAT = £15 000
100% 17.5% = 117.5%

£15 000 ÷ 1.175 = £12 766 (to nearest pound)

Exercise N8.5

1 Find the result of these following percentage increases.
Do these mentally.

a £100 is increased by 25% b £50 is increased by 20%
c 40 m is increased by 15% d 36 cm is increased by 50%

2 Calculate mentally the result of these percentage decreases.

a £60 is decreased by 25% b 80 cm is decreased by 75%
c 720 cm is decreased by 50% d 120 mm is decreased by 15%

3 Write a decimal number that you could multiply a quantity by to find the results of these percentage changes.

a An increase of 20% b A decrease of 15%
c An increase of 6% d A decrease of 5%
e A decrease of 6% f A decrease of 17%

4 Calculate the result of these percentage increases and decreases, by multiplying by the correct decimal number. Show your method.
You may use a calculator.

a £64 decreased by 7% b 45 kg increased by 14%
c 10.4 seconds decreased by 3% d 120 m increased by 65%
e €240 increased by 8.5% f $340 decreased by 11.5%

5 Calculate the original cost of these items, before the percentage changes shown.
Show your method. You may use a calculator.

a A hat that costs £46.50 after a 7% price cut.
b A skirt that costs £32.80 after a price rise of 6%.

6 A computer costs £658, including VAT at 17.5%.
Find the price, before VAT was added.

7 To decrease an amount by 8%, multiply it by 0.92.
For a further decrease of 8%, multiply by 0.92 again, and so on.
Use this idea to calculate

a the final price of an item with an original price of £380, which is given two successive price cuts of 8%.
b the final price of an item with an original price of £2400, which is given three successive price cuts of 10%.

N8 Exam review

Key objectives

- Divide a quantity in a given ratio
- Solve word problems about ratio and proportion
- Reverse percentage problems

Worked solution	Commentary and misconceptions
1	Ensure that students understand using fractions and percentages as operators and can convert between forms before attempting this question. Ensure that students read word problems carefully and understand what they are required to work out. Encourage students to approach word problems methodically, writing down their workings.
a Shop A: ₭30 is 60% of the original cost. 60% = 0.6 So original cost × 0.6 = 30 So original cost = $\frac{30}{0.6}$ = ₭50. Shop B: original cost = ₭42. So, the lowest original price is in shop B.	Ensure that students understand how to use percentage as an operator. Encourage them to use the method they feel most comfortable with. Converting to a decimal or a fraction are common methods. Ensure that students use the correct value from the question.
b Shop B: Sale price = $\frac{3}{4}$ × 42 = ₭31.50. Shop A: sale price = ₭30. So, lowest sale price is in shop A.	Ensure that students understand how to use fraction as an operator. Ensure that students use the correct value from the question.
2 **a** **i** total income = ₭50 + (240 × ₭5) = ₭1250. **ii** $\frac{50}{1250} = \frac{1}{25}$	Ensure that students use the correct order of operations and include all the relevant amounts in their calculation. Ensure that students give their answer in its simplest form as required.
b 60% of ₭1000 = 0.6 × 1000 = ₭600. So, the ratio is 600 : 250 So in its simplest form the ratio is 12 : 5.	Ensure that students understand how to use percentage as an operator. Ensure that students express their answer in its correct simplified form.

Enlargement and similarity

Objectives

F/H Understand that enlargements are specified by a centre and a scale factor

F/H Recognise, visualise and construct enlargements of objects using positive fractional scale factors

F Recognise that enlargements preserve angle but not length

F/H Understand the implications of enlargement for perimeter

F/H Identify the scale factor of an enlargement as the ratio of the lengths of any two corresponding line segments

Unit overview

This unit consolidates the previous work on transformations, extending it to enlargements. It covers using centres of rotation and scale factors, the implication of enlargement on 2-D shapes and what their scale factor represents. Students are taught how to describe and construct enlargements and then how to solve geometric problems involving enlargement by considering similar shapes.

Prior knowledge

Before your students start this unit they should be able to:

- Recognise and plot points in all four quadrants of the coordinate grid
- Understand and use ratio notation, including simplification and its link to fractions
- Set up and solve linear equations
- Understand and use fraction notation
- Add, subtract, multiply and divide with integers, decimals and fractions
- Understand, recall and use angle and geometric properties of triangles and other 2-D shapes
- Understand and use parallel lines

Differentiation

- **Foundation** focuses on reflections, rotations and translations, including congruence, and then extends to enlargements on a grid with positive, integer scale factors, and finding their scale factor and centre of rotation
- **Foundation Plus** begins with maps and scale drawings and then extends the Foundation book work on enlargement to considering similar shapes and the relationships involved
- **Higher Plus** extends the work on the coordinate plane, focusing on vectors and vector notation, including using vectors in geometry and proof

S7.1 Enlargement

Objectives

- Enlarge objects, given a centre of enlargement and scale factor, including fractional scale factors
- Recognise that enlargements preserve angle but not length
- Understand how enlargement affects perimeter

Useful resources

- Geometry tool
- Graph paper
- Class set of interlocking cubes

Mental starter

Give pairs of students 10 interlocking cubes. Make a 3-D shape using interlocking cubes without the class seeing.
Describe your shape and ask students to make an identical shape.
Compare students' results with the original shape.

Introductory activity

How does a slide projector work? Ask a student who has seen or used one to try and explain. Make clear that the bulb at the back shines onto the slide then the light travels onto the screen to produce the enlarged image.
Show students the first diagram and explain the construction lines.

Demonstrate an enlargement where the centre of enlargement is outside the object, and one where the centre of enlargement is inside the object. Show the construction and then measure equivalent lengths on the object and the image.

Draw axes from 0 to 16 on x and y. Plot a triangle at (6, 4) (8, 4) (6, 8). Label it A. Enlarge A by the scale factor 2, with the centre of enlargement at (0, 2). Label the image.
Emphasise counting the squares from the centre of enlargement, rather than measuring with a ruler. Most successful students tend to count from centre of enlargement to the nearest point, then go back to the centre of enlargement and count twice the distance. Once one point of the image is placed, the rest of the shape can be drawn in, using the scale factor.

Coordinates of the enlarged figure are (12, 6) (16, 6) (12, 14).

Exercise commentary and misconceptions

Remind students to leave in the construction lines. They will gain marks for this in the exam.

In questions 1, 3 and 4**b** the centre of enlargement is **outside** the object.

In question 2 the centre of enlargement is **inside** the object.

In question 4**c** the centre of enlargement is **on** the object.

Students often have problems with counting the squares both horizontally and vertically. A common mistake is to count 'one' at the centre of enlargement.

Students should begin to realise that the perimeter enlarges by the scale factor, but not the area. Ask students who are confident with their enlargements how the **area** increases when a shape is enlarged (by the scale factor squared).

Plenary

What effect does the scale factor have on the perimeter of the shape? (Multiplied by the scale factor.)

What effect does the scale factor have on the area of the shape? (Multiplied by the square of the scale factor.)

Enlargement is a type of transformation. **What are the other transformations students have met?** (Translation, reflection and rotation.)

S7.1 Enlargement

This spread will show you how to:

- Enlarge objects, given a centre of enlargement and scale factor, including fractional scale factors
- Recognise that enlargements preserve angle but not length
- Understand how enlargement affects perimeter

Keywords
Centre of enlargement
Enlargement
Scale factor
Similar

To enlarge a shape, you multiply all the lengths by the same scale factor.

- **In an enlargement**
  - corresponding angles are the same
  - corresponding lengths are in the same ratio.

You draw an enlargement from a centre.

In the diagram, △PQR is enlarged by **scale factor** 2 from the centre (0, 1).

All the distances are × 2 so CQ′ = 2CQ, CP′ = 2CP, CR′ = 2CR

Lengths on the image are 2 × corresponding lengths on the object so P′R′ = 2PR and so on.

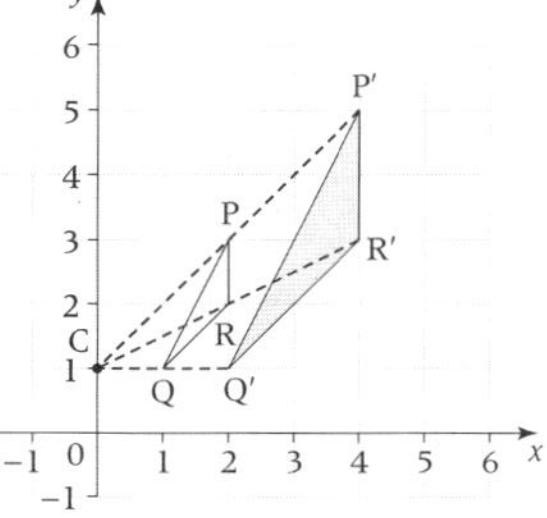

The image, P′Q′R′, is **similar** to the object, PQR.

- **Perimeter of image = scale factor × perimeter of object.**

Example

a Enlarge the white triangle by scale factor 3, centre (2, 1).
b How much larger is the perimeter of the image than the perimeter of the object?

a

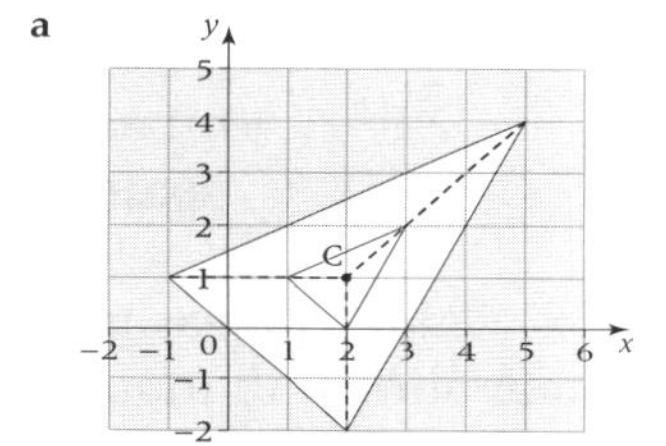

Lengths on the image are 3 × the corresponding lengths on the object.

The distance from C to a point on the image is 3 × the distance from C to the corresponding point on the object.

b Perimeter of image = 3 × perimeter of object.

Exercise S7.1

1 a Copy this diagram, but extend both axes to 16.
b Enlarge triangle T by scale factor 2, centre (0, 0). Label the image U.
c Enlarge triangle T by scale factor 3, centre (0, 0). Label the image V.

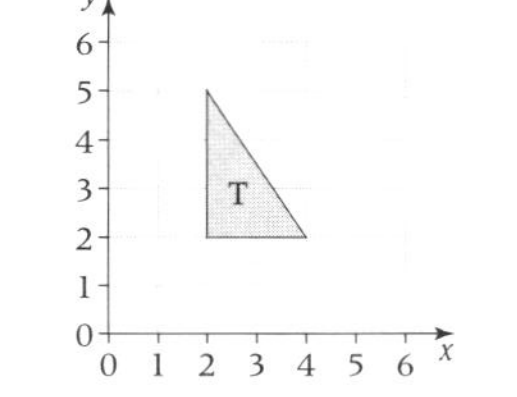

2 a Copy this diagram, but extend both axes from −7 to 7.
b Enlarge kite K by scale factor 2, centre (1, 3). Label the image L.
c Enlarge kite K by scale factor 4, centre (1, 3). Label the image M.

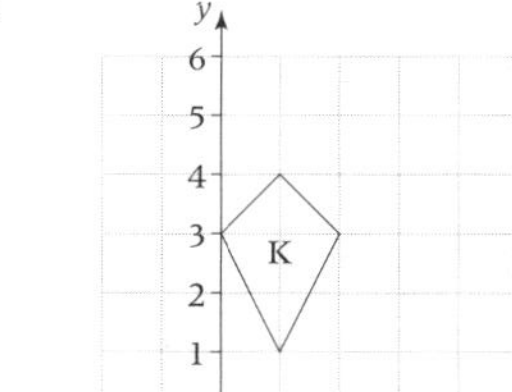

3 a Draw a grid with an x-axis from −6 to 6 and a y-axis from −3 to 15.
Plot the points (1, 3) (1, 5) (3, 7) (4, 6).
Join them to make quadrilateral Q.
b Enlarge quadrilateral Q by scale factor 2, centre (4, 8). Label the image R.
c Enlarge quadrilateral Q by scale factor 2, centre (0, 6). Label the image S.
d How much larger is the perimeter of R than the perimeter of Q?

DID YOU KNOW?
The E. coli. bacteria in this picture has been enlarged by a scale factor of 5000.

4 a Draw a grid with x- and y-axes from 0 to 13.
Plot the points (1, 3) (4, 4) (2, 1).
Join them to make triangle A.
b Enlarge triangle A by scale factor 3, centre (0, 0). Label the image B.
c Enlarge triangle A by scale factor 2, centre (2, 1). Label the image C.
d How much larger is the perimeter of B than the perimeter of A?

S7.2 Fractional scale factors

Objectives
- Enlarge objects, given a centre of enlargement and scale factor, including fractional scale factors

Useful resources
- Geometry tool
- Graph paper

Mental starter
The longest dinosaur ever was 48 m long. The smallest was only 60 cm long. **What fraction of the longest dinosaur's length is the smallest dinosaur's length?** ($\frac{60}{4800} = \frac{1}{80}$)

Introductory activity
Remind students of the previous lesson using scale factors which were positive integers. Enlarging with a scale factor of 2 doubles all the dimensions of the object.
What would enlarging with a factor of $\frac{1}{2}$ do? (Halves all dimensions.) Note that this is still called enlargement, not reduction!

Use the diagram in the middle of the student book page to demonstrate enlargement with fractional scale factors.

Emphasise the importance of counting from the centre of enlargement to points on the object and image.

Exercise commentary and misconceptions
In questions 3 and 4 the construction lines are diagonals so need extra care. Remind students again to count from the centre of enlargement and always count the squares.

In all questions transforming one point then drawing the rest of the shape around this point is acceptable but is much more difficult with irregular shapes, so students should be encouraged to transform all the points unless they are very competent.

Plenary
What effect does the scale factor have on the perimeter of the shape? (Multiplies it by the scale factor.)

What effect does the scale factor have on the area of the shape? (Multiplies it by the square of the scale factor.)

Is it possible to enlarge a cube by a scale factor using coordinates? (Using 3-D coordinates encourage students to explain how this could work.)
When a balloon is expanded (blown up) where is the centre of enlargement? (Inside the shape.)

S7.2 Fractional scale factors

This spread will show you how to:

- Enlarge objects, given a centre of enlargement and scale factor, including fractional scale factors

Keywords
Centre of enlargement
Enlargement
Scale factor

A map or a scale drawing is an enlargement by a fractional scale factor.

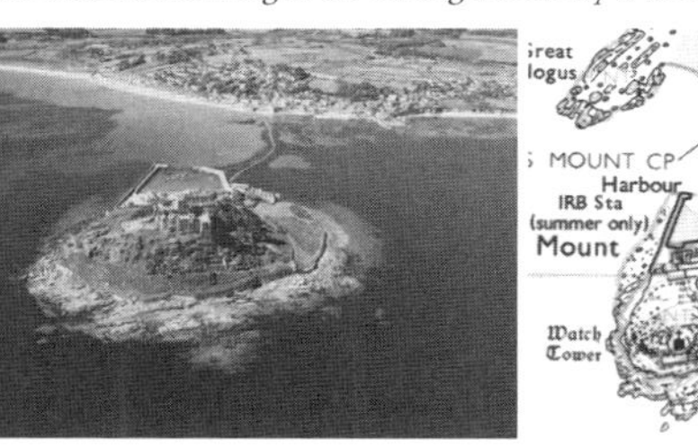

This map is an enlargement of St Michael's Mount by scale factor $\frac{1}{25000}$.

- **Enlargement by a scale factor less than 1 produces a smaller image.**

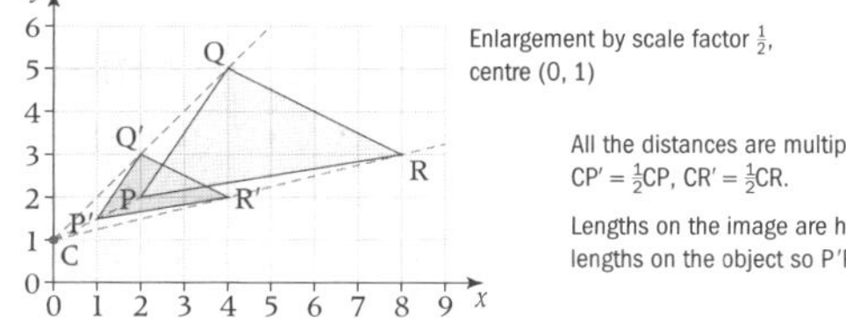

Enlargement by scale factor $\frac{1}{2}$, centre (0, 1)

All the distances are multiplied by $\frac{1}{2}$ so $CQ' = \frac{1}{2}CQ$, $CP' = \frac{1}{2}CP$, $CR' = \frac{1}{2}CR$.

Lengths on the image are half the corresponding lengths on the object so $P'R' = \frac{1}{2}PR$ and so on.

Enlargement by scale factor $\frac{1}{2}$ is the inverse of enlargement by scale factor 2.

Example

Enlarge triangle A by scale factor $\frac{1}{2}$, centre (−4, 2).

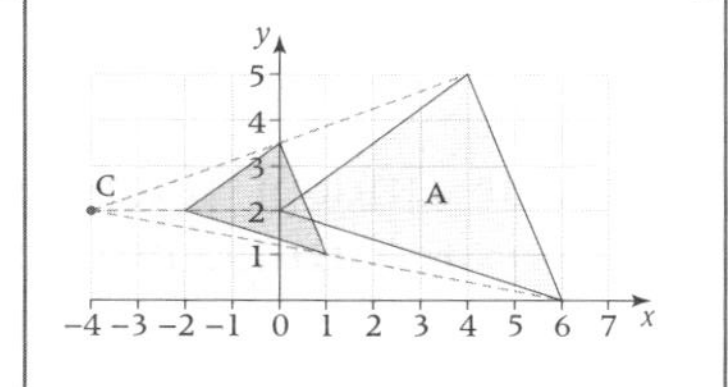

Example

Enlarge triangle B by scale factor $\frac{1}{3}$ centre (−2, −2).

Lengths on the image are $\frac{1}{3}$ corresponding lengths on the object.

Exercise S7.2

1 **a** Copy this diagram.

b Enlarge triangle T by scale factor $\frac{1}{2}$, centre (0, 0). Label the image U.

c Enlarge triangle T by scale factor $\frac{1}{2}$, centre (2, 2). Label the image V.

2 **a** Copy this diagram.

b Enlarge kite K by scale factor $\frac{1}{3}$, centre (0, 0). Label the image L.

c Enlarge kite K by scale factor $\frac{1}{3}$, centre (3, 3). Label the image M.

3 **a** Draw a grid with an x-axis from −3 to 5 and a y-axis from −2 to 10. Plot the points (1, 3) (1, 6) (3, 9) (4, 6). Join them to make quadrilateral Q.

b Enlarge quadrilateral Q by scale factor $\frac{1}{3}$, centre (4, 3). Label the image R.

c Enlarge quadrilateral Q by scale factor $\frac{1}{2}$, centre (−2, −2). Label the image S.

4 **a** Draw a grid with x-axis from −6 to 6 and y-axis from 0 to 11. Plot the points (−6, 6) (0, 6) (3, 3). Join them to make triangle A.

b Enlarge triangle A by scale factor $\frac{1}{3}$, centre (0, 0). Label the image B.

c Enlarge triangle A by scale factor $\frac{1}{2}$, centre (0, 10). Label the image C.

S7.3 Describing an enlargement

Objectives

- Describe an enlargement by giving the scale factor and centre of enlargement
- Understand, identify and use scale factors

Useful resources

- OHT of example in student book

Mental starter

An artist's painting is 4 m by 2 m. Within the picture he paints himself painting himself, painting himself etc. The painting the artist paints within each painting is always 10% of the size of the previous version. The human eye can only see to a width of 0.07 mm. **How many images of the canvas will he paint?** (5 including the original, or 4 smaller ones. Sizes are 4 m by 2 m, 40 cm by 20 cm, 4 cm by 2 cm, 4 mm by 2 mm, 0.4 mm by 0.2 mm, next size is too small.)

Introductory activity

Instead of being asked to perform an enlargement, you could be asked to describe an enlargement. To find the length of a side in the image, you multiply the length in the object by the scale factor. **How could you find the scale factor if you know corresponding lengths in the object and the image?** (Divide the length in the image by the length in the object.)

To describe an enlargement fully you must give the coordinates of the centre of enlargement. Join corresponding points in the object and the image and extend the lines until they meet. This point is the centre of enlargement.

Show this on an OHT of the first diagram.

Exercise commentary and misconceptions

In the phrase 'map P onto Q', P is the object and Q is the image. Students often reverse these, in a similar way to the problem they have with bearings. The scale factor is the fraction formed by dividing a length in the image by the corresponding length in the object.

Students should always use the word 'enlargement' and give the scale factor and the centre of enlargement. It may be helpful to remind students of the detail they had to give when describing other transformations.

In questions 1, 2 and 3 the second transformation is the inverse of the first.
The centre of enlargement is the same but the scale factor is the reciprocal.

Plenary

Describe how you would enlarge a circle, scale factor 2, with the centre of enlargement at the centre of the circle.

What difference would it make if the centre was a point on the circumference of the circle?

S7.3 Describing an enlargement

This spread will show you how to:

- Describe an enlargement by giving the scale factor and centre of enlargement
- Understand, identify and use scale factors

Keywords
Centre of enlargement
Enlargement
Scale factor

To describe an **enlargement** you give the **scale factor** and the **centre of enlargement**.

Construction lines join corresponding vertices on the object and image.

The construction lines meet at the centre of enlargement.

This is an enlargement of scale factor $\frac{1}{2}$, centre (1, 1).

The scale factor of an enlargement is the ratio of corresponding sides.

- Scale factor = $\frac{\text{length of image}}{\text{length of original}}$

You can write this as a ratio length of image : length of object

Example

Describe fully the single transformation that maps triangle PQR onto P′Q′R′.

The construction lines meet at (−3, −2)

Scale factor = $\frac{P'R'}{PR} = \frac{6}{2} = 3$.

The transformation that maps PQR onto P′Q′R′ is an enlargement, centre (−3, −2), scale factor 3.

Exercise S7.3

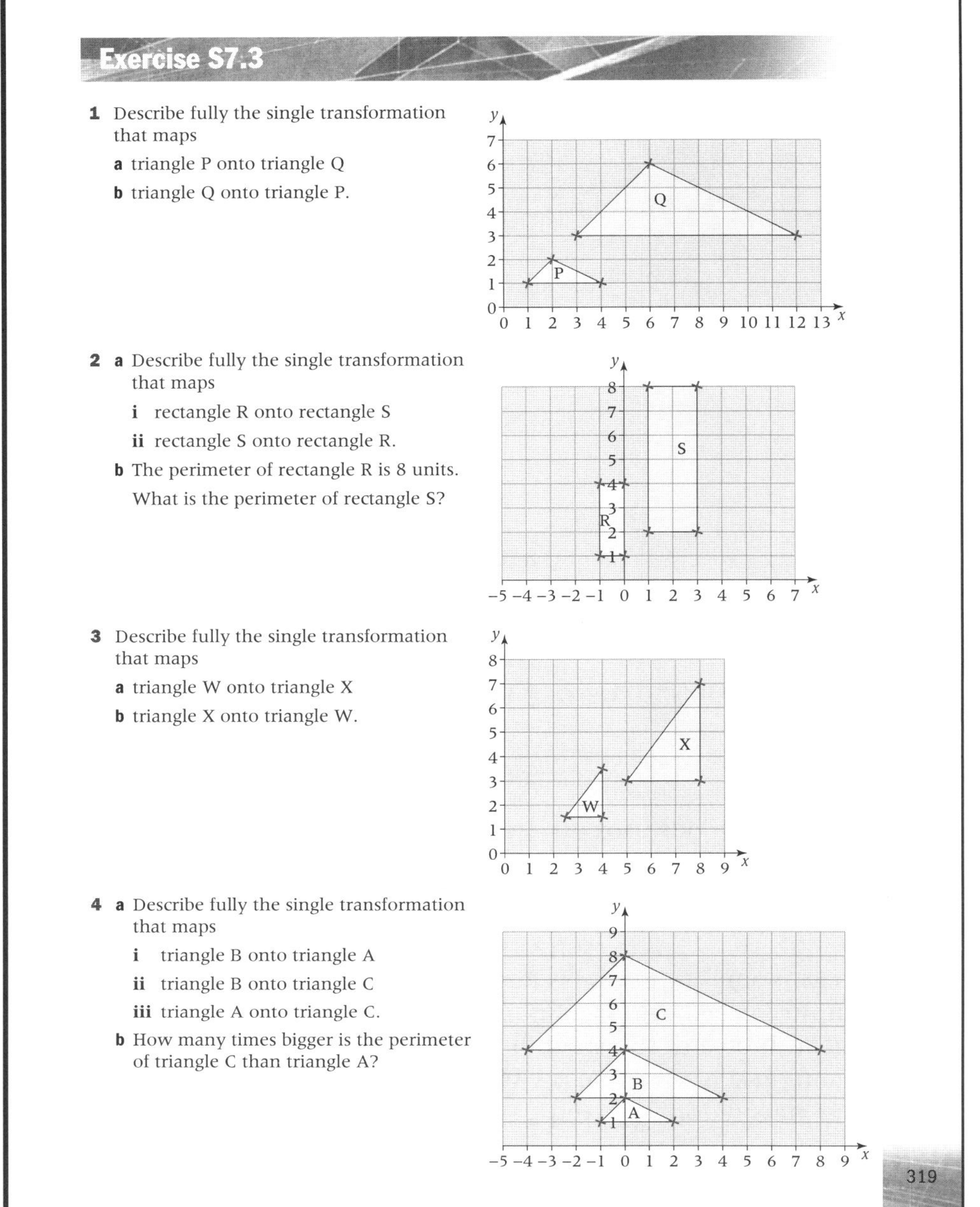

1 Describe fully the single transformation that maps

a triangle P onto triangle Q

b triangle Q onto triangle P.

2 **a** Describe fully the single transformation that maps

i rectangle R onto rectangle S

ii rectangle S onto rectangle R.

b The perimeter of rectangle R is 8 units. What is the perimeter of rectangle S?

3 Describe fully the single transformation that maps

a triangle W onto triangle X

b triangle X onto triangle W.

4 **a** Describe fully the single transformation that maps

i triangle B onto triangle A

ii triangle B onto triangle C

iii triangle A onto triangle C.

b How many times bigger is the perimeter of triangle C than triangle A?

S7.4 Similar shapes

Objectives
- Understand similarity of 2-D shapes, using this to find missing lengths and angles

Useful resources
- OHT of diagrams in student book examples

Mental starter
If I could jump a distance of 52 feet, travel at 42 mph and perform a $1\frac{2}{3}$ turn on my skateboard, I would be $\frac{2}{3}$ as good as the best pro skateboarders in the world.
What are the world records for each one? (78 feet jump, 63 mph, $2\frac{1}{2}$ turns.)

Introductory activity
What does 'similar' mean in the English language? (Variety of responses.)

In maths there is only one meaning. Two shapes are similar if they are exactly the same shape, but one shape is an enlargement of the other. (If they are the same size they are congruent.)
What does this mean about the angles in the two shapes? (Corresponding angles are equal.)
What does this mean about the lengths of the sides? (Corresponding sides are enlarged by the same scale factor.)

When solving problems with similar shapes it is easiest to find the scale factor first. Do this by finding the ratio between corresponding sides in the two shapes.

Use the examples in the student book to show how similarity can be used to find lengths.
It is important that students appreciate that the orientation of the two figures is unimportant and also that it is essential to be given the lengths of at least one pair of corresponding sides.

Emphasise that similarity must be stated, it cannot be assumed just because two figures **look** similar.

Exercise commentary and misconceptions
Questions 1 and 2 have integer scale factors, so most students will find it easy to find the scale factor then either multiply or divide by the scale factor to find enlarged or reduced lengths or sides.

Questions 3–5 have fractional scale factors so students may prefer to use a ratio method. Writing the dimensions in a contingency table, with the two figures in separate columns and corresponding sides in the same row makes it easier to use ratios.

Encourage students to check that their answers are as expected – a side in the larger figure will be larger than the corresponding side in the smaller figure.

Plenary
What other 2-D shapes are always similar to each other?
(Circles, right-angled isosceles triangles and regular polygons including squares and equilateral triangles.)

S7.4 Similar shapes

This spread will show you how to:
- Understand similarity of 2-D shapes, using this to find missing lengths and angles

Keywords
Enlargement
Ratio
Similar

In an **enlargement** the object and the image are mathematically **similar**.

- **In similar shapes**
  - **corresponding pairs of angles are equal**
  - **corresponding pairs of sides are in the same ratio.**

Any two circles are similar.
Any two squares are similar.

You can use the **ratio** between similar shapes to find missing side lengths.

Example

These quadrilaterals are similar.
Find the side lengths s and t.

Side 4.5 cm corresponds to side 9 cm.
As a fraction,
the ratio larger : smaller is $\frac{9}{4.5} = 2$ the ratio smaller : larger is $\frac{4.5}{9} = \frac{1}{2}$

Side s corresponds to side 6.4 cm.   Side t corresponds to side 2.1 cm.

$\frac{s}{6.4} = \frac{1}{2}$ $\frac{t}{2.1} = 2$

$s = \frac{1}{2} \times 6.4 = 3.2$ cm $t = 2.1 \times 2 = 4.2$ cm

Choose the ratio that gives a fraction with the side you want on top.

You need to be able to identify corresponding sides when one shape is upside down.

Example

These two pentagons are similar.
Find the side lengths x and y.

Side 3 cm corresponds to side 12 cm.
As a fraction,
the ratio larger to smaller is $\frac{3}{12} = \frac{1}{4}$ the ratio smaller to larger is $\frac{12}{3} = 4$

Side x corresponds to side 4 cm.   Side y corresponds to side 1.5 cm.

Side x is smaller. $4 \times \frac{1}{4} = 1$, $x = 1$ cm   Side y is larger. $1.5 \times 4 = 6$, $y = 6$ cm

Exercise S7.4

1 These two trapeziums are similar.
Find the lengths a and b

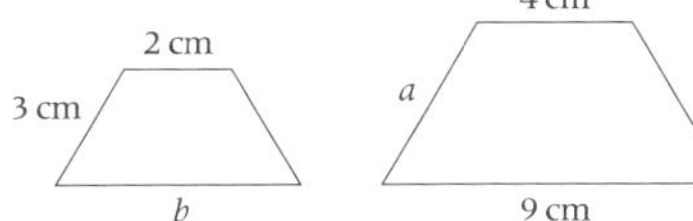

2 These two quadrilaterals are similar.
Find the lengths c and d

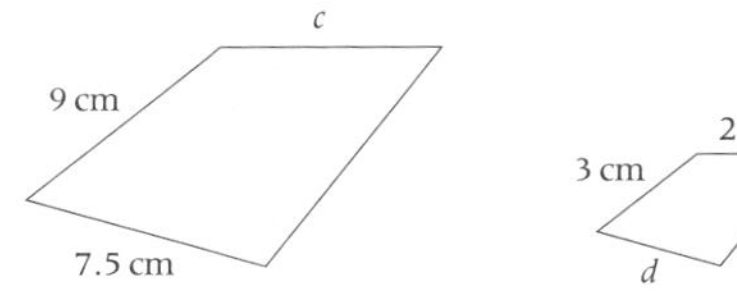

3 These two pentagons are similar.
Find the lengths e and f.

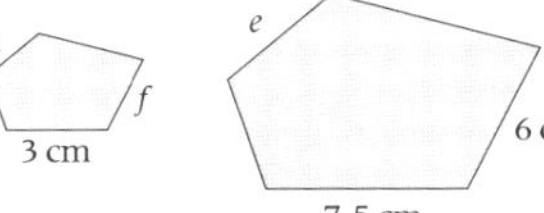

4 These two parallelograms are similar.
Find the perimeter of the smaller parallelogram.

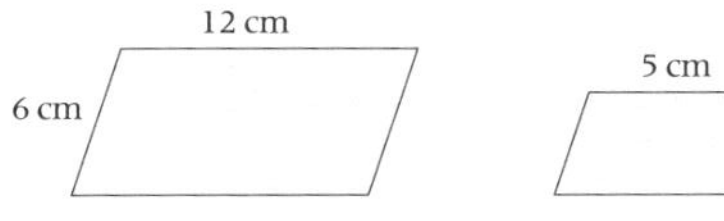

5 These two isosceles triangles are similar.

a Write the ratio
larger : smaller
for these triangles.

b Find the length x.

c Find the perimeter of each triangle.

d Write the ratio of their perimeters.
What do you notice?

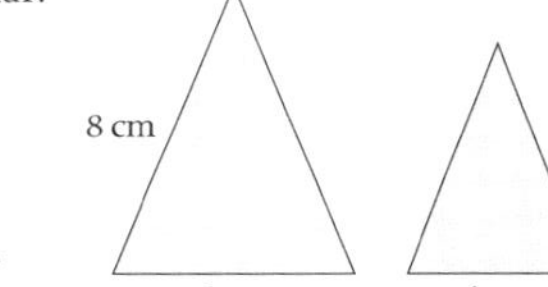

S7.5 Similar triangles

Objectives

- Understand similarity of 2-D shapes, using this to find missing lengths and angles
- Use common sense to check answers to geometric problems

Useful resources

- OHT of diagrams in student book examples

Mental starter

A map is to a scale 1 : 25 000.
Ask students to find the real distances corresponding to these map lengths:
2 cm, 5 cm, 10 cm, 12 cm.

Introductory activity

Look at the first example.
Why are triangles ACD and ABE similar?
(Corresponding angles between parallel lines.)

When comparing similar triangles in one figure, students should always draw the two triangles separately and mark the equal angles before matching up corresponding sides. A contingency table may help students to identify the ratio. When using ratios it is best to ensure that the unknown side is the numerator in one of the fraction. Emphasise to students that an equation of ratios must only contain one unknown side.

Exercise commentary and misconceptions

In all questions ensure that students sketch the two triangles separately. Then match corresponding sides. Then find the scale factor or set up a ratio equation.

Encourage students to think about their work (as always!) so that they know whether they expect the answer to be larger or smaller than the corresponding side.

It may be wise to remind students that the same methods can be used with shapes other than triangles, though triangles with a line drawn parallel to one of the sides occur most often in exams.

Plenary

What 3-D solids are always similar?
(Cubes, spheres, tetrahedrons, octahedrons, dodecahedrons and icosahedrons. Some students may have encountered these solids as the shape of unbiased die.)

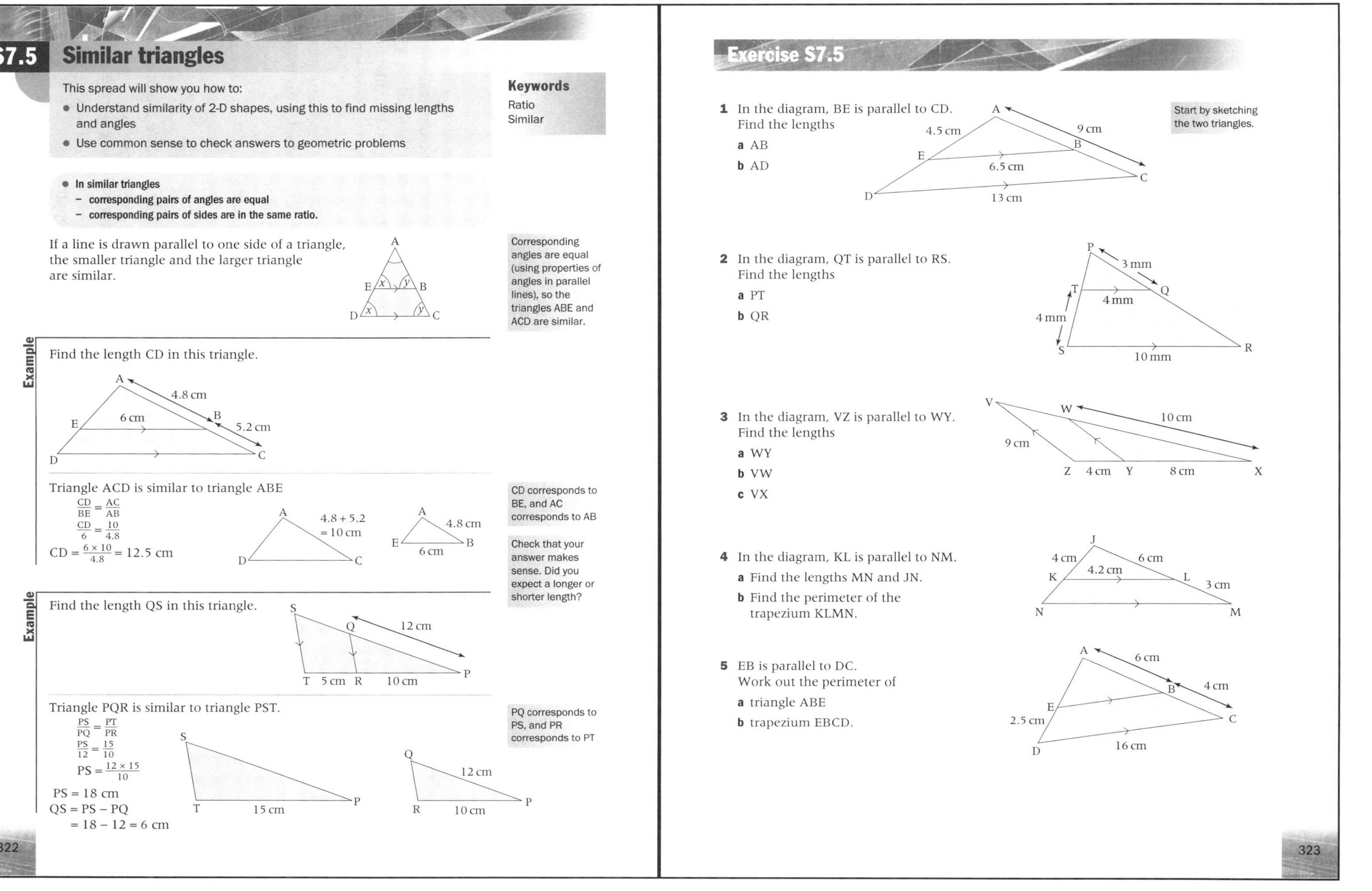

S7.5 Similar triangles

This spread will show you how to:

- Understand similarity of 2-D shapes, using this to find missing lengths and angles
- Use common sense to check answers to geometric problems

Keywords
Ratio
Similar

- **In similar triangles**
  - **corresponding pairs of angles are equal**
  - **corresponding pairs of sides are in the same ratio.**

If a line is drawn parallel to one side of a triangle, the smaller triangle and the larger triangle are similar.

Corresponding angles are equal (using properties of angles in parallel lines), so the triangles ABE and ACD are similar.

Example

Find the length CD in this triangle.

Triangle ACD is similar to triangle ABE

$\frac{CD}{BE} = \frac{AC}{AB}$

$\frac{CD}{6} = \frac{10}{4.8}$

$CD = \frac{6 \times 10}{4.8} = 12.5$ cm

CD corresponds to BE, and AC corresponds to AB

Check that your answer makes sense. Did you expect a longer or shorter length?

Example

Find the length QS in this triangle.

Triangle PQR is similar to triangle PST.

$\frac{PS}{PQ} = \frac{PT}{PR}$

$\frac{PS}{12} = \frac{15}{10}$

$PS = \frac{12 \times 15}{10}$

PS = 18 cm

QS = PS − PQ

= 18 − 12 = 6 cm

PQ corresponds to PS, and PR corresponds to PT

Exercise S7.5

Start by sketching the two triangles.

1 In the diagram, BE is parallel to CD. Find the lengths

a AB

b AD

2 In the diagram, QT is parallel to RS. Find the lengths

a PT

b QR

3 In the diagram, VZ is parallel to WY. Find the lengths

a WY

b VW

c VX

4 In the diagram, KL is parallel to NM.

a Find the lengths MN and JN.

b Find the perimeter of the trapezium KLMN.

5 EB is parallel to DC. Work out the perimeter of

a triangle ABE

b trapezium EBCD.

Exam review

Key objectives

- Recognise, visualise and construct enlargements of objects using positive fractional scale factors
- Identify the scale factor of an enlargement as the ratio of the lengths of any two corresponding line segments and understand the implications of enlargement for perimeter
- Understand similarity of triangles and other plane figures and use this to make geometric inferences

Worked solution	Commentary and misconceptions
1 a 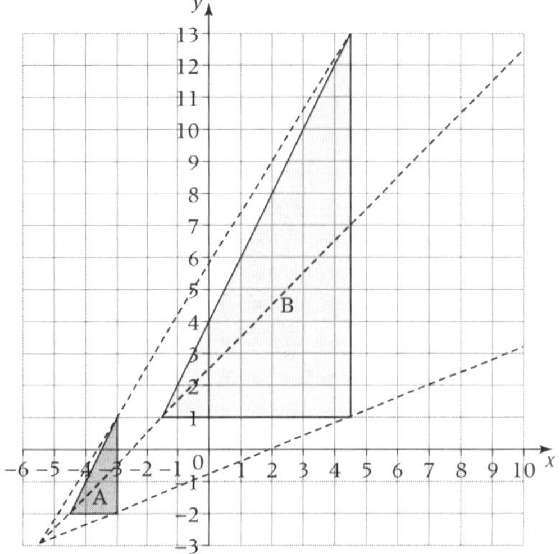	This question tests students' understanding of enlargements, including their centre and scale factor. Ensure that students draw their diagram neatly and accurately. Encourage students to label their diagram clearly and show their workings, including their construction lines.
b An enlargement with centre (−5.5, −3) and scale factor 1/4 maps B back to A.	Students should be able to answer this question without any further working. Ensure that students understand how to reverse an enlargement.
2	Students should recognise that the triangles are similar and understand the implication this has on their side lengths. Ensure that students read the questions carefully and find the correct side length in each case.
a $\frac{\text{length } AB}{\text{length } AC} = \frac{4.5}{6} = \frac{3}{4}$ Triangles ADE and ABC are similar, so $\frac{\text{length } AD}{\text{length } AE} = \frac{3}{4}$ so length AD $= (6 + 4) \times \frac{3}{4} = 7.5$ cm	This is one possible method. Students could use different ratios, for example $\frac{\text{length } BD}{\text{length } AB} = \frac{\text{length } CE}{\text{length } AC}$. Ensure that students state the units of their answer correctly.
b Similarly, $\frac{\text{length } BD}{\text{length } AC} = \frac{\text{length } DE}{\text{length } AE} = \frac{5}{10}$ so length BC $= 6 \times \frac{5}{10}$ So length BC = 3 cm.	Again, students could use different ratios, for example $\frac{\text{length } BC}{\text{length } DE} = \frac{\text{length } AC}{\text{length } AE}$ Ensure that students state the units of their answer correctly.

D6 Tree diagrams

Objectives

F Understand and use the probability scale

F/H Identify different mutually exclusive outcomes and know that the sum of the probabilities of all these outcomes is 1

H Know when to add or multiply two probabilities: if *A* and *B* are mutually exclusive, then the probability of *A* or *B* occurring is P(*A*) + P(*B*), whereas if *A* and *B* are independent events, the probability of *A* and *B* occurring is P(*A*) × P(*B*)

F/H Understand and use estimates or measures of probability from theoretical models, or from relative frequency

H Use tree diagrams to represent outcomes of compound events, recognising when events are independent

Unit overview

This unit consolidates the knowledge of probability and understanding of processing, representing and interpreting data gained in unit D3 and the other previous data units. It covers mutually exclusive and independent events, focusing on drawing and using tree diagrams.

Prior knowledge

Before your students start this unit they should be able to:

- Add, subtract, multiply and divide integers, decimals and fractions
- Understand and use the probability scale
- Understand and use the vocabulary and notation related to probability
- Understand and use estimates or measures of probability from theoretical models, or from relative frequency
- Understand and use tables, charts and diagrams to represent data

Differentiation

- **Foundation** consolidates previous knowledge of probability and using two-way tables. It then focuses on expected and relative frequency and estimates of probability before extending to calculating the probability of two events, listing outcomes in a systematic way
- **Foundation Plus** extends to mutually exclusive events and random sampling
- **Higher Plus** extends the Higher book to sampling with or without replacement and conditional probability

D6.1 Probability revision

Objectives

- Understand and use the probability scale
- Find probabilities of mutually exclusive events
- Calculate theoretical probabilities

Useful resources

- Spinners, dice and coins

Mental starter

Ask students to find the complement of 1 for a variety of proper fractions, for example: $\frac{12}{90}$ $(1 - \frac{12}{90} = \frac{78}{90} = \frac{13}{15})$.
Students should give their answer in its simplest form.

Introductory activity

Recap probability.
Recap the term 'mutually exclusive'.

Consider the example:
Event A is 'rolling a 6 on a dice'.
Event B is 'rolling a 1 on a dice'.
Discuss why these events are mutually exclusive.
Now consider the example:
Event A is 'it will rain tomorrow'.
Event B is 'the sun will shine tomorrow'.
Discuss why these events may not be mutually exclusive.

Focus on the key point:
When you have mutually exclusive events you can add their probabilities together.

Exercise commentary and misconceptions

The questions focus on solving simple probability questions based on earlier work in section D3.
Ensure students write probability as fractions (simplified) or decimals, whichever is the simpler.
Calculate expectation using decimal multiplication, finding the fraction or ratio, whichever method the student is most comfortable with.

Plenary

A spinner has 4 sections labelled, win, lose, lose and draw. The spinner has an equal probability of landing in any one of the 4 sections. You can pay £2 to spin the spinner. If you get 'win' you win your money back and an extra £3. If you get 'draw' you get your money back. If you get 'lose' you get nothing back. **Would you play?** Suppose you played 4 times. You would expect on average to lose twice, draw once and win once. This would cost you £8 and you would win back £7. So it is not worth the risk.

D6.1 Probability revision

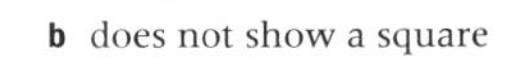
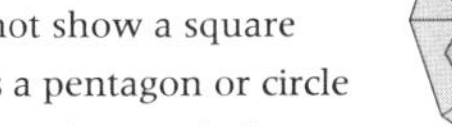
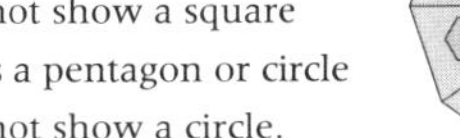

This spread will show you how to:

- Understand and use the probability scale
- Find probabilities of mutually exclusive events
- Calculate theoretical probabilities

Keywords
Event
Expected number
Mutually exclusive
Outcome

You write a probability as a decimal, fraction or percentage.

- **Probability takes a value from 0 to 1.**
- **Probability of an outcome = $\frac{\text{number of favourable outcomes}}{\text{total number of outcomes}}$.**

For an **event** the total probability for all possible outcomes = 1.

- **For an event A, P(not A) = 1 – P(A)**

P(A) means the probability of outcome A.

Two or more events are **mutually exclusive** if they cannot happen at the same time. You use the addition rule to find the probability of mutually exclusive events.

Sometimes called the OR rule.

- **For two mutually excusive events A and B, P(A or B) = P(A) + P(B).**
- **Expected number = total number of outcomes × probability of outcome.**

Example

A spinner has 8 sides, of which 4 show squares, 3 show triangles, and 1 shows a circle.

a The spinner is spun once. Find the probability that the spinner

i shows a triangle
ii does not show a triangle
iii does not show a circle
iv shows a square or a circle
v shows a triangle or a square.

b The spinner is spun 400 times. Find the number of times the spinner is expected to show

i a circle **ii** a triangle.

a i P(triangle) = $\frac{3}{8}$
ii P(not triangle) = $1 - \frac{3}{8} = \frac{5}{8}$
iii P(not circle) = $1 - \frac{1}{8} = \frac{7}{8}$
iv P(circle or square) = P(circle) + P(square) = $\frac{1}{8} + \frac{4}{8} = \frac{5}{8}$
v P(triangle or square) = P(triangle) + P(square) = $\frac{3}{8} + \frac{4}{8} = \frac{7}{8}$
b i Expected number of circles = 400 P(circle) = $400 \times \frac{1}{8} = 50$
ii Expected number of triangles = 400 × P(triangle) = $400 \times \frac{3}{8} = 150$

Exercise D6.1

1 A spinner has ten equal sides, of which 4 show squares, 3 show pentagons, 2 show hexagons and 1 shows a circle. The spinner is spun once. Find the probability that the spinner

a shows a square **b** does not show a square
c shows a square or pentagon **d** shows a pentagon or circle
e does not show a hexagon **f** does not show a circle.

2 A fair coin is thrown 482 times.
How many times would you expect the coin to land on tails?

3 An ordinary fair dice is rolled 612 times.
How many times would you expect it to land on 6?

4 In a class of 28 students, 15 are girls and 13 are boys. 22 students wear glasses. One student is chosen at random.
What is the probability that this student

a is a boy **b** is a girl
c wears glasses **d** does not wear glasses?

5 A spinner has 16 equal sides. The table shows the colours of the sides and the shapes drawn on them. The spinner is spun once.

	Circle	Triangle	Square
Red	3	5	1
Black	4	1	2

a Find the probability that the spinner

i shows red **ii** shows a triangle
iii shows a circle or triangle **iv** does not show a circle
v shows a red triangle **vi** shows a black square.

b Copy and complete this sentence

'The probability of the spinner landing on the red square is the same as the probability of the spinner landing on ______ ______'

6 A fair eight-sided dice has equal sides. Two sides show 2s, three sides show 3s and three sides show 5s.

a The dice is thrown once. Find the probability that the dice shows

i an even number **ii** an odd number
iii a prime number **iv** a multiple of 4.

b The dice is thrown 256 times.
How many times would you expect the dice to land on 2?

D6.2 Independent events

Objectives
- Calculate probabilities of independent events

Useful resources
- Mini-whiteboards for mental starter
- Spinners, dice and coins

Mental starter
Ask students to find products of fractions, for example:
$\frac{1}{2} \times \frac{5}{12}$, $\frac{1}{3} \times \frac{6}{7}$, $\frac{2}{5} \times \frac{3}{4}$
Encourage students to simplify their answer.

Introductory activity
Sometimes the outcome of one event has no effect on the outcome of another event. These events are said to be independent.

Sometimes the outcome of one event does effect the outcome of another event, and in this case the events are said to be conditional or dependent.

If I throw a dice and get a six, and then flip a coin and get tails, **do you think these events are independent or does one event effect the other?** (Independent, because getting a six doesn't make getting tails more or less likely.)

Suppose next Monday it rains and I decide to come to school by car, do you think these two events are independent? (No, I am more likely to use a car if it rains than if it is dry, so one event is conditional on the other.)

Highlight the key point:
For independent events you can find the probability that 2 things happen by multiplying the probabilities together.

P(roll a six and throw a tail) $= \frac{1}{6} \times \frac{1}{2} = \frac{1}{12}$.
P(roll 2 sixes) $= \frac{1}{6} \times \frac{1}{6} = \frac{1}{36}$.

Exercise commentary and misconceptions
Question 1**b** can be done using the table drawn or by multiplying the probabilities. Students may like to check that the two methods give the same answer.

Question 2 can be done the same way as question 1.

The table which was drawn for question 2 could also be used for question 3.

If students want to use a table for question 4, they need to have 10 columns for the spinner and mark four of them with squares, three with pentagons, two with hexagons and one with a circle. This is tedious, but does show that each side has to be considered an equally likely outcome.

Discourage students from drawing a table of outcomes for question 5, but point out that 100 cells would be needed!

Plenary
Discuss how you might evaluate the probability of more that two events. Offer a concrete example, perhaps with a coin, a dice and a spinner.

D6.2 Independent events

This spread will show you how to:

- Calculate probabilities of independent events

Keywords
Event
Independent
Outcome

- **Two or more events are Independent if the outcome of one event has no effect on the outcome of the other.**

When a dice is rolled and a coin is thrown, the number rolled on the dice has no effect on whether the coin lands on a head or a tail. The events are independent.

When two or more coins are thrown, a head or tail on the first coin has no effect on what shows on any of the other coins. All the events are independent.

You use the multiplication rule to find the probability of two or more independent events.

- **For two independent events A and B, P(A and B) = P(A) × P(B)**

Sometimes called the AND rule.

Example

A spinner has 12 equal sides: five green, four blue, two red and one white.

A fair coin is thrown and the spinner is spun.

a What is the probability of getting
i a head and green **ii** a tail and white
iii a tail and red **iv** a head and red?

b Why are the answers to **iii** and **iv** the same?

a i $P(H) = \frac{1}{2}$ $P(\text{green}) = \frac{5}{12}$
$P(\text{head and green}) = P(H) \times P(\text{green}) = \frac{1}{2} \times \frac{5}{12} = \frac{5}{24}$

ii $P(T) = \frac{1}{2}$ $P(\text{white}) = \frac{1}{12}$
$P(\text{head and white}) = P(H) \times P(\text{white}) = \frac{1}{2} \times \frac{1}{12} = \frac{1}{24}$

iii $P(T) = \frac{1}{2}$ $P(\text{red}) = \frac{2}{12}$
$P(\text{head and red}) = P(H) \times P(\text{red}) = \frac{1}{2} \times \frac{2}{12} = \frac{2}{24}$

iv $P(H) = \frac{1}{2}$ $P(\text{red}) = \frac{2}{12}$
$P(\text{head and red}) = P(H) \times P(\text{red}) = \frac{1}{2} \times \frac{2}{12} = \frac{2}{24}$

P(H) = P(T)

b The answers to **iii** and **iv** are the same since the coin is fair.

Example

A dice is rolled twice.

Find the probability that on the first roll the dice shows a four and on the second roll the dice shows an odd number.

$P(4) = \frac{1}{6}$ $P(\text{odd number}) = \frac{3}{6}$ $P(4, \text{odd number}) = \frac{1}{6} \times \frac{3}{6} = \frac{3}{36}$

Exercise D6.2

1 A fair coin is thrown and an ordinary dice is rolled.

a Copy and complete the table to list all the outcomes. One has been done for you.

Coin/Dice	1	2	3	4	5	6
Head		H2				
Tail						

b Find the probability of getting
i a head and a 2 **ii** a tail and a 4 **iii** a tail and a 5.

2 A red dice and a blue dice are rolled.

a Draw a table to show all the possible outcomes.

b Find the probability of getting
i 6 on the red dice and 6 on the blue dice
ii 3 on the red dice and 5 on the blue dice
iii 3 on the red dice and an odd number on the blue dice
iv 5 or greater on the red dice and 1 on the blue dice.

3 An ordinary fair dice is rolled twice. Find the probability that the dice shows an even number on the first roll and a number greater than 4 on the second roll.

4 A spinner has ten equal sides: 4 show squares, 3 show pentagons, 2 show hexagons and 1 shows a circle. A fair coin is thrown and the spinner is spun. Find the probability of getting

a a square and a head **b** a circle and a head
c a pentagon and a tail **d** a hexagon and a tail
e a circle and a tail.

5 A spinner has ten equal sides. 4 show squares, 3 show pentagons, 2 show hexagons and 1 shows a circle. The spinner is spun twice. Find the probability of getting

a a circle on the first spin and a circle on the second spin
b a circle on the first spin and a hexagon on the second spin
c a triangle on the first spin and a square on the second spin
d a square on the first spin and a circle on the second spin
e a hexagon on first spin and a pentagon on second spin.

6 A 10 pence coin and a 2 pence coin are spun.

a Draw a table to show all the possible outcomes.

b Find the probability that the 10p shows tails and the 2p shows heads.

7 One coin is spun twice. Find the probability that both times the coin shows heads.

D6.3 Drawing tree diagrams

Objectives
- Draw and use tree diagrams for the outcomes of several events

Useful resources
- Blank tree diagram template

Mental starter
A clown brings 20 balloons to a party.
The probability a balloon pops when being blown up is 0.01. There are 18 children at the party. **Should the clown feel confident that every child will get a balloon?** (Yes, unlikely one will pop and extremely unlikely three will pop and leave only 17 balloons.)

Introductory activity
When you list all possible outcomes of a series of events, especially when all the events are not equally likely, a good way to do this is on a probability tree.
Decide what possible outcomes there are for the first event and mark each of these on a branch with the probability that it will happen.
Do the same for the second event, repeating the set of outcomes at the end of each of the branches.

Draw the tree diagram of 7 yellow and 3 blue marbles used as the example in the students' book, explaining the technique as you draw it.

Repeat for another example, question 2 from exercise D6.3, asking students to suggest the branches and the probabilities. This is much easier using a template to show where to write the outcomes and where to write the probabilities. Explain carefully that the values on the branches show the probability of that event occurring if the previous event has occurred, not the overall probability of the previous and current events occurring.

Note that in this session, no probabilities are calculated.

Exercise commentary and misconceptions
The first three questions are all based on similar scenarios, so should not cause any problems.

In question 4, some students may wish to make the first set of outcomes 'getting a tail with the 10 pence coin' and 'getting a tail with the 2 pence coin' and in this case the probabilities would indeed add to '1'. Pointing out that there would be a problem with three coins should help in persuading them to use each set of branches for a single coin.

Trees look much better, and will be easier to use in subsequent sessions, if they are drawn with a ruler.

Students will need the diagrams they draw for questions 2, 3 and 4 in the next session.

Plenary
Using the two tree diagrams drawn in the introduction, write the combined outcomes of the two events alongside the ends of the second set of branches.
The upper branch of the first diagram will then be '2 yellow marbles'.

D6.3 Drawing tree diagrams

This spread will show you how to:
- Draw and use tree diagrams for the outcomes of several events

Keywords
Random
Replaced
Tree diagram

- **You can use a tree diagram to show the possible outcomes of two events.**

- **In a tree diagram,**
  - **write the outcomes at the end of each branch**
  - **write the probability on each branch**
  - **the probabilities on each set of branches should add to 1.**

Example

A bag contains 7 yellow and 3 blue marbles.

A marble is chosen at **random** from the bag, its colour is noted and then it is **replaced** in the bag.

The bag is shaken and then a second marble is chosen at random.

Draw a tree diagram to show all the possible outcomes.

Label the sets of branches with each choice.

First choice **Second choice**

First set of branches for first choice.

$\frac{7}{10}$ Yellow — $\frac{7}{10}$ Yellow, $\frac{3}{10}$ Blue

$\frac{3}{10}$ Blue — $\frac{7}{10}$ Yellow, $\frac{3}{10}$ Blue

Second sets of branches for second choice.

Note each branch from first choice must have its own set of second choice branches.

Write the probability on each branch.

Write the outcome at the end of each branch.

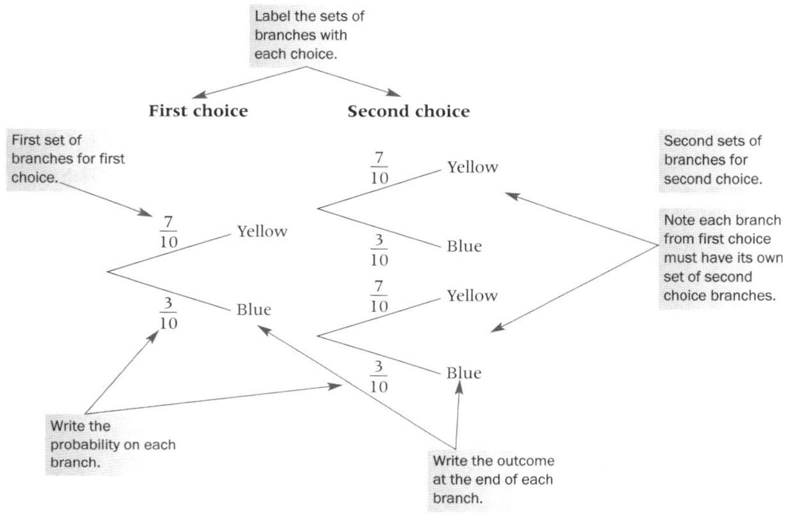

Exercise D6.3

You will need the tree diagrams drawn for questions **2**, **3** and **4** for Exercise D6.4.

1 A bag contains 6 purple counters and 11 orange counters. A counter is chosen at random from the bag, its colour noted and then it is replaced in the bag. The bag is shaken and a second counter is chosen at random. Copy and complete the tree diagram to show all the possible outcomes.

First choice **Second choice**

___ Purple — ___ Purple, ___ Orange

___ Orange — ___ Purple, ___ Orange

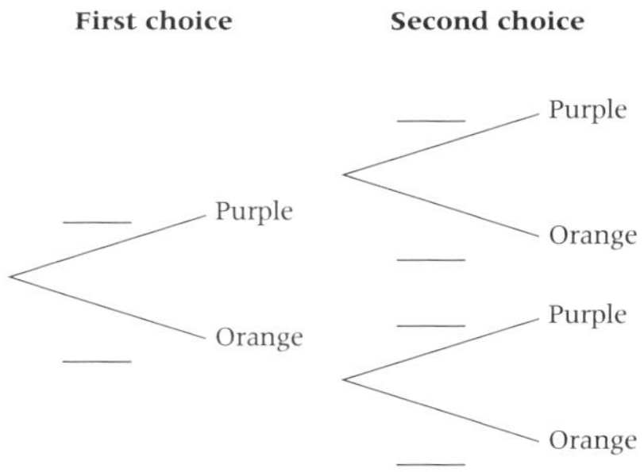

2 A bag contains 4 green and 5 red marbles. A marble is chosen at random from the bag, its colour noted and then it is replaced in the bag. The bag is shaken and a second marble is chosen at random. Draw a tree diagram to show all the possible outcomes.

3 A bag contains 3 white counters and 8 black counters. A counter is chosen at random from the bag, its colour noted and then it is replaced in the bag. The bag is shaken and then a second counter is chosen at random. Draw a tree diagram to show all the possible outcomes.

4 A 10 pence and a 2 pence coin are thrown. Draw a tree diagram to show all the possible outcomes.

5 A bag contains 2 white and 5 black counters. A counter is chosen from the bag and a fair coin is thrown. Draw a tree diagram to show all the possible outcomes.

6 Dave owns 13 CDs. Three of the CDs are by his favourite group, Cloudplay. Dave chooses one of the CDs at random, notes whether it is a Cloudplay CD, and replaces it. He then chooses another one of the 13 CDs at random. Copy and complete the probability tree diagram that has been started.

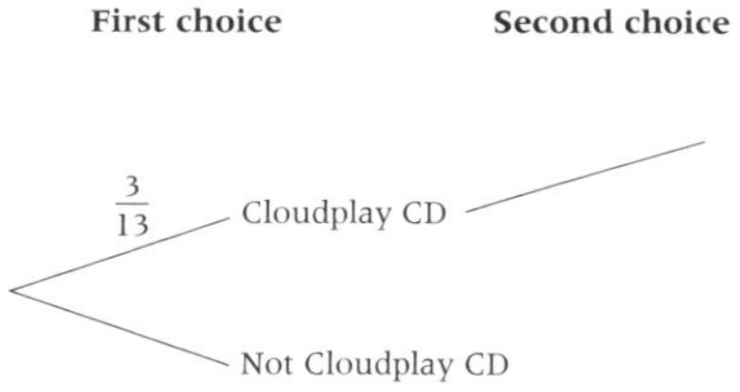

First choice **Second choice**

$\frac{3}{13}$ Cloudplay CD

Not Cloudplay CD

D6.4 Using tree diagrams to find probability

Objectives

- Draw and use tree diagrams for the outcomes of several events

Useful resources

- OHTs of the tree diagrams for 'yellow and blue marbles' and 'green and red marbles' used in D6.3
- Students' tree diagrams drawn in D6.3

Mental starter

Ask students to multiply products of decimal smalller than 1, for example:

0.6×0.4 $\quad$ 0.3×0.8

0.25×0.6 $\quad$ 0.15×0.9

Introductory activity

Recap the previous lesson (D6.3).
The students' book revisits the scenario portrayed in the earlier lesson, of the bag containing 7 yellow and 3 blue marbles.

Using an OHT of 'yellow and blue marbles' calculate the probabilities of the four combined outcomes, YY, YB, BY and BB.

Repeat with an OHT of 'green and red marbles', as outlined in question 2 of the exercise.

If you trace the path along the branches, you multiply the probabilities together.

Exercise commentary and misconceptions

If the tree diagrams were correctly drawn in the previous session, students should find it easy to calculate the probabilities.

Question 5 is the first tree diagram where the probabilities are different on the two sets of branches.

This session is good revision for multiplying fractions – remind students of the need to simplify, preferably by cancelling before multiplying. Every answer should be less than 1, because they are calculating probabilities.

Plenary

Question 4 can be done with a table, as in D6.2, or with a probability tree. Discuss the advantages and disadvantages of the two methods.

D6.4 Using tree diagrams to find probability

This spread will show you how to:

- Draw and use tree diagrams for the outcomes of several events

Keywords
Tree diagram

- **You can use a tree diagram to find the probability of an outcome of an event.**

- **To find the probability of an outcome, multiply the probabilities along the branches leading to that outcome.**

Example

A bag contains 7 yellow and 3 blue marbles.

A marble is chosen at random from the bag, its colour noted and it is replaced in the bag.

The bag is shaken and then a second marble is chosen at random.

The tree diagram shows all the possible outcomes.

First choice **Second choice**

$\frac{7}{10}$ Yellow — $\frac{7}{10}$ Yellow, $\frac{3}{10}$ Blue

$\frac{3}{10}$ Blue — $\frac{7}{10}$ Yellow, $\frac{3}{10}$ Blue

Use the tree diagram to find the probability of choosing

a two yellow marbles
b two blue marbles
c a yellow marble then a blue marble, in that order.

a $P(YY) = \frac{7}{10} \times \frac{7}{10} = \frac{49}{100}$
$= \frac{49}{100}$

b $P(BB) = \frac{3}{10} \times \frac{3}{10} = \frac{9}{100}$
$= \frac{9}{100}$

c $P(YB) = \frac{7}{10} \times \frac{3}{10} = \frac{21}{100}$
$= \frac{21}{100}$

Exercise D6.4

1 A bag contains 3 green and 4 white marbles. A marble is chosen at random from the bag, its colour noted and then it is replaced in the bag. The bag is shaken and a second marble is chosen at random. The tree diagram shows all the possible outcomes. Use the tree diagram to find the probability of choosing

a two green marbles

b two white marbles

c a white marble then a green marble in that order.

First choice **Second choice**

$\frac{3}{7}$ Green — $\frac{3}{7}$ Green, $\frac{4}{7}$ White

$\frac{4}{7}$ White — $\frac{3}{7}$ Green, $\frac{4}{7}$ White

2 A bag contains 4 green and 5 red marbles. A marble is chosen at random from the bag, its colour noted and then it is replaced in the bag. The bag is shaken and a second marble is chosen at random. Use the tree diagram from Exercise D6.3 question **2** to find the probability of choosing

a two green marbles **b** two red marbles.

3 A bag contains 3 white counters and 8 black counters. A counter is chosen at random from the bag, its colour noted and then it is replaced in the bag. The bag is shaken and a second counter is chosen at random. Use the tree diagram from Exercise D6.3 question **3** to find the probability of choosing

a two white counters **b** two black counters.

4 A 10 pence and a 2 pence coin are thrown. Use the tree diagram from Exercise D6.3 question **4** to find the probability of getting

a two heads **b** two tails **c** a head on the 10p and a tail on the 2p.

5 Bag A contains 5 red counters and 7 black counters. Bag B contains 2 yellow counters and 8 red counters. A counter is chosen from each bag.

a Copy and complete the tree diagram to show the possible outcomes.

b Find the probability of choosing

i a black and a yellow counter

ii two red counters.

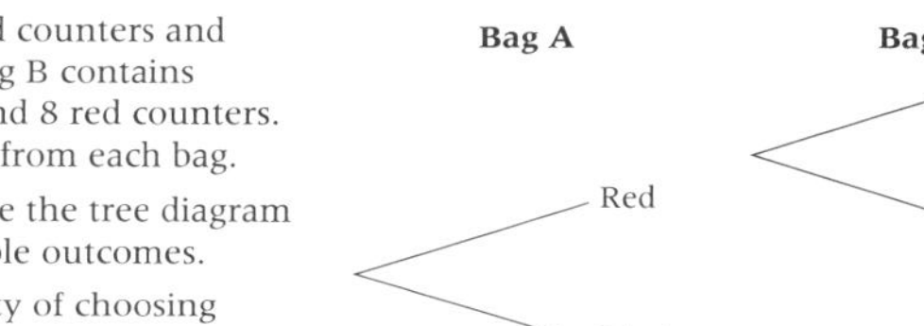
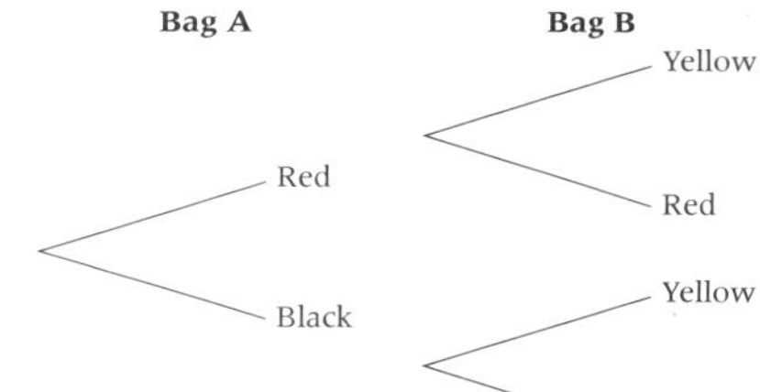

D6.5 Using tree diagrams to find harder probabilities

Objectives

- Use a tree diagram to calculate the probability of an event that can happen in more than one way

Useful resources

- OHTs as in previous lesson

Mental starter

Ask students to add fractions smaller than 1, for example:

$\frac{2}{5}+\frac{1}{5}$, $\frac{3}{4}+\frac{1}{8}$, $\frac{1}{3}+\frac{2}{5}$, $\frac{3}{7}+\frac{1}{4}$

Extend to adding decimals smaller than 1.

Introductory activity

Remind students how to add fractions, that they need to have the same denominator. When calculating probabilities with tree diagrams, it is best **not** to cancel or simplify fractions when multiplying as this makes the addition easier. Students should simplify fractions which are used in the final answer though.

Using the OHT from the previous lesson, highlight the fact that two different routes through the branches give the same result of one yellow and one blue marble. To find the probability of choosing one yellow and one blue marble they must add the two probabilities at the ends of the branches together.

Reinforce this point with the red and green marble problem from the previous lesson.

Emphasise that students will always need to multiply probabilities when tracing outcomes along branches, but that they will add when finding the probability of an outcome that can happen in more than one way.

Remind students that no probability answer can be greater than 1.

Exercise commentary and misconceptions

Students should be familiar with multiplying both decimals and fractions from the previous lesson. Encourage students to consider the reasonableness of their answer, especially if it is over 1!

In question 1 (and later questions) emphasise the difference between '1' and 'at least 1'. Encourage students to think what outcome is excluded from 'at least 1' as this gives the alternative way of calculating.
P(at least one) = 1 – P(none).

In all the questions, there are two events; two cars in question 1, two counters in question 2 and so on.

Plenary

Imagine a bag with three red and two blue balls in it. I take out 2 balls.
What is the probability they are the same colour?

This is a very tough question without a tree diagram. Notice that this time the probabilities of the colour of the second ball depend on what is picked for the first ball; this is dependent or conditional probability. You can still answer the question as long as you take care to change the probabilities for the second ball.

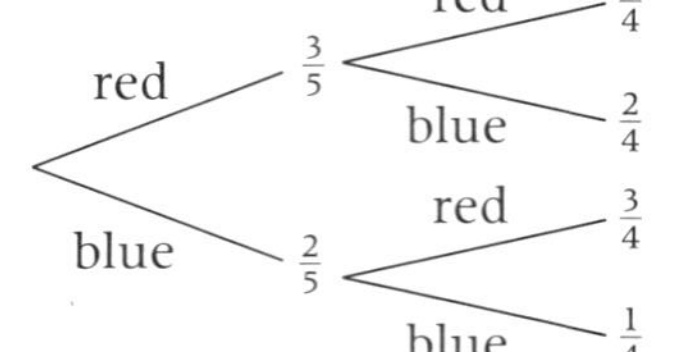

P(both are red) $= \frac{3}{5}\times\frac{2}{4}=\frac{6}{20}$.

P(both are blue) $= \frac{2}{5}\times\frac{1}{4}=\frac{2}{20}$.

P(both are same colour) $= \frac{6}{20}+\frac{2}{20}=\frac{8}{20}=\frac{2}{5}$.

D6.5 Using tree diagrams to find harder probabilities

This spread will show you how to:

- Use a tree diagram to calculate the probability of an event that can happen in more than one way

Keywords
Event
Independent
Mutually exclusive
Outcome

You can use a tree diagram to find the probability of an event that can happen in more than one way.

- **To find probabilities when an event can happen in different ways**
  - **multiply the probabilities along the branches**
  - **add the probabilities for the different ways of getting the chosen event.**

Example

Pamela makes two pottery vases.

Each vase is made independently.

The probability that a vase cracks while it is in the kiln is 0.2.

a Find the probability that the vase does not crack while it is in the kiln.
b Draw a tree diagram to show all the possible outcomes for two vases.
c Find the probability that one of the vases will crack while it is in the kiln.

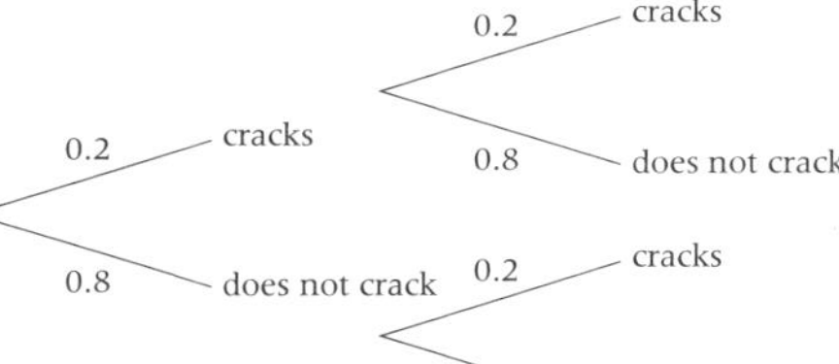

a P(does not crack) = 1 − P(cracks)
P(cracks) = 0.2
So, P(does not crack) = 1 − 0.2 = 0.8

The outcomes 'cracks' and 'does not crack' are **mutually exclusive**.

b **First Vase** **Second Vase**

First vase: 0.2 cracks; 0.8 does not crack
- cracks → 0.2 cracks; 0.8 does not crack
- does not crack → 0.2 cracks; 0.8 does not crack

The state of the first vase has no effect on the state of the second vase – the events are **independent**.

c P(one will crack) = P(cracks, does not crack) + P(does not crack, cracks)
= (0.2 × 0.8) + (0.8 × 0.2)
= 0.16 + 0.16
= 0.32

Two ways of getting one cracked vase.

Exercise D6.5

1 Batteries are placed in two toy cars, a red car and a blue car. The probability that a battery lasts for more than 20 hours is 0.8.

a Find the probability that a battery does not last for more than 20 hours.

b Draw a tree diagram to show the outcomes for the batteries in the two cars.

c Find the probability that the batteries last more than 20 hours

i in both cars **ii** in only one car

iii in at least one car.

2 A bag contains 6 white and 2 black counters. A counter is chosen at random from the bag, its colour noted and then it is replaced. A second counter is chosen at random.

a Draw a tree diagram to show all the outcomes of choosing two counters.

b Find the probability of choosing

i two black counters **ii** one counter of each colour

iii at least one black counter.

3 Two delicate glasses are placed in a dishwasher. The probability that a glass breaks in the dishwasher is $\frac{1}{20}$.

a Find the probability that a glass does not break in the dishwasher.

b Draw a tree diagram to show all the outcomes for the two glasses.

c Find the probability that while in the dishwasher

i neither glass breaks **ii** only one glass breaks

iii at least one glass breaks.

4 A spinner has five equal sectors. Three are coloured orange and two are coloured black. The spinner is spun twice.

a Draw a tree diagram to show all the outcomes of two spins on the spinner.

b Find the probability that on two spins the spinner lands

i on black both times **ii** on black at least one time

iii once on each colour.

5 Josh makes two model aeroplanes, a grey plane and an orange plane. He flies both of them. The probability that one crashes is 0.1.

a Find the probability that a model aeroplane does not crash.

b Draw a tree diagram to show all the outcomes for the two model aeroplanes.

c Find the probability that both model aeroplanes will crash.

d Find the probability that only one of the aeroplanes will crash.

D6 Exam review

Key objectives

- Identify different mutually exclusive outcomes and know that the sum of the probabilities of all these outcomes is 1
- Know when to add or multiply two probabilities
- Use tree diagrams to represent outcomes of compound events, recognising when events are independent

Worked solution	Commentary and misconceptions
1	This question tests students' understanding of probability, the probability scale and mutually exclusive events.
a **i** P(red ball) $= \frac{2}{12} = \frac{1}{6}$	Ensure that students show their working, explaining their reasoning. Encourage students to express their answer in its simplest form.
ii P(not a red ball) $= 1 -$ P(red ball) $= 1 - \frac{1}{6} = \frac{5}{6}$	Encourage students to use this formula directly rather than calculate all the individual probabilities.
iii P(red ball or white ball) = P(red ball) + P(white ball) $= \frac{1}{6} + \frac{3}{12} = \frac{5}{12}$	Students should recall this formula. Encourage students to write down this intermediate step.
b P(purple ball) = 0 There are no purple balls in the bag so it is impossible for Anna to pick a purple ball and so the probability of this event is 0.	This question tests students' basic understanding of the probability scale. Ensure that they can recall this fact directly. They must explain their reasoning and understand that impossible events have a probability of 0 whilst certain events have a probability of 1.
2 **a** The expected number of sixes from a fair dice would be 100. This suggests that the dice is not fair, however it is impossible to know for sure due to the unpredictable nature of such experiments.	Students should be familiar with expected frequency and comparing experimental and theoretical results. They should also know how to draw tree diagrams. Ensure students answer the question, giving a coherent explanation.
b (tree diagram below)	Ensure students know that each branch should add to 1. They can use this fact to check their diagram.

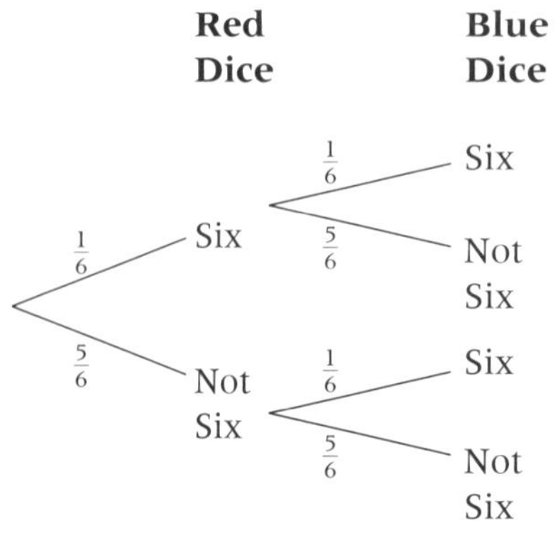

b

Red Dice → Blue Dice

- Six ($\frac{1}{6}$)
  - Six ($\frac{1}{6}$)
  - Not Six ($\frac{5}{6}$)
- Not Six ($\frac{5}{6}$)
  - Six ($\frac{1}{6}$)
  - Not Six ($\frac{5}{6}$)

A8 Using graphs

Objectives

H Discuss and interpret graphs modelling real situations

F/H Understand and use compound measures, including speed and density

F/H Construct linear functions and plot the corresponding graphs arising from real-life problems

F Use the conventions for coordinates in the plane; plot points in all four quadrants

H Understand that the form $y = mx + c$ represents a straight line and that m is the gradient of the lines and c is the value of y-intercept

F/H Draw lines of best fit by eye, understanding what these represent

F/H Draw and produce scatter graphs

Unit overview

This unit consolidates students' knowledge of the coordinate grid and plotting and interpreting graphs. It covers distance–time graphs, rates of flow and other linear and quadratic graphs. Students are given the opportunity to use the understanding they have gained through the graphical work in previous algebra units to analyse real-life situations.

Prior knowledge

Before your students start this unit they should be able to:

- Understand and use compound measures, including speed
- Recognise the equation form of a straight line and interpret its gradient and y-intercept
- Find the gradient and y-intercept of a straight line
- Plot points in all four quadrants of the coordinate grid
- Draw and interpret linear graphs
- Label axes clearly, using appropriate units and range
- Use formulae from mathematics and other subjects
- Recognise, draw and interpret quadratic graphs

Differentiation

- **Foundation** focuses on drawing and interpreting conversion graphs and simple distance–time graphs
- **Foundation Plus** extends to calculating average speed from a distance–time graph and considers other graphs modelling real-life situations
- **Higher Plus** focuses on transforming graphs

A8.1 Distance–time graphs

Objectives

- Draw and interpret distance–time graphs
- Understand and use compound measures, including speed

Useful resources

- Square paper
- Graph plotting tool

Mental starter

I cycle to work a distance of 30 miles there and 30 miles back. I travel at 30 mph on the way to work and 15 mph on the way back. What is my average speed? (You could let students know that the answer is not $(30 + 15)/2 = 22.5$ mph.)

Remind students that speed = distance ÷ time. In this case distance = 60 miles and time is 3 hours. So speed is $60 \div 3 = 20$ mph.

Introductory activity

Show Dan's journey on his bike, as portrayed in the student book. You could use an OHT of the graph.
Discuss each phase of the journey, and discuss the significance of positive and negative gradients, and also of zero gradient.
Ask for Dan's average speed ($8\frac{1}{3}$ mph).
Discuss the example in the student book (Janine's journey).

Exercise commentary and misconceptions

Question 1 involves a car journey with a number of phases. Refer students to Dan's bike journey as previously discussed if they are unsure of the significance of different parts of the graph.
Question 3 requires students to construct distance–time graphs from a written description of journeys.
Question 4 extends to a speed–time graph.

Plenary

Discuss question 4, **or**:
An Olympic athlete runs 100 m in 10 sec. What speed is this in km/h? (Hint, what does speed in km/h mean? How far you can go in 1 hour?) So use ratio to change 100 m : 10 sec into 1 hour.)

$$\begin{aligned} 100\text{ m} &: 10\text{ sec} \\ \times 6 \quad &\quad \times 6 \\ 600\text{ m} &: 1\text{ min} \\ \times 60 \quad &\quad \times 60 \\ 36\,000\text{ m} &: 1\text{ hour} \\ 36\text{ km} &: 1\text{ hour} \end{aligned}$$

So 36 km/h. (Note, ÷8 and ×5 to change into mph ⟶ 22.5 mph.)

A8.1 Distance–time graphs

This spread will show you how to:
- Draw and interpret distance–time graphs
- Understand and use compound measures, including speed

Keywords
Distance
Gradient
Speed

- **You can represent a journey on a distance–time graph.**
- **Time is always plotted on the horizontal axis.**
- **Distance is plotted on the vertical axis.**

This graph shows Dan's journey on his bike.

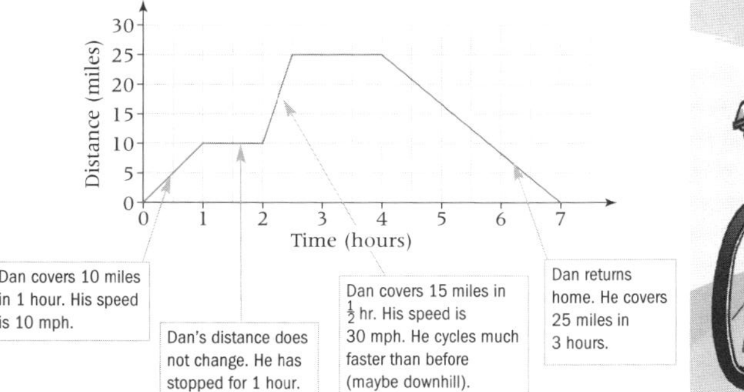

Example

Janine leaves home at 1 p.m. and cycles to her friend's house, 30 km away, at a speed of 20 km/h. She stays for 2 hours, then cycles home, arriving at 6 o'clock.

Draw a distance-time graph to represent the journey and determine her speed on the way home and her average speed for the entire journey.

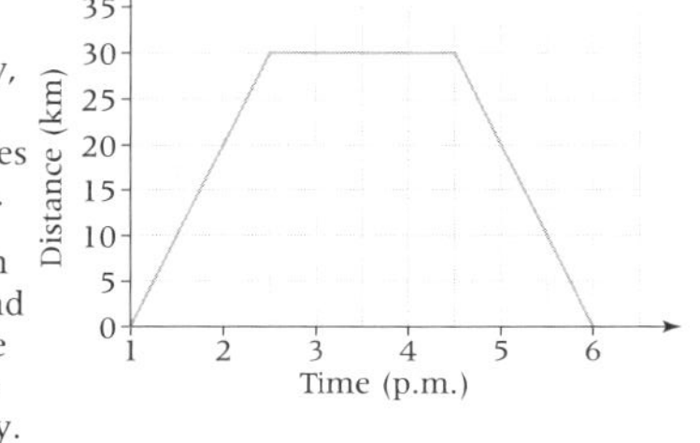

Janine returning home, is shown by the graph going down, back to the x-axis

On the journey home, Janine covers the 30 km in $1\frac{1}{2}$ hours. This means that she has covered 10 km in each half hour and, hence, 20 km in each hour. Her speed is 20 km/h.

Since she covers 60 km in 5 hours, her average speed for the whole journey, including the stop, is 12 km/h.

average speed = $\frac{\text{total distance}}{\text{total time}}$

Exercise A8.1

1 The distance–time graph shows the journey of a car between Birmingham and Stoke-on-Trent.

Distance (km): 20, 40, 60; Time (hours): 9 a.m., 10 a.m., 11 a.m., 12 noon, 1 p.m.; Birmingham; Stoke-on-Trent

a How far is it from Birmingham to Stoke-on-Trent?

b For how long did the car stop?

c What was the speed of the car for the first part of the journey?

d Between which two times was the car travelling fastest?

e What was the average speed of the car for the whole journey?

2 Three students have drawn distance–time graphs. Two have made mistakes. Which two students have made a mistake and what mistake is it?

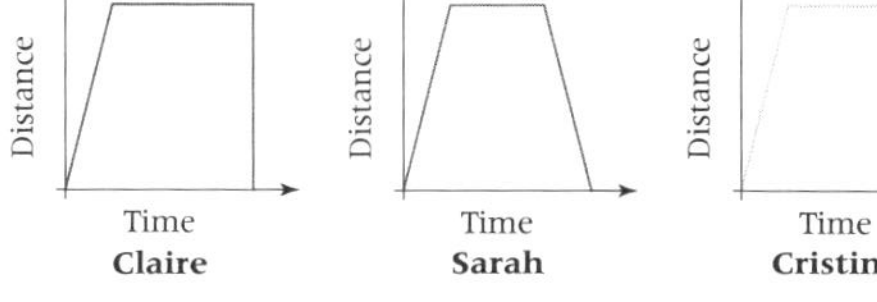

3 Construct a distance–time graph to show each journey.

a A car travels between Bristol and London. On the outward journey, it travels the 120 miles to London in $2\frac{1}{4}$ hours. The driver remains in London for $1\frac{1}{2}$ hours. The car travels half way back to Bristol at 40 miles per hour, as the motorway is busy, then the remaining distance at 80 miles per hour.

b Two brothers both went to see each other on the same day. Henry left his home at 2 p.m. to go and see Leo, who lives 5 miles away. Henry walked at an average speed of 4 miles per hour, but he stopped half way for a 15 minute rest. At 2:30 p.m., Leo set out on his bicycle from his home in order to go and visit Henry. He cycled straight there in $\frac{1}{2}$ hour.

Draw the two journeys on one graph.

4 The following graph represents the journey of a car. Construct a graph of speed (km/h) against time (h) for this journey.

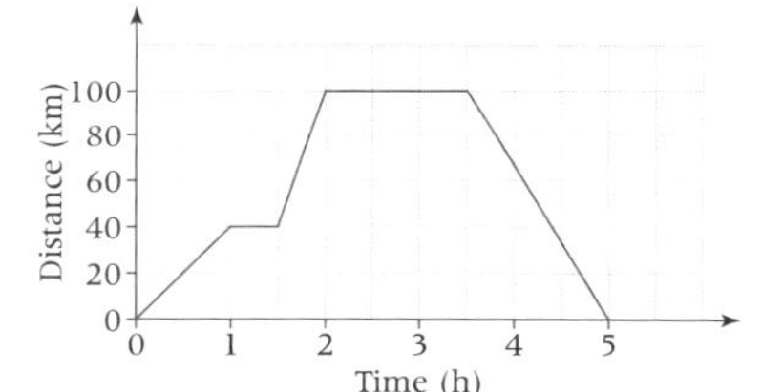

A8.2 Other real-life graphs

Objectives

- Draw and interpret graphs modelling real-life situations

Useful resources

- Different shaped perspex containers and a jug of water

Mental starter

A robot is given these instructions for running a bath.

1 Turn on the taps.
2 When the bath is $\frac{3}{4}$ full the task is complete.

What will happen if the robot follows these instructions? What is wrong with these instructions? (Did not mention putting in the plug, does not tell you to turn off the taps, does not take any notice of the temperature of the bath.)

Introductory activity

Draw a number of different-shaped containers – you can use the containers in question 1 for inspiration, or the container shown at the top of the student book page.

In each case ask students to imagine filling the container with a constant flow of water; as the water pours in the depth will increase, but in some containers the speed at which the depth is increasing changes. Think of filling a water bottle at the sink. The depth increases steadily, then suddenly the water rushes to the top.

With the whole class, sketch the shape of each graph on the board.

Ask what is happening at each part of the graph.

Exercise commentary and misconceptions

Emphasise that time is always plotted on the horizontal axis.

Ensure students appreciate the difference between a sudden change in height, weight or temperature which is shown by a steep gradient and a gradual change which is shown by a shallow gradient.

The shapes in question 1 are standard ones which often occur in exam questions.

In question 5 the graph of acceleration is a difficult one for most students to understand. Emphasise that a slowing down in speed will be caused by negative acceleration, and that acceleration is the gradient of the speed–time graph. Students may find it helpful to draw the acceleration graph immediately below the speed graph.

Plenary

Discuss question 5, in particular the acceleration–time graph in part **b**.

A8.2 Other real-life graphs

This spread will show you how to:
- Draw and interpret graphs modelling real-life situations

Keywords
Model

You can use graphs to **model** the depth of water flowing in or out of a container at a constant rate.

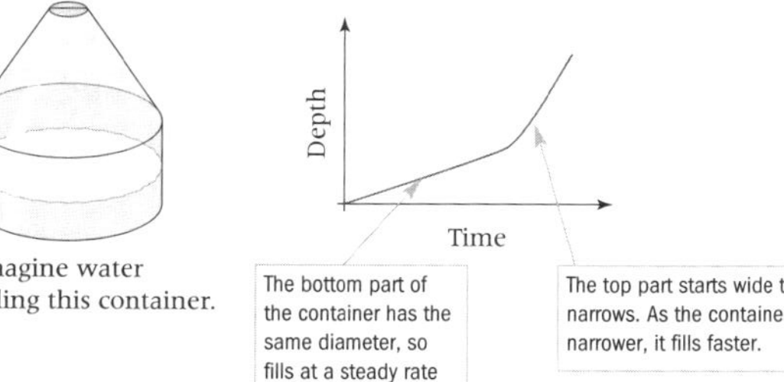

Imagine water filling this container.

- **In a real-life graph involving time, time is usually represented by the *x*-axis.**

Example

Sketch a graph to show what happens as

1. Sam fills a bath with both taps running
2. He realises it is too hot so turns off the hot tap
3. He turns off the cold tap
4. He gets in
5. Has a long soak
6. Gets out
7. Pulls out the plug.

Label the axes with the quantities they represent. A scale is not needed for a sketch graph.

Exercise A8.2

1 Match the four sketch graphs with the containers.

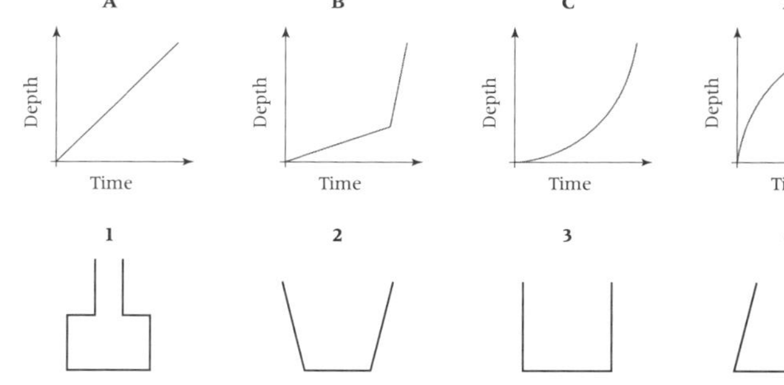

2 The sketch graph shows Andrew's height from 3 to 23 years of age. Explain what the graph shows at each stage and explain why this might be.

3 Sketch a graph of depth against time as this container is filled.

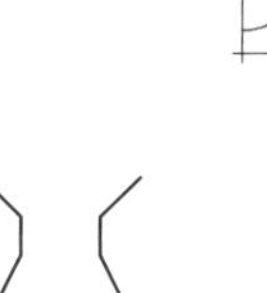

4 Construct sketch graphs to represent these situations.

a A woman is pregnant and puts on weight. When her baby arrives, she finds it difficult to lose any of the weight for six months, but then joins an exercise class and eats healthily. She is back to her natural weight a year and a half after becoming pregnant. (Graph is weight against time.)

b A frozen chicken is taken out of the freezer and left to defrost. Two hours later it is put in the microwave briefly to speed up and finish off the process. The chicken is then put in the oven to roast for Sunday lunch. (Graph is temperature against time.)

5 A skateboarder at a skate park uses a ramp, as shown. For one go on the ramp, construct a sketch graph of

a speed against time

b acceleration against time.

A8.3 Linear graphs in real life

Objectives
- Draw and interpret graphs modelling real-life situations
- Form linear functions, using the corresponding graphs to solve real-life problems

Useful resources
- 2 mm graph paper
- Graph plotting tool

Mental starter
There are approximately 2.2 pounds (lb) in 1 kg. Ask students to convert these weights: 10 kg, 11 lb, 8 lb, 7 kg and 3.8 lb.

Pounds	(22)	11	8	(15.4)	3.8
kg	10	(5)	(3.6)	7	(1.7)

What might make this easier, apart from a calculator?

(A graph of the relationship).

Introductory activity
A delivery company is fed up with offering free deliveries and decides to charge. The further away you live from the company the more you pay. For every mile you pay £2. But if you live less than a mile the company still wants to charge you. **How can they do this?** Discuss the options the company might have. Settle on paying a standard charge plus £2 per mile.

Ask students to copy and complete this table:

Miles	0	2	4
Cost			

Draw a graph of this charge using horizontal axes for the miles from the company and the vertical axes for the cost of the delivery.

What is the y-intercept (c) and what does it represent? (The fixed cost of a delivery.)

What is the gradient and what does it represent? (2, represents the cost per mile.)

Exercise commentary and misconceptions
In question 1 emphasise that the graph can be used to change miles to km and vice versa.

The axes in question 2 need to go from £0 to £400, and from €0 to €600. Students may find it difficult to decide on efficient scales to use for the axes. The origin and the point representing £400 and €580 are sufficient to draw an accurate graph, but calculating and plotting a third point, perhaps for £200, should safeguard against any errors.

The graph in question 3 has a positive y-intercept – students often assume that all graphs will go through the origin. Suggest to students that they check what values they will need to read off the finished graph to answer question **b**; they can then make sure that the axes have a sufficient range.

Ask students what is the meaning of the constant terms in the equations in question 4. They are often unfamiliar with the concept of a standing charge.

Plenary
£1 = \$1.80 and \$1 = €0.80. Ask students to draw a conversion graph for pounds to euros.

£	0	5	10
$	0	9	18
€	0	7.20	14.20

A8.3 Linear graphs in real life

This spread will show you how to:

- Draw and interpret graphs modelling real-life situations
- Form linear functions, using the corresponding graphs to solve real-life problems

Keywords
Formula
Gradient
Graph
Intercept

- You can use a graph to represent a real-life situation.

Suppose you have a mobile phone. You pay £10 line rental each month, then 50p for every minute you spend making calls.

Work in £
50p = £0.5

Time on calls in minutes (x)	0	1	2	3	4
Price of calls in pounds	0	0.50	1.00	1.50	2.00
Line Rental (£)	10	10	10	10	10
Total cost in pounds (y)	10	10.50	11	11.50	12

To get the total cost, you add £10 (the line rental) to the call cost.
The call cost is the number of minutes on the phone multiplied by 0.50.
Hence,

Total cost = 0.50 × time on phone + 10

$y = 0.5x + 10$

This formula is linear and its graph is a straight line. The y-axis intercept is 10 and the gradient is 0.5.

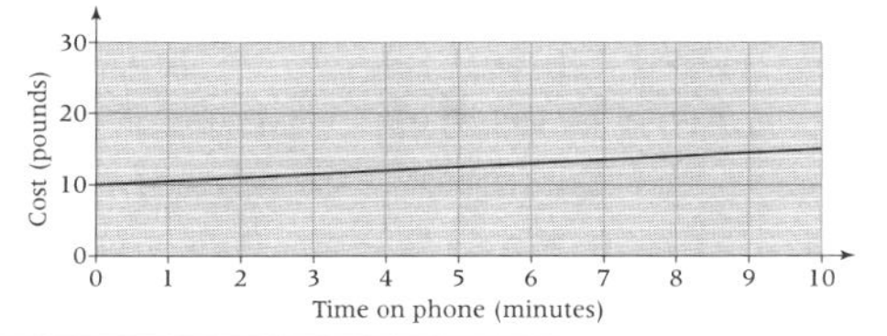

Example

Plot a graph to represent the total cost of hiring a party venue, if the owner charges £100 hire fee and £5 per guest.
Use the graph to estimate the number of guests if the total bill is £285.

If x is the number of guests and y is the cost (£s), then

$y = 5x + 100$

Draw a horizontal line from £285 to the graph. Draw a vertical line from the graph to the horizontal axis.

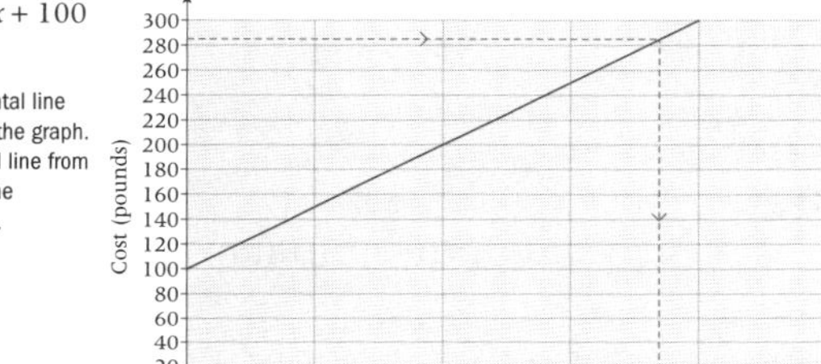

Label axes with the quantities they represent and the units in which they are measured.

For cost of £285, number of guests = 37.

Exercise A8.3

1 **a** This graph is a conversion graph for miles to kilometres and vice versa.
Use the graph to convert

i 20 miles into kilometres

ii 60 kilometres into miles.

b If Dan ran 30 miles and Charlie ran 50 kilometres, who ran further?

c By finding the gradient of the line, give a formula to connect the number of miles (x) with the number of kilometres (y).

2 Pauline and her family are going on holiday and exchange £400 spending money into euros (€) before they go. At the bank, the exchange rate is £1 = €1.45.

£1 = €1.45

a Construct a graph that the family can take on holiday to convert any amount of their spending money from pounds to euros or vice versa.

b Use the graph to find

i the cost, in euros, of a side trip which is advertised for £95

ii the cost, in pounds, of a meal in a restaurant that comes to €85.

3 A campsite charges £15 per night per tent, plus an extra £3 per person.

a Construct a table of charges and, hence, a graph to show the cost of the campsite depending upon how many people stay in the tent. (The largest tent available is one that sleeps 15 people.)

b Use your graph to calculate how many people stayed in the tent if the total cost was £36.

c Explain why the total cost could never be £50.

d Suggest an equation for your graph, stating clearly the meaning of any letters you use.

e Use your equation to work out the cost of pitching a new Supertent that sleeps 27 people.

4 Two competing electricity companies use these formulae to work out customers' bills.

POWER UP! $y = 3x + 5$

SPARKS ARE US! $y = 2x + 15$

The number of units of electricity used is x.
The price of the electricity is £y.
Using graphs, compare the pricing policies of the two companies and advise householders from which company they should buy their electricity.

A8.4 Further linear graphs in real life

Objectives

- Form linear functions, using the corresponding graphs to solve real-life problems
- Draw and use scatter diagrams and lines of best fit

Useful resources

- Graph plotting tool

Mental starter

You can add the squares of all the numbers from 1 to 10 either directly as $1^2 + 2^2 + 3^2 + 4^2 + 5^2 + 6^2 + 7^2 + 8^2 + 9^2 + 10$ **or** $5 \times 10^2 + 5^2 - 2(1 \times 9 + 2 \times 8 + 3 \times 7 + 4 \times 6)$.
Which is faster? Set half the class to try each of the methods. For the second method you have $500 + 25 - 2 \times 70 = 385$, the correct answer.

Introductory activity

Ask students to draw axes from 0 to 10 in the x direction and −4 to 15 in the y direction, and draw a line through the points (3, 3) and (6, 9).
How can you find the equation of the line? Remind students that they must find the gradient m and the y-intercept c (covered in section A4.4). Recap the method of finding the gradient.
In this case $m = 2$ and $c = -3$ so the equation is $y = 2x - 3$.
Recapping the work form A6.3, display the graph from the top of the student book page. The graph relates to a hire car company charges.
Discuss how to find the equation of the graph, and discuss the practical significance of the gradient and the intercept.
Discuss the example in the student book – you may need to revise students' knowledge of scatter diagrams.

Exercise commentary and misconceptions

When using linear graphs for real-life situations students need to have a feel for what they look like. When using a fractional gradient point out that the fraction can appear in front of the x or be absorbed by the x, for example $y = \frac{2}{3}x + 3$ or $y = \frac{2x}{3} + 3$. Remind students of TUBA, and that a steeper gradient has a larger value for m.

The 'driving test' graph in question 2 is invalid for x values less than 17 in the UK.

Encourage students to sketch the graphs in question 3, but not plot them on graph paper.

Plenary

All the graphs drawn in this section and the previous one have used only positive values for x and y. **Could any of them have been extended to include negative values? Which ones? Why couldn't the others?** (Could go through them, only the temperature conversion graph could include negatives.)

A8.4 Further linear graphs in real life

This spread will show you how to:

- Form linear functions, using the corresponding graphs to solve real-life problems
- Draw and use scatter diagrams and lines of best fit

Keywords
Gradient
Graph
Intercept

You can use a straight line graph to represent real-life information.
For example, this graph shows a hire car company's charges.

You can find the gradient and the y-axis intercept.

The line intercepts the y-axis at (0, 50).

For every 10 squares you move across, you travel 5 squares up:

Gradient $= \frac{\text{rise}}{\text{run}} = \frac{5}{10}$ *or* $\frac{1}{2}$

The equation of the graph is $y = \frac{1}{2}x + 50$.

- Graphs help to give you more information.
  - The intercept of 50 tells you that you are charged £50 for a car.
  - The gradient of $\frac{1}{2}$ tells you that for every 2 miles you travel, you are charged an extra £1.

Example

Interpret the line of best fit on this scatter diagram of 18 students' heights and weights.

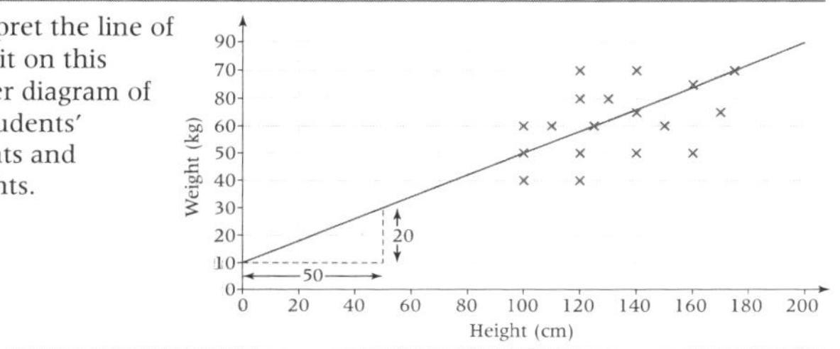

The gradient, $m = \frac{20}{50} = \frac{2}{5}$
The y-axis intercept, $c = 10$
The equation of the graph is $y = \frac{2}{5}x + 10$.

The gradient tells you that for every 5 cm you grow you gain 2 kg.
The intercept tells you that at 0 cm height, you weigh 10 kg – in other words, it does not always make sense to interpret the intercept on a straight line graph for a big age range.

Exercise A8.4

1 Match each graph with an equation.

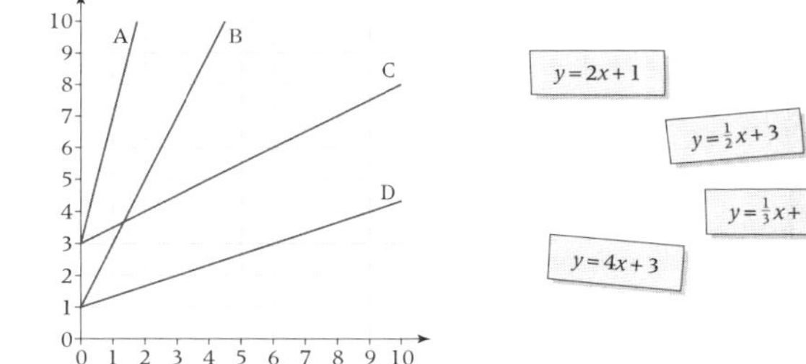

2 For each graph, find its equation in the form $y = mx + c$ and interpret the meaning of m and c, deciding if it is sensible to interpret c.

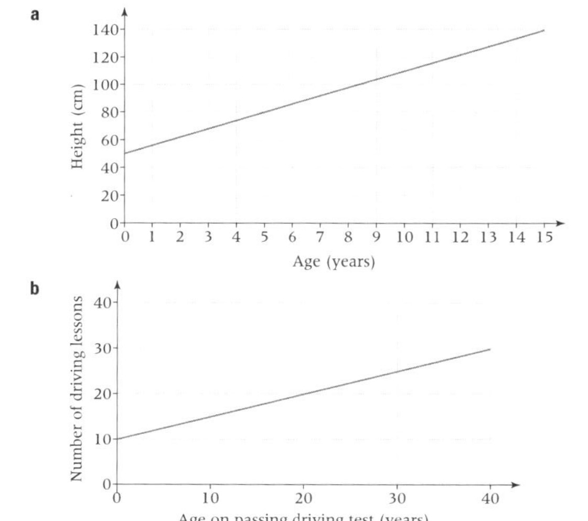

3 Interpret these equations representing real-life situations and discuss their limitations.

a $y = 0.3x + 2$ ... x is age in years, y is amount of pocket money in £.

b $y = 1\frac{4}{5}x + 32$ ... x is temperature in degrees Celsius (°C) and y is temperature in degrees Fahrenheit (°F).

A8.5 Using quadratic graphs

Objectives

- Plot graphs of simple quadratic functions

Useful resources

- Graph plotting tool
- 2 mm graph paper

Mental starter

What is the value of x^2 when $x = -5$? (25.)
What is the value of $2x^2$ when $x = 5$? ($2 \times 5^2 = 50$.)
What is the value of $-x^3$ if $x = -2$? (8, cubing -2 gives -8, so $- -8$ is positive.)
Invent other questions of this type.

Introductory activity

A ball is fired vertically from the ground at an initial speed of 30 m/s. The height, h, in metres and the time, t, in seconds after the ball is fired are governed by the formula $h = 5t(6 - t)$.
Draw a graph of height against time.

Time, t	0	1	2	3	4	5	6
Height, h	0	25	40	45	40	25	0

Should you extend the graph for times past 6 seconds? (No, because the height can't be negative.) Notice that the graph is symmetrical and has a maximum point (vertex). This type of graph is known as a quadratic or parabola. Some parabolas have a 'U' shape. **What is it about the equation that tells you this would be a ∩ shape?** (Negative coefficient of t^2 when multiplied out, see A7.1.) **How high does the ball reach?** (45 m.)

Emphasise the difference between this graph and that for the javelin in the students' book. Here the x-axis is time, not horizontal distance, so it doesn't show the path of the ball.

Exercise commentary and misconceptions

When completing the table of values for question 2 students will need to take care with the negative values, especially when squaring. Encourage them to fill in the table of values, or to put brackets round negative numbers if using a calculator to square them.
A few calculators actually put brackets around the negative number automatically when you raise it to a power, but most calculators will give students an incorrect answer when they attempt to square a negative number.

When calculating $4x^2$ in question 3, remind students of the rule of BIDMAS, indices first.

Plenary

Can you find values of x and y that fit the equation $x^2 + y^2 = 25$? (Hint that the answers are integers and lie between −5 and 5 inclusive.)

Encourage students to start with a table:

x	−5	−4	−3	0	3	4	5
y	0	3 or −3	4 or −4	5 or −5	4 or −4	3 or −3	0

Can you draw it? (A circle, centre at the origin with a radius of 5 units.)

A8.5 Using quadratic graphs

This spread will show you how to:

- Plot graphs of simple quadratic functions

Keywords
Maximum
Model
Parabola
Quadratic

- You can model some real-life situations with **quadratic** graphs.

This **parabola** shows the height of a javelin as it is thrown.

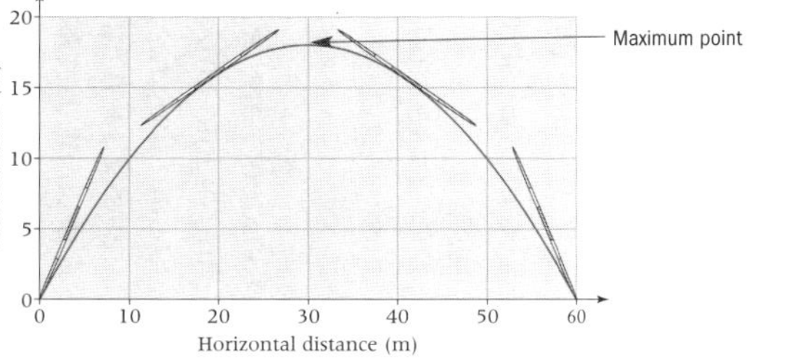

The **maximum** point on the graph shows that the javelin reaches a height of 18 m when it is 30 m away from the throwing line.

Example

One part of a roller coaster is modelled using the equation

$y = \dfrac{x^2 - 20x}{10} + 10$ where x is the horizontal distance and y is the vertical distance from the start of the section.

a Complete this table of values and use it to plot the graph that shows the path of the roller coaster.

x	0	5	10	15	20
x^2	0	25			
$-20x$	0	-100			
$\dfrac{x^2 - 20x}{10}$	0	-7.5			
y	10	2.5			

b How long is this part of the roller coaster ride?

a

x	0	5	10	15	20
x^2	0	25	100	225	400
$-20x$	0	-100	-200	-300	-400
$\dfrac{x^2 - 20x}{10}$	0	-7.5	-10	-7.5	0
y	10	2.5	0	2.5	10

b 20 m

Exercise A8.5

1 The graph $y = 2x^2 + x - 4$ is shown. What are the coordinates of the minimum point of the graph?

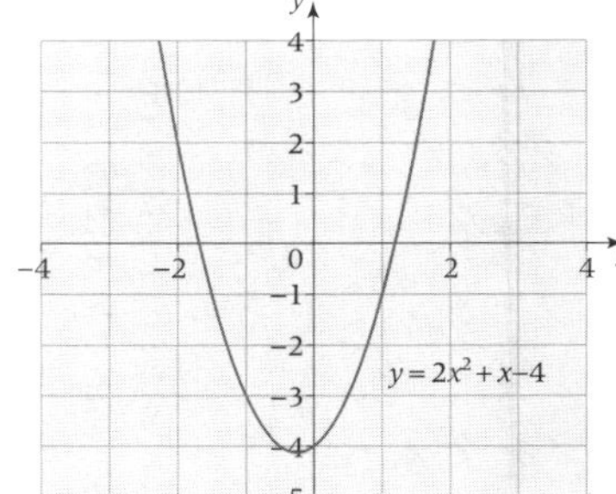

2 **a** Copy and complete the table of values for the graph $y = x^2 + x + 1$.

x	-3	-2	-1	0	1	2	3
x^2	9						
y	7				3		

b Plot the points for x and y and join them to form a smooth parabola.

c What is the approximate minimum value of $x^2 + x + 1$ and for what value of x does it occur?

3 A ball is thrown into the air.
The formula, $y = 20x - 4x^2$, shows its height, y metres, above the ground x seconds after it is thrown.

a Copy and complete the table of values to show the height of the ball during its first five seconds.

Time (x)	0	1	2	3	4	5
$20x$						
$4x^2$						
Height (y)						

b Use the table to plot a graph to show the ball's height against time.

c Use your graph to find

i the maximum height reached by the ball and the time at which it reaches this height

ii two times when the ball is 12 metres above the ground

iii the interval of time when the ball is above 15 metres.

A8 Unit review

Key objectives

- Construct linear functions and plot the corresponding graphs arising from real-life problems
- Discuss and interpret graphs modelling real situations

Worked solution	**Commentary and misconceptions**

1

a

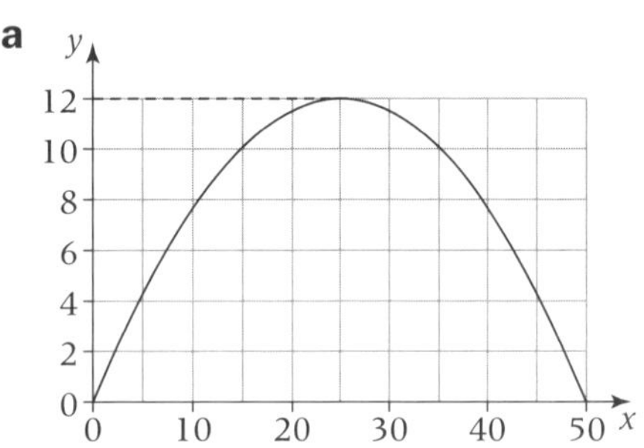

Encourage students to either sketch the graph with a line to show their working or explain their method.

Maximum height = 12 metres.

Ensure that students state the correct units with their answer.

b Distance from start position = 50 metres.

Encourage students to explain their method.
Ensure that students state the correct units with their answer.

2

a

$$\text{Speed} = \frac{\text{distance}}{\text{time}} = \frac{20}{0.5} = 40 \text{ km/h}$$

Ensure that students use the correct units when answering this question. They need to convert 30 minutes to $\frac{1}{2}$ hour before completing the calculation.
Revision of converting units may be useful.
Ensure that students realise that dividing by 0.5 is equivalent to multiplying by 2.
Ensure that students state the correct units.
Encourage students to think about their answer and decide if it seems feasible.

b

Distance in km from Sian's house
20
15
10
5
0
0
20
40
60
80
Time in minutes

Students are required to convert 60 km/hr into km/minutes.
Ensure they realise it takes 20 minutes to travel 20 km at this speed.
A prompt that 60 km/hr is equivalent to 1 km/minute may help.

S8 Pythagoras and trigonometry

Objectives

H Understand, recall and use trigonometrical relationships in right-angled triangles

H Understand similarity of triangles and of other plane figures, and use this to make geometric inferences

H Use an extended range of function keys, including trigonometrical and statistical functions

F/H Understand, recall and use Pythagoras' theorem

F Understand angle measure using the associated language

F Use and interpret maps and scale drawings

Unit overview

This unit consolidates knowledge of Pythagoras' theorem from unit S4 and of bearings and scale drawings from unit S5. It introduces trigonometry in right-angled triangles, using this to find missing sides and angles. Students are then encouraged to use their gained knowledge to solve 2-D problems.

Prior knowledge

Before your students start this unit they should be able to:

- Change the subject of algebraic formulae
- Use the rules of arithmetic to manipulate algebraic expressions
- Recall, understand and use Pythagoras' theorem
- Recall, understand and use angle properties of 2-D shapes
- Use a scientific calculator effectively, including brackets
- Add, subtract, multiply and divide integers, decimals and fractions
- Understand and use angle measure and the associated language
- Give answers to an appropriate level of accuracy
- Know when and when not to round in calculations

Differentiation

- **Foundation** focuses on geometric properties of 2-D shapes and tessellations. It then extends to 2-D representations of 3-D shapes and properties of cubes and cuboids
- **Foundation Plus** extends the work on tessellations and 2-D shapes, and then introduces Pythagoras' theorem in 2-D
- **Higher Plus** extends the Higher book to consider trigonometric equations and graphs

S8.1 Tangent ratio

Objectives

- Understand, recall and use trigonometry in right-angled triangles
- Understand and use similarity and ratio to find missing angles and lengths in triangles

Useful resources

- Calculators
- Geometry tool

Mental starter

Lightning travels at about 435 600 km/h. A typical lightning storm is 12.1 km high. **How long does it take lightning to travel the length of the storm to earth?** (0.1 sec.)

Introductory activity

Remind students that calculators must be in degree mode.

Recap how in similar triangles there is a constant ratio between sides. In this lesson we are looking at the ratio between the two short sides in a right-angled triangle. The first step in any problem is to label the sides of the triangle as hypotenuse, opposite (to the marked angle) and adjacent (to the marked angle). It is worth giving three examples with differing orientations to reinforce this. This lesson and S8.2 only cover finding an unknown side.

The first worked example in the students' book finds an opposite side and the second finds an adjacent side. Some students find the algebraic manipulation of the equation difficult in the second example and may be helped by using a triangle similar to that used for speed, distance and time, with O at the top and T and A below.

Exercise commentary and misconceptions

In question 1 the first four parts require finding opposite sides and the next four require finding adjacent sides.
Encourage students always to label the triangle as the first step to solving any trigonometric problem. Students often find this quite difficult when the adjacent side is not at the base of the triangle.

Plenary

Discuss different angle measures: degrees, radius and gradians. Focus on the different ways in which they divide up a circle. Discuss the merits of using 360 (you may wish to ask students to list the factors).

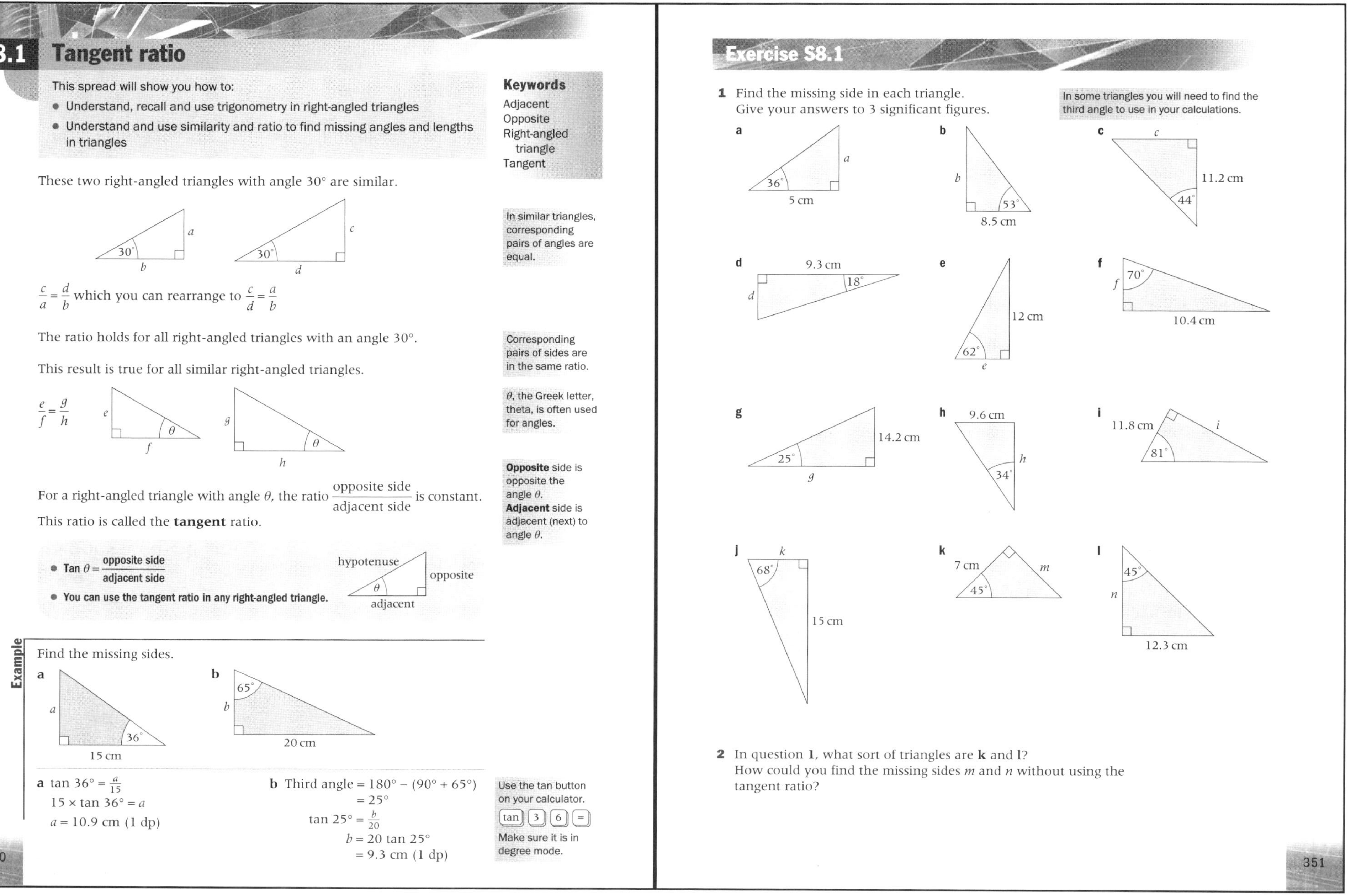

S8.1 Tangent ratio

This spread will show you how to:

- Understand, recall and use trigonometry in right-angled triangles
- Understand and use similarity and ratio to find missing angles and lengths in triangles

Keywords
Adjacent
Opposite
Right-angled triangle
Tangent

These two right-angled triangles with angle 30° are similar.

In similar triangles, corresponding pairs of angles are equal.

$\frac{c}{a} = \frac{d}{b}$ which you can rearrange to $\frac{c}{d} = \frac{a}{b}$

The ratio holds for all right-angled triangles with an angle 30°.

Corresponding pairs of sides are in the same ratio.

This result is true for all similar right-angled triangles.

$\frac{e}{f} = \frac{g}{h}$

θ, the Greek letter, theta, is often used for angles.

For a right-angled triangle with angle θ, the ratio $\frac{\text{opposite side}}{\text{adjacent side}}$ is constant.

This ratio is called the **tangent** ratio.

Opposite side is opposite the angle θ.
Adjacent side is adjacent (next) to angle θ.

- **Tan θ = $\frac{\text{opposite side}}{\text{adjacent side}}$**
- **You can use the tangent ratio in any right-angled triangle.**

Example

Find the missing sides.

a

b

a $\tan 36° = \frac{a}{15}$
$15 \times \tan 36° = a$
$a = 10.9$ cm (1 dp)

b Third angle = 180° − (90° + 65°)
= 25°
$\tan 25° = \frac{b}{20}$
$b = 20 \tan 25°$
= 9.3 cm (1 dp)

Use the tan button on your calculator.
tan 3 6 =
Make sure it is in degree mode.

Exercise S8.1

1 Find the missing side in each triangle.
Give your answers to 3 significant figures.

In some triangles you will need to find the third angle to use in your calculations.

a

b

c

d

e

f

g

h

i

j

k

l

2 In question **1**, what sort of triangles are **k** and **l**?
How could you find the missing sides *m* and *n* without using the tangent ratio?

S8.2 Sine and cosine ratios

Objectives

- Understand, recall and use trigonometry in right-angled triangles
- Use the trigonometric functions of a scientific calculator

Useful resources

- Calculators
- Geometry tool

Mental starter

Write equations on the board:

$0.5 = \frac{x}{16}$ $\quad$ $12x = 108$

$\frac{m}{2} = 0.72$ $\quad$ $\frac{r}{9.7} = 0.1$

Ask students to solve them.

Extend to harder examples of the type:

$\frac{3}{p} = 0.6$

Introductory activity

This lesson introduces the other two major trig ratios that can be used to solve right-angled triangles.

As before, the first step in any problem-solving is to label the sides in the diagram. The other two ratios are:
sine, which uses opposite and hypotenuse sides,
and cosine which uses adjacent and hypotenuse sides.

It is worth highlighting the facts that dividing by a number less than 1 always increases a value, that both sine and cosine ratios cannot be greater than 1 and that the hypotenuse must be longer than any other known side. Finding a hypotenuse always involves dividing by a sine or cosine ratio.

Work through the worked examples in the students' book. You may use formula triangles to help.

Exercise commentary and misconceptions

The first three parts use the sine ratio to find a short side, the next two use the cosine ratio to find a short side.
Parts **f**, **g** and **i** use the cosine ratio for the hypotenuse and **h** and **j** use the sine ratio for the hypotenuse.

If students always label the triangles with the names of the sides they are more likely to make the right choice of ratio.

This is a good opportunity to remind students to use an appropriate level of accuracy in their answers.

Plenary

Discuss different ways of remembering the three trigonometric ratios.
Formula triangles provide one easy way, but may not work for all students as it is not always clear where the letters go. Suggest a mnemonic, like SOHCAHTOA for example.

S8.2 Sine and cosine ratios

This spread will show you how to:
- Understand, recall and use trigonometry in right-angled triangles
- Use the trigonometric functions of a scientific calculator

Keywords
Cosine
Hypotenuse
Right-angled triangle
Sine
Tangent

The tangent ratio is:

$$\text{Tan } \theta = \frac{\text{opposite side}}{\text{adjacent side}}$$

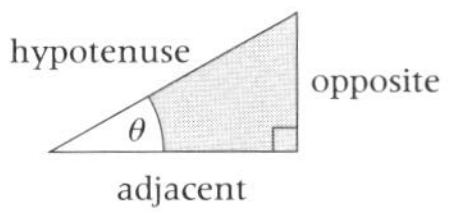

hypotenuse
opposite
adjacent
θ

The **hypotenuse** is the longest side, opposite the right angle.

There are two other ratios you can use in right-angled triangles.

- **Sine ratio**

$$\sin \theta = \frac{\textbf{opposite side}}{\textbf{hypotenuse}}$$

- **Cosine ratio**

$$\cos \theta = \frac{\textbf{adjacent side}}{\textbf{hypotenuse}}$$

hypotenuse
opposite
adjacent
θ

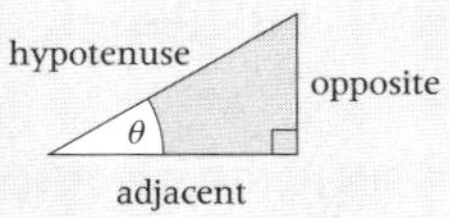

Label the sides you want to find and the side you know.
Remember that opposite and adjacent refer to the angle.

Example

Find the missing sides.

a a, 14 cm, 32°

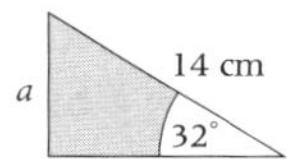

b 8 cm, 22°, b

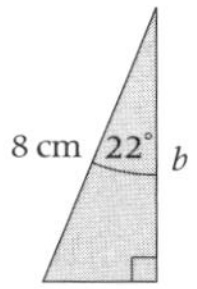

c 28 cm, h, 65°

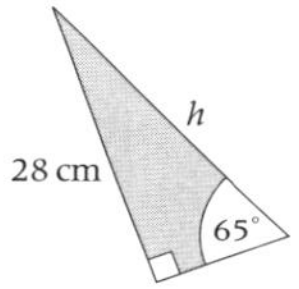

a $\sin 32° = \frac{a}{14}$
$14 \times \sin 32° = a$
$a = 7.42$ cm (3 sf)

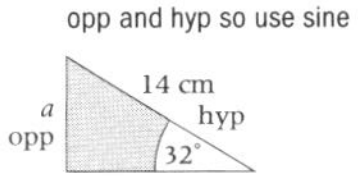

opp and hyp so use sine
a opp, 14 cm hyp, 32°

Use the (sin) key on your calculator.

b $\cos 22° = \frac{b}{8}$
$8 \times \cos 22° = b$
$b = 7.42$ cm (3 sf)

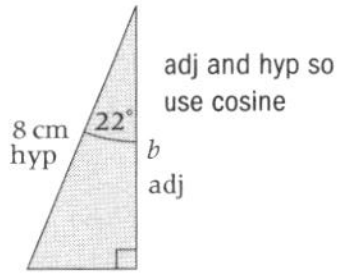

8 cm hyp, 22°, b adj
adj and hyp so use cosine

Use the (cos) key on your calculator.

c $\sin 65° = \frac{28}{h}$
$h = \frac{28}{\sin 65}$
$h = 30.9$ cm (3 sf)

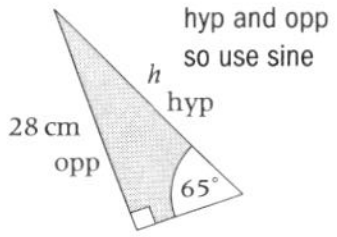

hyp and opp so use sine
h hyp, 28 cm opp, 65°

To find the hypotenuse, you will always need to divide by either sin or cos.

Exercise S8.2

1 Find the missing side in each of these right-angled triangles.
Give your answer to 3 significant figures.

a 9 cm, 50°, a

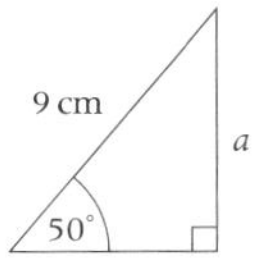

b 35°, 15 cm, b

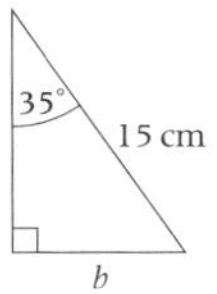

c c, 10.6 cm, 48°

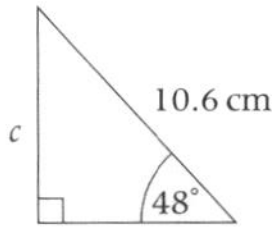

d 7.5 cm, 52°, d

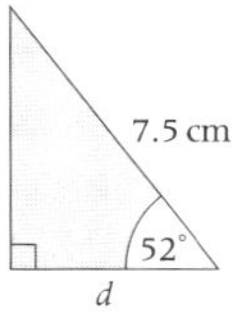

e e, 67°, 12.2 cm

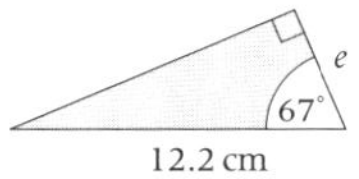

f f, 80°, 6.9 cm

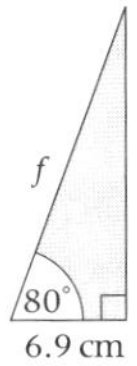

g 9.3 cm, 29°, g

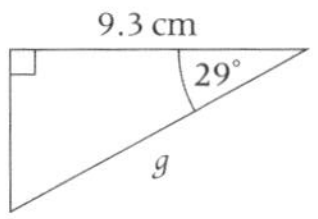

h h, 42°, 6.4 cm

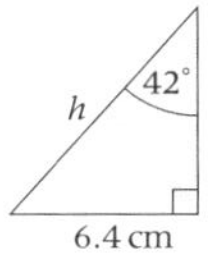

i i, 66°, 14 cm

j 72°, 12.8 cm, j

k k, 16.7 cm, 45°

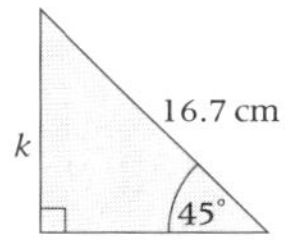

l l, 12.3 cm, 30°

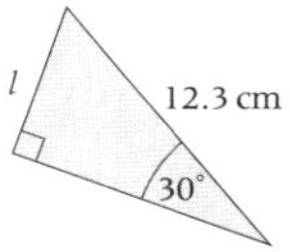

m m, 30°, 12.3 cm

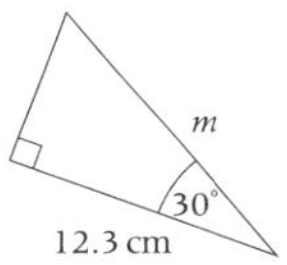

n 57°, p, 9.4 cm

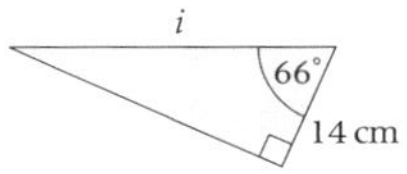

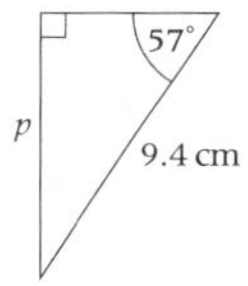

o q, 9.4 cm, 33°

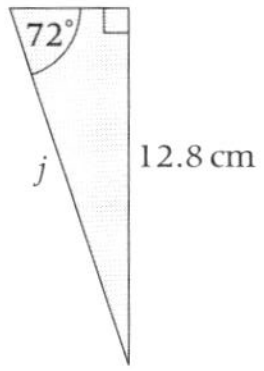

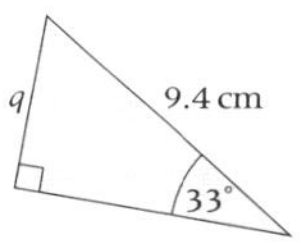

S8.3 Finding angles in right-angled triangles

Objectives

- Understand, recall and use trigonometry in right-angled triangles
- Use the trigonometric functions of a scientific calculator

Useful resources

- Calculators

Mental starter

A birthday cake is in the shape of a prism. Its cross-section is the base, a right-angled triangle, size 16 by 12 by 20 cm. It is 5 cm deep. It is cut into two pieces along a vertical line through the centre points of the two shorter sides of the cross-section. The larger piece is eaten. **What fraction is left?** ($\frac{1}{4}$)

Introductory activity

In this lesson, all three trigonometrical ratios are used to find angles.

Check that all students have their calculators in degree mode and that they can find the inverse trigonometry functions.

As before, students should label the sides of the triangle, identify the ratio which is needed and write down the equation to find the angle.

Encourage students to use the bracket function on their calculators so that to find x from the equation $\cos x = \frac{6}{10.5}$, the key operations are

$\cos^{-1}(6 \div 10.5) =$

Recap the steps in solving trigonometry questions.

1 Label the sides, hyp, opp, adj.
2 Identify the two sides that are involved in the question.
3 Select the formula that uses the two sides identified.
4 Write the equation using the numbers and letters given.
5 Solve the equation.

Remind students that they always require an inverse function when finding an angle.

Exercise commentary and misconceptions

In this exercise, all the questions involve finding an angle and therefore the inverse function is always used.

The first two parts use the sine ratio, the next two the cosine ratio and the next four use the tangent ratio.

The two most likely problem areas are that the student has not identified the correct trigonometrical ratio for the question, or has forgotten to use the inverse function. Ensure students know how their calculators work.

Plenary

Ask students to use a calculator to find values of $\cos \theta$ and $\sin \theta$ up to 360°, using 20° intervals. Plot the values on a graph of $\sin \theta$ against $\cos \theta$.
Ask students what they notice.
This plenary may be spread over two lessons.

S8.3 Finding angles in right-angled triangles

This spread will show you how to:

- Understand, recall and use trigonometry in right-angled triangles
- Use the trigonometric functions of a scientific calculator

Keywords
Cosine
Right-angled triangle
Sine
Tangent

You can use the **sine**, **cosine** and **tangent** ratios in a right-angled triangle.

$$\sin\theta = \frac{\text{opp}}{\text{hyp}} \qquad \cos\theta = \frac{\text{adj}}{\text{hyp}} \qquad \tan\theta = \frac{\text{opp}}{\text{adj}}$$

- **You can use the inverse functions $\sin^{-1}$, $\cos^{-1}$ and $\tan^{-1}$ to find the angle if you know two sides.**

Always start you calculation by labelling opposite side and adjacent side in relation to the angle.

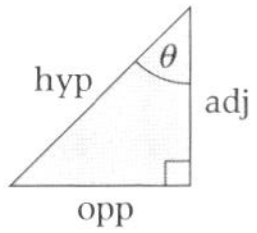

Find the [$\sin^{-1}$], [$\cos^{-1}$] and [$\tan^{-1}$] keys on your calculator. They may be 2nd functions, or you may need to use the [INV] key.

hyp, adj, opp, θ

Example

Find the missing angles.
Give your answers to the nearest degree.

a 10.5 cm, x

b 7.5 cm, 11.7 cm, y

c 11.1 cm, 5.2 cm, z

a You have adjacent and hypotenuse, so use cosine.

$\cos x = \frac{6}{10.5}$

$x = \cos^{-1}\frac{6}{10.5}$

$x = 55°$ (to the nearest degree)

On your calculator, use the brackets keys for the fraction.

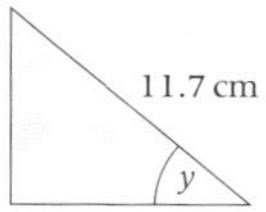

b You have opposite and hypotenuse, so use sine.

$\sin y = \frac{7.5}{11.7}$

$y = \sin^{-1}\frac{7.5}{11.7}$

$y = 40°$ (to the nearest degree)

c You have opposite and adjacent, so use tan.

$\tan z = \frac{5.2}{11.1}$

$z = \tan^{-1}\frac{5.2}{11.1}$

$z = 25°$ (to the nearest degree)

Tan can be a fraction >1. sin and cos are always fractions <1.

Exercise S8.3

1 Find the missing angle in each triangle.
Give your answers to 3 significant figures.

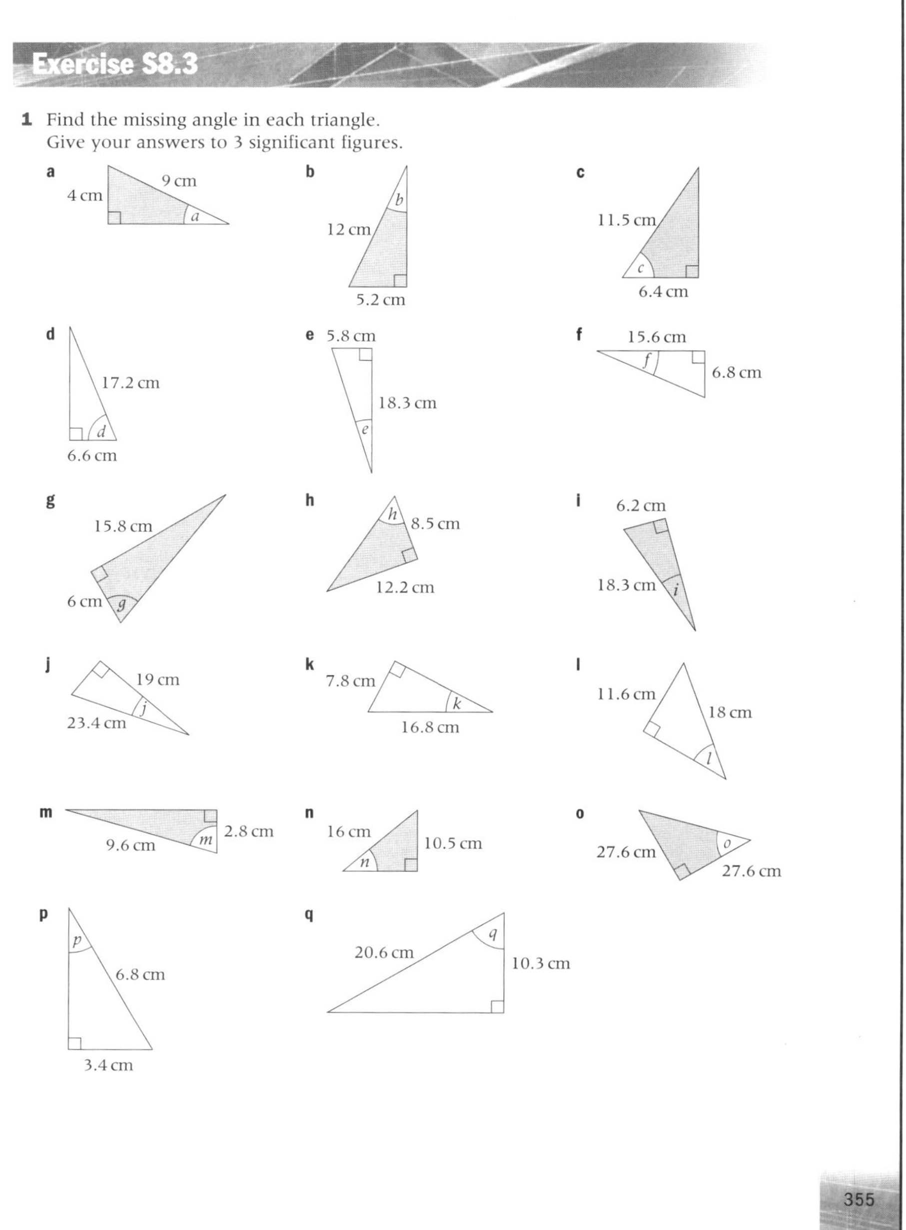

S8.4 Pythagoras' theorem and trigonometry

Objectives
- Understand, recall and use Pythagoras' theorem and trigonometry in 2-D problems

Useful resources
- Calculators

Mental starter
Write these lengths of triangles on the board:

a 5, 12, 13 **b** 6, 7, 8
c 5, 8, 11 **d** 16, 30, 34
e 13, 15, 23

Ask students to find which are right-angled. Are the others acute- or obtuse-angled? Students should not construct the triangles, but should try to sketch them.

Introductory activity
Students may need reminding of the Pythagoras formula.

Exam questions often combine trigonometry questions and Pythagoras questions because they are both used in right-angled triangles. Discuss the examples in the student book.

In the second example in the students' book it would have been possible to use the value for the side found in **a** for part **b**, but students should be discouraged from doing this in case they have made an error in part **a**.

Exercise commentary and misconceptions
Encourage students to always draw a sketch and divide the shape into convenient right-angled triangles.

In question 1**a** students need to use the value they have found for **a** in the second part of the questions. Students may be tempted to assume that angle ABC is a right angle because it looks like one! Remind them that right angles must be labelled as such, unless there is a mathematical explanation.

In question 2**a** students must find the length of BD first, even though this length is not asked for in the question. Triangle ABC is not right-angled!

In question 2**b**, FH must be found before the required angle.

In question 2**c**, KM must be found.

Question 3**a** uses Pythagoras, but question 3**b** uses the tan ratio.

In question 4, JL is the hypotenuse in triangle JKL, but the opposite side in triangle JLM.

In question 5, PR is the opposite side in triangle PRS, but the hypotenuse in triangle QPR.

Plenary
Students may continue with the plenary from S8.3, by completing their graphs.

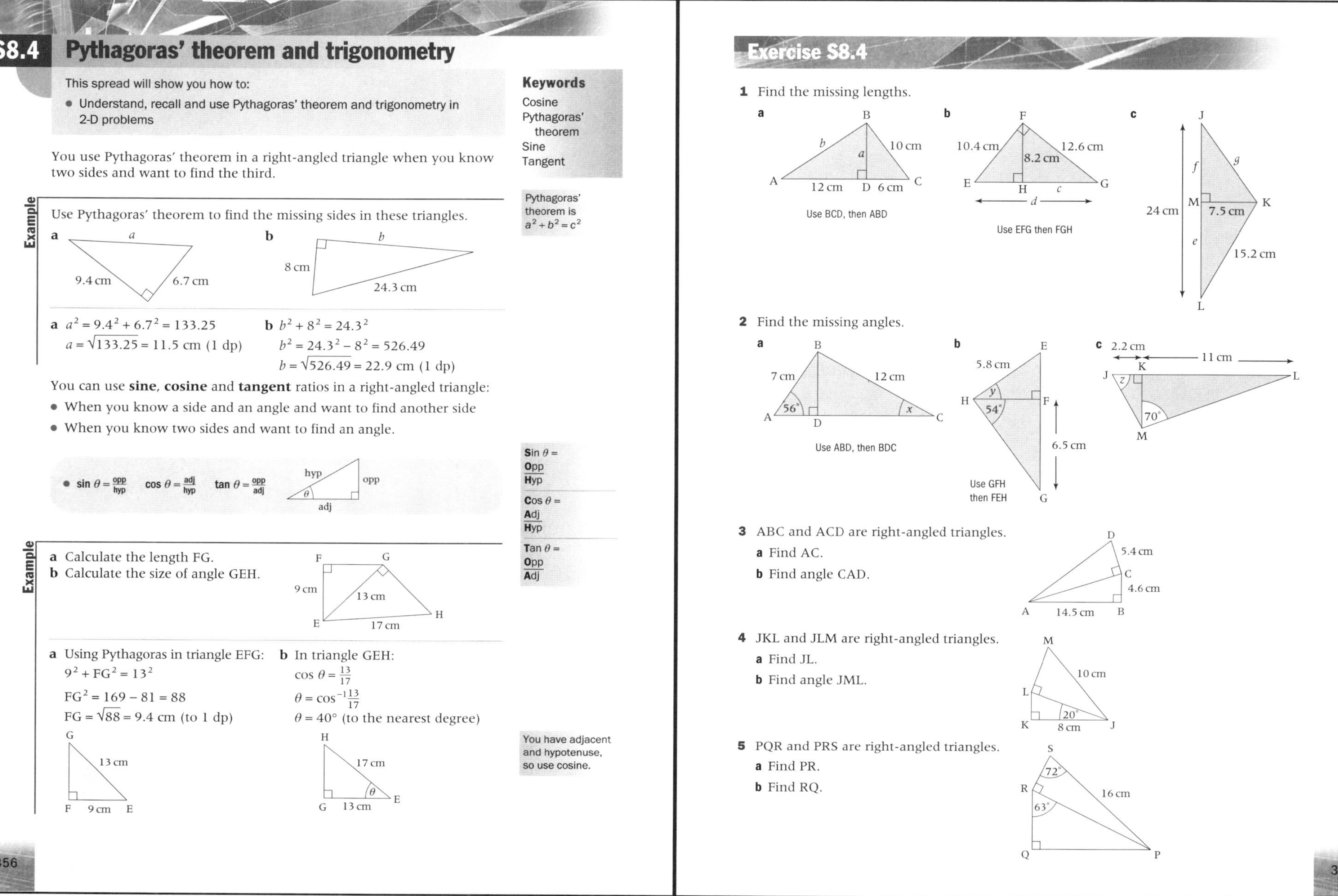

S8.4 Pythagoras' theorem and trigonometry

This spread will show you how to:
- Understand, recall and use Pythagoras' theorem and trigonometry in 2-D problems

Keywords
Cosine
Pythagoras' theorem
Sine
Tangent

You use Pythagoras' theorem in a right-angled triangle when you know two sides and want to find the third.

Pythagoras' theorem is $a^2 + b^2 = c^2$

Example

Use Pythagoras' theorem to find the missing sides in these triangles.

a

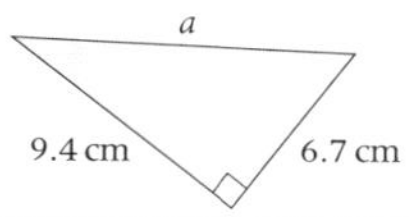

b

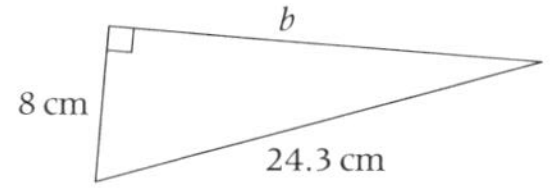

a $a^2 = 9.4^2 + 6.7^2 = 133.25$
$a = \sqrt{133.25} = 11.5$ cm (1 dp)

b $b^2 + 8^2 = 24.3^2$
$b^2 = 24.3^2 - 8^2 = 526.49$
$b = \sqrt{526.49} = 22.9$ cm (1 dp)

You can use **sine**, **cosine** and **tangent** ratios in a right-angled triangle:
- When you know a side and an angle and want to find another side
- When you know two sides and want to find an angle.

- $\sin\theta = \frac{\text{opp}}{\text{hyp}}$ $\cos\theta = \frac{\text{adj}}{\text{hyp}}$ $\tan\theta = \frac{\text{opp}}{\text{adj}}$

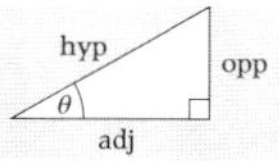

Sin θ = Opp / Hyp

Cos θ = Adj / Hyp

Tan θ = Opp / Adj

Example

a Calculate the length FG.
b Calculate the size of angle GEH.

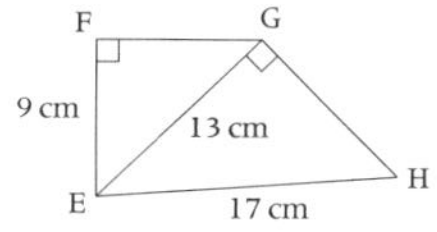

a Using Pythagoras in triangle EFG:
$9^2 + FG^2 = 13^2$
$FG^2 = 169 - 81 = 88$
$FG = \sqrt{88} = 9.4$ cm (to 1 dp)

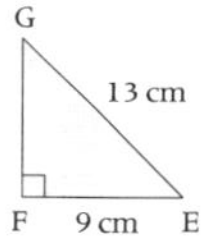

b In triangle GEH:
$\cos\theta = \frac{13}{17}$
$\theta = \cos^{-1}\frac{13}{17}$
$\theta = 40°$ (to the nearest degree)

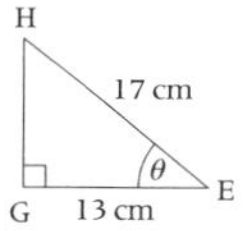

You have adjacent and hypotenuse, so use cosine.

Exercise S8.4

1 Find the missing lengths.

a

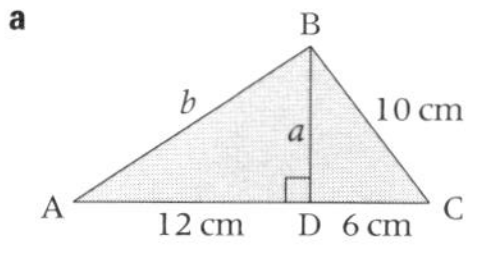

Use BCD, then ABD

b

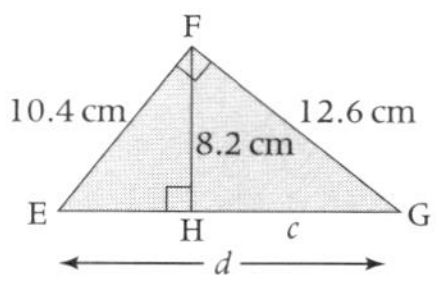

Use EFG then FGH

c

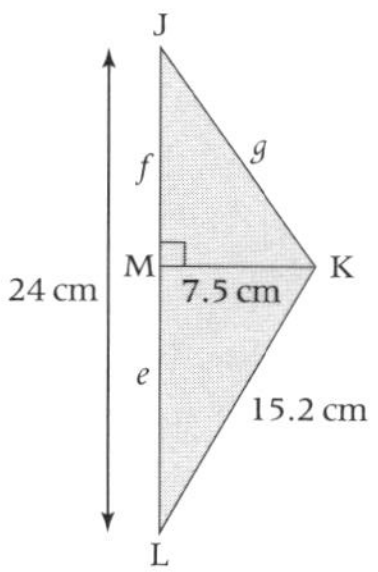

2 Find the missing angles.

a

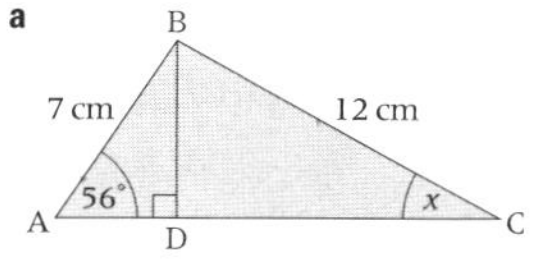

Use ABD, then BDC

b

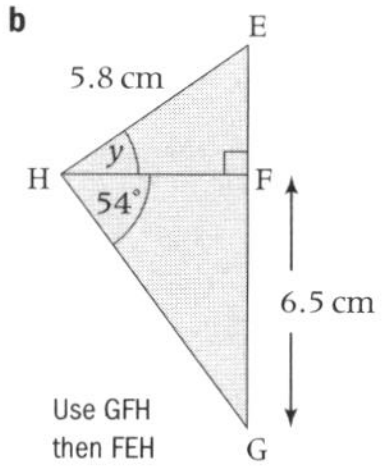

Use GFH then FEH

c

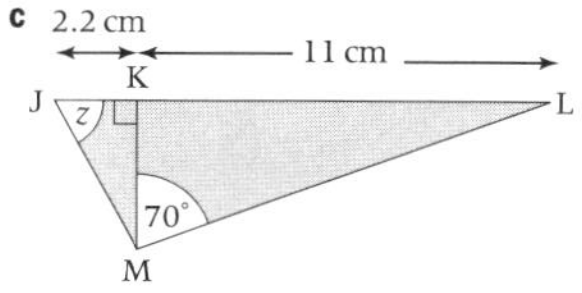

3 ABC and ACD are right-angled triangles.
a Find AC.
b Find angle CAD.

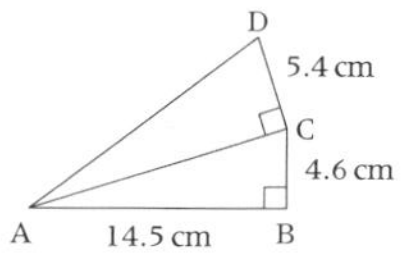

4 JKL and JLM are right-angled triangles.
a Find JL.
b Find angle JML.

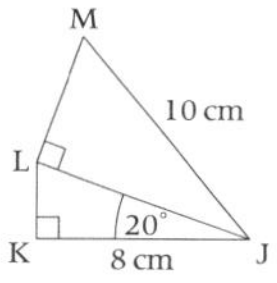

5 PQR and PRS are right-angled triangles.
a Find PR.
b Find RQ.

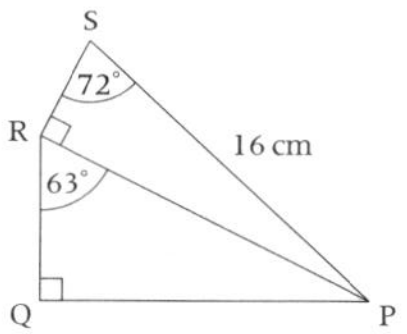

S8.5 Trigonometry in problem solving

Objectives
- Understand, recall and use Pythagoras' theorem and trigonometry in 2-D problems
- Use bearings, maps and scale drawings in problem-solving

Useful resources
- Calculators
- Mini-whiteboards for starter activity

Mental starter
Write these problems on the board:

$5^2 + 8^2 =$

$10^2 - 3^2 =$

$16^2 - 2^2 =$

$9^2 + 9^2 =$

$1.5^2 + 1.5^2 =$

Ask students to work these out mentally.

Introductory activity
Sometimes the necessary right angle has to be inferred from the information given in the question, or from a construction which the question suggests.

Common situations are with compass directions, angle in a semicircle, the height of a triangle, especially one which is isosceles or equilateral, a rectangle of square or the diagonals of a rhombus or kite.

In any such situation, students should draw a diagram and carefully label any right angles.

In the first example in the students' book, the height of the parallelogram is necessary to find the area, so a perpendicular has been drawn from R to PQ.

The right angle in the second example is between the compass directions North and East.

Exercise commentary and misconceptions
The questions combine trigonometry and Pythagoras with practical problems. Ensure students draw a sketch and label all the information they have. Always look for right-angled triangles and draw them in if necessary. Try to keep intermediate work either as full decimals or in terms of roots and trig values.

Plenary
Sketch an equilateral triangle with sides 2 cm. Divide the triangle into 2 right-angled triangles.

Show that $\sin 60° = \frac{\sqrt{3}}{2}$, that $\cos 60° = \frac{1}{2}$ and $\tan 60° = \sqrt{3}$.

S8.5 Trigonometry in problem solving

This spread will show you how to:

- Understand, recall and use Pythagoras' theorem and trigonometry in 2-D problems
- Use bearings, maps and scale drawings in problem-solving

Keywords
Bearing
Cos
Pythagoras' theorem
Sin
Tan

In more complex problems involving lengths and angles, it helps to sketch the situation.

Example

PQRS is a parallelogram. PQ = 9.4 cm. QR = 7.8 cm. Angle PQR = 47°.
Find the area of the parallelogram.

Draw a diagram.

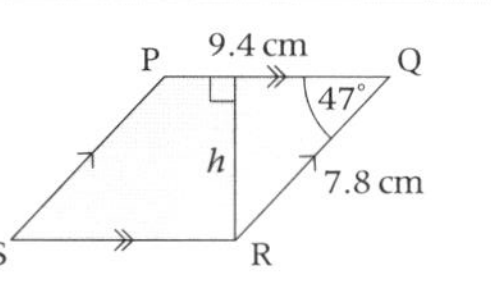

First find the vertical height, h, between PQ and RS.

h is at right-angles to PQ and RS.

Do not round intermediate values in the calculation.

$\sin 47° = \frac{h}{7.8}$

$h = 7.8 \times \sin 47° = 5.704 \ldots$

Area of PQRS $= h \times b = 5.704 \ldots \times 9.4 = 53.6\ \text{cm}^2$ (to 3 sf)

Example

A boat sails from a harbour on a bearing of 072° to a buoy 12 km away. Then it changes direction and sails 20 km to a lighthouse due east of the harbour.

a On what bearing does the boat sail from the buoy to the lighthouse?
b How far is it from the lighthouse back to the harbour?

Draw a diagram.

H is the harbour.
B is the buoy.
L is the lighthouse.

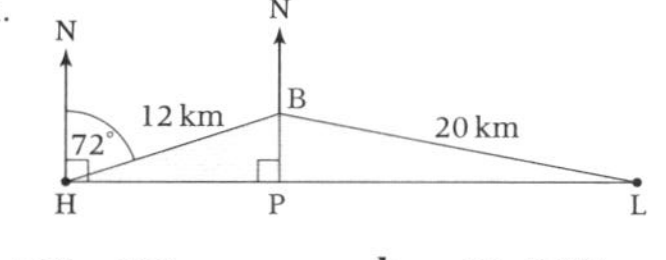

Draw in BP to divide HBL into two right-angled triangles.

a $\angle BHP = 90° - 72° = 18°$

$\sin 18° = \frac{BP}{12}$

$BP = 12 \times \sin 18° = 3.708 \ldots$

$\cos PBL = \frac{3.708 \ldots}{20}$

$\angle PBL = \cos^{-1}\left(\frac{3.708 \ldots}{20}\right)$

$= 79.314 \ldots°$

Bearing of L from B
$= 180° - 79.314 \ldots°$
$= 101°$ (nearest degree)

b BP = 3.708...

$PL^2 = 20^2 - 3.708 \ldots^2$ Pythagorean theorem

$PL = \sqrt{386.249 \ldots}$

$= 19.653 \ldots$

$PH^2 = 12^2 - 3.708 \ldots^2$

$PH = \sqrt{140.291 \ldots}$

$= 11.412 \ldots$

$HP = PL + PH$
$= 11.412 \ldots + 19.653 \ldots$
$= 31.1$ km (to 3 sf)

Exercise S8.5

1 Find the area of a parallelogram with side lengths 5 cm and 11 cm and smaller angle 64°.

2 A rhombus has side lengths 9 cm and smaller angle 52°.
Find the area of the rhombus.

3 Find the area of a rhombus with sides 7 cm and smaller angle 40°.

4 A chord AB with length 10 cm is drawn inside a circle with centre O and radius 7 cm.
Find the angle AOB.

5 A chord PQ is drawn inside a circle with centre O and radius 8 cm such that angle POQ = 80°.
Find the length of the chord PQ.

6 A chord ST is drawn inside a circle with centre O and radius 9 cm such that angle SOT = 110°.
Find the length of the chord ST.

7 An isosceles triangle has side lengths 9 cm, 9 cm and 6 cm.
Find the angle between the two equal sides.

8 Jenny walks 4 km on a bearing 052°. She changes direction and walks a further 5 km to finish due east of her starting point.
Find how far Jenny is from her starting point.

9 Liz leaves home and cycles 16 km on a bearing 215° to a lake.
She changes direction and cycles 12 km to a wood which is due south of her home.

a On what bearing does she cycle from the lake to the wood?
b How far does she have to cycle home?

10 A flag pole TP, with T at the top, is held upright by two ropes, TX and TY, fixed on horizontal ground at X and Y.
Angle PXT = 23°. Angle PYT = 36°. TX = 10 m.
Find the length of TY.

11 Ali and Pete are estimating the height of a phone mast.
Ali stands 15m from the mast and measures the angle of elevation to the top as 60°.
Pete stands 25m from the mast and measures the angle of elevation to the top as 46°.
Can they both be correct? Discuss.

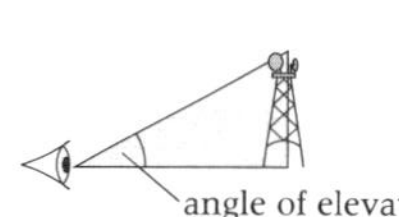

S8 Exam review

Key objectives

- Understand, recall and use Pythagoras' theorem in 2-D problems
- Understand, recall and use trigonometrical relationships in right-angled triangles, and use these to solve problems

Worked solution	Commentary and misconceptions
1 a 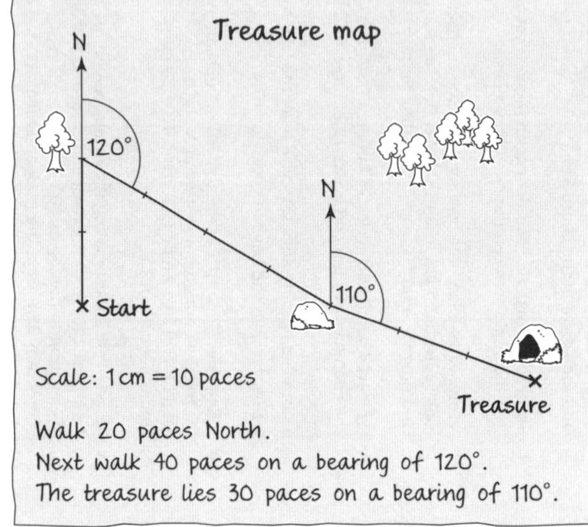	Ensure that students understand angle measure, bearings and scale drawings before attempting this question. Ensure that students draw the diagram accurately, using the correct scale.
b i Distance of treasure from starting point = 63.7 paces.	Students should recall Pythagoras' theorem for right-angled triangles.
ii Bearing = 279.3°.	Ensure students use the correct scale conversion to give their final answers. Answers should be given in paces and degrees.
2 a Triangle EDG is right-angled. Length DG^2 = length ED^2 + length EG^2 $= 6^2 + 10^2 = 136$ So length DG $= \sqrt{136} = 11.66190379$ $= 11.7$ m to 3 sf	Encourage students to look at the diagram carefully. Students should recall Pythagoras' theorem and how to use it to find missing sides. Ensure students realise that DG is the hypotenuse. Ensure that students remember to calculate $\sqrt{136}$. Ensure they round correctly and express their answer in the units requested.
b Triangle EFG is right-angled. Angle $\cos x = \frac{8}{10} = \frac{4}{5}$ So, $x = \cos^{-1}(\frac{4}{5})$ $= 36.86989765 = 36.9°$ to 1 dp	Students should recall how to use trigonometrical relationships in right-angled triangles to find missing angles. Ensure they realise that FG is adjacent to *x* and that EG is the hypotenuse. Ensure that students use the cosine rule. Ensure students can use the inverse trig function of their calculator correctly. Ensure students state their answer in degrees to 1 dp as required.

GCSE formulae

In your students' Edexcel GCSE examination they will be given a formula sheet like the one on this page.

Volume of a prism = area of cross-section × length

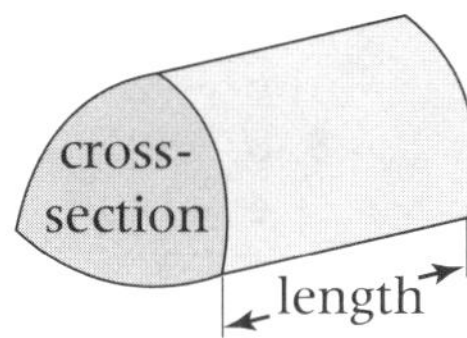

Volume of sphere = $\frac{4}{3}\pi r^3$

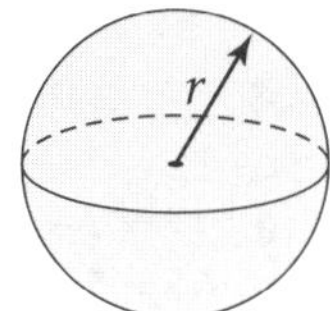

Surface area of sphere = $4\pi r^2$

Volume of cone = $\frac{1}{3}\pi r^2 h$

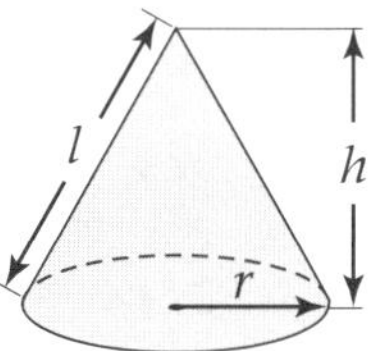

Curved surface area of cone = $\pi r l$

In any triangle ABC

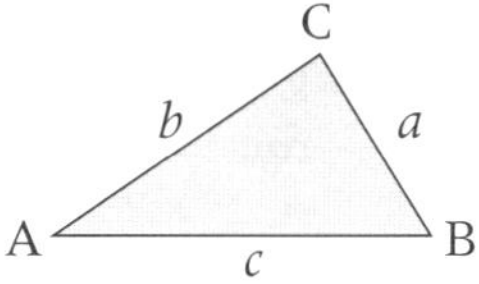

Sine rule $\frac{a}{\sin A} = \frac{b}{\sin B} = \frac{c}{\sin C}$

Cosine rule $a^2 = b^2 + c^2 - 2bc \cos A$

The Quadratic Equation
The solutions of $ax^2 + bx + c = 0$ where $a = 0$, are given by

$$x = \frac{-b \pm \sqrt{(b^2 - 4ac)}}{2a}$$

N1 Before you start ...

1 a Four thousand **b** Four hundred
c Four tenths **d** Four thousandths

2 a

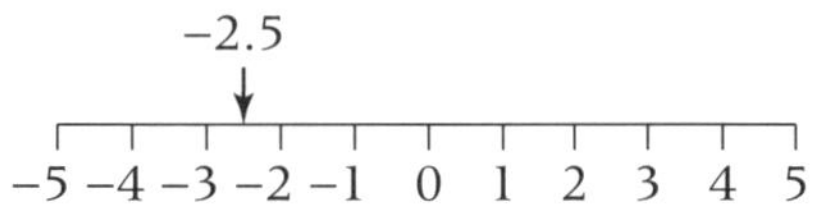

−2.5
−5 −4 −3 −2 −1 0 1 2 3 4 5

b −3, −2.4, −1.8, 0, +1.5, +5

3 ai 1, 2, 3, 6
aii 1, 2, 3, 4, 6, 12
aiii 1, 2, 4, 7, 14, 28
aiv 1, 2, 3, 4, 6, 9, 12, 18, 36
b 2, 3, 5, 7, 11, 13, 17, 19, 23, 29, 31, 37, 41, 43, 47

N1.1

1 a 0.1, 0.3, 1, 1.3, 2, 3.1
b 6.07, 7.06, 27.6, 77.2, 607
2 a 682.8, 862.6, 6000.8, 6008, 8000.6
b 47.9, 49.7, 74.9, 79.4, 94.7, 97.4
3 a 167 **b** 248 **c** 7.16
d 10.95 **e** 2430 **f** 2813
4 a 21.4 **b** 6.73 **c** 410.6
d 20.07 **e** 0.6025 **f** 8.6
5 a 4.52 **b** 5.5 **c** 16.8 **d** 16.8
6 a 7.03, 7.08, 7.3, 7.38, 7.8, 7.83
b 2.18, 2.4, 4.18, 4.2, 8.24, 8.4
7 a 18.7, 18.16, 17.6, 17.16, 16.7, 16.18
b 13.2, 13.145, 2.5, 2.38, 1.1, 1.06
8 ai 3050 **aii** 3000 **bi** 1760 **bii** 1800
ci 290 **cii** 300 **di** 50 **dii** 100
ei 40 **eii** 0 **fi** 740 **fii** 700
9 a 3000 **b** 1000 **c** 0
d 25 000 **e** 16 000 **f** 168 000
10 ai 39.1 **aii** 39.11 **bi** 7.1 **bii** 7.07
ci 5.9 **cii** 5.92 **di** 512.7 **dii** 512.72
ei 4.3 **eii** 4.26 **fi** 12.0 **fii** 12.01
gi 0.8 **gii** 0.83 **hi** 26.9 **hii** 26.88
11 ai 0.1 **aii** 0.07 **aiii** 0.070
bi 15.9 **bii** 15.92 **biii** 15.918
ci 128.0 **cii** 128.00 **ciii** 127.998
di 887.2 **dii** 887.17 **diii** 887.172
ei 55.1 **eii** 55.14 **eiii** 55.145
fi 0.0 **fii** 0.01 **fiii** 0.007
12 a 1306 **b** 2.085 **c** 1085 **d** 2.487
e 0.0008 **f** 6.19 **g** 0.04513 **h** 0.0045

N1.2

1 a −2 **b** −1 **c** −2 **d** −2 **e** −5 **f** −5
2 a +1 **b** +3 **c** +4 **d** +9 **e** +12 **f** +5
3 a −5 **b** −2 **c** −4 **d** +2 **e** +4 **f** +7
4 a −2 **b** + **c** +, − **d** −, −
5 a +22 **b** −12 **c** −2 **d** −14
e +12 **f** +12 **g** +19 **h** −15
i +11 **j** −61 **k** +344 **l** +49
6 a −10.8 **b** +6.3 **c** +13.5 **d** −0.7
e −38.2 **f** +112.7
7 a £5.49 **b** £172.38
8 a −13 °C **b** 15 °C **c** 41 °C

N1.3

1 a −2 **b** −1 **c** −2 **d** −2 **e** −5 **f** −5
1 a −15 **b** −18 **c** −21 **d** −56 **e** −36 **f** −12
2 a +16 **b** +16 **c** +15 **d** +42 **e** +56 **f** +81
3 a −25 **b** −32 **c** −72 **d** −20 **e** +30 **f** +49
g +16 **h** −20 **i** −18 **j** +26 **k** −42 **l** −48
4 a −3 **b** −4 **c** −7 **d** −2 **e** +19 **f** +5
5 a −2 **b** −5 **c** +5 **d** +4 **e** −22 **f** −1
g +40 **h** +4 **i** +5 **j** −17 **k** −3 **l** +27
6 a + **b** −6 **c** +45 **d** −12
7 a +450 **b** −150 **c** +63 **d** −25
e −0.73 **f** +0.092
8 a −49 **b** +63 **c** +3.77 **d** −619.7
e +140.9 **f** +0.09
9 a −36 **b** +0.1 **c** −0.98 **d** −0.0087
e +0.0073 **f** −0.00006
10 a −55 **b** +0.48 **c** +5.266 **d** −156
e +0.0082 **f** +0.50005
11 a +0.18 **b** −27 **c** −7 **d** −380

N1.4

1 a 12 **b** 8 **c** 4 **d** 6
2 a $77 = 7 \times 11$ **b** $51 = 3 \times 17$
c $65 = 5 \times 13$ **d** $91 = 7 \times 13$
e $119 = 7 \times 17$ **f** $221 = 13 \times 17$

3

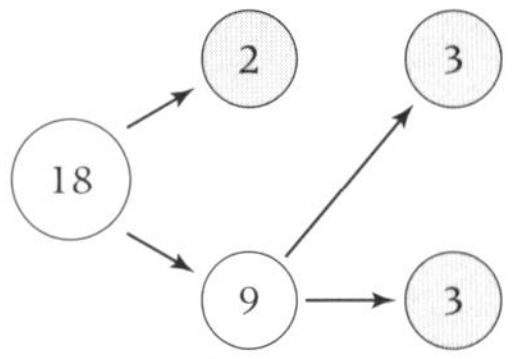

18
2
9
3
3

4 a $2^2 \times 3^2$ **b** $2^3 \times 3 \times 5$
c 2×17 **d** 5^2
e $2^4 \times 3$ **f** $2 \times 3^2 \times 5$
g 3^3 **h** $2^2 \times 3 \times 5$
5 a $2^2 \times 263$ **b** $2^9 \times 5$
c $2 \times 3^2 \times 5 \times 7$ **d** $3 \times 5^2 \times 11$
e $5 \times 11 \times 13$ **f** $7 \times 11 \times 13$
g 3×73 **h** 17^2
i $2^3 \times 5 \times 71$ **j** $5 \times 7^2 \times 11$
k $7 \times 13 \times 19$ **l** $2 \times 3^2 \times 11 \times 17$
m $2^2 \times 11 \times 13 \times 17$ **n** $2 \times 5 \times 7 \times 13^2$
o $2^2 \times 23 \times 31$ **p** $3^3 \times 13 \times 29$
6 a $2 \times 3 \times 35$, $2 \times 5 \times 21$, $2 \times 7 \times 15$, $3 \times 7 \times 10$, $3 \times 5 \times 14$, $5 \times 6 \times 7$
b $1 \times 1 \times 121$, $2 \times 3 \times 35$, $2 \times 5 \times 21$, $2 \times 7 \times 15$, $3 \times 7 \times 10$, $3 \times 5 \times 14$, $5 \times 6 \times 7$

N1.5

1 a $6^2 = 36$, so 6 joins to itself.
b 1, 2, 3, 4, 6, 8, 12, 16, 24, 48

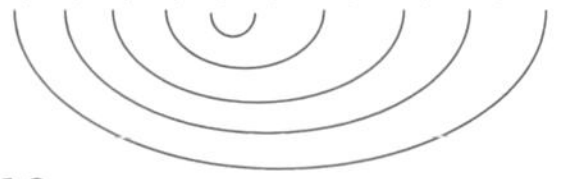

c 12
2 a 1 **b** 1 **c** 3 **d** 2 **e** 8 **f** 10
3 a Multiples of 12 = 12, 24, 36, 48, ...
Multiples of 9 = 9, 18, 27, 36, 45, ...
b 36
4 a 20 **b** 36 **c** 30 **d** 60 **e** 70 **f** 40
5 8
6 a 5600 **b** 4432 **c** 12 720
d 14 168 **e** 5105 **f** 10 220

N1 Exam review

1 ai −25.9 **aii** −25.9 **aiii** 25.9
bi −53.2 **bii** −0.532 **biii** 5.32
2 ai $2^2 \times 3 \times 5$ **aii** $2^5 \times 3$
b 12 **c** 480

S1 Before you start ...

1 ai 16 cm aii 160 mm aiii 12 cm^2
bi 18 cm bii 180 mm biii 20.25 cm^2
ci 19.1 cm cii 191 mm ciii 14 cm^2
2 54 cm^2

S1.1

1 a 28 cm^2 b 22.26 cm^2 c 26.1 cm^2
d 7440 mm^2 e 10 290 mm^2
2 a 7.5 cm^2 b 11.76 cm^2 c 126 mm^2
d 6 cm^2 e 14 cm^2 f 7.2 cm^2
g 10.8 cm^2
3 ai 40 cm aii 72 cm^2
bi 33 cm bii 32 cm^2
ci 36 cm cii 44 cm^2
4 479 cm^2

S1.2

1 a 15 cm^2 b 33.48 cm^2 c 45.6 cm^2
d 31.5 cm^2 e 13.34 cm^2
2 a 12 cm^2 b 20 cm^2 c 20.5 cm^2
d 600 mm^2 e 1250 mm^2
3 205.5 cm^2

S1.3

1 a 25.1 cm b 239 mm c 50.3 cm
d 47.1 cm e 75.4 mm f 132 mm
g 82.9 cm
2 a 75.4 mm b 145 cm c 330 mm
d 3.77 cm e 22.6 cm f 393 cm
3 a 50.3 cm^2 b 4540 mm^2 c 201 cm^2
d 177 cm^2 e 452 mm^2 f 1390 mm^2
g 547 cm^2
4 a 452 mm^2 b 1660 cm^2 c 8660 mm^2
d 1.13 cm^2 e 40.7 cm^2 f 12 300 cm^2
5 18.8 m
6 7.0 cm to 1 dp
7 a 13.9 cm^2 b 3.79 cm^2

S1.4

1 a 39.3 cm^2 b 81.4 cm^2 c 905 mm^2
d 127 cm^2 e 402 mm^2 f 373 cm^2
g 422 cm^2 h 103 cm^2
2 a 226 cm^2 b 8.31 cm^2 c 65.3 m^2
d 195 cm^2 e 142 mm^2 f 4.51 cm^2
3 a 25.7 cm b 37.0 cm c 123 mm
d 46.3 cm e 82.3 mm f 79.2 cm
g 84.3 cm h 33.5 cm
4 a 61.7 cm b 11.8 cm c 33.2 m
d 57.3 cm e 48.8 mm f 8.71 cm
5 a 6.28 m^2
b 20 flowers; he has 0.28 m^2 left
6 113 m
7 2510 cm^2

S1.5

1 a 142 cm^2 b 98 cm^2 c 114 cm^2
d 65.6 cm^2 e 96 cm^2 f 102 mm^2
2 a 44.0 cm^2 b 165 cm^2 c 75.4 cm^2
d 66.4 cm^2
3 a 330 cm^2 b 468 cm^2

S1 Exam review

1 32 cm
2 88.4 cm^2

A1 Before you start ...

1 a 45 b 52 c −26 d 196
e 7 f 13 g −50 h 30
2 a 15 b 21 c 15 d 15
3 a 1, 2, 3, 4, 6, 8, 12, 24
b 1, 2, 3, 4, 6, 9, 12, 18, 36
c 1, 2, 4, 5, 10, 20, 25, 50, 100
d 1, 11, 121
4 a 3 b 4 c 10 d 6
e 4 f 25 g 33 h 7

A1.1

1 a $5w$ b $\frac{6}{k}$ c y^2 d $6ab$ e $8k^3$
2 a 20 b 4 c 36 d 22 e 108
3 a $11a+6b$ b $2t+26$ c $x-12y$
d p^2+14p e $20xy$ f $7ab$
4 Abdul
5 a $28mn$ b $12m^2$ c $10p$
d 2 e $24abc$ f $6k^3$
g $4b$ h $9c$
6

$4p+7q$	$4p+7q$	**$4p+2q$**
$6mn$	**$10mn$**	$6mn$
$2d$	**2**	$2d$
$2n-8$	**$2n$**	$2n-8$

7 ai $8p+16$ aii $32p$ b $3x$, $2y$
8

$9b-2a$	$12a^2$	$4b$
$2ab$	$5p^3+7p^2+10p$	$13abc$
$5m-4$	$60m^3$	$\frac{2}{a^2}$

A1.2

1 a $4n+20$ b $6b-42$
c a^2+3a d $ab-ac$
e $8x+12y-16z$ f $2h^2+18h$
2 a $-3k-27$ b $-2h+10$
c $-w+4$ d $-t+p$
e $-k^2-7k$ f $-18m+9k-36$
g $-x^2+x+8$ h $-2x^2-6$
i $-3+3x$
3 a $10c+62$ b $23x+67$
c $2x^2+10x$ d $17t^2+32t$
e $7x-45$ f $2x-11$
g $2m-26$ h $-11g+33$
i $p+14$ j $2q-7$
4 $-5x^2+44x$
5 a $3(2x-1)$ b $6x-3$
c $6x-3=15$, which gives $6x-18=0$
6 a For example, $8(3x+2)$
b For example, $2(2x+3)+5(4x+2)$
7 $y^2(y+7)$, y^3+7y^2

A1.3

1 a x^2+5x+6 b $p^2+11p+30$
c w^2+5w+4 d $c^2+10c+25$
e x^2+2x-8 f $y^2+5y-14$
g $t^2+4t-12$ h $x^2-7x+10$
i $y^2-14y+40$ j w^2-3w+2
k $p^2-10p+25$ l $q^2-24q+144$
2 a $6x^2+17x+7$ b $10p^2+19p+6$
c $6y^2+11y+4$ d $4y^2+24y+36$
e $10t^2+12t-16$ f $15w^2+42w-9$
g $6x^2-6y^2$ h $9m^2-24m+16$
i $6p^2-pq-40q^2$ j $4m^2-12mn+9n^2$
3 a $(x-3)(x+6)=x^2+3x-18$
b $(2m-3)^2=4m^2-12m+9$

4 a $a^2 + 2ab + b^2$ b 16
c For example, $2.5^2 + 2 \times 2.5 \times 3.5 + 3.5^2 = 6^2$
5 a $(3x - 1)(2x + 3) = 75$, which gives $6x^2 + 7x - 78 = 0$
b $(3x - 4)(5x + 2) = 2 \times 75$, which gives $15x^2 = 14x^2 + 158$

A1.4

1 a $2(x + 2)$ b $3(y - 2)$
c $12(p + 3q)$ d $5(5w - 1)$
e $x(6y + w)$ f $b(a - 2c)$
g $q(pr + rt - sw)$ h $x(5y - 1)$
i $2x(y + 3)$ j $2a(2b - 3a)$
k $5p(5p - 2)$ l $7x(1 + 2y)$
m $2a(c + 2a - 4)$ n $5m(3n - 1 + 2m^2)$
o $6p(p^3 - 2)$
2 Correct factorisations are: Clare $5x(1 + 2y)$, Ben $3p(2q + 1)$, Vicky $7p(3 + 2q)$
3 a $23(x + y)$ b $(a - b)(a - b + 5)$
c $(q + r)(6 - (q + r)^2)$ d $7(pt - w)$
4 a $(a + b)(x + y)$ b $(c + b)(d + m)$
c $(a + b)(a + 2)$ d $(c - m)(d + e)$
5 a $4(x - 1)$ b $20(b + 2)$
6 a 6 b 16.5 c 58.6 d 33.2
7 $4(3x + 2) - 2(2x - 1) = 8x + 10 = 2(4x + 5)$

A1.5

1 a $(x + 2)(x + 4)$ b $(x + 3)(x + 7)$
c $(x + 4)(x + 7)$ d $(x + 3)(x + 8)$
e $(x - 2)(x - 6)$ f $(x - 3)(x - 6)$
g $(x - 9)(x - 4)$ h $(x + 4)(x - 3)$
i $(x - 7)(x + 5)$ j $(x + 9)(x - 3)$
k $(x - 16)(x + 2)$ l $(x + 20)(x - 2)$
2 a $x^2 + 9x - 22$ b $(x + 11)(x - 2)$
3 a $(x + 12)(x - 6)$ b $(x - 12)(x + 2)$
c $(x - 15)(x - 5)$ d $(x + 16)(x - 4)$
e $(x - 8)(x + 8)$ f $(x - 4)(x - 25)$
4 a $(x + 2)(x + 19)$ b $x(5x + 5 + y)$
c $(x + 11)^2$ d $(x + 9)(x - 2)$
e $(p + 3)(p + 11)$ f $x(2x + 3y)$
5 $(x + 5)(x + 4) = 12$
$x^2 + 9x + 20 = 12$
$x^2 + 9x + 8 = 0$
$(x + 1)(x + 8) = 0$
6 a $2(x + 4)(x + 7)$ b $x(x - 8)(x + 3)$
c $x(x - 4)(x + 4)$
7 $(2.3 + 1.7)^2 = 4^2 = 16$

A1 Exam review

1 a $2x(2 + x)$ b $x^2 + 3x - 10$ c $12a^2b^5$
2 a $4x + 8$ b 15.5

N2 Before you start ...

1 a 120 b 138 c 90 d 265
2 a 4438 b 1977 c 857 d 14 224
3 a 147 b 1515 c 66 560 d 51 450
4 a 11 700 b 78 408 c 205 d 67 564

N2.1

1 a 30 b 30 c 50 d 210
e 780 f 23 780
2 a 6 b 4 c 22 d 39
e 18 f 454
3 a 200 b 200 c 100 d 700
e 1400 f 134 600
4 a 2000 b 13 000 c 8000 d 11 000
e 78 000 f 156 000
5 a 0.3 b 0.7 c 0.3 d 0.2
e 4.6 f 105.4
6 a 0.32 b 0.46 c 15.30 d 104.68
e 16.45 f 0.00
7 a 480 b 1200 c 490 d 14 000
e 530 f 15 000
8 a 0.36 b 0.42 c 0.057 d 0.0047
e 1.4 f 0.0000042
9 a 200 b 2000 c 5 d 10
e 0.0005 f 100 000
10 a 0.62 b 0.57 c 0.56 d 380
e 550 f 7 300 000
11 a $400 \div 20$ b 40×40 c $1000 \div 90$
d $4000 + 10\,000$ e $100 + (2000 \div 50)$
12 a 20 b 1600 c 11 d 14 000 e 140
13 a 16.9047619, estimate is slightly high
b 1677, estimate is close
c 11.44565217, estimate is close
d 16 040, estimate is slightly low
e 153.3846154, estimate is close

N2.2

1 a 0.8 b 0.5 c 0.4 d 1.1
e 0.4 f 1
2 a 5.8 b 6.5 c 7.4 d 4.1
e 11.4 f 10
3 a 10 b 10.1 c 10.3 d 11.1
e 16.4 f 7.5
4 a 7.77 b 5.25 c 3.6 d 3.9
e 2.13 f 13.04
5 a 4.15 b 6.85 c 4.58 d 10.42
e 12.14 f 2.04
6 a 0.7 b 0.6 c 0.3 d 3.4
e 10.2 f 14.1
7 a 0.8 b 0.7 c 0.8 d 4.1
e 10.7 f 14.4
8 a 0.74 b 1.05 c 1.51 d 7.14
e 1.08 f 0.64
9 a 1.62 b 3.51 c 0.45 d 4.37
e 14.52 f 11.66
10 a 1.72 b 3.61 c 0.65 d 4.67
e 15.02 f 12.36

N2.3

1 a 12.4 b 12 c 13.1 d 22.5
e 23.1 f 26.1
2 See Q1
3 a 13 b 1.871 c 201.321 d 45
e 38.97 f 21.69
4 a 140.33 b 242.3 c 98.807 d 203.59
e 161.002 f 102
5 See Q4
6 a 2.2 b 9.1 c 15 d 4.9
e 8.4 f 2.9
7 See Q6
8 a 2.6 b 7.86 c 5.9 d 92.54
e 0.97 f 24.27
9 a 13.896 b 19.45 c 359.79 d 7.683
e 0.326 f 11.42
10 See Q9
11 a 4.1 b 40.2 c 11.288 d 5.968
e 0.892 f 0.469

N2.4

1 **a** 4.8 **b** 4.8 **c** 0.48 **d** 48 **e** 48
2 **a** 9.1 **b** 9.1 **c** 0.91 **d** 91 **e** 91
3 **a** 42 **b** 16 **c** 7 **d** 104 **e** 2 **f** 28
4 **a** 4.2 **b** 1.6 **c** 0.7 **d** 10.4 **e** 2 **f** 0.28
5 **a** $9 \times 7 = 63$, $63 \div 7 = 9$, $63 \div 9 = 7$
b $8 \times 6 = 48$, $48 \div 6 = 8$, $48 \div 8 = 6$
c $7 \times 13 = 91$, $91 \div 13 = 7$, $91 \div 7 = 13$
d $18 \times 15 = 270$, $270 \div 15 = 18$, $270 \div 18 = 15$
e $3.5 \times 5 = 17.5$, $17.5 \div 3.5 = 5$, $17.5 \div 5 = 3.5$
f $3.9 \times 2.4 = 9.36$, $9.36 \div 2.4 = 3.9$, $9.36 \div 3.9 = 2.4$
7 **a** 4.5 **b** 4500 **c** 45 **d** 0.45
e 5 **f** 90 **g** 500
8 **a** 345.8 **b** 345.8 **c** 38 **d** 345.8
e 9.1 **f** 0.0091
9 **a** 10 185 **b** 101.85 **c** 0.10185 **d** 0.35
e 3.5 **f** 0.00291
10 **a** 29.61 **b** 0.2961 **c** 6300

N2.5

1 **a** 98 **b** 152 **c** 273 **d** 323 **e** 308
2 See Q1
3 **a** 9.8 **b** 15.2 **c** 2.73 **d** 0.0323 **e** 0.308
4 **a** 80 **b** 12 **c** 12 **d** 21 **e** 22
5 See Q4
6 **a** 0.8 **b** 1.2 **c** 1.2 **d** 0.21 **e** 0.22
7 **a** 24.91 **b** 4.284 **c** 105.84 **d** 130.8985
e 42.9442
8 See Q7
9 **a** 3.87 **b** 0.775 **c** 0.916 **d** 7.53 **e** 18.13
10 See Q9
11 **a** 3.45 **b** 4.15 **c** 7.74 **d** 4.08 **e** 2.35
12 See Q11
13 **a** 5.26 **b** 28.88 **c** 1384.29 **d** 175.56 **e** 28.65
14 See Q13

N2 Exam review

1 **a** 2.7 **b** 20 **c** 49 **d** 4.8
2 **a** 119.31 **b** 119 310 **c** 1.23

A2 Before you start ...

1 **a** 9 **b** 4 **c** 30 **d** 10
2 **a** I think of a number and multiply it by 6.
b I think of a number and multiply it by itself.
c I think of a number, multiply it by 2 and then subtract 3.
d I think of a number, subtract 4 from it and then multiply by 4.
e I think of a number and divide it by 7.
f I think of a number, square it and then multiply by 2.
g I think of a number, multiply it by 2 and then square.
h I think of a number, multiply it by 3 and then subtract the result from 10.
3 **a** $3x + 27$ **b** $8x - 4$ **c** $6 - 12y$
d $x^2 - 7x$
4 **a** $5 > 3$ **b** $-9 < 1$ **c** $-2 > -5$
d $0.9 > 0.85$ **e** $-\frac{1}{4} > -\frac{1}{2}$

A2.1

1

20, −8, x, 40 → ÷ 4 → + 5 → 10, 3, $\frac{x}{1.5} + 5$, 15

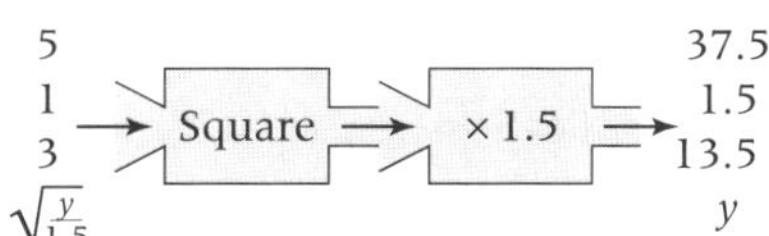

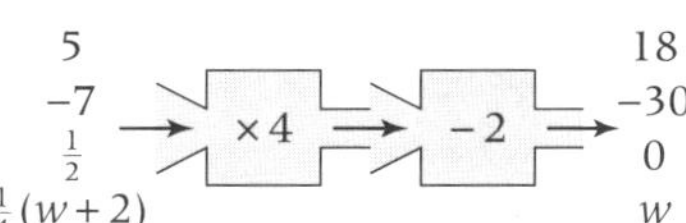

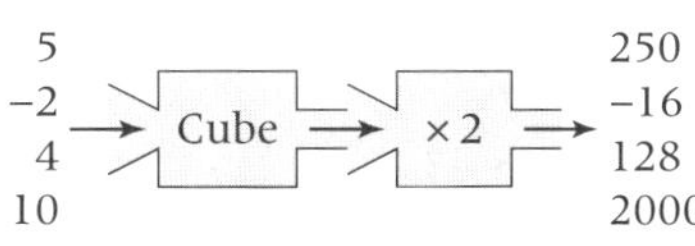

2 The starting numbers are different, so 10% of the final answer is different from 10% of the starting value.
3 **a** 5.5 **b** 5 **c** −3 **d** 40 **e** 100
4 Yes, −5, as 25 has two possible square roots
5 **ai** Add 3
aii Multiply by 2, subtract 1
aiii Add 5, divide by 2
aiv Square, multiply by 3
bi 14 **bii** 9 **biii** 43 **biv** 4 or −4
6 **a** −2 **bi** 7 **bii** 10 **biii** $-\frac{3}{16}$
7 **a** Many possibilities, e.g. ×2, ×2, ×2, +20, −7
b Depends on **7a**

A2.2

1 **a** 5 **b** 6 **c** 10 **d** −2 **e** 5 or −5
f 13 **g** $8\frac{1}{3}$ **h** $\sqrt[3]{16}$ **i** 100

2

[1] 1	3	■	[2] 4
0	■	[3] 4	0
[4] 5	[5] 1	■	■
■	[6] 3	2	5

3 sides = 1 and 14
4 51
5 **a** 5 **b** 1
6 **a** $2x - 10$ **b** $5x - 20 = 180$ **c** 40°, 70°, 70°

A2.3

1 **a** −1 **b** 2 **c** 3 **d** $-\frac{13}{16}$
2 **a** $6a - 2 = 2a + 6$ **b** $3b - 14 = b$
c $1 - 4c = 2 - 8c$ **d** $5d + 7d - 3 = 10 - d$
3 **a** $8x - 2 = 2x + 10$, $x = 2$ **b** $5x + 3 = 24 - 2x$, $x = 3$
c $11 - 2x = 14 - 3x$, $x = 3$
4 **a** 40°, 60°, 80°
b Square: 8 × 8, rectangle: 10 × 6
c Mark: 160 cm, Miranda: 144 cm

A2.4

1 **a** $\frac{1}{3}$ **b** $\frac{1}{3}$ **c** $1\frac{1}{3}$ **d** $\frac{7}{8}$
e −5 **f** $\frac{5}{11}$ **g** $-2\frac{1}{3}$ **h** $\frac{1}{3}$
2 **a** 5 **b** 10 **c** 28 **d** −1
3 **a** −8 **b** $\frac{3}{4}$ **c** $\frac{2}{11}$ **d** $-1\frac{1}{4}$
4 **a** −7 **b** 2 **c** −4 **d** 2

5 a $\frac{16}{x} = 10$, $x = 1.6$ b $\frac{12}{x+4} = 7$, $x = -2\frac{2}{7}$
c $\frac{11}{x-3} = \frac{8}{x}$, $x = -8$
6 a 36 b 14 c 80
7 $x = 37\frac{2}{3}$
Values for Set 1: $74\frac{1}{3}$, 115, $192\frac{1}{3}$, $263\frac{2}{3}$, 222, $-65\frac{1}{3}$
Values for Set 2: 106, $196\frac{1}{3}$, $-24\frac{2}{3}$, 228, $162\frac{2}{3}$

A2.5

1 a $x \leqslant 3$ b $2 \leqslant x \leqslant 8$ c $-5 < x < 12$
2 a $x \leqslant 2$ b $x > -1$ c $x \geqslant -1$
d $1 \leqslant x < 5$ e $-7 < x \leqslant 0$
3 True
4 a $x \leqslant 7$ b $x > 11$ c $p \leqslant -16$
d $x > -3$ e $y \leqslant \frac{2}{3}$ f $y < -3$
g $x \leqslant 2$ h $x > -5$ i $x \leqslant 10$
j $p \leqslant -3$ k $x > -18$ l $x \geqslant \frac{1}{2}$
5 a $6(x-2) > 12 + 2(x-2)$, $x > 5$ b 6

A2 Exam review

1 a $x \geqslant -3$ b $-3 < x \leqslant 2$

−5 −4 −3 −2 −1 0 1 2 3 4 5

2 a $y = 3\frac{1}{2}$ b $x = 7$

D1 Before you start ...

1 a A census is a survey where every member of a population is questioned.
b A sample looks at a fraction of the population.
2 a Primary: Nicola, Secondary: Maddy
b Primary data is data collected by yourself. Secondary data has already been collected by someone else.
3 ai There are two groups containing 4 hours.
aii There are no groups containing 2 hours.
b Use inequalities for the groups, e.g. $0 \leqslant t < 2$, $2 \leqslant t < 4$, 4 or more.

D1.1

1 a Assumes you visit cinema. Answers may differ at different times of year so could say on average. No answer choices given.
b On average how many times do you go to the cinema in one month? Once or less often, 2 or 3 times a month, 4 times or more often.
c On average how much do you spend when you go to the cinema? Less than £5, £5 to £10, more than £10.
2 ai Leading.
aii What is your favourite sport? Tennis, swimming, football, rugby, cricket, other.
b On average, how many times a week do you play sport? Never, 1 or 2 times, 3 or 4 times, more than 4 times.
3 a May not like either/not enough choices.
b What is your favourite flavour of crisps? Plain, cheese and onion, salt and vinegar, smokey bacon, ketchup, other.
ci 'lots' is vague; options do not cover all possible answers; needs a time frame.
cii How many times have you visited the tuck shop in the last month? Never, once or twice, three to five times, more than five times.
4 ai Does not cover all possible answers.
aii How far would you travel to see your favourite band? Less than 1 mile, 1 mile to 5 miles, between 5 and 10 miles, 10 miles or more.
b How much would you pay for a ticket to see your favourite band? Less than £5, £5 to £10, £10.01 to £15, more than £15.

D1.2

1 Obviously visit cinema so not typical of population.
2 People in an athletics club will probably play sport more often than those who aren't.
3 a It is not representative of the people who use the school tuck shop.
b It is not representative of the whole school.
c Pick names out of a hat, or take every 20th person on a list of all the people in the school.
4 a His friends are not representative of the whole population, they might particularly like (or dislike) travelling to see bands.
b People listening to MP3 players may be more interested in music than is typical.
5 All girls/all friends so may have same taste in music/small sample.
6 Cars passing at similar time/small sample.
7 a People at bus stops are more likely to take the bus to work.
b Biased against people who are ex-directory, don't have a landline or aren't in when phoned.

D1.3

1 Two-way table with number of visits per week and amount spent.
2 Two-way table with number of times per week play sport and gender.
3 Two-way table with crisp flavour and Year groups.
4 Two-way table with distances and money.
5 Two-way table with favourite bands and numbers of CDs.
6 Two-way table with colour and make of car.
7 Two-way table with mode of travel and time taken.
8 a 500
bi 20.8% bii 24% biii 39.2%

D1.4

1 ai 7 aii 6 aiii 5.82 aiv 6 av 2
bi 75 bii 63 biii 60.1 biv 63 bv 27
ci 8 cii 96 ciii 95.6 civ 96 cv 2
di 71 dii 22, 37 diii 40.4 div 37 dv 38
ei 26 eii 88, 89 eiii 84.2 eiv 87 ev 7
fi 72 fii 27 fiii 46.9 fiv 34 fv 37
gi 8 gii 105 giii 105.2 giv 105 gv 3
2 Range is unduly affected by one extreme value, which IQR ignores.
3 Mode is the lowest value.
4 a 1, 6, 8, 2, 8, 5, 6, 9, 3, 5, 7, 4, 4, 5, 5
bi 8 bii 5 biii 5.2 biv 5 bv 3
c Range and IQR stay same.
d Answers are as for Q1 less 100.
e Adding 100 does not affect spread of values but does affect averages.
5 ai 200, 200 aii 100, 100 aiii 100, 100
aiv 100, 100 av 2, 2
b All measures the same although sets of numbers are different.

D1.5

1 80.4 minutes
2 7.325 hours
3 43 lessons
4 78.6%
5 7.33
6 8.29

D1 Exam review

1 $\frac{30a + 20b}{50}$
2 Two-way table with crisp flavour and gender.

N3 Before you start ...

1 a 6 squares shaded b 8 squares shaded
c 9 squares shaded d 10 squares shaded
e 7 squares shaded
2 a $\frac{1}{2}$ b $\frac{3}{4}$ c $\frac{4}{5}$ d $\frac{19}{20}$ e $\frac{3}{4}$
3 a 0.75 b 0.4 c 0.7 d 0.45 e 0.17
4 a $\frac{1}{2}$ b $\frac{1}{4}$ c $\frac{3}{10}$ d $\frac{4}{5}$ e $\frac{9}{20}$
5 a 50% b 25% c 10% d 20% e 5%

N3.1

1 a 4 b 10 c 24 d 12
e 20 f 35 g 60 h 60
2 a 6 squares, 3 shaded
b 6 squares, 4 shaded
c 15 squares, 9 shaded
d 20 squares, 15 shaded
3 a $\frac{20}{60}$ b $\frac{15}{60}$ c $\frac{40}{60}$ d $\frac{24}{60}$
4 a $\frac{18}{24}$ b $\frac{8}{24}$ c $\frac{9}{24}$ d $\frac{10}{24}$
5 a $\frac{20}{30}$ b $\frac{18}{42}$ c $\frac{35}{45}$ d $\frac{25}{40}$
6 a $\frac{2}{10}$ and $\frac{3}{10}$, $\frac{3}{10}$ b $\frac{8}{12}$ and $\frac{9}{12}$, $\frac{3}{4}$
c $\frac{6}{15}$ and $\frac{5}{15}$, $\frac{2}{5}$ d $\frac{21}{30}$ and $\frac{20}{30}$, $\frac{7}{10}$
7 a Two sets of 10 squares; 2 and 3 shaded
b Two sets of 12 squares; 8 and 9 shaded
c Two sets of 15 squares; 6 and 5 shaded
d Two sets of 30 squares; 21 and 20 shaded
8 a 30 b 12 c 24 d 28
9 a $\frac{6}{12}, \frac{8}{12}, \frac{9}{12}$ b $\frac{4}{20}, \frac{15}{20}, \frac{7}{20}$
c $\frac{3}{24}, \frac{14}{24}, \frac{16}{24}$ d $\frac{56}{84}, \frac{64}{84}, \frac{24}{84}$
10 a $\frac{2}{15}, \frac{1}{5}, \frac{2}{3}$ b $\frac{1}{4}, \frac{7}{20}, \frac{2}{5}$
c $\frac{5}{14}, \frac{3}{8}, \frac{3}{7}$ d $\frac{2}{7}, \frac{2}{3}, \frac{5}{6}$
11 a $\frac{1}{2}, \frac{2}{5}, \frac{3}{10}, \frac{1}{4}$ b $\frac{4}{5}, \frac{1}{4}, \frac{3}{20}, \frac{1}{10}$
c $\frac{3}{4}, \frac{17}{40}, \frac{2}{5}, \frac{3}{8}$ d $\frac{5}{6}, \frac{5}{8}, \frac{7}{12}, \frac{11}{24}$

N3.2

1 a 3 squares, 1 and 2 shaded different colours
b 5 squares, 1 and 3 shaded different colours
2 a $\frac{3}{5}$ b 1 c $\frac{5}{7}$ d 1
3 a 5 squares, 1 and 2 shaded different colours
b 4 squares, 1 and 3 shaded different colours
c 7 squares, 2 and 3 shaded different colours
d 8 squares, 3 and 5 shaded different colours
4 a $\frac{1}{3}$ b $\frac{3}{5}$ c $\frac{2}{3}$ d $\frac{3}{5}$
5 a 6 squares, 2 and 3 shaded different colours
b 10 squares, 6 and 3 shaded different colours
6 a $\frac{3}{10}$ b $\frac{5}{6}$ c $\frac{11}{20}$ d $\frac{3}{8}$
7 a 10 squares, 2 and 1 shaded different colours
b 6 squares, 4 and 1 shaded different colours
c 20 squares, 8 and 3 shaded different colours
d 8 squares, 1 and 2 shaded different colours
8 a $\frac{1}{8}$ b $\frac{1}{2}$ c $\frac{3}{4}$ d $\frac{5}{16}$
9 a $\frac{5}{21}$ b $\frac{2}{15}$ c $\frac{1}{18}$ d $\frac{7}{20}$
10 a $1\frac{1}{10}$ b $1\frac{7}{12}$
11 a $1\frac{1}{6}$ b $1\frac{3}{10}$ c $1\frac{2}{15}$ d $1\frac{1}{14}$
12 a $1\frac{1}{4}$ b $1\frac{4}{5}$ c $1\frac{5}{8}$ d $4\frac{1}{4}$
13 a $\frac{7}{4}$ b $\frac{23}{16}$ c $\frac{14}{9}$ d $\frac{18}{7}$
14 a $3\frac{14}{15}$ b $4\frac{7}{12}$ c $7\frac{13}{14}$ d $7\frac{55}{63}$
e $1\frac{7}{20}$ f $\frac{3}{4}$ g $1\frac{19}{20}$ h $4\frac{13}{14}$

N3.3

1 a $\frac{1}{4}$ b $\frac{1}{6}$ c $\frac{1}{10}$ d $\frac{1}{12}$
2 a 5 b 9 c 2 d 3
3 a $8 \times \frac{1}{5}$ b $6 \times \frac{1}{4}$ c $9 \times \frac{1}{5}$ d $17 \times \frac{1}{3}$
4 a $\frac{1}{4}$ b 2 c 3 d $\frac{1}{5}$
5 a

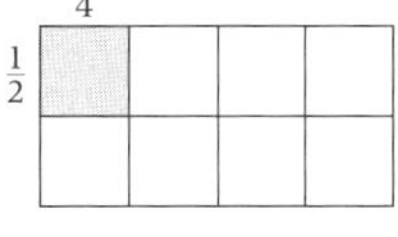

b

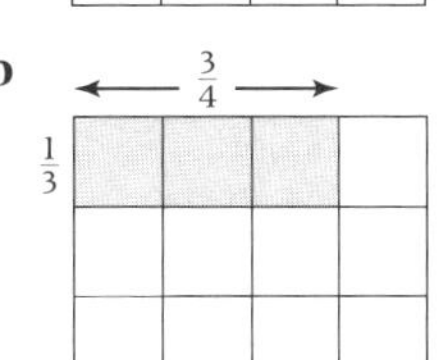

c

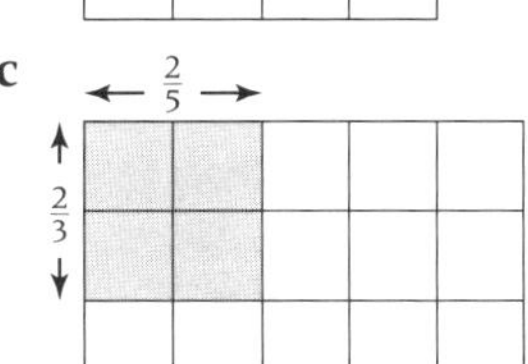

d

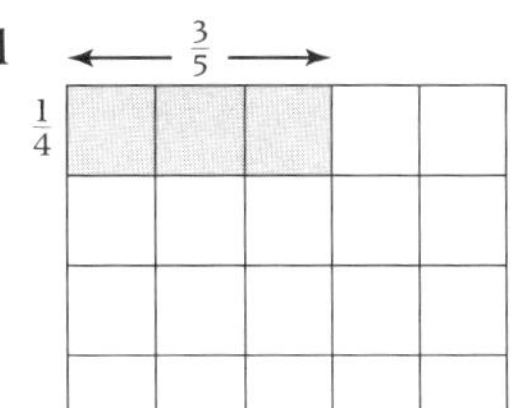

6 a $\frac{3}{20}$ b $\frac{4}{27}$ c $\frac{1}{14}$ d $\frac{1}{4}$
e $\frac{20}{63}$ f $\frac{1}{12}$ g $\frac{12}{65}$ h $\frac{1}{15}$
7 a $\frac{3}{4}$ b $\frac{3}{4}$ c $\frac{4}{5}$ d $\frac{4}{5}$
e $\frac{3}{7}$ f $1\frac{4}{5}$ g $3\frac{3}{7}$ h $2\frac{3}{4}$
8 a $\frac{5}{32}$ b $\frac{1}{8}$ c $\frac{2}{21}$ d $\frac{1}{48}$
e 20 f 9 g $12\frac{1}{2}$ h $25\frac{2}{3}$
9 a $\frac{1}{3}$ b $\frac{1}{4}$ c $\frac{1}{6}$ d $\frac{1}{4}$
e $\frac{8}{9}$ f $\frac{45}{56}$ g $\frac{2}{9}$ h $\frac{8}{21}$
10 a $1\frac{1}{8}$ b $1\frac{1}{10}$ c 3 d 3
e $4\frac{19}{24}$ f $1\frac{3}{14}$ g $1\frac{21}{22}$ h $4\frac{4}{27}$

N3.4

1 a 0.5 b 0.75 c 0.4
d 0.1 e 0.2 f 0.25

2 a 50% b 75% c 40%
d 10% e 20% f 25%

3 a 62.5% b 80% c 87.5%
d 60% e 37.5% f 12.5%

4 See Q3

5 a 0.0625 b 0.28 c 0.056
d 0.075 e 0.4375 f 0.03125

6 a 6.25% b 28% c 5.6%
d 7.5% e 43.75% f 3.125%

7 a $0.\dot{3}$ b $0.1\dot{6}$ c $0.\dot{6}$
d $0.\dot{1}4285\dot{7}$ e $0.\dot{1}$ f $0.8\dot{3}$

8 a $33.\dot{3}\%$ b $16.\dot{6}\%$ c $66.\dot{6}\%$
d $14.\dot{2}8571\dot{4}\%$ e $11.\dot{1}\%$ f $83.\dot{3}\%$

9 See Q7 and Q8

10 a $0.\dot{4}2857\dot{1}$ b 0.1875 c 0.2125
d $0.\dot{5}$ e $0.1\dot{6}$ f $0.\dot{7}1428\dot{5}$

11 a Terminating (only prime factor of denominator is 5)
b Terminating (only prime factors of denominator are 2 and 5)
c Recurring (denominator has a prime factor of 11)
d Recurring (denominator has prime factors of 3 and 7)
e Terminating (only prime factor of denominator is 5)
f Terminating (only prime factor of denominator is 2)

12 The decimal does recur. The restricted number of digits on the calculator display, and the rounding of the final digit, obscure the recurring pattern.

N3.5

1 a 0.43 b 0.86 c 0.94
d 0.455 e 0.0375 f 1.05

2 a $\frac{1}{2}$ b $\frac{1}{4}$ c $\frac{1}{5}$ d $\frac{1}{8}$ e $\frac{3}{4}$ f $\frac{9}{10}$

3 a $\frac{51}{100}$ b $\frac{43}{100}$ c $\frac{413}{1000}$ d $\frac{719}{1000}$ e $\frac{91}{100}$ f $\frac{871}{1000}$

4 a $\frac{49}{100}$ b $\frac{53}{100}$ c $\frac{73}{100}$ d $\frac{81}{100}$ e $\frac{37}{100}$ f $\frac{19}{100}$

5 a $\frac{8}{25}$ b $\frac{11}{20}$ c $\frac{11}{25}$ d $\frac{31}{200}$ e $\frac{16}{25}$ f $\frac{53}{200}$

6 a $\frac{11}{20}$ b $\frac{31}{50}$ c $\frac{21}{25}$ d $\frac{13}{20}$ e $\frac{18}{25}$ f $\frac{37}{200}$

7 $\frac{8}{11}$, 11 not multiple of 2 or 5

8 a $0.\dot{1}$ b $0.\dot{5}$ c $0.7\dot{5}$
d $0.\dot{3}4\dot{6}$ e $0.7\dot{6}\dot{5}$

9 a $\frac{2}{9}$ b $\frac{2}{3}$ c $\frac{25}{99}$ d $\frac{3}{11}$ e $\frac{545}{999}$ f $\frac{605}{999}$

10 a $\frac{47}{90}$ b $\frac{1}{18}$ c $\frac{249}{550}$ d $\frac{752}{9000}$ e $\frac{2}{3}$ f $\frac{8197}{99\,900}$

11 $\frac{13\,717\,421}{1\,111\,111\,111}$

N3 Exam review

1 ai $\frac{3}{5}$ aii $\frac{13}{25}$
bi $\frac{97}{100}$ bii $\frac{13}{100}$ biii $\frac{9}{14}$

2 a 0.067, 0.56, 0.6, 0.605, 0.65
b −10, −6, −4, 2, 5
c $\frac{2}{5}, \frac{1}{2}, \frac{2}{3}, \frac{3}{4}$

S2 Before you start ...

1 $a = 56°$, $b = 112°$
2 $c = 235°$, $d = 160°$
3 $e = 72°$, $f = 63°$
4 $g = 118°$, $h = 75°$

S2.1

1 ai

aii

bi

bii

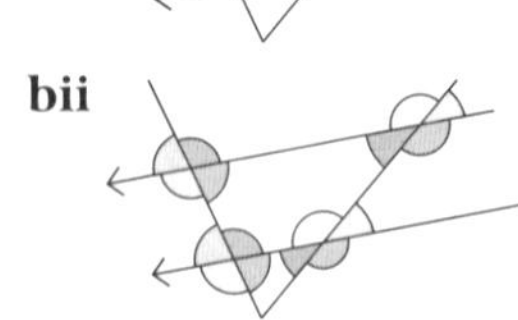

ci

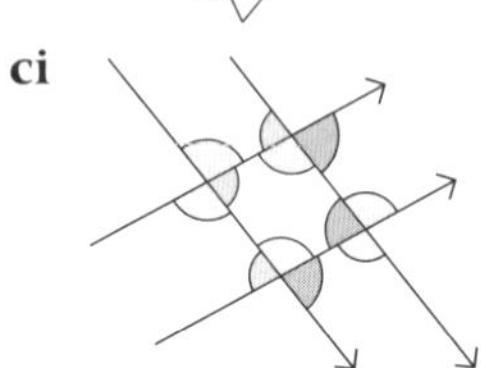

cii

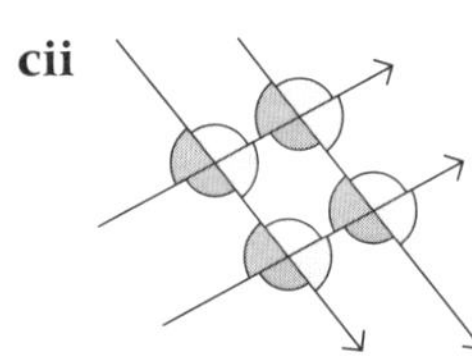

2 a $a = 17°$, angles on a straight line; $b = 17°$, alternate angles; $c = 163°$, corresponding angles
b $d = 125°$, alternate angles; $e = 105°$, vertically opposite angles
c $f = 134°$, vertically opposite angles; $g = 134°$, corresponding angles; $h = 24°$, angles in a triangle

3 a 180°
b x (bottom left) and z (bottom right)
c angles in the triangle are x, y, z (part **b**) and add up to 180° (part **a**)

4 a $a = 25°$, $b = 155°$, $c = 25°$, $d = 155°$
b Parallelogram

5 $a = 80°$, $b = 110°$, $c = 70°$, $d = 30°$, $e = 150°$

S2.2

1

Shape	△	□	⬠	⬡	⯃	⬭
Number of sides	3	4	5	6	8	10
Number of triangles the shape splits into	1	2	3	4	6	8
Sum of the interior angles in the shape	180°	360°	540°	720°	1080°	1440
Size of one interior angle	60°	90°	108°	120°	135°	144
Size of one exterior angle	120°	90°	72°	60°	45°	36

2 a $x = 80°$ b $x = 77°$, $2x = 154°$
c $x = 55°$, $3x = 165°$

3 a $x = 125°$, $y = 50°$ b $x = 110°$, $y = 65°$

4 ai 120° aii 108° aiii 118°
b The sum of the two opposite angles

5 Drawing a diagonal splits the quadrilateral into two triangles, so the sum of the interior angles is $2 \times 180° = 360°$

S2.3

1 a $a = 63°$, angle at centre is double the angle at the circumference
   b $b = 46°$, angle at centre is double the angle at the circumference
   c $c = 118°$, angle at centre is double the angle at the circumference
   d $d = 32°$, angles in same arc
   e $e = 78°$, angles in same arc
   f $f = 103°$, angle at centre is double the angle at the circumference
   g $g = 122°$, angles in same arc
   h $h = 21°$, angles in same arc
   i $i = 140°$, angle at centre is double the angle at the circumference
   j $j = k = l = 38°$, angles in same arc
   k $m = 47°$, $n = 62°$, angles in same arcs
   l $p = 75°$, $q = 25°$, angles in same arcs
   m $r = 90°$, angle in same arc
   n $s = 49°$, $t = 22°$, angles in same arcs
   o $u = 54°$, $v = 45°$, angles in same arcs
   p $w = 180°$, angle at centre is double the angle at the circumference
   q $x = 45°$, angles in isosceles triangle
   r $y = 45°$, angle at centre is double the angle at the circumference

S2.4

1 a $a = 90°$, angle in a semicircle
   b $b = 61°$, angles in a triangle
   c $c = 43°$, angles in a triangle
   d $d = 135°$, opposite angles in cyclic quadrilateral
   e $e = 76°$, opposite angles in cyclic quadrilateral
   f $f = 15°$, angles in a triangle
   g $g = 143°$, $h = 100°$, opposite angles in cyclic quadrilateral
   h $i = 124°$, $j = 54°$, opposite angles in cyclic quadrilateral
   i $k = 56°$, $l = 124°$, angles in a triangle, angles on a straight line
   j $m = 125°$, $n = 250°$, angles at centre are double the angles at the circumference
   k $p = q = 106°$, opposite angles in cyclic quadrilateral, angles on same arc
   l $r = 50°$, angle in a semicircle
   m $s = 35°$, angle in a semicircle
   n $t = 62°$, $u = 118°$, angles on a straight line, opposite angles in a cyclic quadrilateral
   o $v = 74°$, angles on a straight line, opposite angles in a cyclic quadrilateral
   p $w = x = 90°$, opposite angles in a cyclic quadrilateral
   q $y = 45°$, angles in isosceles triangle
   r $z = 100°$

S2.5

1 a $a = 90°$, angle between tangent and radius
   b $b = 33°$, angle between tangent and radius, $c = 57°$, angles in a triangle in a semicircle
   c $d = 42°$, angle between tangent and radius, $e = 48°$, angles in a triangle in a semicircle
   d $f = 26°$, angle between tangent and radius, $g = 26°$, angles in a triangle in a semicircle
   e $h = 51°$, angle between tangent and radius, angles in a triangle in a semicircle
   f $i = 16°$, $j = 4$ cm; 2 tangents drawn from a point to a circle
   g $k = 44°$, angles in a quadrilateral
   h $l = 40°$, $m = 70°$, angles in a quadrilateral and angle at centre is double the angle at the circumference
   i $n = 16°$, $p = 82°$, angles in a quadrilateral and angle at centre is double the angle at the circumference
   j $q = 30°$, $r = 105°$, angles in a quadrilateral and angle at centre is double the angle at the circumference
   k $s = 100°$, $t = 50°$, angles in a quadrilateral and angle at centre is double the angle at the circumference
   l $u = 116°$, $v = 58°$, angles in a quadrilateral and angle at centre is double the angle at the circumference

S2 Exam review

1 60°
2 a 27°, angle between tangent and radius
   b 63°, angles in a triangle in a semicircle

A3 Before you start ...

1 a 10, 12   b 70, 64   c 16, 22
   d −2, −5   e 48, 96   f $\frac{1}{6}, \frac{1}{7}$
2 a $4(n-1)$, $2n+7$, $15-n$, $2n^2 = \frac{9}{n} + 15$
3 Square numbers. They form a pattern of squares when drawn.

A3.1

1 a 29, 34   b 65, 58
   c 16, 21   d 999 999, 9 999 999
   e 13, 21   f 3.375, 1.6875
2 a 7, 13   b 8, 16
   c 93, 91, 89   d 8, 125
3 a 3, 6, 9, 12, 15
   b 2, 4, 8, 16, 32
   c 2, 3, 5, 7, 11
   d 121, 144, 169, 196, 225
4 a 6, 10, 15
   b 1, 3, 6, 10, 15, 21, 28, 36, 45, 55
   c They form a pattern of squares when drawn.
   d They form a pattern of cubes when drawn.
5 a 110   b $\frac{10}{11}$ c 100   d 100 000
6 a $6667^2 = 44\,448\,889$, $66\,667^2 = 4\,444\,488\,889$
   b 4 444 444 444 488 888 888 889
   c 666 666 667

A3.2

1 a 10, 18, 26, 34, 42   b 1, 6, 11, 16, 21
   c 7, 14, 21, 28, 35   d 8, 6, 4, 2, 0
   e −2, 1, 6, 13, 22   f 2, 8, 18, 32, 50
2 a 8, 7, 6, 5, 4   b 6, 12, 20, 30, 42
   c $1, \frac{1}{2}, \frac{1}{3}, \frac{1}{4}, \frac{1}{5}$ d −1, −8, −27, −64, −125
   e −4, −9, 0, 35, 108   f 1, 16, 81, 256, 625
3 a n^2, n^3 b $18 + 2n$, $n(n+5)$
   c n^3, $n(n+5)$
4 a 10, 14, 18, 22, 26, 30, 34, 38; $T(n) = 4n + 6$ (for $n \leqslant 7$)
   b $T(n) = n^2 + 1$, until $T(n)$ is over 49
5 a For example, n^2, $6n - 5$
   b For example, n^2, $6n - 5$
   c No

A3.3

1 a $5n - 1$ b $2n - 1$ c $2n + 8$
d $0.5n + 0.5$ e $2n - 6$ f n
g $13n$ h $10n - 6$ i $12 - 2n$
j $105 - 5n$ k $50\frac{1}{4} - \frac{1}{4}n$ l $79 - 4n$

2 Many possibilities

3 a False b False c False

4 a $8n - 20$ b $3n + 2$ c $2n - 46$

5 a $\frac{2n+1}{3n+4}$ b $\frac{2n+8}{33-3n}$ c $\frac{n+6}{n^2}$ d $\frac{n^3}{13-2n}$

6 a $\frac{1}{n}$ b

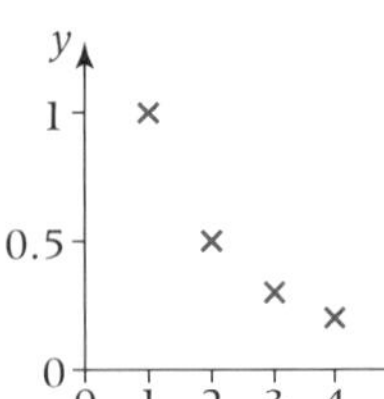

c It gets closer and closer to zero without ever actually reaching it.

A3.4

1 a For each square added, 3 sides added, and in first case there is an extra side
b Number of black tiles is matched by number of white tiles in middle of figure, with 4 extra whites at side
c Length of rectangle is always one more than width

2 a $W = 2B + 6$
bi $B = 2W + 2$ bii $L = 3W + 1$
ci $P = n(n - 1)$ cii $H = \frac{n(n-1)}{2}$

3 a $C = 4$, $E = 4(n - 2)$, $M = (n - 2)^2$
b $C = 4$, $E = 2(m - 2) + 2(n - 2)$, $M = (m - 2)(n - 2)$

A3.5

1 a 6, 9, 14, 21, 30 b −1, 2, 7, 14, 23
c 3, 12, 27, 48, 75 d 3, 8, 15, 24, 35
e 1, 7, 17, 31, 49 f 4, 13, 28, 49, 76

2 a $n^2 + 3$ b $n^2 - 3$ c $10n^2$
d $3n^2$ e $n^2 + n$ f $0.5n^2$

3 a

*n*th term	First five terms	First differences	Second differences
$2n^2$	2, 8, 18, 32, 50, …	6, 10, 14, 18, …	4, 4, 4, …
$3n^2$	3, 12, 27, 48, 75, …	9, 15, 21, 27, …	6, 6, 6, …
$4n^2$	4, 16, 36, 64, 100, …	12, 20, 28, 36, …	8, 8, 8, …
$5n^2$	5, 20, 45, 80, 125, …	15, 25, 35, 45, …	10, 10, 10, …
$10n^2$	10, 40, 90, 160, 250, …	30, 50, 70, 90, …	20, 20, 20, …

b Second difference is double the coefficient of the n^2 term
ci $6n^2$ cii $2n^2 - 1$ ciii $3n^2 + n$

4 a $T = h^2 + h$
b Width of rectangle is one more than the height.

5 a $A = (n + 1)(n + 2)$
bi The height of the rectangle is always one more than the term, the width is always two more.
bii Formula equals $n^2 + 3n + 2$, which is quadratic.

A3 Exam review

1 a $2n - 1$ b 1, 3 2 a $m = 6n$

D2 Before you start ...

1

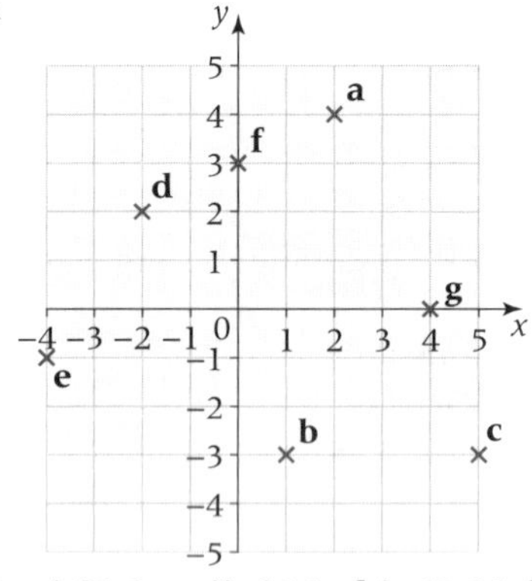

2 ai €2.8 aii €6.3 bi £2.85 bii £5

3 a 40 b 60 c 80 d 15 e 50
f 30 g 75 h 150 i 120 j 630

D2.1

1 a

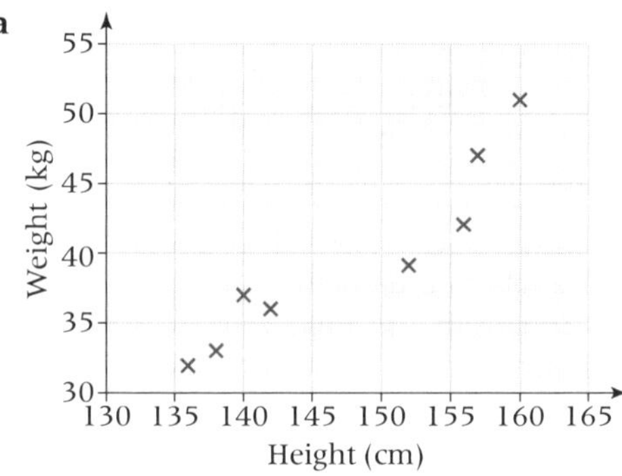

b Positive correlation
c As weight increases, so does height.

2 a

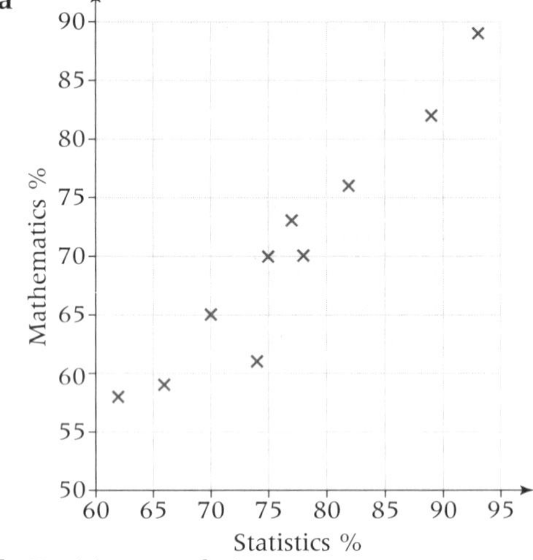

b Positive correlation
c As percentage achieved in statistics increases, so does percentage achieved in maths.

3 a

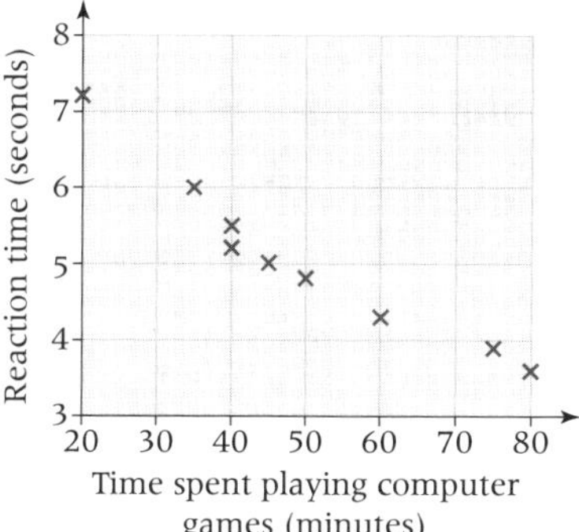

b Negative correlation

c As time spent playing computer games increases, reaction time decreases.

4 a

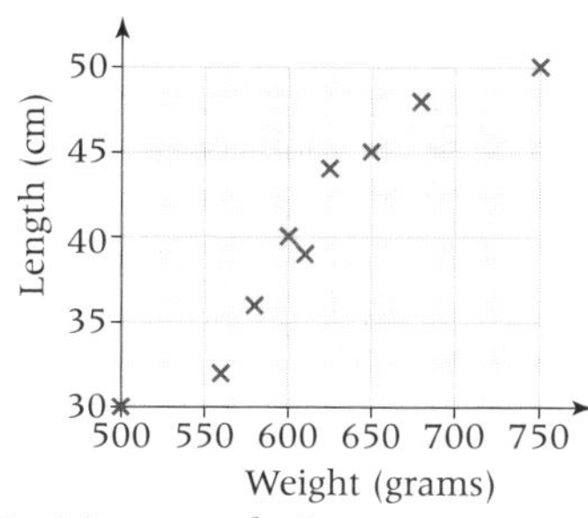

b Positive correlation

c As weight of fish increases, so does length.

D2.2

1 a

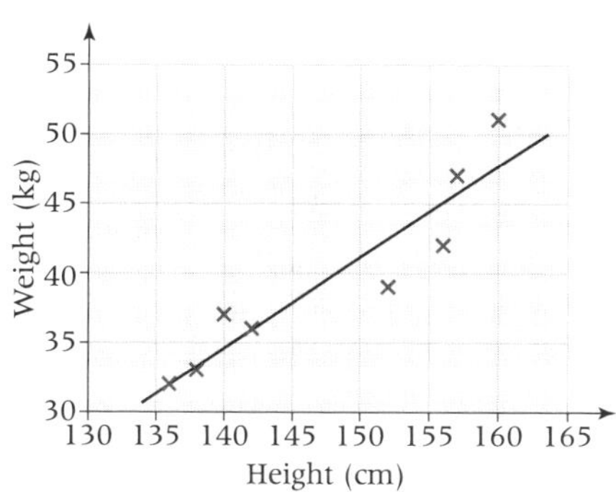

bi 38 kg **bii** 156 cm

2 a

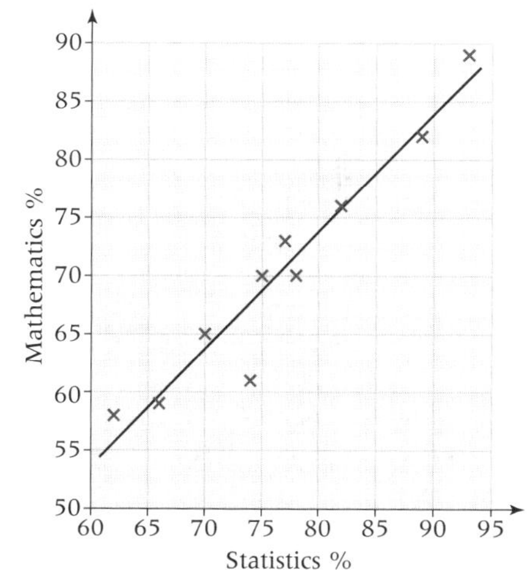

bi 74% **bii** 86%

c Score of 46% outside range of data collected

3 a

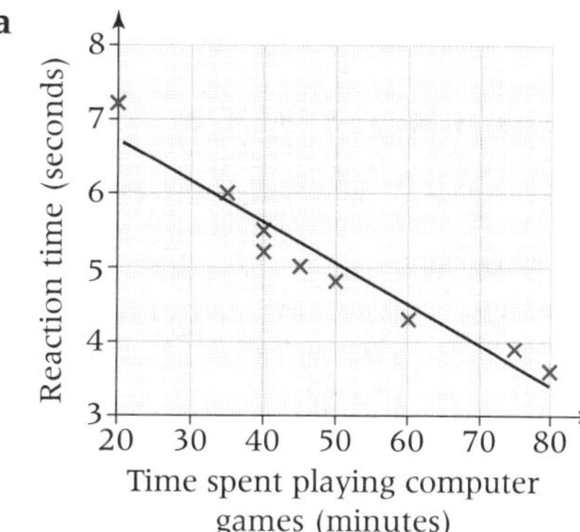

bi 3.9 seconds **bii** 26 minutes

c Outside range of data collected

4 a

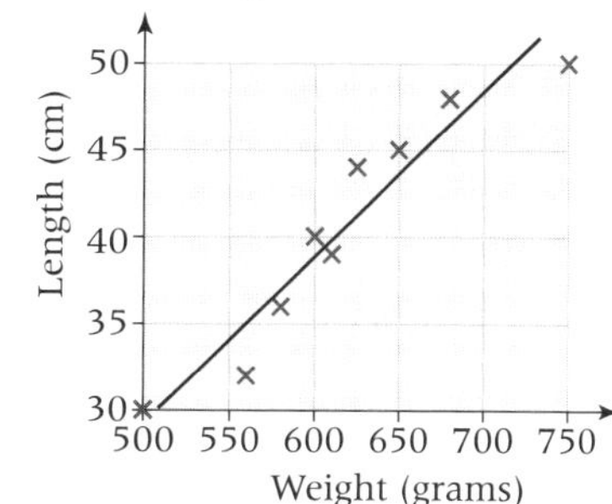

bi 48 cm **bii** 635 g

D2.3

1 a

Stem	Leaves
4	5 8
5	1 1 2 4 9 9
6	1 2 3 6 9
7	0 1 4 5 7 8
8	1 2 9
9	3

Key: 4|5 means 45%

b Most students scored between 50% and 80%.

ci 48% **cii** 66%

ciii 54%, 77% **civ** 23%

2 a

Stem	Leaves
2	6 8 9
3	0 3 7 8 9
4	0 1 3 3 4 7 9
5	2 5 8 9
6	0 1 3 6 7 9
7	1 3 5 7
8	0 4

Key: 3|4 means 34 minutes

b Most people spend 30 to 70 minutes playing computer games each day.

ci 58 minutes

cii 52 minutes

ciii 39 minutes, 67 minutes

civ 28 minutes

3 a

Stem	Leaves
0	3 5 6 7 8 9
1	0 1 2 2 5 6 7 8 9
2	0 1 2 4 6 7 7 9 9
3	1 2 3 4 5 7
4	0 1

Key: 2|7 means 27 minutes

b Most people take between 10 and 30 minutes to complete the crossword.

ci 38 minutes

cii 20.5 minutes

ciii 11.5 minutes, 30 minutes

civ 18.5 minutes

4 a

Stem	Leaves
3	2 6 7 9
4	2 5 7 8 9
5	1 2 4 6 6 6 7 8 9
6	0 1 3 6
7	0

Key: 3|6 means 36 kg

b Most boys weigh between 40 kg and 60 kg.
ci 38 kg **cii** 54 kg
ciii 45 kg, 59 kg **civ** 14 kg

5 a

Stem	Leaves
14	6 7 9
15	0 0 2 2 3 4 5 5 7 8 9
16	0 2 3 5 7 8
17	1 2 2

Key: 14|2 means 142 cm

b Most girls are between 150 cm and 160 cm tall.
ci 26 cm **cii** 157 cm
ciii 152 cm, 165 cm **civ** 13 cm

6 a

Stem	Leaves
8	8 9
9	1 2 4 8
10	1 3 4 5 6 7 8 8 9
11	0 2 4 6 6 7 7 8 9
12	1 5 6 7 9
13	1 3

Key: 11|4 means IQ of 114

b Most students have an IQ between 100 and 120.
ci 45 **cii** 110
ciii 103, 119 **civ** 16

7 a

Stem	Leaves
0	3 6 8 9 9
1	0 1 2 5 6 7 8 9
2	1 1 2 3 3 5 7
3	0 2 3

Key: 3|2 means 32 minutes

b Most people took 10 to 20 minutes
ci 30 minutes
cii 18 minutes
ciii 10 minutes, 23 minutes
civ 13 minutes

D2.4

1 a

A		B
8 7	3	
9 7 5 4 4 3 2 1	4	1 4 8
8 6 6 5 3 2 1	5	3 3 4 6 7 7 7
2 1	6	1 3 5 6 9
	7	0 2 2 5 9

Key: 2|5|3 means 52% in A, 53% in B

bi A: 49%, B: 61%
bii A: 10%, B: 17%
biii A: 24%, B: 38%
c Results for test A are generally lower but less varied than for test B.

2 a

Men		Women
	14	8 9
9 8	15	0 1 2 3 4 5 5 6 7 8 8
9 9 8 7 7 5 0	16	1 2 2 5 6 7 9
8 8 7 6 5 4 2 1	17	2 4 8
8 4 3 3 2 0	18	

Key: 2|16|3 means 162 cm for men, 163 cm for women

bi Men: 174 cm, women: 158 cm
bii Men: 13 cm, women: 13 cm
biii Men: 30 cm, women: 29 cm
c Men are taller than women and heights are evenly spread in both groups.

3 a

X		Y
9	8	7
9 8 6	9	3 7 9
9 8 6 5 5 4 3	10	2 3 4 6 7 7 9
8 8 7 7 6 5 5 2 0 0	11	3 4 4 4 6 8 8 8 9
9 6 4 3 3 2 1	12	1 1 4 6 6 8 9
4 1 0	13	0 1 2

Key: 7|9|6 means for 97 in X, 96 in Y

bi X: 116, Y: 116
bii X: 18, Y: 18
biii X: 45, Y: 45
c Average IQ and spread are same in X and Y.

4 a

Boys		Girls
8 8 8 6 2 2	3	
7 5 4 2	4	0 3 4 6 8
9 8 7 6 3 2	5	2 2 5 6 9
8 6 4 2 1	6	2 3 3 5 6 7
2 1	7	2 3 4 6 7
	8	0 2

Key: 5|4|6 means 4.5 minutes for boys, 4.6 seconds for girls

bi Boys: 5.3 seconds, girls: 6.3 seconds
bii Boys: 2.4 seconds, girls: 2.1 seconds
biii Boys: 4 seconds, girls: 4.2 seconds
c Boys have a faster reaction time than girls, variation in times is similar.

5 a

P		Z
9 8 8 6	0	8 9
9 8 7 7 4 3 2 1 0	1	1 2 4 5 6 7 7 8 9
9 7 4 2	2	1 3 4 5 8 9
7 1	3	2 3

Key: 4|3|6 means 34 minutes for P, and 36 minutes for Z

bi P: 17 minutes, Z: 18 minutes
bii P: 14 minutes, Z: 11 minutes
biii P: 31 minutes, Z: 25 minutes
c The average time taken is similar but times for P are more varied than those for Z.

D2.5

1

40 50 60 70 80 90 100
(%)

2

20 30 40 50 60 70 80 90 100
(Minutes per day)

3

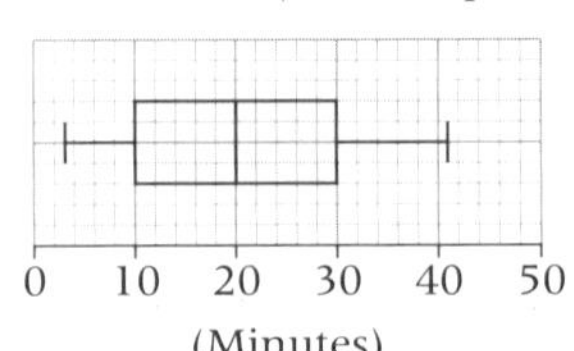

4

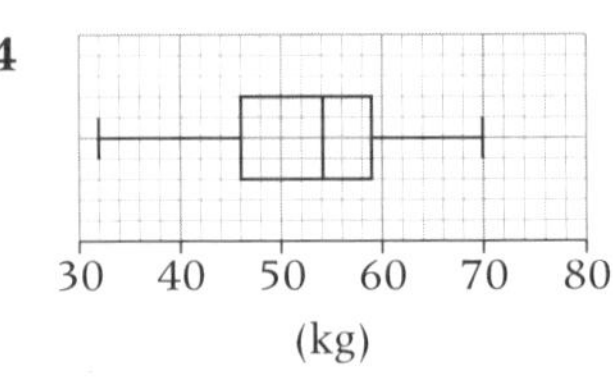

5

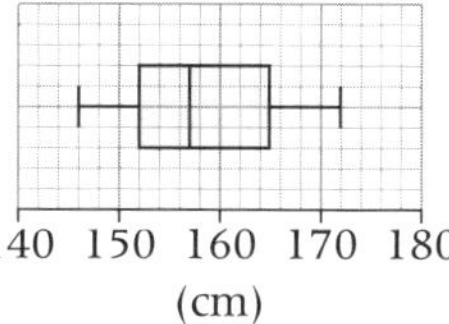

6

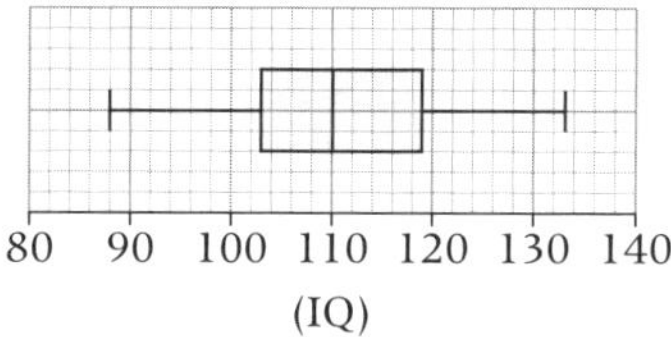

7

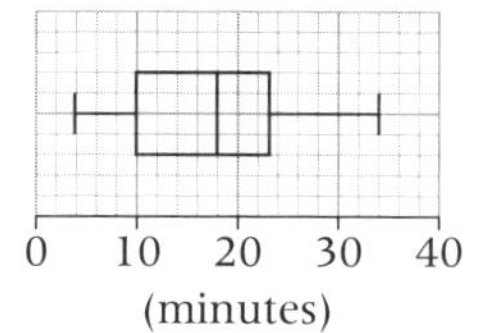

D2 Exam review

1 a

Key: 5|2 means 52 minutes

bi 58 minutes
bii 9 minutes
biii 23 minutes

2 **a** and **2 c**

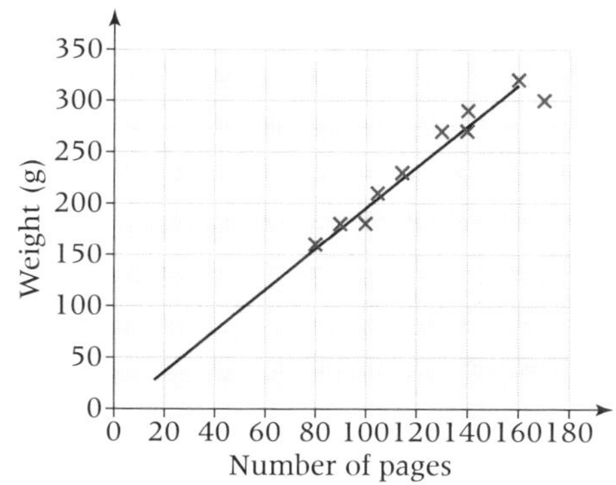

b The weight increases with the number of pages.

di 143 pages **dii** 235 g

A4 Before you start ...

1

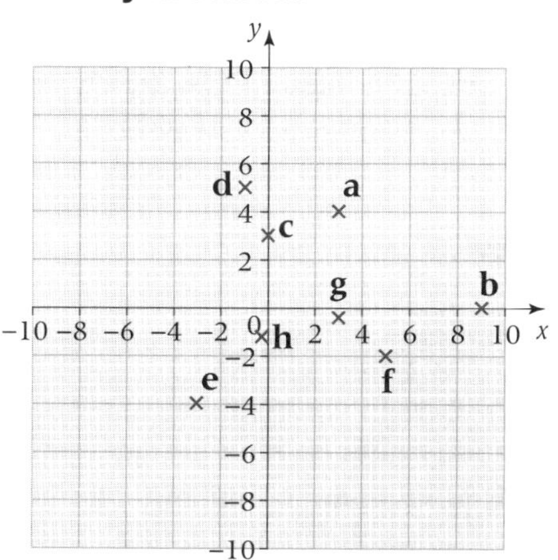

2

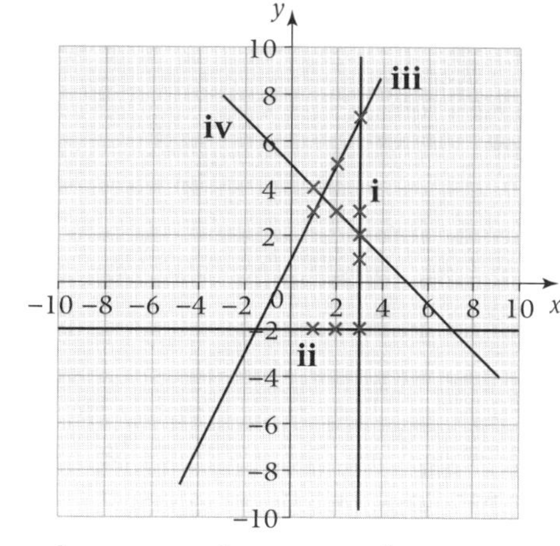

3 **a** $6\frac{1}{2}$ **b** $5\frac{1}{4}$ **c** $4\frac{5}{7}$ **d** $-9\frac{1}{11}$

e $\frac{7}{3}$ **f** $\frac{15}{2}$ **g** $\frac{23}{4}$ **h** $-\frac{35}{8}$

4 **a** $n = 2$ **b** $m = 1$ **c** $p = \frac{1}{2}$

A4.1

1 $y = 2x + 3$, $y = 7 - 3x$, $y = 5x$, $y = 7$, $2x + 7y = 8$, $x = -2$

2 **ai**

x	0	1	2
y	2	5	8

aii

x	0	2	3
y	−2	0	2

aiii

x	0	5	1
y	2	0	1.6

b

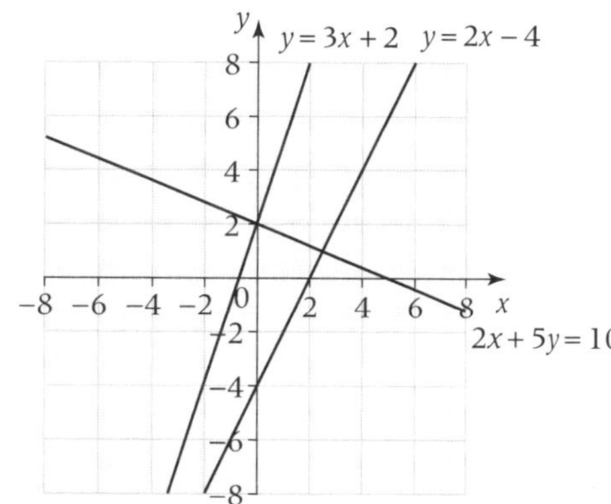

3 **a**

x	1	2	3	4	5
y	12	19	26	33	40

b

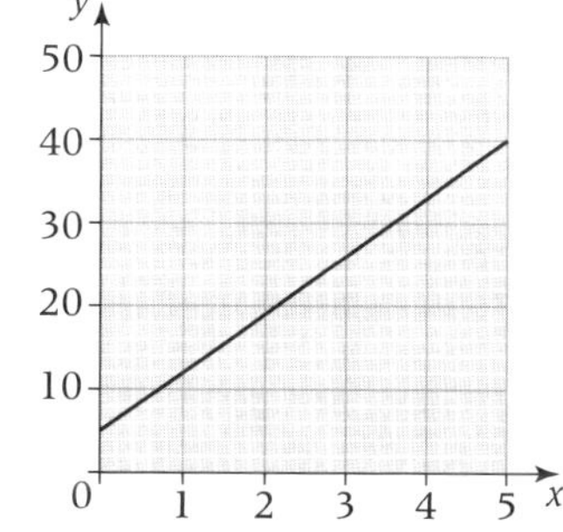

c £26

d $y = 7x + 5$

4 **a** No, $2 \times 3 + 1 \neq 8$

b For example (3, 7)

5 a

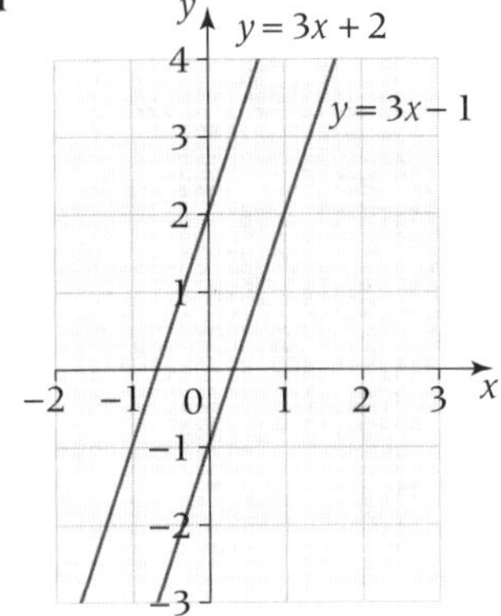

b They are parallel so will never intersect.

c $y = 3x + k$, where k is any number other than −1 or 2.

6 Need to do 40 chores to receive the same amount under both options. If you do less than 40 chores, better to take £5 option; if more, take £3 option.

A4.2

1 a $y = 7$ is horizontal, $x = 9$ and $x = -0.5$ are vertical, $y = 2x - 1$ is diagonal, $y = x^2 + x$ is none of these

2 a $y = 3$ **b** $y = \frac{3}{4}$ **c** $y = -3$

d $x = -2$ **e** $x = \frac{1}{4}$ **f** $x = 2.5$

3 a Vertical line cutting x-axis at $x = 5$

b Horizontal line cutting y-axis at $y = 2$

c Vertical line cutting x-axis at $x = 1.6$

d Horizontal line cutting y-axis at $y = -3$

e Horizontal line cutting y-axis at $y = 1$

f Vertical line cutting x-axis at $x = -1\frac{1}{4}$

4 a (5, 2) **b** (4, −3) **c** (−2, 9) **d** (−2, −4)

5 a For example $x = 4$, $x = 7$, $y = 1$, $y = 6$

b For example $x = 4$, $x = 7$, $y = 3$, $y = 6$

c For example $x = 4$, $y = 3$, $4y + 3x = 36$

6 a (1, 2)

b For example, below $y = 5$, above $y = 3$, right of $x = 2$, on the line $y = x + 1$

A4.3

1 a −2, 1 **b** 3, 2 **c** 1, −3 **d** $\frac{1}{2}$, −1

2 a Any line sloping up from left to right.

b Any line through (0, 3).

c Any line that goes up 1 unit for every 4 across.

d Any line that slopes down and goes through (0, −2).

e The line $y = 3x + 5$.

f The line $y = \frac{2}{3}x + 1$.

3 a

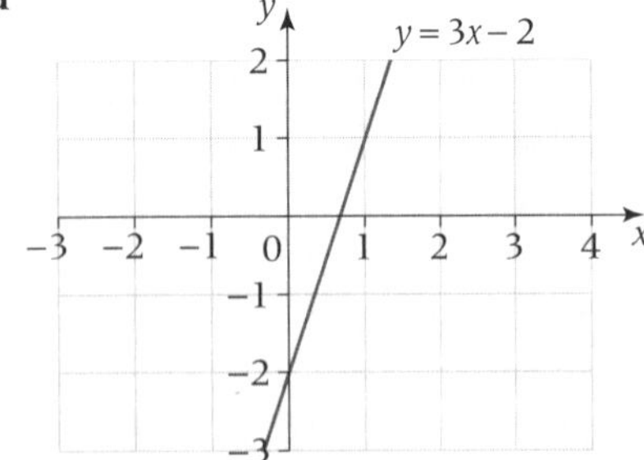

b 3, −2

c Number before x gives gradient, number after x term gives intercept.

d

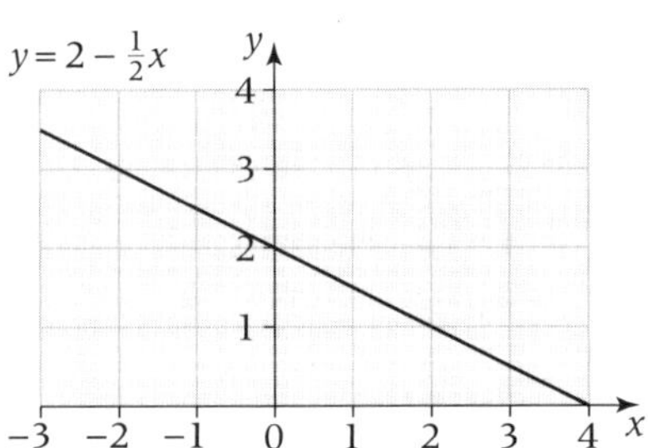

Gradient $-\frac{1}{2}$, intercept 2

4 a $\frac{2}{3}$

b Divide difference in y-coordinates by difference in x-coordinates.

c 2

A4.4

1 a $y = 2 - 4x$ **b** $y = 3x + 1$

c $y = 4x - 2$ **d** $y = x$

2

Equation	Gradient	Direction	Intercept
$y = 4x + 3$	4	Positive	3
$y = 3x + 4$	3	Positive	4
$y = 9x - 2$	9	Positive	−2
$y = 4x - 5$	4	Positive	−5
$2y = 8x + 6$	4	Positive	3

3 ai Gradient = 1 Intercept = 2

aii $y = x + 2$

bi Gradient = 2 Intercept = 3

bii $y = 2x + 3$

ci Gradient = 3 Intercept = 0

cii $y = 3x$

di Gradient = −2 Intercept = 3

dii $y = -2x + 3$

4 a $y = 3x + k$, where k is any number

b $y = 3 - 2x$ **c** $y = -3x + 1$

5 a $y = 3x + 4$ **b** $y = 3x + 4$

A4.5

1 $y = 3x + 5$, $y = 5x - 2$, $y = 7 - 2x$, $2y = x + 18$, $4y = -x - 12$, $y = 4$, $y = x$

2 a $y = 4x$, $y = 12 - 2x$ **b** $y = 4x$, $y = 5x - 1$

3 a $y = 7x + 5$ **b** $y = \frac{1}{2}x + 3$ **c** $y = 4x - 4$

d $y = 3x - 5$ **e** $y = 5 - 2x$ **f** $y = \frac{1}{4}x - 2$

g $y = 4x + 1$ **h** $y = x + 2$ **i** $y = 8x - 6$

4 a 3 **b** $y = 3x + 5$ **c** −2, $y = 16 - 2x$

5 a (0, −2.5) **b** $(\frac{5}{9}, 0)$ **c** (2, 6.5)

6 $3x + 2y = 12$

A4 Exam review

1 a

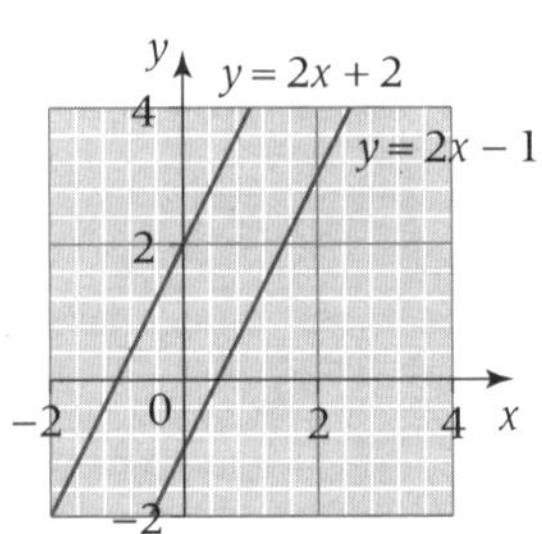

b No, they are parallel

2 $y = 2x + 6$

D3 Before you start ...

1 a $\frac{3}{5}$ b $\frac{3}{7}$ c $\frac{5}{8}$ d $\frac{7}{11}$

2 a $\frac{5}{6}$ b $\frac{9}{20}$ c $\frac{23}{24}$ d $\frac{5}{8}$

3 a $\frac{1}{6}$ b $\frac{3}{20}$ c $\frac{2}{15}$ d $\frac{5}{12}$

4 a 0.7 b 0.75 c 0.375
d 0.4 e $0.\dot{3}$ f 0.0625

D3.1

1 0.53
2 0.65
3 a $\frac{5}{18}$ b 0 c $\frac{5}{9}$ d $\frac{1}{6}$ e $\frac{5}{6}$
4 a $\frac{7}{25}$ b $\frac{18}{25}$ c $\frac{9}{25}$ d $\frac{16}{25}$ e $\frac{2}{5}$ f $\frac{17}{25}$
5 0.27
6 a Gerry will not be chosen.
b 0.4
7 a The chance of Rangers winning is double that of Rovers winning.
b 0.2

D3.2

1 a $\frac{7}{20}$ b $\frac{13}{20}$ c $\frac{1}{20}$ d $\frac{19}{20}$ e $\frac{11}{20}$
f $\frac{9}{20}$ g $\frac{8}{20}=\frac{2}{5}$ h $\frac{15}{20}=\frac{3}{4}$ i $\frac{16}{20}=\frac{4}{5}$
2 a $\frac{1}{9}$ b $\frac{8}{9}$ c $\frac{5}{9}$ d $\frac{1}{3}$ e $\frac{1}{9}$
f $\frac{8}{9}$ g $\frac{2}{9}$ h $\frac{7}{9}$ i $\frac{4}{9}$
3 a 30
bi $\frac{7}{30}$ bii $\frac{1}{2}$ biii $\frac{2}{5}$
biv $\frac{3}{10}$ bv $\frac{3}{10}$ bvi $\frac{7}{10}$
4 a $\frac{1}{15}$ b $\frac{1}{3}$ c $\frac{23}{30}$ d $\frac{1}{2}$

D3.3

1 a

	France	Spain	UK	Total
Girls	3	4	11	18
Boys	3	8	3	14
Total	6	12	14	32

bi $\frac{3}{16}$ bii $\frac{13}{16}$ biii $\frac{7}{16}$
biv $\frac{2}{3}$ bv $\frac{9}{16}$ bvi $\frac{5}{6}$

2 a

	Orienteering	Paintballing	Quadbiking	Total
Girls	11	8	4	23
Boys	5	8	14	27
Total	16	16	18	50

bi $\frac{8}{25}$ bii $\frac{17}{25}$ biii $\frac{16}{25}$ biv $\frac{17}{25}$
ci 128 cii 184

3 a

	Science	Humanities	Other subjects	Total
Girls	29	18	32	79
Boys	27	3	11	41
Total	56	21	43	120

bi $\frac{79}{120}$ bii $\frac{7}{40}$ biii $\frac{77}{120}$ biv $\frac{9}{40}$
ci 280 cii 105

D3.4

1 140
2 120
3 $\frac{5}{8}$
4 a Frequency of 3 is double frequency of 1.
b $\frac{17}{60}$ c 50
5 ai $\frac{11}{100}$ aii $\frac{8}{25}$ aiii $\frac{27}{100}$
aiv $\frac{41}{100}$ av $\frac{87}{100}$
bi 80 bii 120
6 a 104 b 114 c 25

D3.5

1 Expected number of tails is 160, which is not close to 114, so coin is probably not fair.
2 133 is close to expected number of black (140), so spinner is fair.
3 Yes, each outcome occurs with similar frequency.
4 No, 2 occurs more than twice as often as other outcomes.
5 a Relative frequency: $\frac{4}{10}, \frac{7}{20}, \frac{11}{30}, \frac{13}{40}, \frac{18}{50}, \frac{22}{60}, \frac{26}{70}, \frac{29}{80}, \frac{32}{90}, \frac{34}{100}$
b $\frac{34}{100}$
c Expected number of heads is 50, which is not close to 34, so coin could be biased.

D3 Exam review

1 ai $\frac{10}{100}$ aii $\frac{23}{100}$ aiii $\frac{33}{100}$
b Might not be fair as a 6 was rolled more than twice as often as a 2.
2 a

	France	Germany	Spain	Total
Female	2	23	9	34
Male	15	2	9	26
Total	17	25	18	60

b $\frac{5}{12}$

N4 Before you start ...

1 a 40 b 630 c 90 d 200
2 £8.50
3 a 40 mph b 42.5 mph c 54 mph
d 32 mph
4 ai £49.50 aii 66 mm
aiii 52.8 km aiv 4.4 hours
bi 4.5 miles bii 43.5 minutes
biii 10.5 kg biv £46.5

N4.1

1 a 75 g b 4.5 g c 100 g d 125 g
2 a 90 ml b 228 ml c 1.2 litres d 168 cm^3
3 a 60 g b 36 mm c 380 g d 31.2 km
4 a $\frac{1}{10}$ b $\frac{3}{40}$ c $\frac{1}{5}$ d $\frac{3}{16}$
5 a 10% b 7.5% c 20% d 18.75%
6 ai $\frac{3}{20}$ aii 15% bi $\frac{1}{25}$ bii 4%
ci $\frac{2}{5}$ cii 40% di $\frac{8}{125}$ dii 6.4%
7 a 73.536 mm b £315 c £9.88 d 840 kg
8 a $\frac{3}{25}$ b 12%
9 a 200 ml b 3 litres c 1.8 litres
10 a £2100 b $\frac{8}{15}$, 53.3%

N4.2

1 a £8.25 b £12.38 c £18.84 d £75.76
2 a £17.25 b £11.90 c £7.76 d £16.73
3 12 kg
4 75 kg
5 a 60p b £1.80
6 a 40p b £2.80
7 a 3p b 0.52 g
8 a £1.88 b £1.48 c £1.73
d £2.04 e £1.86
9 £3.64
10 a Pack A = 1.21p per pin. Pack B = 1.15p per pin. Pack B is the better value.
11 Regular, at 0.7p per sheet, is better value than Super, at 0.77p per sheet.

N4.3

1 a $31 b $54.25 c $16.28 d $59.75
2 a €7.45 b €44.70 c €87.91 d €393.36
3 a £1.07 b £19.22 c £127.67 d £305.31
4 a £2 b £10 c £1.46 d £3.66
e £31.71 f £2.90
5 a CAN$40.91 b $61.11
6 a €501 b €760.87
7 1.760 pints
8 a 12.701 kg b 1016.047 kg
c 170.324 kg d 17.293 kg
9 a 40.2 km to 3 sf b 6.21 miles to 3 sf
c 24 900 miles to 3 sf d 7.24 km to 3 sf
10 a 157.5 litres b 11.1 gallons to 3 sf
c 56.25 litres to 3 sf d 8.62 gallons to 3 sf

N4.4

1 9 mph
2 £2.10 per metre
3 32 mph
4 89.25 mph
5

Speed (km/h)	Distance (km)	Time
105	525	5 hours
48	106	2 hours 12.5 minutes
$37\frac{1}{3}$	84	2 hours 15 minutes
86	215	2 hours 30 minutes
37.1	65	1 hour 45 minutes

6 a 5 g/cm^3 b 87.88 g
7 a 7.59 g/cm^3 to 3 sf b 105
8 a 9.46 kg to 3 sf b 6.15 litres to 3 sf
9 1358 m

N4.5

1 a 25 g b 200 cm c 7.5 h d 625 kg
2 a 400 g b 20 sec c 30° d 48 h
3

Original number	Proportional change	Result
42	Decrease by $\frac{1}{4}$	31.5
110	Increase by $\frac{1}{5}$	132
250	Increase by $\frac{1}{10}$	275
450	Decrease by $\frac{2}{5}$	270
965	Increase by $\frac{1}{10}$	1061.5

4 90 g butter, 4.5 tsp sugar, 300 ml milk
5 a 280 g b 350 g c $233\frac{1}{3}$ g d 875 g
6 a £126 b £252 c £525 d £75.60
7 a £495 b £510 c £954 d £658.60
8 Andrew's pay increases by a larger amount (£10.80) than Bella's (£10.40)
9 £203.34 to 2 dp

N4 Exam review

1 a 750 g b 187.5 g
2 £9720

S3 Before you start ...

1 A(3, 2), B(5, 5), C(4, −1), D(−2, −3), E(1, 6)
2 a $x = 3$ b $x = 5$ c $y = 2$
d $y = -1$ e $y = x$
3

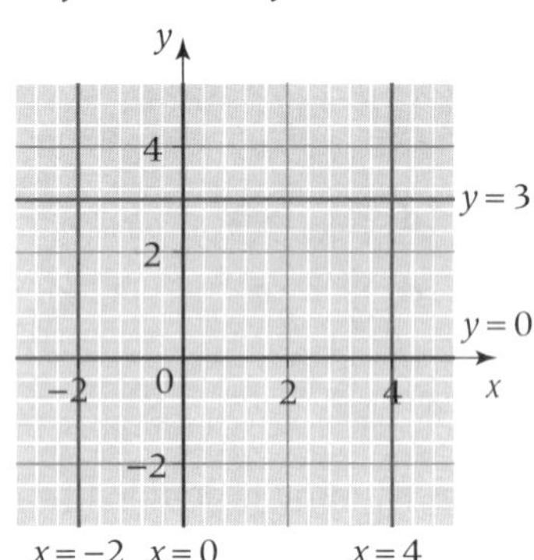

S3.1

1

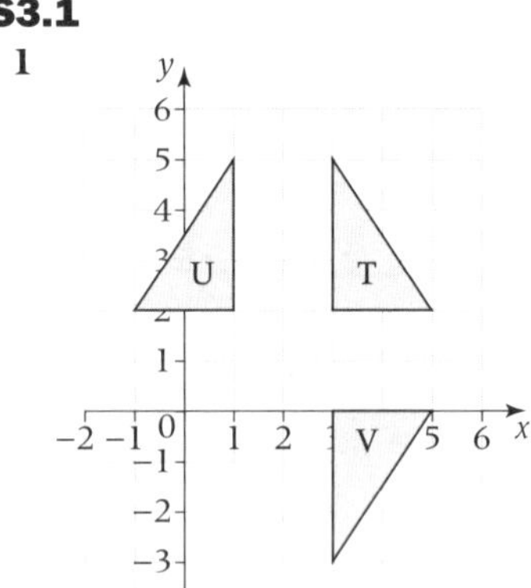

2

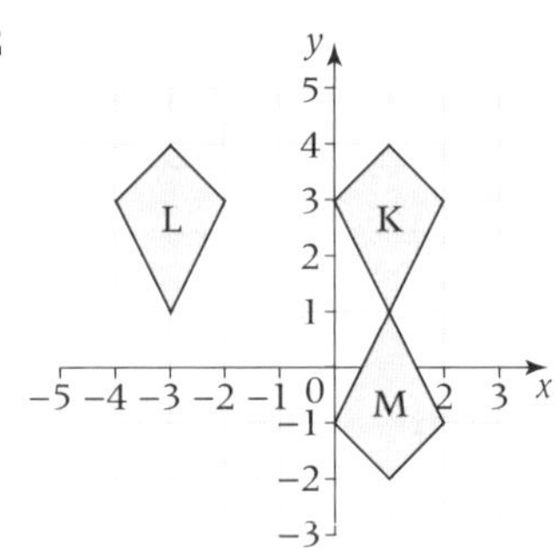

3

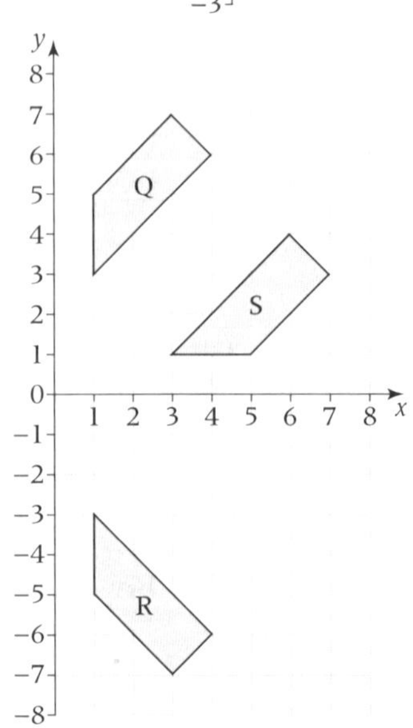

4 a and **b**

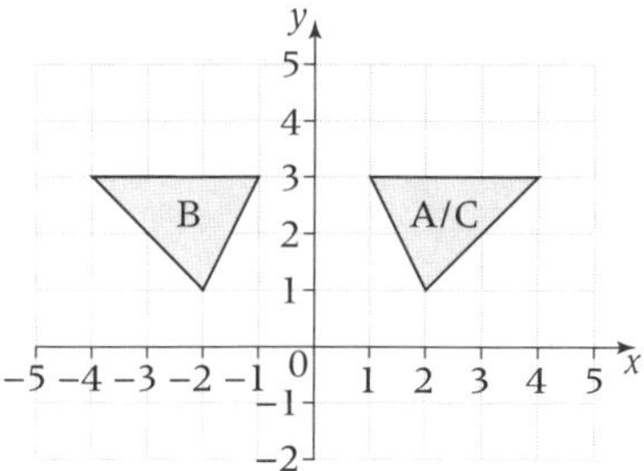

c A and C are the same.

S3.2

1

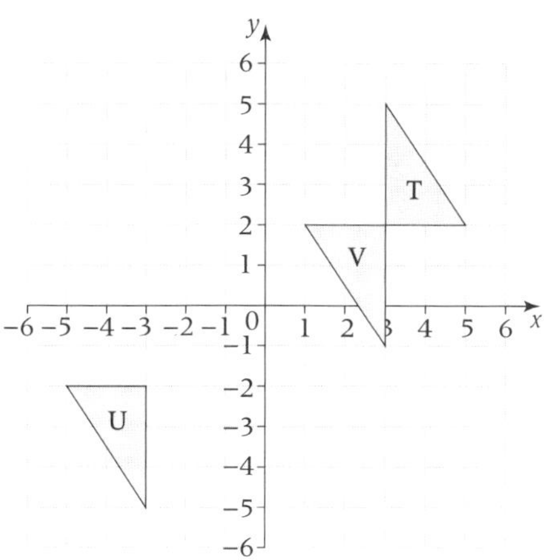

2

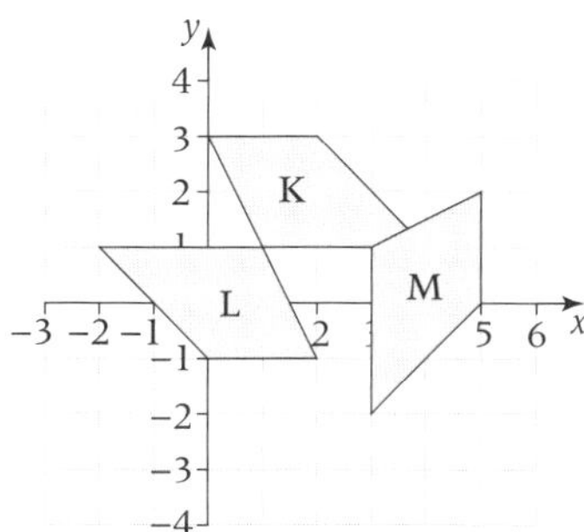

3

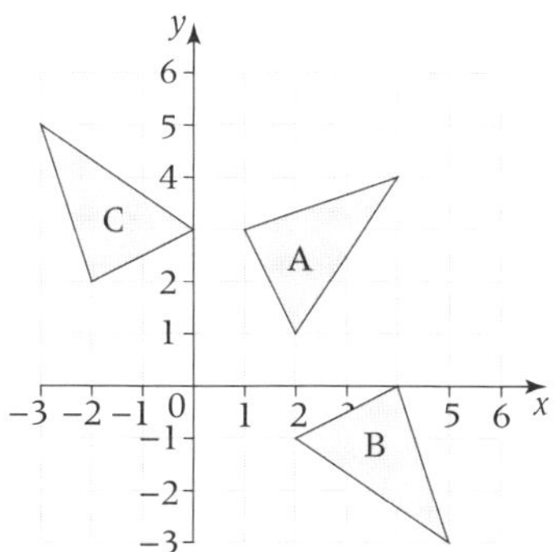

4 a and **b**

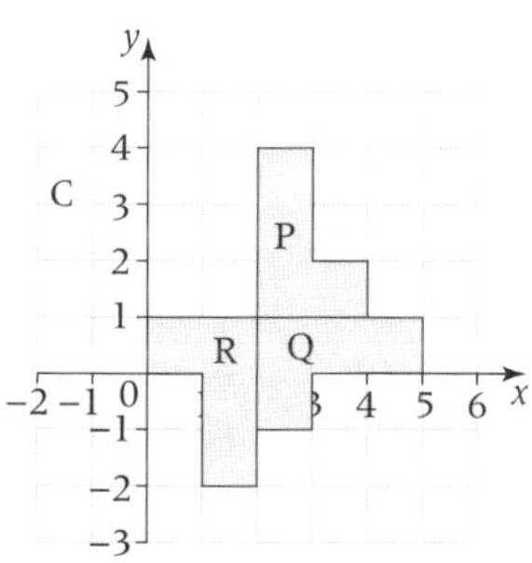

c It maps to R. Two 90° rotations are the same as one 180° rotation.

S3.3

1

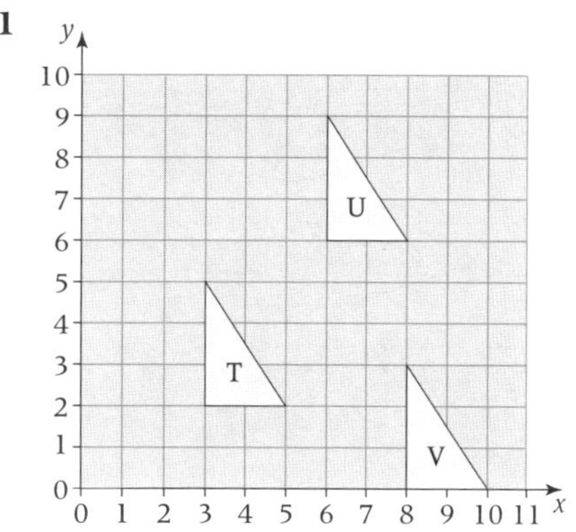

2

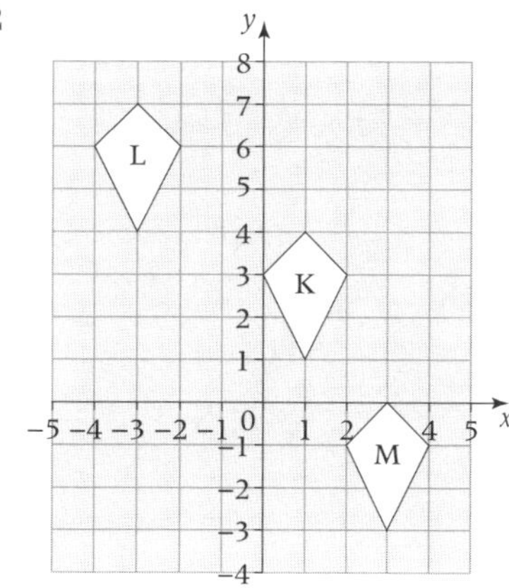

3 a

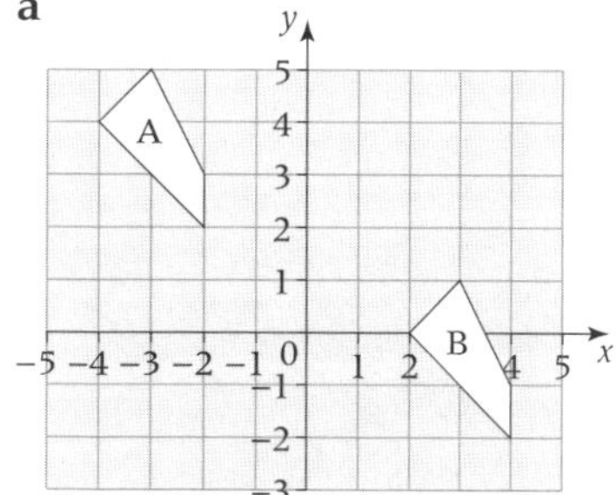

b B translates back to A. The translation has been reversed.

S3.4

1 a Reflection in *y*-axis
b Rotation by 180° about (0, 0)
c Reflection in *x*-axis
d Rotation by 180° about (0, 0)
e Reflection in *y*-axis
f Reflection in *y*-axis

2 a Translation by $\begin{pmatrix}16\\2\end{pmatrix}$ **b** Translation by $\begin{pmatrix}5\\3\end{pmatrix}$

c Translation by $\begin{pmatrix}9\\8\end{pmatrix}$ **d** Translation by $\begin{pmatrix}-4\\-5\end{pmatrix}$

e Translation by $\begin{pmatrix}7\\-6\end{pmatrix}$ **f** Translation by $\begin{pmatrix}-7\\6\end{pmatrix}$

3 a Rotation by 90° clockwise about (0, 0)
b Rotation by 180° about (0, 0)
c Rotation by 90° anticlockwise about (0, 0)
d Rotation by 90° clockwise about (0, 0)
e Rotation by 90° clockwise about (0, 0)
f Rotation by 180° about (0, 0)

S3.5

1

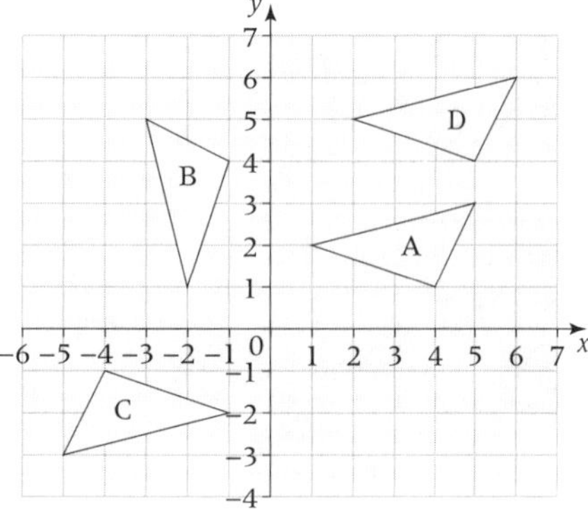

c Rotation 180° about (0, 0)

e Translation by $\begin{pmatrix} 1 \\ 3 \end{pmatrix}$

2

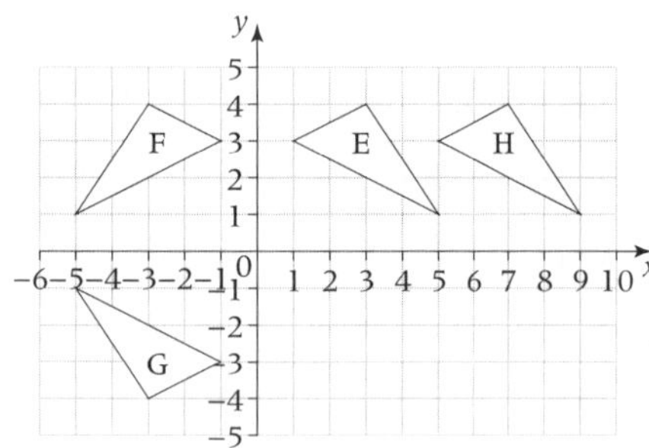

c Rotation 180° about (0, 0)

e Translation by $\begin{pmatrix} 4 \\ 1 \end{pmatrix}$

3

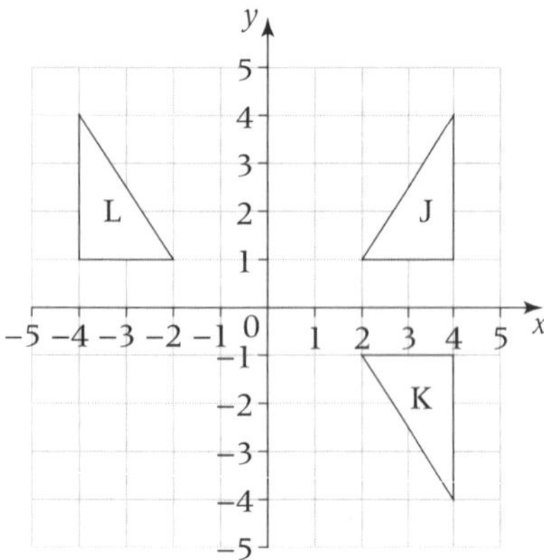

c Reflection in y-axis

4 **a** and **b**

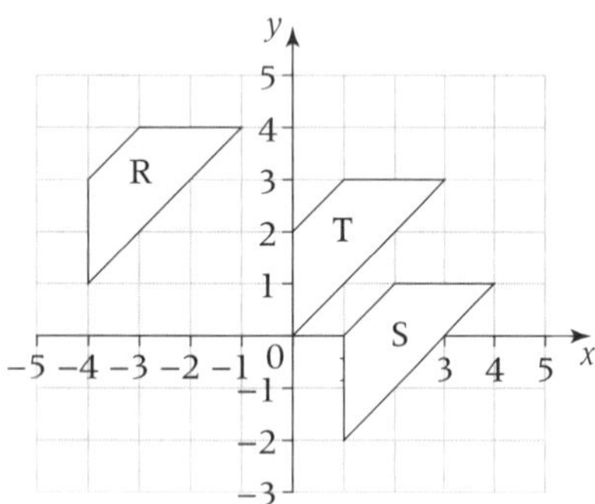

c Translation by $\begin{pmatrix} -4 \\ 1 \end{pmatrix}$

S3 Exam review

1

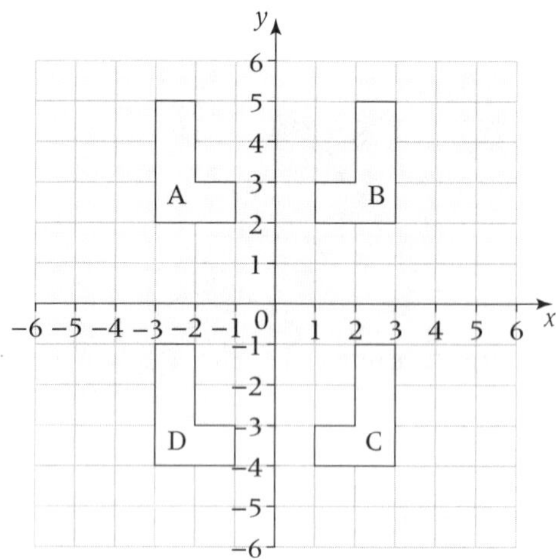

b Translation by $\begin{pmatrix} -2 \\ -6 \end{pmatrix}$

d Reflection in the line $x = 0$

2 a

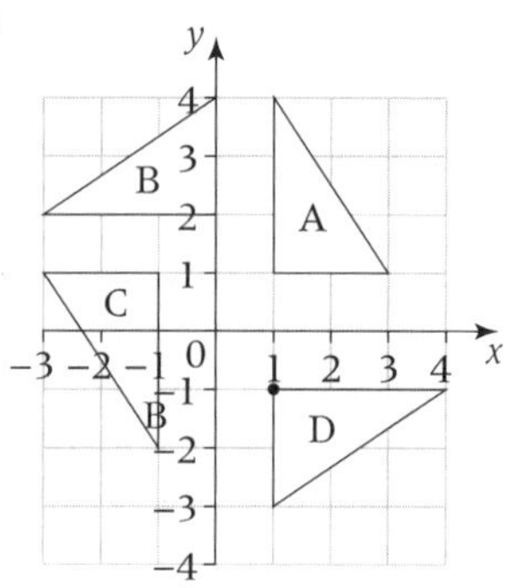

b Rotation by 180° about (0, 1)

N5 Before you start ...

1 **a** 3^2 **b** 4^5 **c** 6^3 **d** 5^4

2 **a** $3 \times 3 \times 3$ **b** 6×6
c $4 \times 4 \times 4 \times 4 \times 4$ **d** $8 \times 8 \times 8$
e $7 \times 7 \times 7 \times 7$ **f** $5 \times 5 \times 5 \times 5 \times 5$

3 **a** 16 **b** 27 **c** 16
d 125 **e** 128 **f** 7

N5.1

1 **a** 3^2 **b** 2^3 **c** 3^3 **d** 5^4
e 7^3 **f** 10^3 **g** 6^4 **h** 5^3

2 **a** $3 \times 3 \times 3 \times 3$
b 5×5
c $7 \times 7 \times 7 \times 7$
d $10 \times 10 \times 10 \times 10 \times 10$
e $4 \times 4 \times 4 \times 4 \times 4 \times 4 \times 4 \times 4 \times 4$
f $6 \times 6 \times 6$
g $2 \times 2 \times 2 \times 2 \times 2$
h $9 \times 9 \times 9$

3 **a** 16 **b** 64 **c** 32 **d** 100
e 1000 **f** 27 **g** 8 **h** 9

4

Index form	Product	Value
10^6	$10 \times 10 \times 10 \times 10 \times 10 \times 10$	1 000 000
10^5	$10 \times 10 \times 10 \times 10 \times 10$	100 000
10^4	$10 \times 10 \times 10 \times 10$	10 000
10^3	$10 \times 10 \times 10$	1 000
10^2	10×10	1 00
10^1	10	1 0

5

Index form	Product	Value
2^8	$2\times2\times2\times2\times2\times2\times2\times2$	256
2^7	$2\times2\times2\times2\times2\times2\times2$	128
2^6	$2\times2\times2\times2\times2\times2$	64
2^5	$2\times2\times2\times2\times2$	32
2^4	$2\times2\times2\times2$	16
2^3	$2\times2\times2$	8
2^2	2×2	4
2^1	2	2

6 **a** 9^2 **b** 5^3 **c** 2^7 **d** 10^5
e 3^4 **f** 7^3
7 **a** 3^6 **b** 2^9 **c** 4^8 **d** 5^5
e 8^8 **f** 3^8 **g** $2^7\times3^4$ **h** $5^4\times7^7$
8 **a** 2^4 **b** 11^2 **c** 3^3 **d** 5^3
e 13^2 **f** 5^4 **g** 3^5 **h** 2^8
9 **a** $2^2\times13$ **b** $2^2\times3^2$ **c** 2×5^2 **d** $2^3\times3$
e 2×3^2 **f** $2^4\times3$ **g** $2^2\times3\times5$
j $2^4\times3^2$
10 **a** 16, 448 **b** 19, 266

N5.2

1 **a** 7^3 **b** 3^3 **c** 5^3 **d** 6^4
e 5^2 **f** 8^5 **g** 9^6 **h** 8^8
2 **a** 6^5 **b** 4^9 **c** 2^{13} **d** 11^7
e 1^{30} **f** 7^{12} **g** 3^{12} **h** 9^{10}
3 **a** 7^2 **b** 8^4 **c** 3 **d** 9^3
e 4^6 **f** 1 **g** 12^2 **h** 1
4 **a** 8^5 **b** 5^5 **c** 2^6 **d** 9^2
e 8^8 **f** 7^7 **g** 4^{10} **h** 11
5 **a** 3 **b** 5^{10} **c** 4^2 **d** 7^2
e 8^{11} **f** 9^{10}
6 **a** 4^2 **b** 6^2 **c** 9^2 **d** 8
e 5^3 **f** 6^5 **g** 8^2 **h** 10^0
7 **a** $3^3\times4^4$ **b** $7^2\times8^3$ **c** $5^6\times6^4$ **d** $3^6\times4^9$
e $2^3\times5^2$ **f** $7^4\times9^7$ **g** $5^3\times8^5$ **h** $3^8\times8^7$
i $2^2\times9^5$
8 **a** $5^2\times8^3$ **b** $6^2\times7^2$ **c** $5^2\times6^2$ **d** $5^3\times7^6$
e $3^3\times8^2$ **f** $4^5\times5^2$ **g** $6^4\times7^2$ **h** $4^4\times7^5$

N5.3

1 **a** 5 **b** 6 **c** 1 **d** 1
e 1 **f** 41 **g** 1 **h** 0
2 **a** 10 **b** 4 **c** 7 **d** 2
e 8 **f** 3 **g** 11 **h** 8
i 12 **j** 2 **k** 3 **l** 10
3 **a** 6 **b** 36 **c** 9 **d** 81
4 **a** 3^{-1} **b** 5^{-1} **c** 7^{-1} **d** 11^{-1}
e 2^{-1} **f** 5^{-1} **g** 10^{-1} **h** 3^{-1}
5 **a** $\frac{1}{9}$ **b** 1 **c** 3 **d** 9
e 81 **f** 729 **g** 59 049 **h** $\frac{1}{729}$
6 **a** 7^{-2} **b** 9^{-2} **c** 2^{-2} **d** 2^{-3}
e 2^{-5} **f** 3^{-4} **g** 5^{-3} **h** 6^{-4}
7 **a** $\frac{12}{8}$ **b** $\frac{13}{7}$ **c** $\frac{12}{5}$ **d** $\frac{14}{9}$
e $\frac{12}{3}$ **f** $\frac{13}{9}$ **g** $\frac{15}{4}$ **h** $\frac{16}{6}$
8 **a** $\frac{1}{16}$ **b** $\frac{1}{4}$ **c** 1 **d** 2
e 4 **f** 16 **g** 64 **h** $\frac{1}{2}$
9 **a** $3^{-\frac{1}{2}}$ **b** $5^{-\frac{1}{2}}$ **c** $7^{-\frac{1}{2}}$ **d** $11^{-\frac{1}{2}}$
10 **a** 2^4 **b** 2^6 **c** 3^6 **d** 4
e 5^8 **f** 4^{-6} **g** 7^{12} **h** 5^{-4}
11 **a** 2^{-2} **b** 3^{-1} **c** 4^{-6} **d** $3^{-1}\times5^{-1}$
e $5^4\times7^{-4}$

N5.4

1 **a** 10^2 **b** 10^1
c 10^5 **d** 10^0
2 **a** 2×10^2 **b** 8×10^2 **c** 9×10^3
d 6.5×10^2 **e** 6.5×10^3 **f** 9.52×10^2
g 2.358×10 **h** 2.5585×10^2
3 **a** 500 **b** 3000 **c** 100 000
d 250 **e** 4900 **f** 3 800 000
g 750 000 000 000
h 8 100 000 000 000 000 000
4 **a** 6×10^2 **b** 4.5×10^4
c 6.5×10^0 **d** 5×10^6
5 **a** 4×10^5 **b** 9×10^7
c 2.5×10^8 **d** 2.4×10^{13}
6 **a** 2×10^2 **b** 2×10^4
c 5×10 **d** 7.5×10^2
7 **a** 9.75×10^9 **b** 1.37×10^4
c 4.01×10^{11} **d** 2.06×10^8
8

Planet	Mean distance from Sun (m)	Light travel time
Mercury	5.79×10^{10}	3 minutes 13 seconds
Earth	1.50×10^{11}	8 minutes 20 seconds
Mars	2.28×10^{11}	12 minutes 40 seconds
Jupiter	7.78×10^{11}	43 minutes 13 seconds
Pluto	5.90×10^{12}	5 hours 27 minutes 47 seconds

N5.5

1 **a** 3×10^{-1} **b** 4.7×10^{-3}
c 7.8×10^{-5} **d** 4.485×10^{-1}
2 **a** 2.8×10^{-1} **b** 4×10^{-2}
c 1.35×10^{-3} **d** 1.2×10^{-7}
3 **a** 1×10^{-2} km **b** 2×10^{-3} g
c 5×10^{-6} m **d** 1.1×10^{-2} litre
4 **a** 5×10^{-1} **b** 9.2×10^{-8}
c 2×10^{-2} **d** 4.2×10^{-8}
5 **a** 1.15×10^{-7} **b** 1.83×10^{-1}
c 4.85×10^{-6} **d** 5.01×10^{-1}
6 **a** 5.2×10^{-1} **b** 4.6×10^{-2}
c 2.09×10^{-2} **d** 1.3×10^{-2}
7 See Q6
8 9.11×10^{-8} m^3
9 5.3×10^{-3} kg
10 About 385 atoms

N5 Exam review

1 **a** $\frac{1}{3}$ **b** 3 **c** 1
2 **a** 1×10^{-9} **b** 2×10^8

A5 Before you start ...

1 $A=lw$, area of a rectangle; $A=\pi r^2$, area of a circle; $a^2+b^2=c^2$, length of sides in a right-angled triangle; $V=lwh$, volume of a cuboid; $A=\frac{1}{2}bh$, area of a triangle
2 **a** $x=5\frac{2}{3}$ **b** $x=2\frac{1}{2}$ **c** $x=\pm5$ **d** $x=16$
e $x=2$ **f** $x=2$
3 **a** $2+4=6$ **b** $(-2)^2=4$
c $2+3=5$ **d** $3\times2=6$

A5.1

1

Identities	Equations	Formulae
$3x(x+1)=3x^2+3x$	$3x+1=10$	$C=2\pi r$
$y+y=y^2$	$2x+5=3-7x$	$A=\frac{1}{2}(a+b)h$
$20-x=-(x-20)$	$2x^2=50$	$a^2+b^2=c^2$

2 **a** £17 **b** 9
c You can't have half a person.

3 **a** 53.6 °F **b** 0 **c** 4
d 108, 5 or −5 **e** −14 **f** 1

4 **a** 33 cm^2
b Any a, b such that both are positive and $a + b = 10$

A5.2

1 **a** $P = 4s$ **b** $A = \frac{1}{8}\pi r^2$
c $P = 26 + 2y$ **d** $T = 10 + 35p$

2 a, b, c are lengths of the base of the triangle, height of the triangle and length of the prism. Correct as multiplies area of cross-section by length.

3 **a** $V = \pi r^2 h$
b $2\pi r^2$ represents the combined area of the top and bottom circles. The $2\pi rh$ term represents the curved section; if opened out flat it would be a rectangle measuring $2\pi r$ in length (the circumference of the circle) and h in height.

A5.3

1 **a** $x = \frac{C - b}{a}$ **b** $x = M + b + c$
c $x = Kt + qt$ **d** $x = \frac{W - t}{y}$
e $x = Hp - z$ **f** $x = q + \frac{D}{p}$ or $\frac{D + pq}{p}$
g $x = AB + ct$ **h** $x = \frac{Y - c}{m}$

2 Sebastian expanded the brackets to begin with but James didn't.

3 **a** $y = \sqrt{c}$ **b** $y = 4(k + 2)$
c $y = \frac{M + t}{xz}$ **d** $y = 4x^2$
e $y = \sqrt[3]{p - 2}$ **f** $y = \sqrt{\frac{T}{k}}$
g $y = \frac{3R}{az}$ **h** $y = p^3$

4 He should have divided by m before taking cube root.

5 $\frac{8(D + k)}{ab} = c$
$8(D + k) = abc$
$D + k = \frac{1}{8}abc$
$d = \frac{1}{8}abc - k$

6 **a** 86 °F **b** $C = \frac{5(F + 40)}{9} - 40$ **c** $-35\frac{5}{9}$ °C

A5.4

1 **a** $p = \frac{m + q}{x}$ **b** $p = \sqrt{w + r}$
c $p = (m - h)^3$ **d** $p = t(h + g)$
e $p = 2(q - r)$ **f** $p = \sqrt{\frac{k}{b}}$
g $p = \frac{z}{aw}$ **h** $p = (2x + y)^2$

2 **a** $x = k - w$ **b** $x = \frac{t - p}{a}$
c $x = \frac{b - y}{t}$ **d** $x = a - \frac{m}{n}$ or $\frac{an - m}{n}$
e $x = \frac{k}{w}$ **f** $x = \frac{t}{m}$
g $x = \frac{h}{g - p}$ **h** $x = \sqrt{\frac{p}{k}}$

3 **a** 36 mph **b** $t = \frac{d}{S}$
c 1 hour 26 minutes to nearest minute

4 $k = \frac{t}{p - q}$

5 D $x = \frac{2(p - y)}{ab}$
A $abx = 2(p - y)$
E $\frac{1}{2}abx = (p - y)$
B $y + \frac{1}{2}abx = p$
C $y = p - \frac{1}{2}abx$

6 **a** $g = \dfrac{4\pi^2 p}{T^2}$ **b** 9.8

7 **a** $h = \dfrac{V}{\pi r^2}$ **b** $r = \sqrt{\frac{V}{\pi h}}$

8 $b = \frac{2A}{h} - a$

A5.5

1 **a** 21 − 7 = 14, which is even
b 5 squared is 25, which is odd
c 3 × 2 = 6, which is even, and 2 is prime
d 3 × 4 = 12, which is even

2 **a** For example, 2 + 3 + 4 + 5 + 6 = 20 which is 4 × 5
b $n + (n + 1) + (n + 2) + (n + 3) + (n + 4)$
$= 5n + 10 = 5(n + 2)$, which is clearly a multiple of 5 for any n

3 **a** Even numbers $2m$ and $2n$, then $2m + 2n = 2(m + n)$ which is a multiple of 2 and hence even for any m, n
b $(2n)^2 = 4n^2$, which is always in the four times table for any n
c $2n + (2m + 1) = 2(m + n) + 1$, which is odd for any m, n
d $k(k + 1) - k = k^2 + k - k = k^2$, which is a square number for any k

4 **a** Any number between 0 and 1 inclusive
b Any number between 0 and 1 inclusive

5 **a** $k(k + 2) = k^2 + 2k$,
$(k + 1)^2 = k^2 + 2k + 1 = k(k + 2) + 1$ for any k
b $(n + 4)(n + 7) - (n + 3)(n + 8)$
$= n^2 + 11n + 28 - n^2 - 11n - 24 = 4$
c $m^2 - n^2 = m^2 - mn + mn - n^2 = (m - n)(m + n)$ for any m, n

6 If angle ACB is x then angle OAC is also x (isosceles triangle) so angle AOC is 180 – $2x$ (angles in a triangle).
If angle ABC is y then angle OAB is also y (isosceles triangle) so angle AOB is 180 – $2y$ (angles in a triangle).
Hence, 180 – $2x$ + 180 – $2y$ = 180 (angles on a straight line)
$180 = 2x + 2y$
$90 = x + y$
i.e. angle CAB is 90°

A5 Exam review

1 **a** $P = 3x - 1$
bi $x = \frac{P + 1}{3}$ **bii** 3 cm

2 When $n = 2$, $n^2 + 3 = 7$ which is not even.

S4 Before you start ...

1

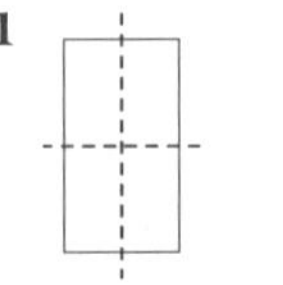
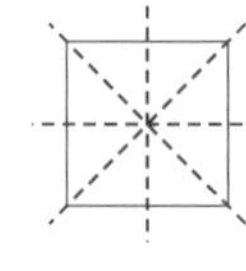
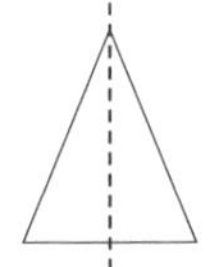

2 **a** 49 **b** 52 **c** 34 **d** 48 **e** 45 **f** 80

S4.1

1 **a** Yes **b** No **c** Yes

2 **a**

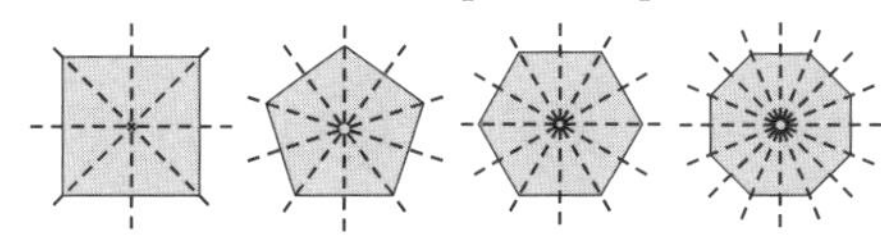

2 **b** First is a line of symmetry, second is not.

3

4 Cuboid: planes through middle of sides and through diagonally opposite edges.
Triangular prism: planes vertically through middle and from edges through opposite side.

5 One horizontal plane of symmetry halfway up the cylinder and infinitely many vertical planes of symmetry through the diameters of the circle on top of the cylinder.

S4.2

1 22.5 cm^2

2 **a** Square with sides 7.1 cm **b** 50 cm^2

3 **a** All 42 cm^2
b Halve the product of the diagonals.

4 60 cm^2

5 Both shapes have been split into congruent triangles; all sides equal in both cases.

6 **a** True; all squares have 2 pairs of sides of equal length and 4 equal angles.
b False; not all kites have diagonals which bisect each other.
c False; not all rhombuses have 4 equal angles.

S4.3

1 **a** 5 cm **b** 17 cm **c** 13 cm **d** 10.3 cm
e 10.8 cm **f** 9.90 cm **g** 12.5 cm **h** 7.3 cm

2 3, 4, 5
8, 15, 17
5, 12, 13

3 **a** 8 cm **b** 19.8 cm **c** 10 cm **d** 9 cm
e 7.5 cm **f** 5.4 cm **g** 5.5 cm **h** 7.1 cm

4 **a** 6, 8, 10; 10, 24, 26; 9, 12, 15
b 6, 8, 10 is 3, 4, 5 doubled; 10, 24, 26 is 5, 12, 13 doubled; 9, 12, 15 is 3, 4, 5 trebled

S4.4

1 **a** 8.6 cm **b** 7.9 cm

2 9.6 cm

3 11.3 cm

4 5.7 cm

5 16.7 cm

6 13.6 cm

7 5.4 m

S4.5

1 **a** (4, 5.5) **b** (0.5, 1) **c** (2, 3) **d** (1.5, 2)

2 **a** 5 units **b** 8.6 units **c** 10.0 units
d 10.6 units

3 J (2, 4, 5), K (3, 1, 5), L (−1, 2, 5), M(4, −3, 0)
N (0, 2, 0), P (−3, 2, −1), Q (0, 0, 2), R (2, 0, −3)

4 **ai** (1, 3, 3.5) **aii** (3.5, 3, 2.5)
aiii (1.5, 3.5, 1.5) **aiv** (3, 2.5, 4.5)
b ai Length of AB = 3.61 (2 dp)
aii Length of CD = 4.24 (2 dp)
c BC = $\sqrt{3}$ = 1.72 (2 dp)
If rounded, $\sqrt{2}$ = 1.41 so BC = 1.73 (2 dp)

S4 Exam review

1 **a** 4.1 units **b** (2.5, 5)

2 **ai** D **aii** B **aiii** C **b** D

N6 Before you start ...

1 **a** 14 **b** 2 **c** 15 **d** 16

2 **a** 2 **b** 16 **c** 5 **d** 10

3 **a** 45.6 **b** 9.9 **c** 104.0 **d** 16.1

4 **a** $20 \times 30 = 600z$ **b** $350 + 60 = 410$
c $1200 - 800 = 400$ **d** $7000 + 6000 = 13\,000$

N6.1

1 **a** 59 **b** 37 **c** 2 **d** 26

2 **a** 1 **b** 10 **c** 43 **d** 110

3 **a** + **b** − **c** × **d** ×

4 **a** $(11 - 1) \times 5 = 50$ **b** $(12 + 3) \div 3 = 5$
c $12 - (4 - 1) = 9$ **d** $8 \div (4 + 4) + 1 = 2$

5 **a** 2 **b** 5 **c** 110 **d** 14

6 **a** 196 **b** 4 **c** 18 **d** 289

7 **a** 121 **b** 39 **c** 18 **d** 5.29

8 **a** 10 **b** 18

9 **a** ≈14.72 **b** ≈102.38 **c** ≈286.63 **d** ≈9.90

N6.2

1 **a** $2\sqrt{3}$ **b** $2\sqrt{5}$ **c** 5 **d** $3\sqrt{7}$

2 **a** 2π **b** 3π **c** $2 + 2\pi$ **d** 10π
e 18π **f** 16π

3 **a** $4 + \sqrt{3}$ **b** $3 + \sqrt{2}$ **c** $7 + 2\sqrt{7}$ **d** $14 + \sqrt{17}$

4 **a** 32π **b** $7 + 2\pi$ **c** $28 + 4\sqrt{2}$ **d** 62π

5 **a** 2 **b** 5 **c** $3 + 3\sqrt{3}$ **d** $8 + 2\sqrt{3}$

6 **a** 5.66 **b** 3.24 **c** 4.24 **d** 106.10

7 **a** 20.9 **b** 44.7 **c** 12.3 **d** 32.2

8 **a** $2\sqrt{5}$ **b** $5\sqrt{5}$ **c** $3\sqrt{2}$ **d** $7\sqrt{2}$

9 **a** 3.146 **b** 3.162 **c** 11.18 **d** 4.897

10 **a** 3.146 **b** 3.162 **c** 11.18 **d** 4.899

11 **a** $14 + 7\sqrt{2}$ **b** $17 + 7\sqrt{5}$ **c** $15 - 3\sqrt{3}$ **d** 20

N6.3

1 **a** 1 **b** 1.2 **c** 1.5 **d** 8
e 1.4 **f** 7.5 **g** 16 **h** 15

2 **a** 40 **b** 10 **c** 20 **d** 80
e 0.2 **f** 10 **g** 10 **h** 5000

3 **a** 0.8 **b** 0.8 **c** 21 **d** 0.21
e 4.8 **f** 40 **g** 0.9 **h** 4

4 as Q3

5 **a** $4 \div 2$ **b** $6 \div 5$ **c** $12 \div 5$ **d** $2 \div 1000$

6 **a** 4×2 **b** 6×4 **c** 16×100 **d** 15×20

7 **a** 32 **b** 30 **c** 35 **d** 1
e 0.32 **f** 200 **g** 1 **h** 720
i 0.375 **j** 28 **k** 1 **l** 7

8 **a** 80 **b** 1000 **c** 520 **d** 500

9 **a** 78.72 to 4 sf **b** 1063 to 4 sf
c 509.2 **d** 548.4 to 4 sf

10 **a** 9.3 **b** 700 **c** 12.2 **d** 10.6
e 500 **f** 130 **g** 570 **h** 0.65

11 **a** 9.3 **b** 700 **c** 12.2 **d** 10.6
e 463.3 to 4 sf **f** 130 **g** 570
h 0.65

N6.4

1 **a** 38.76 **b** 2.61 **c** 8.201 **d** 22.52

2 **a** 2.12 **b** 4.19 **c** 2.04 **d** 5.7

3 **a** 155.09 **b** 0.45 **c** 24.32 **d** 14.72

4 **a** 3.97 **b** 0.095 **c** 12.44 **d** 58.34

5 **a** 2.81 **b** 0.757 **c** 88.01 **d** 1126.72

6 See Q1 to Q5

7 **a** 63.6 **b** 5.37 **c** 13.52 **d** 0.003 36
8 **a** 1.7 **b** 0.43 **c** 48 **d** 269
9 **a** 16.72 **b** 10.26 **c** 793.8 **d** 7.918
10 **a** 8.82 **b** 203 **c** 886 **d** 21.9
11 See Q7 to Q10

N6.5

1 **a** 5.6 **b** 73 **c** 0.85 **d** −35
e 110 **f** 38
2 **a** 141 kg **b** 8.4 m^2 **c** £13.63 **d** 45 sec
3 **a** 4.745 m, 4.755 m **b** 12.55 s, 12.65 s
c 149.5 cm, 150.5 cm **d** 24.45 kg, 24.55 kg
e 8.065 g, 8.075 g **f** 4.325 s, 4.335 s
4 **a** 2.545 m, 2.555 m **b** 1.65 s, 1.75 s
c 1.548 m/s to 4 sf **d** 1.454 m/s to 4 sf
5 **a** 56.3 m^2, 72.3 m^2 **b** 40.3 cm^2, 41.6 cm^2
c 1.09 m^2, 1.11 m^2 **d** 8.70 mm^2, 9.30 mm^2
e 14.0 m^2, 14.1 m^2 **f** 99.7 cm^2, 99.9 cm^2
6 **a** 36.765, 38.645 **b** 0.3, 0.5
c 0.3618, 0.3864 **d** 0.1075, 0.1485 to 4 sf

N6 Exam review

1 **ai** 1250 **aii** 125 **b** 500
2 **a** 17.9867 **b** $(1.6 + 3.8 \times 2.4) \times 4.2$

A6 Before you start ...

1 **a** $x = 9$ **b** $x = 3\frac{1}{2}$ **c** $x = -14\frac{1}{2}$
d $x = \frac{1}{4}$ **e** $x = 3$ **f** $x = 14$
2 **a** $x^2 + 7x + 10$ **b** $y^2 + 4y - 21$
c $w^2 + 8w + 16$ **d** $y^2 - 15y + 54$
e $6w^2 + 14w + 4$ **f** $16y^2 - 40y + 25$
3 **a** $2(2x + 1)$ **b** $y(x + y)$
c $5a(1 + 2b)$ **d** $4(x - 2y)$
e $3x^3(3 + x^2)$ **f** $2x(y + 2z - 3x)$
4 **a** $(x + 2)(x + 3)$ **b** $(x - 6)(x + 4)$
c $(x - 3)(x - 4)$ **d** $(x - 3)(x - 3)$
e $(x + 10)(x - 10)$ **f** $2(x + 2)(x + 1)$

A6.1

1 **a** $x = 6$ **b** $y = 14$ **c** $x = 6$ **d** $p = 8$
e $t = 2$ **f** $b = 2$
2 **a** $x = 1.1$ **b** $y = -1\frac{5}{22}$ **c** $z = 1\frac{3}{4}$ **d** $a = 5\frac{1}{8}$
e $a = 4.3$ **f** $x = 5$ **g** $x = -13$ **h** $y = 1$
3 **a** $x = 2$ **b** $y = 3$ **c** $z = \frac{3}{8}$ **d** $a = -\frac{5}{7}$
e $b = -1$ **f** $c = -2.5$
4 **a** $40 - 3x = 22$, $x = 6$
b $2(x + 6) = 6(x - 5)$, $x = 10.5$
c $(x + 7)^2 = (x - 5)^2$, $x = -1$
5 **a** $(x - 4)(x + 11) = (x + 3)^2$, $x = 53$
b $26 - 4x = 15 - 2x$, $x = 5.5$

A6.2

1 **a** $2x^2 = 50$, $x^2 = 100$, $x^2 = -49$, $x^2 = 169$
b $2x^2 = 50$, $x = 5$ or -5
$x^2 = 100$, $x = 10$ or -10
$x^2 = 169$, $x = 13$ or -13
2 **a** $(x + 2)(x + 6)$ **b** $(x + 3)(x + 8)$
c $(x + 9)(x + 4)$ **d** $(x + 11)(x + 5)$
e $(x - 3)(x - 4)$ **f** $(x - 4)(x - 6)$
g $(x + 7)(x - 4)$ **h** $(x + 3)(x - 5)$
3 **a** $x = -3$ or -4 **b** $x = -2$ or -6
c $x = -5$ **d** $x = -7$ or 2
e $x = 5$ or -1 **f** $x = 2$ or 3
g $x = 3$ or 7 **h** $x = 8$ or -5
4 **a** $x = 0$ or 8 **b** $x = 0$ or -4
c $x = 0$ or 6 **d** $y = 0$ or -5
e $x = 0$ or 9 **f** $x = 0$ or 12
g $x = 0$ or -4 **h** $x = 0$ or 6
5 No two integers add up to 7 and multiply to give 11.
6 **a** She should make the right-hand side equal to zero first.
b $(x + 4)(x - 2) - 7 = 0$, $x^2 + 2x - 15 = 0$, $(x + 5)(x - 3) = 0$, $x = -5$ or 3
7 $x = 1$ or 5

A6.3

1 6.2
2 **a** $x(x + 4) - 77 = 0$, length and width are 7 cm and 11 cm
b $h(h + 1)(h + 2) - 990 = 0$; 9 cm by 10 cm by 11 cm
3 **a** $x = -5$ or 3
b Factorisation is more efficient
4 **a** 8.8 **b** 4.6 **c** 9.51 **d** 4.28
5 **a** $x(x + 1)(x + 2) = 85\,140$; 43, 44, 45
b $x(x + 2) = 57$; 6.6 cm
c $3x^3 = 500$, 5.5 mm
6 0 or 6.03

A6.4

1 **a**, **b**, **e**
2 **a** $x = 4, y = 3$ **b** $x = 2, y = -3$
c $x = -1, y = 4$ **d** $x = 3, y = 0.5$
e $m = 5, n = 2$ **f** $x = -2, y = 1$
3 **a** $x = 7, y = -1$ **b** $x = 3, y = 0.5$
c $a = 3, b = -1$ **d** $x = 2, y = 5$
e $x = 3.25, y = -\frac{2}{7}$ **f** $p = 2, q = 7$
4 **a** $x = 5, y = -2$ **b** $x = 2, y = 3$
c $a = -2, b = 3$ **d** $v = 4, w = 2$
e $p = 2, q = 2$ **f** $x = 5, y = 2$
5 **a** $a = 9\frac{2}{3}, b = 1\frac{1}{3}$ **b** $v = 7, w = 4$
6 **a** 17p **b** 6.4 cm

A6.5

1 **a** $x = 12, y = -16$ **b** $x = 7, y = -1$
c $a = 3, b = -2$ **d** $v = 3, w = 2$
e $p = 0, q = 3$ **f** $x = 5, y = 2$
2 $2x + y = 12$, $y - x = 15$; $x = -1$, $y = 14$
$2x + y = 12$, $3x - 4y = 7$; $x = 5$, $y = 2$
$y - x = 15$, $3x - 4y = 7$; $x = -67$, $y = -52$
3 **a** $x = 6, y = 2$ **b** $a = 8, b = -1$
c $p = 8, q = -1$
4 **a** 17, 24 **b** 17, 23
c 4 large, 1 small **d** 105°, 37.5°, 37.5°
e £5 for a paperback, £10 for a hardback

A6 Exam review

1 $x = 4, y = 2$ **2** $x = 4.2$

D4 Before you start ...

1 **a** 10 **b** 25 **c** 20 **d** 12.5
2 **a** 16 **b** 22.5 **c** 14 **d** 18.5
3 **a** 6 **b** 4.5 **c** 7 **d** 10.5
4 **a** 12 **b** 33 **c** 21 **d** 10.5

D4.1

1 **ai** 5 **aii** 6 **aiii** 5.8 **aiv** 4 **av** 2
bi 5 **bii** 5 **biii** 4.96 **biv** 4 **bv** 2
ci 4 **cii** 6 **ciii** 6.59 **civ** 6 **cv** 4
di 6 and 7 **dii** 6 **diii** 5.77 **div** 6 **dv** 3
2 42.5
3 2.46

D4.2

1 **ai** $10 < t \leqslant 15$ **aii** $15 < t \leqslant 20$ **aiii** 16.1
bi $10 < t \leqslant 20$ **bii** $20 < t \leqslant 30$ **biii** 23.5
ci $5 < t \leqslant 10$ **cii** $10 < t \leqslant 15$ **ciii** 14.7
di $15 < t \leqslant 25$ **dii** $25 < t \leqslant 35$ **diii** 30

2 **a** $20 < b \leqslant 30$ **b** £24.17 to 3 sf

3 **a** $165 \leqslant h < 170$ **b** 164.3 cm **c** $165 \leqslant h < 170$

D4.3

1 **a**

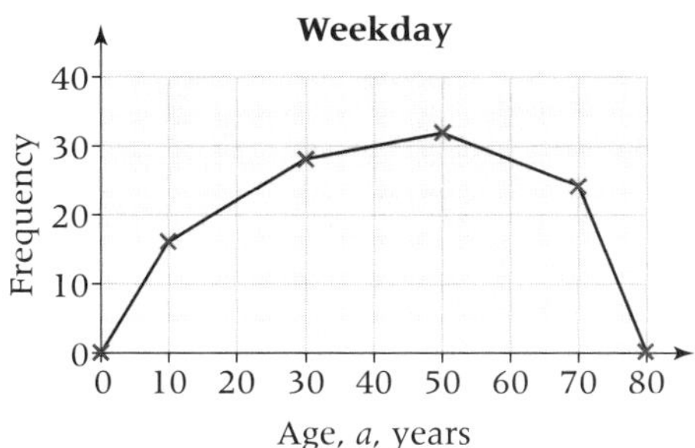

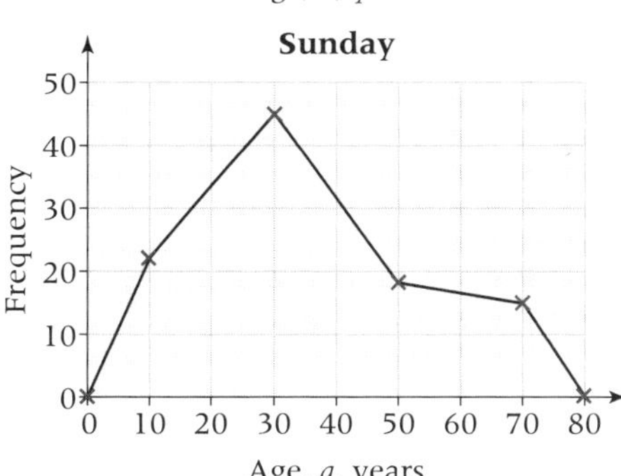

b Weekday: $40 < a \leqslant 60$, Sunday: $20 < a \leqslant 40$

c Generally, people visiting on a weekday are older.

2 **a**

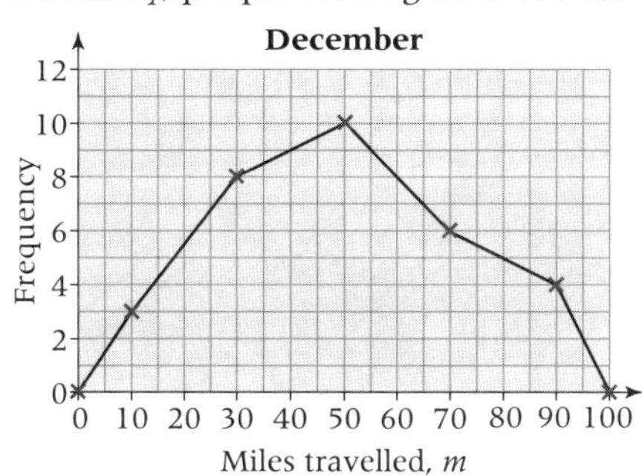

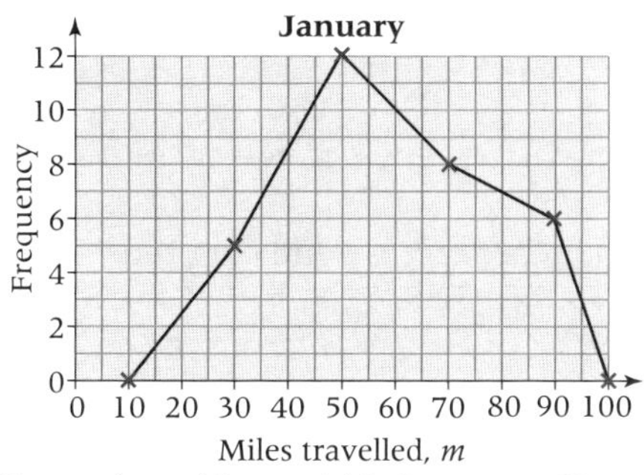

b December: $40 < m \leqslant 60$, January: $40 < m \leqslant 60$

c Less variation in miles travelled in January, less short journeys. The most common journey length does not change.

3 **a**

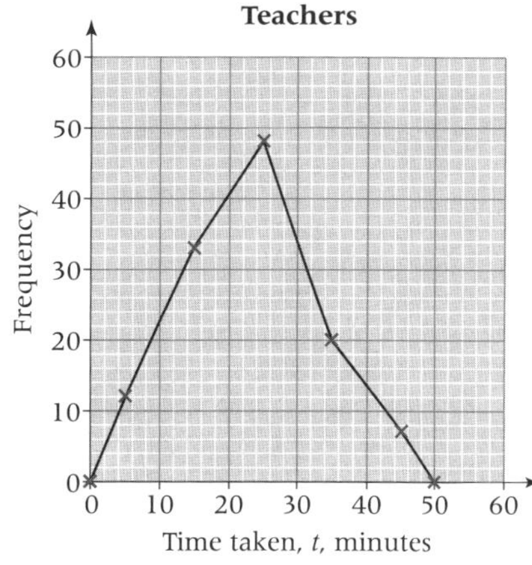

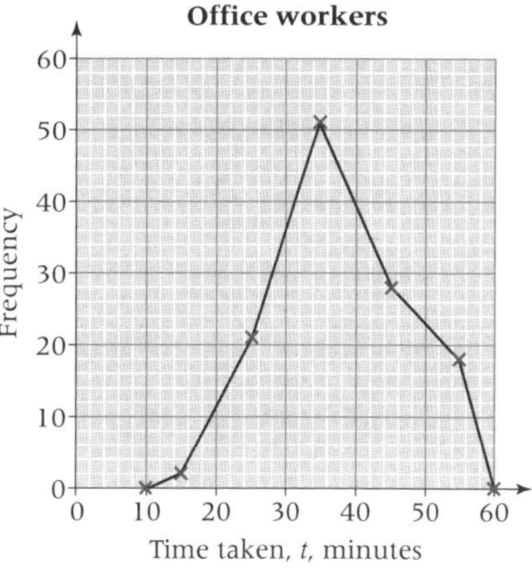

b Teachers: $20 < t \leqslant 30$, office workers: $30 < t \leqslant 40$

c On average, office workers take longer travelling home.

D4.4

1

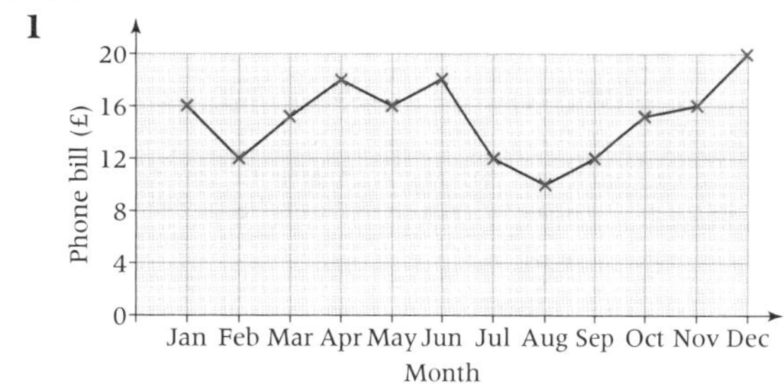

2

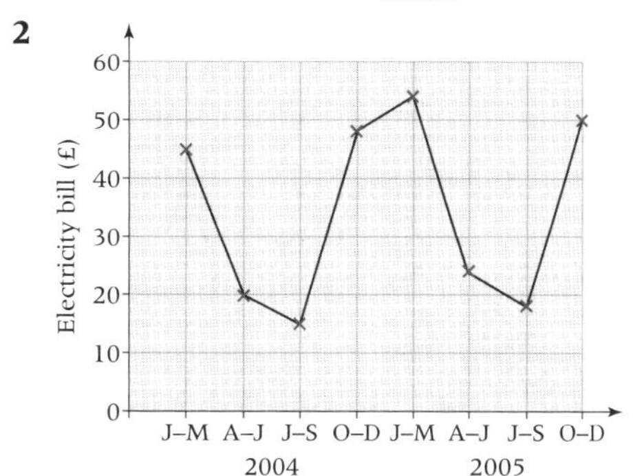

3

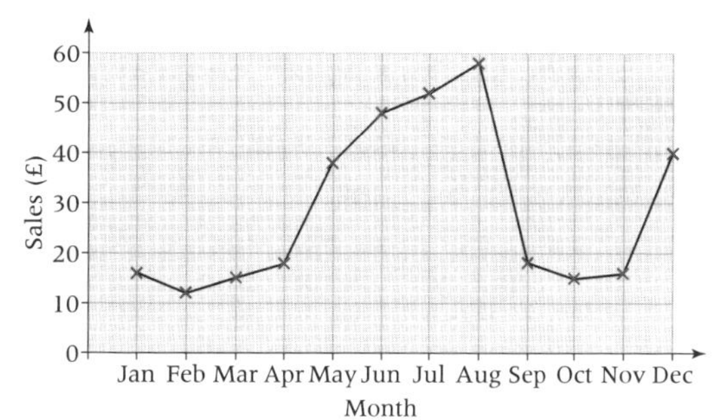

4

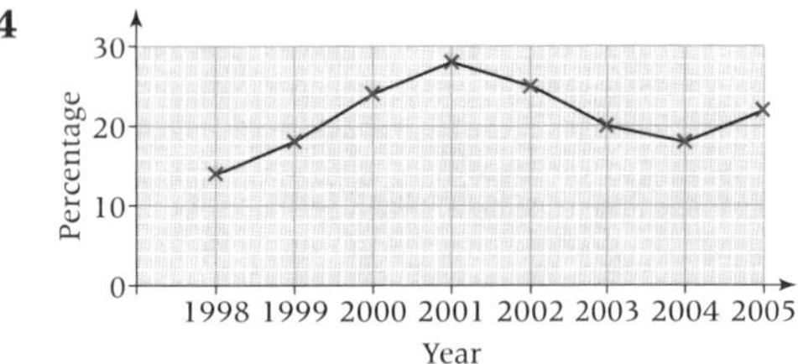

5

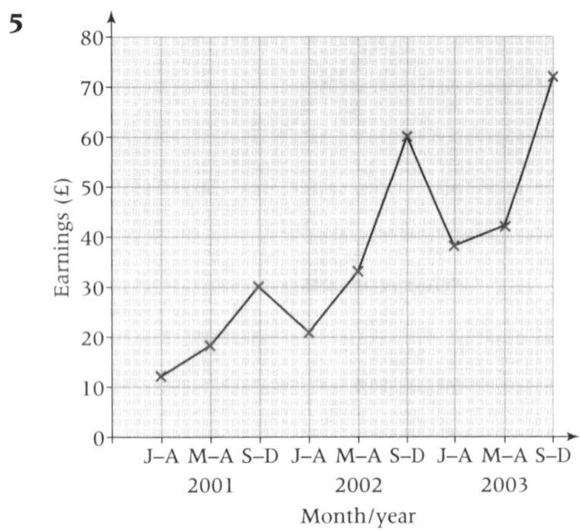

6

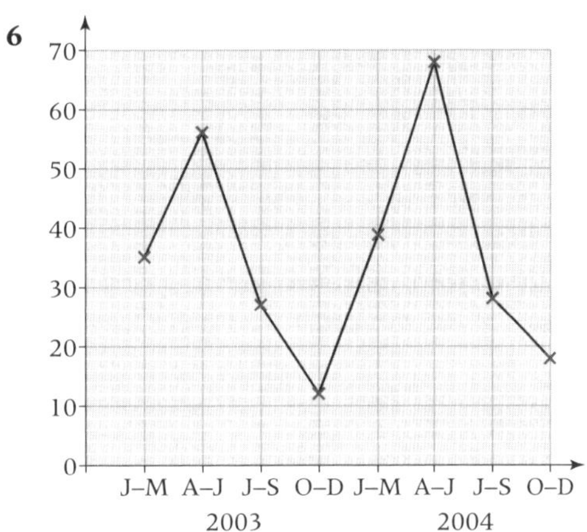

D4.5

1 £14.33, £15, £16.33, £17.33, £15.33, £13.33, £11.33, £12.33, £14.33, £17

2 a £32, £34.25, £35.25, £36, £36.50

b

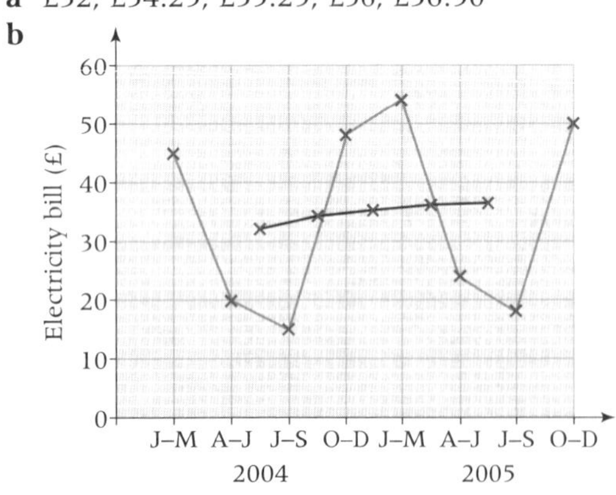

c Bills are gradually increasing

3 £15.25, £20.75, £29.75, £39, £49, £44, £35.75, £26.75, £22.25

4 a 16%, 21%, 26%, 26.5%, 22.5%, 19%, 20%

b 21.8%, 23%, 23%, 22.6%

5 a £20, £23, £28, £38, £44, £47, £51

b

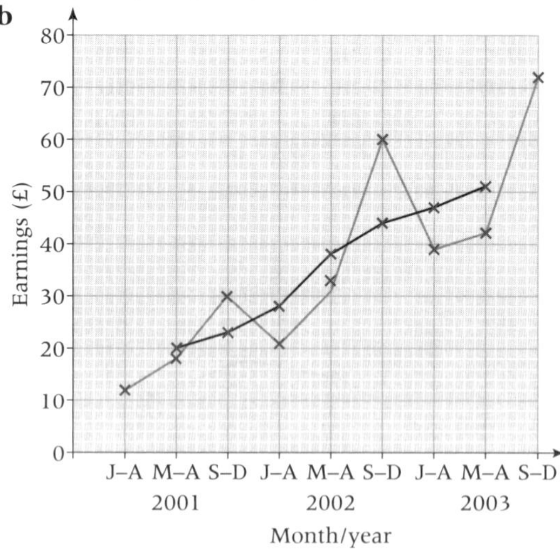

c Money earned is increasing over time.

6 a £32.50, £33.50, £36.50, £37, £38.50

b

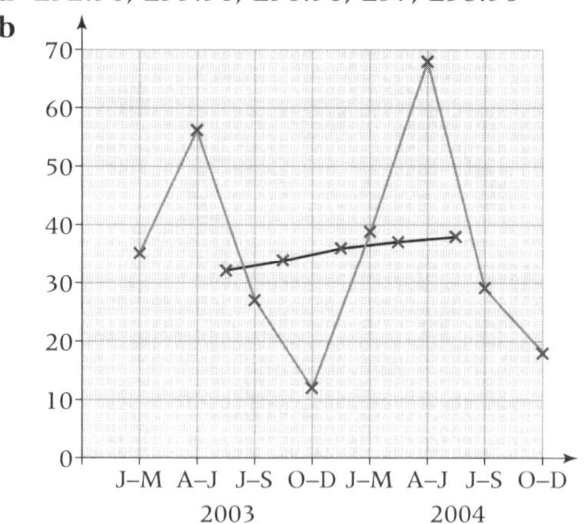

c Expenses are gradually increasing.

7 a

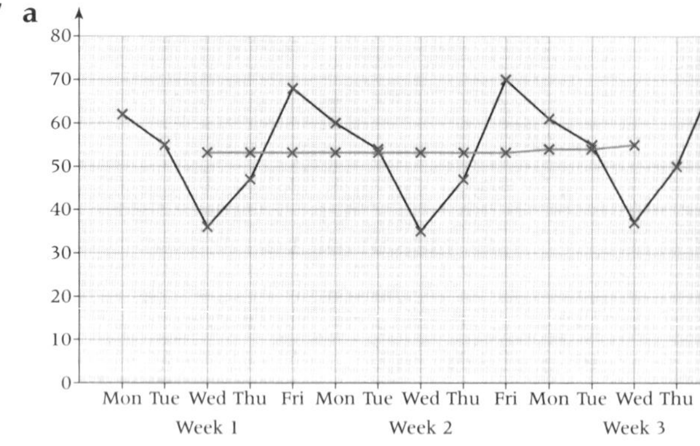

b It will help her spot a trend over time

c 53.6, 53.2, 53, 52.8, 52.8, 53.2, 53.4, 53.6, 54, 54.6, 55.2

d See part **a**

e The number of cups of coffee sold is changing very little over time, perhaps a slight increase.

D4 Exam review

1 1.65

2 a $150 < C \leqslant 250$

b Incorrect, the median (21st customer) is in $150 < C \leqslant 250$

c £6500

S5 Before you start ...

1 34°, 72°, 105°

2 a Circle radius 3 cm b Arc radius 5 cm

3

S5.1

1 a 050° b 320°

2 a Bearing of 070° b Bearing of 155°
  c Bearing of 340° d Bearing of 260°

3 a 316° b 265° c 068°

4 a 284° b 263° c 117°

5 b 170°

6 aii 328° bii 357°

S5.2

1 a Triangle with sides 8 cm, 4 cm, 7 cm
  b Triangle with sides 3 cm and 4 cm, and 30° angle
  c Triangle with sides 10 cm, 7.5 cm, 6 cm
  d Triangle with sides 8 cm and 2 cm, and 90° angle
  e Triangle with sides 6 cm, 9 cm, 5 cm
  f Triangle with two 45° angles and a 4 cm side

2 a $4 + 3 < 9$, two short sides will never meet.
  b $4 + 5 = 9$ so triangle is a straight line.

3 a Yes b Yes c Yes d No
  e No f No g Yes h Yes

4 a Triangle with sides 7 cm, 7 cm, 5 cm
  b Triangle with sides 5 cm, 5 cm, 7 cm

5 a Equilateral triangle with sides 5 cm
  b Rhombus with sides 5 cm
  c Rhombus with sides 3.5 cm

6 a Triangle with sides 5 cm, 12 cm, 13 cm
  b Right-angled triangle

7 a Triangle with sides 3 cm, 4 cm, 5 cm
  b Rectangle with sides 3 cm, 4 cm
  c Rectangle

8 a, b

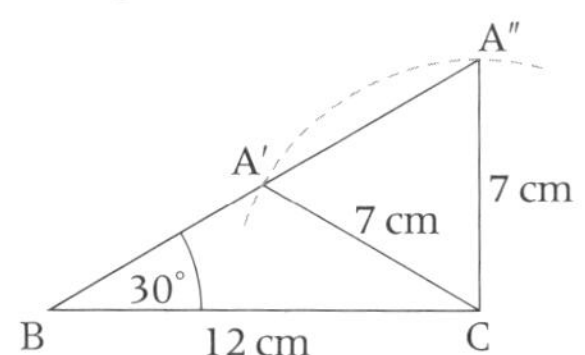

  c No, SSA triangles are not unique since you can construct two different triangles with the given side and angle measurements.

S5.3

1 Angle bisectors

2 a Equilateral triangle with sides 5 cm
  b Angles bisectors of triangle in part **a**
  c Bisectors meet at a point and cut the midpoints of the opposite sides.

3 a Equilateral triangle with sides 4 cm
  b Angles bisector of triangle in part **a**

4 a Perpendicular bisector of 6 cm line
  b Perpendicular bisector of 9 cm line
  c Perpendicular bisector of 5.6 cm line
  d Perpendicular bisector of 10 cm line
  e Perpendicular bisector of 11.2 cm line

5 a Equilateral triangle with sides 5 cm
  b Perpendicular bisectors of sides of triangle in part **a**
  c Perpendicular bisectors and angle bisectors of equilateral triangles are the same.

6 d Both sets of lines meet at common points but not the same points.

7 c 90°, 135°, 180°, 225°, 270°, 315°

S5.4

1 Perpendiculars from points to lines

2 Perpendiculars from points on lines

S5.5

These sketches not drawn to scale.

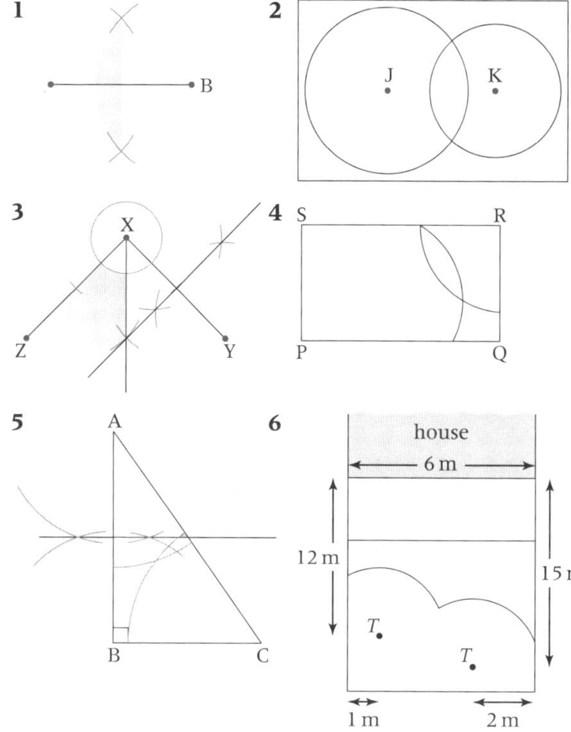

S5 Exam review

1 a 197°

  bi

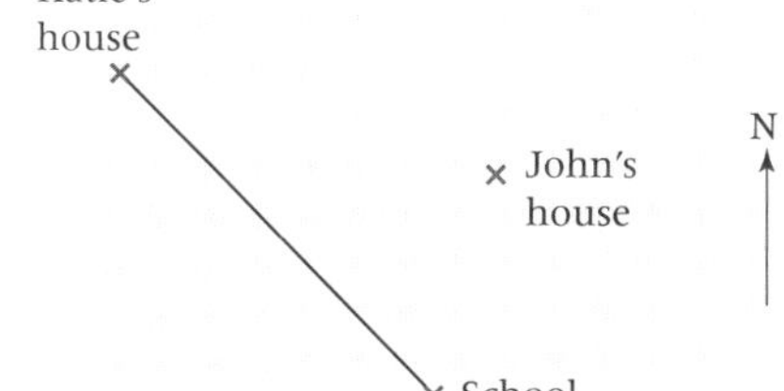

  bii 105°

2

C
A
B

N7 Before you start ...

1 a 30% b 65% c 72.5%
  d 105% e 6%

2 a 0.15 b 0.065 c 0.125
  d 0.975 e 1.08

3 a £35 b 20 m c 54°
  d 120 cm e £54 f 90 mm

N7.1

1 a $3\frac{1}{2}$ b $1\frac{3}{5}$ c $3\frac{1}{3}$ d $\frac{2}{3}$ e $2\frac{2}{5}$
2 a 10 b 21 c 12 d 13 e 22
3 a 24 m b 36 m
4 a 15 litres b 180 cl c 48 cl d 250 ml
5 a 30 m b 8 km c 16 mm d 90 m
6 a $1\frac{2}{3}$ b $5\frac{1}{2}$ c $4\frac{1}{2}$ d $\frac{2}{3}$ e $9\frac{1}{3}$
7 a $2\frac{2}{3}$ miles b $33\frac{1}{3}$ miles
c $12\frac{1}{2}$ miles d $3\frac{3}{4}$ miles
8 a 6.67 m b 5.33 mm
c 8.73 cm d 22.3 km
9 40.83 m^3 to 2 dp
10 a 5 hours 15 minutes b 13 hours 45 minutes
c 16 hours 48 minutes d 24 hours 30 minutes
11 a £4.67 b £16.67 c £12.75 d £29.33.

N7.2

1 a 10.5 b 126 c 240 d 300
e 132 f 99
2 a £225 b 990 m c 2520 kg d €144
3 a 135 b 91 c 744 d 18
e 387 f 192
4 a 1104 mm b 1952 kg
c €1443 d £493
5 a 0.5 b 0.6 c 0.25 d 0.51
e 0.64 f 0.22 g 0.15 h 0.7
i 0.07 j 0.085
6 a 5.7 b 200 c 15.93 d 99.84
e 16.81 f 20
7 a £3.84 b £53.55 c £13.95 d £170.10
e £28.16 f £15.90
8 £425
9 £73.70
10 649
11 £49 090.60

N7.3

1 a 108 b 72 c 49.5 d 618
e 600 f 700
2 a 25.2 b 36 c 51 d 37.5
e 228 f 64.35
3 a 342 b 264 c 1008 d 325
e 1372.5 f 1109.2
4 a 672 b 532 c 272 d 301.5
e 136.5 f 124.8
5 a 1.2 b 1.3 c 1.45 d 1.85
e 1.065
6 a 0.6 b 0.4 c 0.65 d 0.28
e 0.815
7 a 56.71 b 40.32 c 719.2 d 246
8 a £33 600 b £19 317.15
c £27 348 d £56 021
9 a £351 b £559 c £725 d £714.10
10 They could have more than doubled in price, which gives a price increase of more than 100%.
11 A price fall by more than 100% leads to a negative cost, which is impossible.

N7.4

1 a £5 b £24
c £315 d £227.50
2 a £262.50 b £416.16
c £1348.32 d £1061.21
3 a 1.06 b 1.1236
4 a £530 b £561.80
5 a 1.157 625 b 1.370 086 663
6 a £5788.13 b £1096.07
7 Option **a**
8 a 0.85 b Multiply by 0.85 again

N7.5

1 a 80% b 28% c 12% d 50% e 50%
2 a 60% b 48% c 60%
3 a 51.35% b 28.57% c 3.83%
4 a Science: 88.3%, Maths: 86.7% (both to 3 sf)
b Jason dropped a greater proportion of the total marks in maths than he did in science.
5 £5
6 a 50 b 40 c 80 d 110
7 a £9 b £80
8 a Francesca b £78.03
9 She needs to find 20% of the original price, which is not the same as 20% of the sale price.

N7 Exam review

1 a £32 b £12 ci £36 cii $\frac{9}{20}$ ciii 45%
2 a 6 hours b £255.84

D5 Before you start ...

1 a 62 b 70 c 60 d 45 e 102 f 108
2 a 350 b £70
3 2 days

D5.1

1 a

Height, h, cm	Cumulative frequency
$h < 145$	0
$h < 150$	7
$h < 155$	32
$h < 160$	78
$h < 165$	95
$h < 170$	100

b

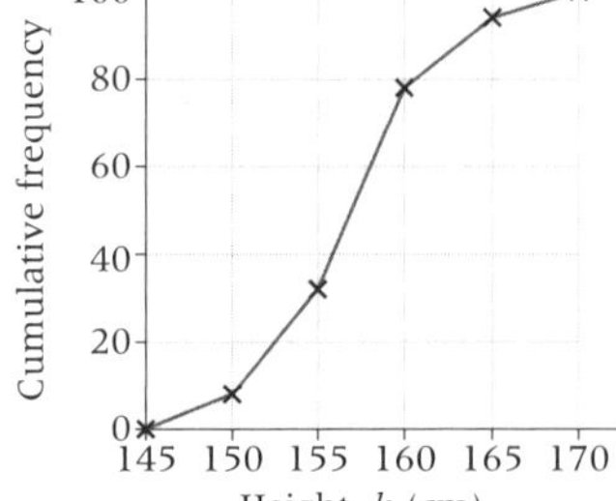

c $155 \leqslant h < 160$

2 a

Age, A, years	Cumulative frequency
$A < 20$	0
$A < 30$	18
$A < 40$	55
$A < 50$	106
$A < 60$	134
$A < 70$	150

b

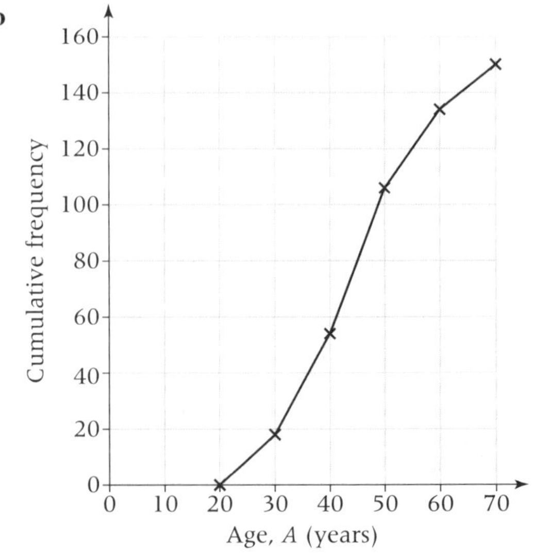

c $40 \leqslant A < 50$

3 a

Time, t, minutes	**Cumulative frequency**
$t<10$	4
$t<20$	15
$t<30$	44
$t<40$	81
$t<50$	108
$t<60$	120

b

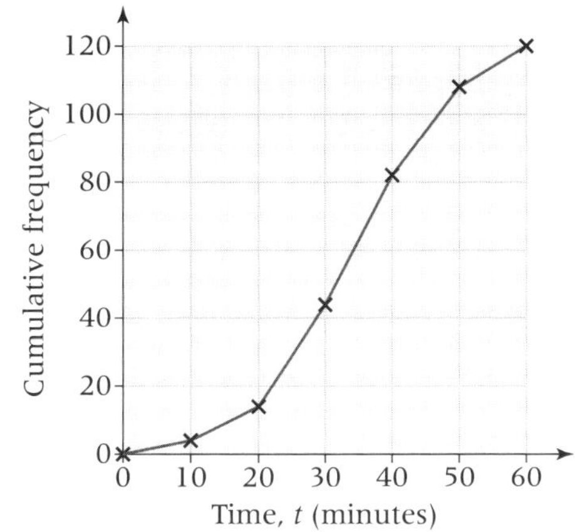

c $30 \leqslant t < 40$

4 a

Weight, w, grams	**Cumulative frequency**
$w<1500$	0
$w<2000$	9
$w<2500$	31
$w<3000$	68
$w<3500$	88
$w<4000$	100

b

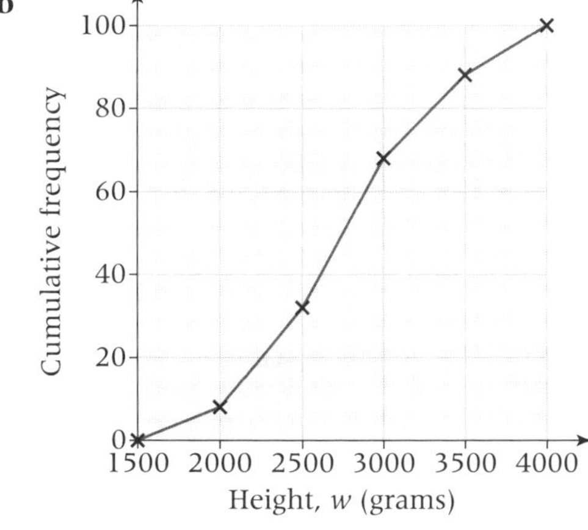

c $2500 \leqslant w < 3000$

5 a

Height, h, cm	**Cumulative frequency**
$h<40$	0
$h<60$	2
$h<80$	19
$h<100$	47
$h<120$	86
$h<140$	110
$h<160$	120

b

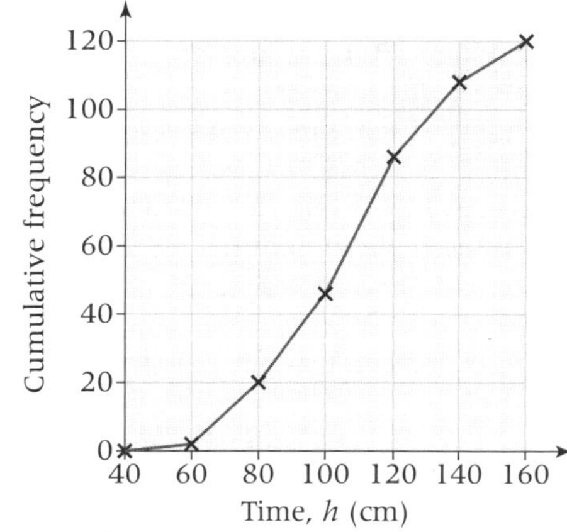

c $100 \leqslant h < 120$

6 a

Amount, £p	**Cumulative frequency**
$p<10$	16
$p<20$	30
$p<30$	53
$p<50$	70
$p<70$	85
$p<100$	100

b

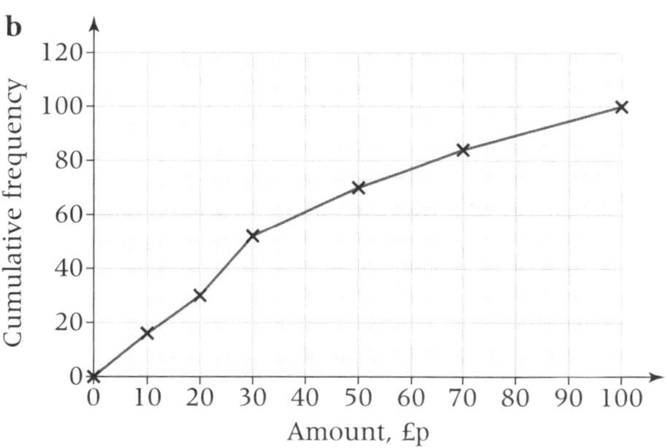

c $20 \leqslant p < 30$

D5.2

1	**a** 157 cm	**b** 6 cm	**ci** 18	**cii** 12
2	**a** 44 cm	**b** 17 cm	**ci** 36	**cii** 30
3	**a** 34 minutes	**b** 18 minutes	**ci** 30	**cii** 26
4	**a** 2750 g	**b** 750 g	**ci** 18	**cii** 10
5	**a** 107 cm	**b** 36 cm	**ci** 98	**cii** 86
6	**a** £28	**b** £39	**ci** 30	**cii** 10

D5.3

1 The boys' results are higher than the girls', on average. The middle half of the girls' results is less varied than that of the boys. The range is the same for the girls and the boys.

2 Farmer Jenkins' sunflowers are shorter than Farmer Giles', on average.
The middle half of Farmer Jenkins' sunflowers vary in height more than those of Farmer Giles.
The range of heights of Farmer Jenkins' sunflowers is greater than that of Farmer Giles'.

3 Boys have higher mobile phone bills, on average. The middle half of the mobile phone bills varies the same for girls and boys. The range of mobile phone bills is the same for girls and boys.

D5.4

1 Min 145 cm, LQ 153.5 cm, median 157 cm, UQ 159.5 cm, max 170 cm
2 Min 20 cm, LQ 35 cm, median 44 cm, UQ 52 cm, max 70 cm
3 Min 0 minutes, LQ 25 minutes, median 34 minutes, UQ 43 minutes, max 60 minutes
4 Min 1500 g, LQ 2350, median 2750 g, UQ 3100 g, max 4000 g
5 Min 40 cm, LQ 88 cm, median 107 cm, UQ 124 cm, max 160 cm
6 Min £0, LQ £17, median £28, UQ £56, max £100
7 a Min 11 years, LQ 13.1 years, median 14.3 years, UQ 15.4 years, max 18 years
  b Min 0 minutes, LQ 23 minutes, median 31 minutes, UQ 42 minutes, max 70 minutes

D5.5

1 On average, waiting times are higher at the dentist. The range of waiting times is greater at the doctor. The middle half of waiting times varies more at the doctor than at the dentist. The waiting times at the doctor are symmetrical, but those at the dentist are negatively skewed.
2 The average reaction time is the same for boys and girls. The range of reaction times is greater for girls. The middle half of reaction times varies less for girls than for boys. Reaction times for girls are symmetrical, but for boys they are negatively skewed.
3 On average, results are the same in the English and French tests. The ranges of results are the same. The middle half of results varies more in the English test. The English test results are negatively skewed, but the French test results are symmetrical.
4 On average, 17-year-old girls make longer phone calls than 13-year-olds. The range of the length of calls made is greater for 17-year-olds.
The middle half of the calls made varies more for 13-year-olds than for 17-year-olds. The lengths of calls made by 13-year-olds is negatively skewed, but those for 17-year-olds are symmetrical.

D5 Exam review

1 a

Height	Cumulative frequency (girls)	Cumulative frequency (boys)
$h<140$	0	0
$h<145$	4	2
$h<150$	9	6
$h<155$	12	11
$h<160$	14	14
$h<165$	15	17

b

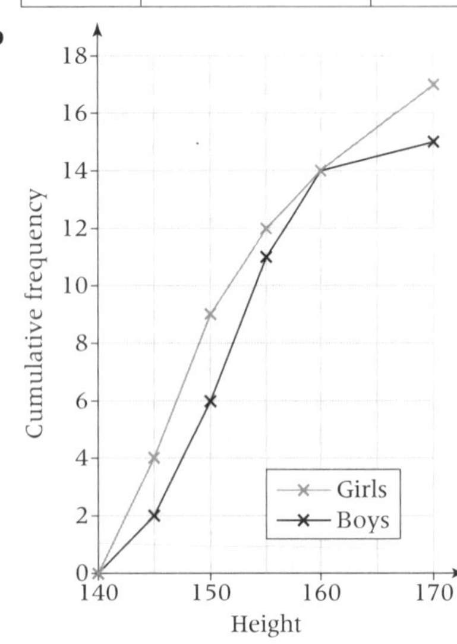

c Girls are on average shorter (lower median). The range of heights are the same. The interquartile ranges are the same.

2 a Median 33 seconds
  b Min 10 seconds, LQ 16 seconds, median 33 seconds, UQ 44.5 seconds, max 60 seconds
  c The average times for boys and girls are similar. The range of times is greater for boys.

S6 Before you start ...

1 a Cuboid b Square-based pyramid
  c Triangular prism
2 For example,

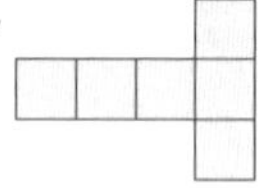

3 a 21 cm^2 b 59.3 cm^2 c 40 cm^2

S6.1

1 ai

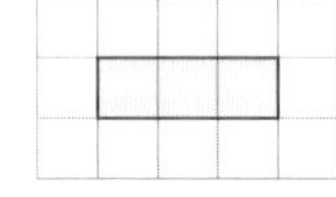

aii

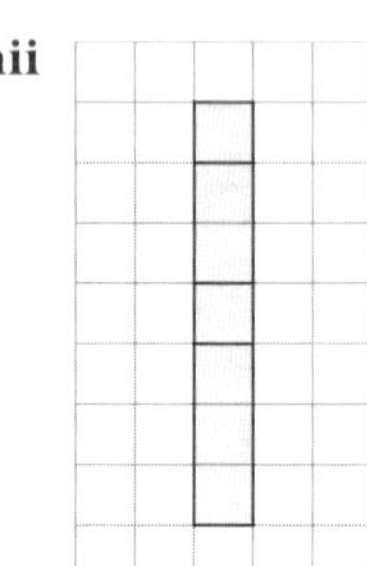

bi

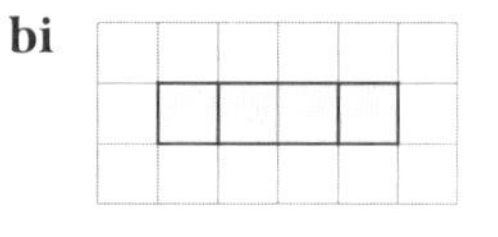

bii

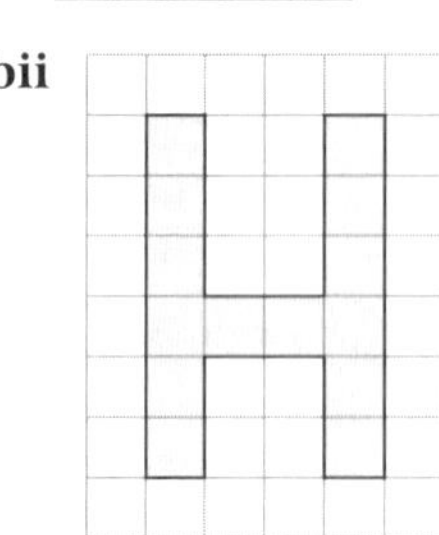

ci

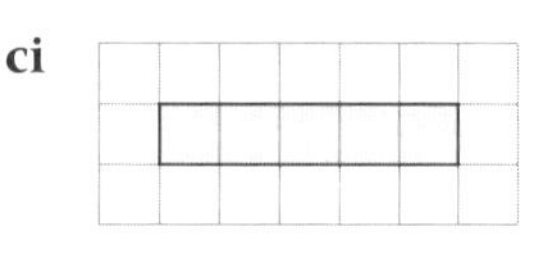

cii

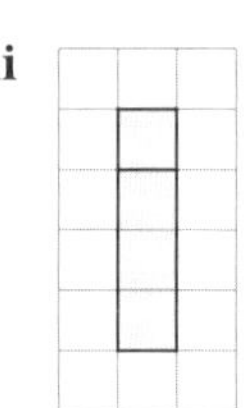

di

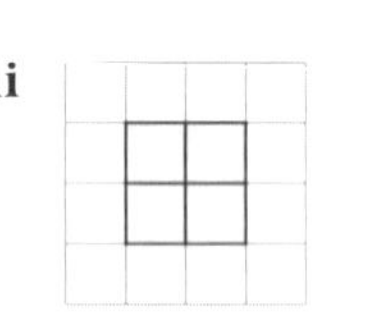

dii

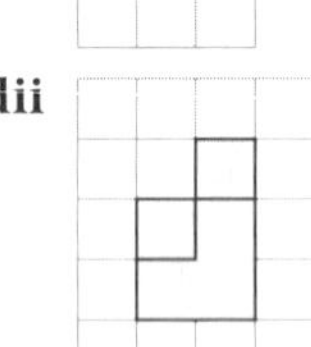

ei

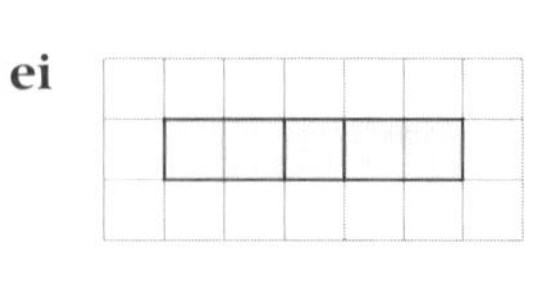

eii

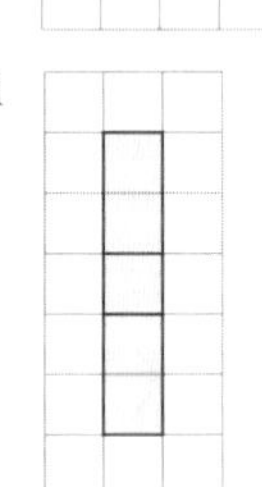

fi

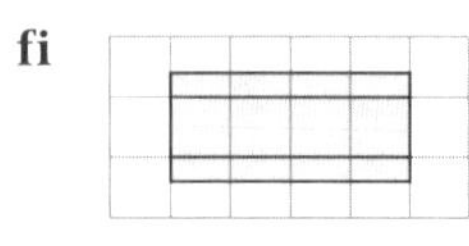

fii

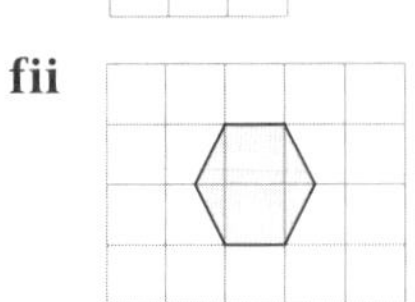

2 a 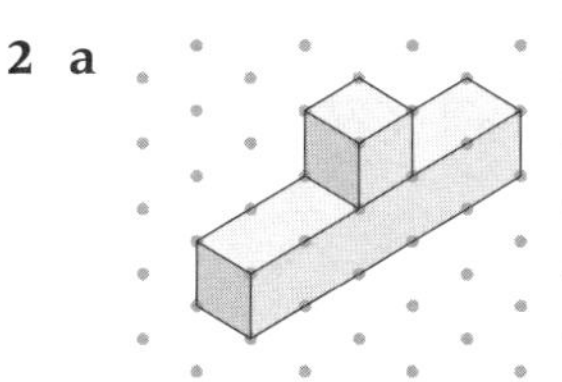
b 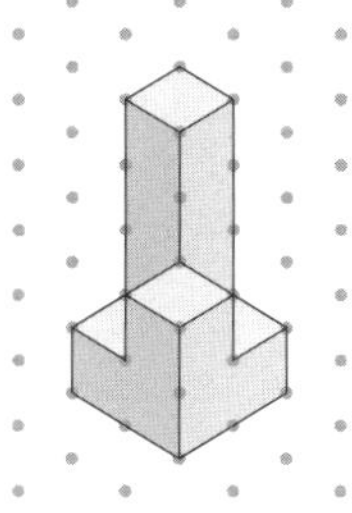
c 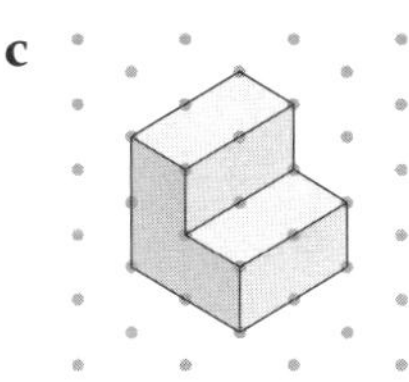
d 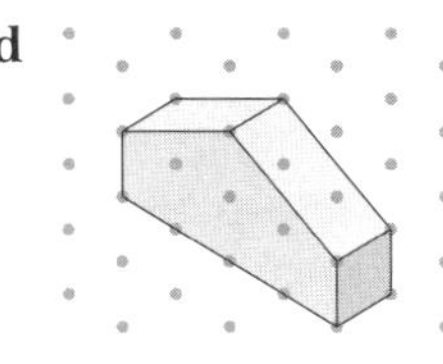
e 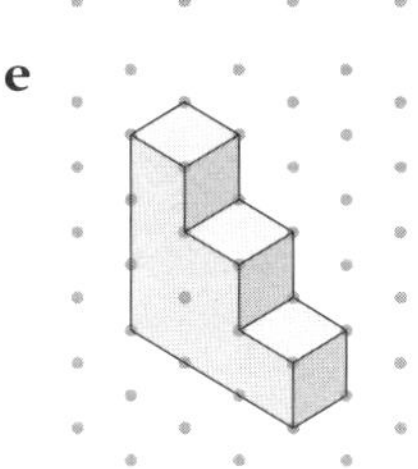
f 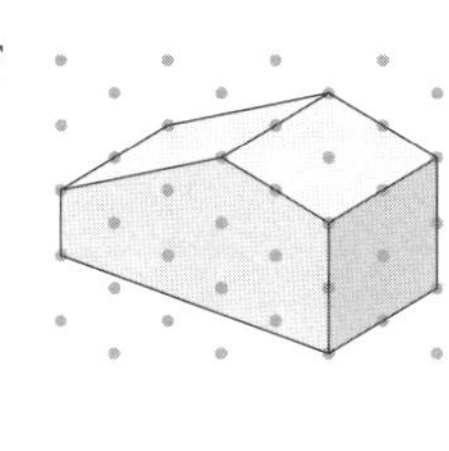

S6.2

1 a 105 cm³ b 60 cm³ c 72 cm³
d 36 cm³ e 64 cm³ f 28 mm³
2 a 75.4 cm³ b 628.3 cm³ c 201.1 cm³
d 160.8 cm³
3 a 270 cm³ b 600 mm³ c 1080 cm³

S6.3

1 a 27 cm³ b 343 cm³
c 64 cm³ d 0.125 m³
2 a 155.9 cm³ b 86.6 cm³
c 304.8 cm³ d 125.7 mm³
3 a 384 cm² b 600 cm²
c 216 cm² d 6 cm²
4 a 1 cm × 1 cm × 80 cm, 2 cm × 2 cm × 20 cm, 4 cm × 4 cm × 5 cm
bi 112 cm² bii 322 cm²
5 a 1 cm × 1 cm × 24 cm, 2 cm × 2 cm × 6 cm; 56 cm²; 98 cm²
b 1 cm × 1 cm × 64 cm, 2 cm × 2 cm × 16 cm, 4 cm × 4 cm × 4 cm; 96 cm²; 258 cm²

S6.4

1 a 400 m b 630 m c 4200 cm
d 12 m e 80 km f 45 m
g 50 mm h 300 mm i 60 cm
j 70 mm k 4 cm l 18 050 m
2 a 2.6 m² b 70 000 m² c 4.5 cm²
d 1200 cm² e 80 cm² f 4.5 cm²
g 8 400 000 mm² h 300 m²
i 20 000 cm² j 1 000 000 m²
3 a 3000 mm³ b 200 cm³ c 4800 cm³
d 3000 cm³ e 10 m³ f 50 m³
4 a Area b Area c Area
d Volume e Length f Length
g Area h None of these i Area
j Volume k None of these l Length
5 a The formula has only two dimensions; it should have three.
b Area
6 a $\pi r + 2h + 2r$ b $2hr + \frac{1}{2}\pi r^2$
7 a $a^2 + 2ab + a\sqrt{a^2 + b^2}$ b $\frac{1}{2}a^2 b$

S6.5

1 a 2.5 litres/s b 1.6 litres/s
2 1200 litres
3 a 30 ml b 24 ml c 12.6 ml
4 ai 96 km aii 12 km aiii 16 km
bi 3 hours bii 1 hour 30 minutes
biii 10 minutes
5

Distance	Time taken	Average speed
120 km	1.5 hours	80 km/h
250 miles	4 hours	62.5 mph
4 km	15 minutes	16 km/h
120 m	24 seconds	5 m/s
300 m	15 seconds	20 m/s
0.4 km	160 seonds	2.5 m/s or 12 km/h
3 km	125 seconds	24 m/s
20 km	20 minutes	60 km/h

6 a 57 040 kg b 34 000 kg c 61 824 kg
d 125 cm³ e 2500 cm³ f 0.002 m³
7 a 8000 kg/m³ b 4500 kg/m³ c 2700 kg/m³

S6 Exam review

1 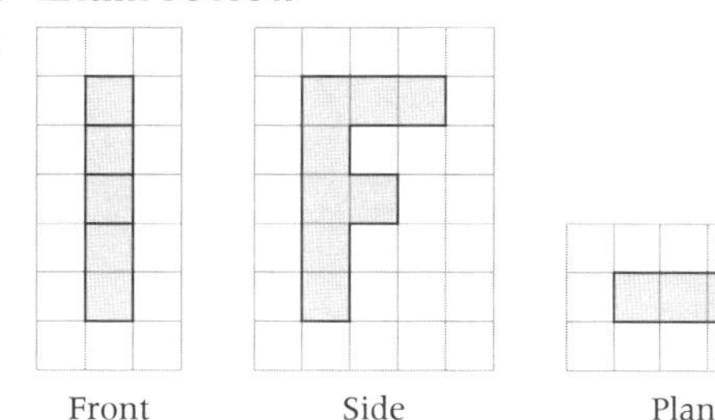

b 8 cm³
2 170 g

A7 Before you start ...

1 a 9 b 27 c 18 d 30
e 24 f 60 g 63 h 87
2 a 4 b −8 c 8 d −12
e 14 f −20 g 8 h 2
3 a Vertical, cutting x-axis at (4, 0)
b Horizontal, cutting y-axis at (0, 5)
c Vertical, cutting x-axis at (−3, 0)
d Gradient 2, cuts y-axis at (0, −1)
4 a

x	1	2	3
y	2	8	11

b

x	−1	0	6
y	14	12	0

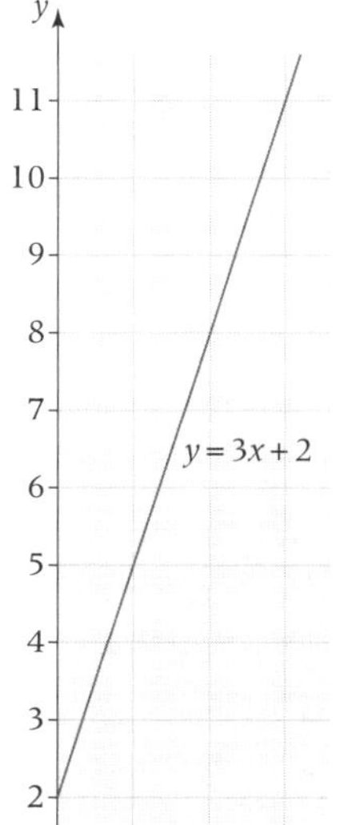

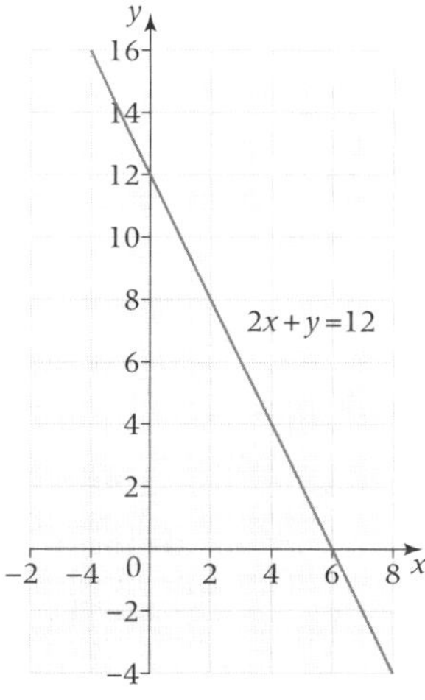

A7.1

1 a

Straight line	Parabola
$y=3x-2$	$y=x^2-2$
$3x+2y=8$	$y=10+x^2$
$y=x$	$y=x^2+2x+1$

2 b

x	−4	−3	−2	−1	0	1	2	3	4
x^2	16	9	4	1	0	1	4	9	16
$y=x^2-2$	14	7	2	−1	−2	−1	2	7	14

c

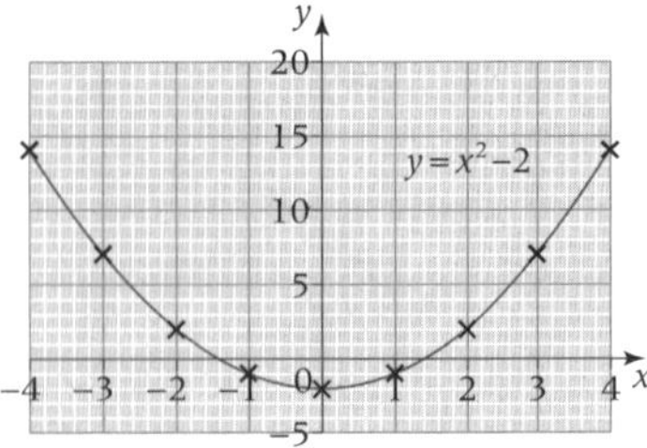

3 ai

x	−4	−3	−2	−1	0	1	2	3	4
$y=x^2+3$	19	12	7	4	3	4	7	14	19

aiii

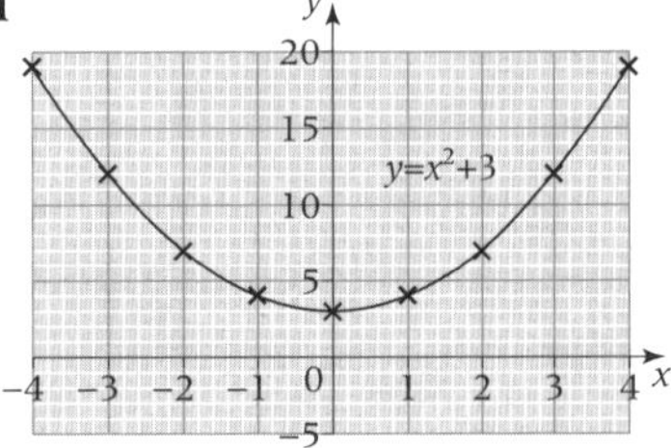

aiv (0, 3)

bi

x	−4	−3	−2	−1	0	1	2	3	4
$y=2x^2$	32	18	8	2	0	2	8	18	32

biii

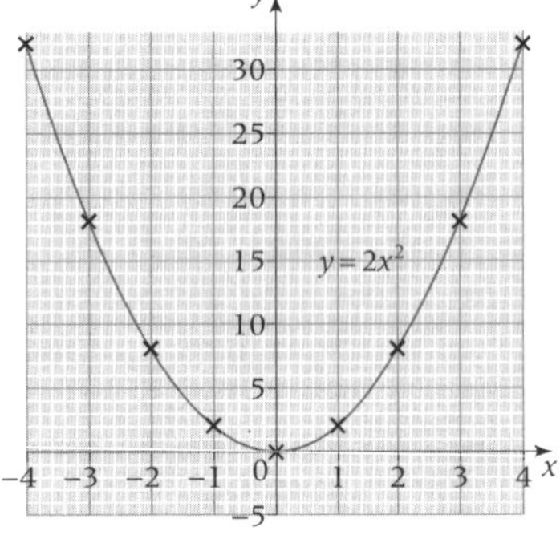

biv (0, 0)

ci

x	−4	−3	−2	−1	0	1	2	3	4
$y=3x^2-1$	47	26	11	2	−1	2	11	26	47

ciii

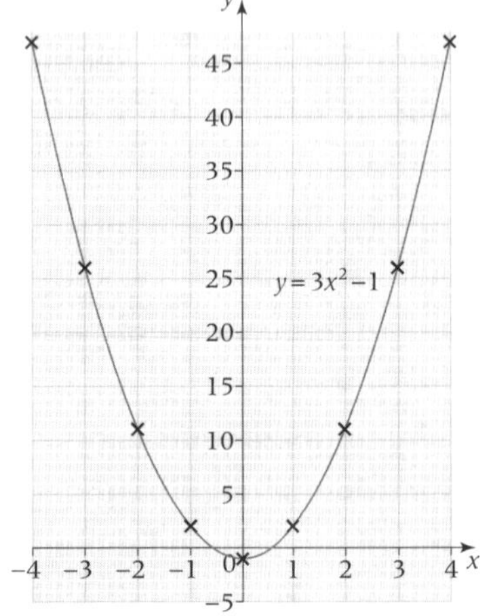

civ (0, −1)

di

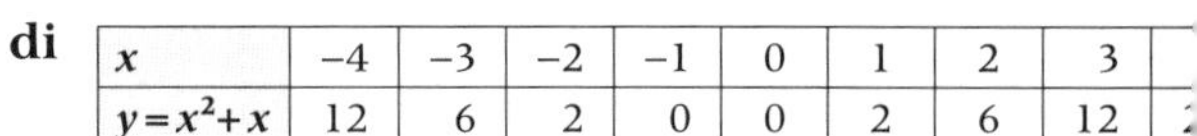

x	−4	−3	−2	−1	0	1	2	3
$y=x^2+x$	12	6	2	0	0	2	6	12

diii

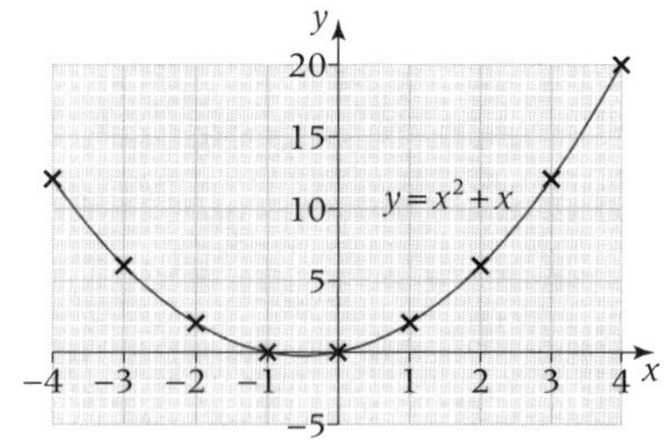

div (−0.5, −0.25)

4 False, since $4^2-5 \neq 10$.

5 a graph will be other way up

b

x	−3	−2	−1	0	1	2	3
$y=10-x^2$	1	6	9	10	9	6	1

c

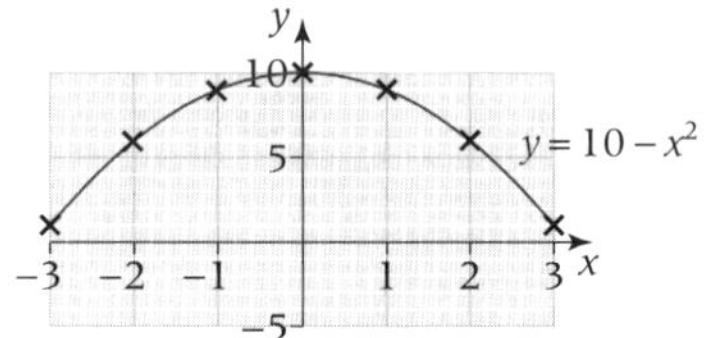

6 a

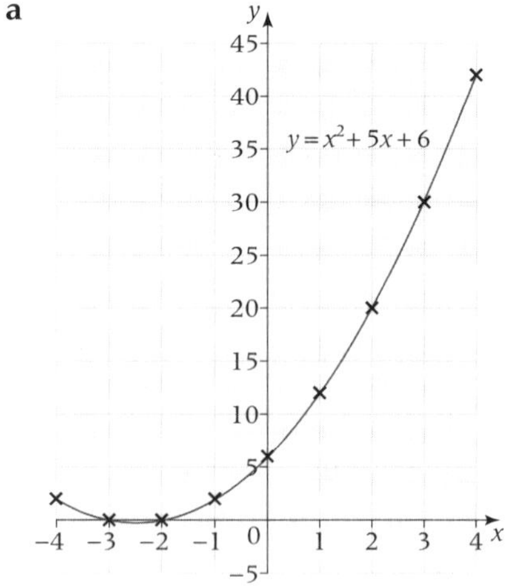

b (−2, 0), (−3, 0)

c $x=-2$ or -3

d The coordinates give the solutions of the equation, as x-axis is the line $y=0$

A7.2

1 From left to right: $y=x^3$, $y=3-2x$, $y=x^2-x-6$, $x=\frac{1}{2}$

Remaining graphs:

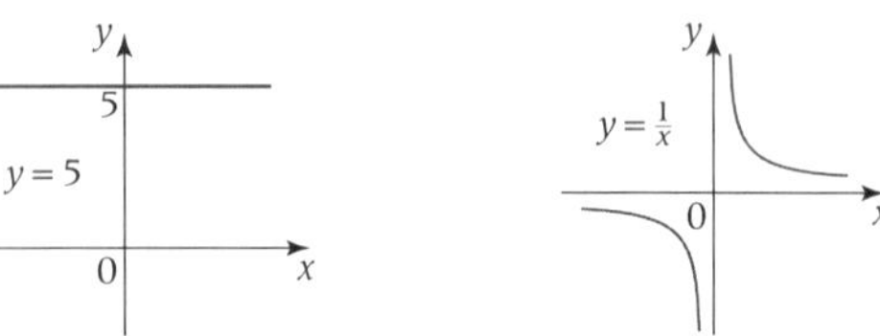

2 b

x	−3	−2	−1	0	1	2	3
y	−26	−7	0	1	2	9	28

c

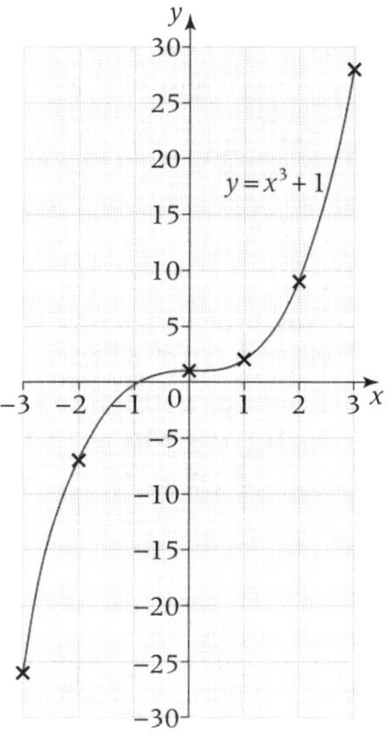

di 4.375 **dii** 1.125

3 a

x	−2	−1	0	1	2	3
x^3	−8	−1	0	1	8	27
$x^3 - 4$	−12	−5	−4	−3	4	23

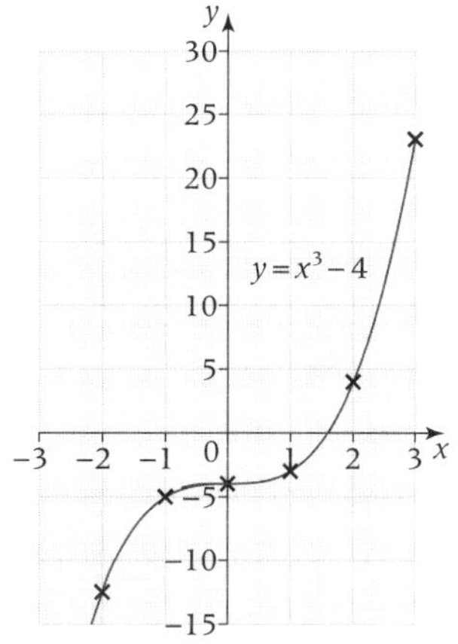

b

x	−2	−1	1	2	3
$\frac{1}{x}$	$-\frac{1}{2}$	−1	1	$\frac{1}{2}$	$\frac{1}{3}$
y	−1	−2	2	1	$\frac{2}{3}$

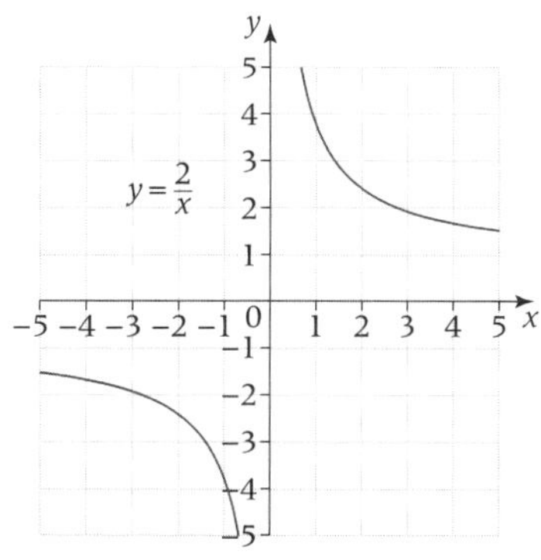

c

x	−2	−1	0	1	2	3
x^3	−8	−1	0	1	8	27
$x + 1$	−1	0	1	2	3	4
y	−9	−1	1	3	11	31

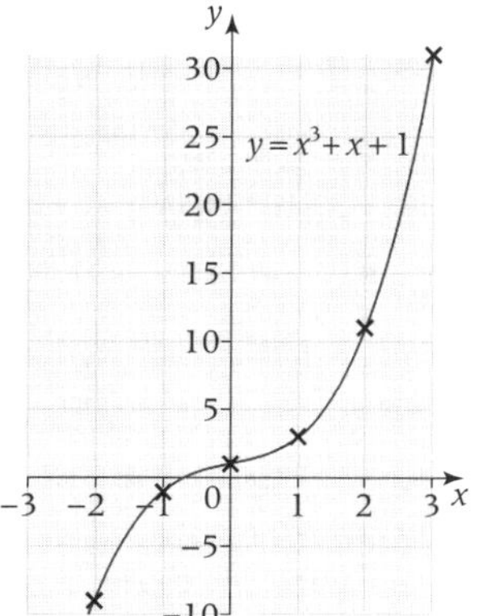

d

x	−2	−1	1	2	3
$\frac{3}{x}$	$-\frac{3}{2}$	−3	3	$1\frac{1}{2}$	1
y	$-\frac{1}{2}$	−2	4	$2\frac{1}{2}$	2

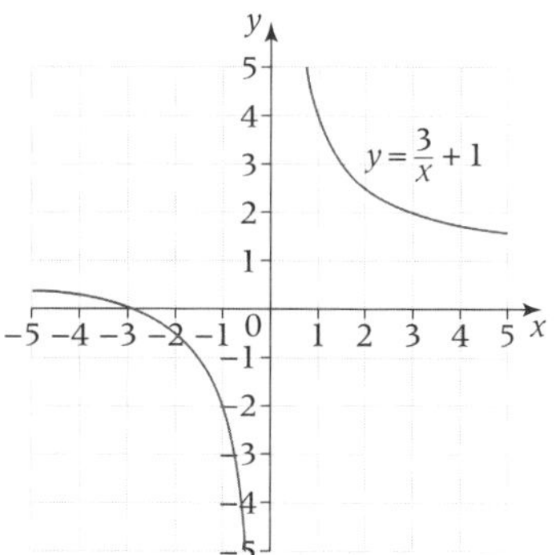

4 a

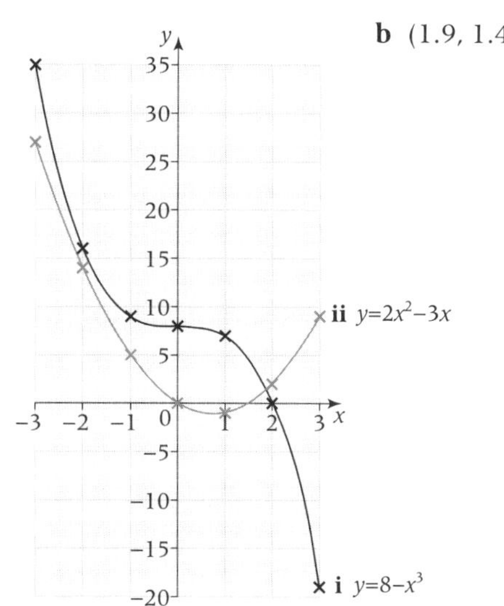

b (1.9, 1.4)

A7.3

1 ai $x = 2, y = 3$ **aii** $x = 1, y = 4$

aiii $x = 1, y = 1$ **aiv** $x = \frac{1}{4}, y = 1\frac{3}{4}$

b The lines are parallel so they never intersect.

2 a $x = 3, y = 7$ **b** $x = 1, y = 1$ **c** $x = 3, y = -1$

3 a $x = 1, y = \frac{1}{2}$ **b** $x = 1, y = \frac{1}{2}$

c They are the same

4 a 2, 0 **b** James is 3, Isla is 1

5 a The lines are parallel so they never intersect.

b Not if both are lines, but you could have a curve and a line intersecting twice or more.

6 a $y = 4x + 3, y = 6x + 1$ **b** $x = 1, y = 7$ **c** (1, 7)

A7.4

1 a $x = -1.5$ or 2.5 b $x = -0.6$ or 1.6
 c $x = -1$ or 3
2 a $x = -1.2$ or 3.2 b $x = 0$ or 2
 c $x = 1$ d $x = -0.3$ or 3.3
 e $x = -1.6$ or 2.6 f $x = -1$ or 3
3 a $x = -2.6$ or 2.6 b $x = -1.3$
 c $x = 0$ or 0.5 d $x = 1.7$

A7.5

1 a $x = -3$ or 1 b $x = -2.6$ or 1.6
 c $x = -3.4$ or 1.4 d $x = -1$ or 0
2 a $y = 3$ b $y = 0$ c $y = 2x + 1$
 d $y = 4$ e $y = 2$ f $y = 10x - 4$
3 a $x = 1.1$ b $x = 1.2$ c $x = -1.5$
 d $x = -1.6$, 0 or 0.6 e $x = -2.5$ f $x = 0.8$

A7 Exam review

1 a

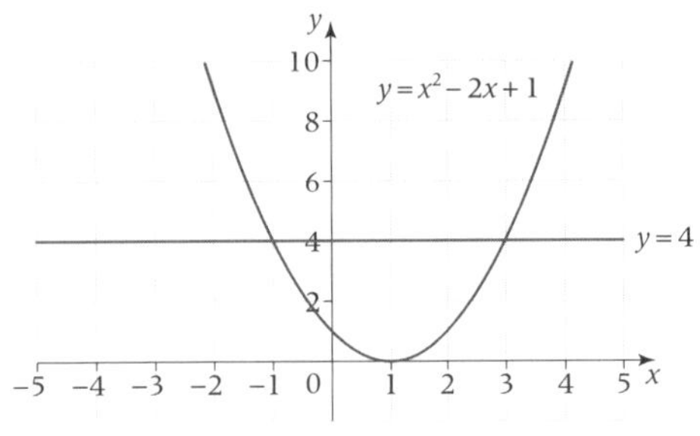

 b $x = 3$, $x = -1$; the x-coordinate of where the graphs cross gives the solutions

2 a

x	−2	−1	0	1	2	3
y	−1	1	3	5	7	9

 b

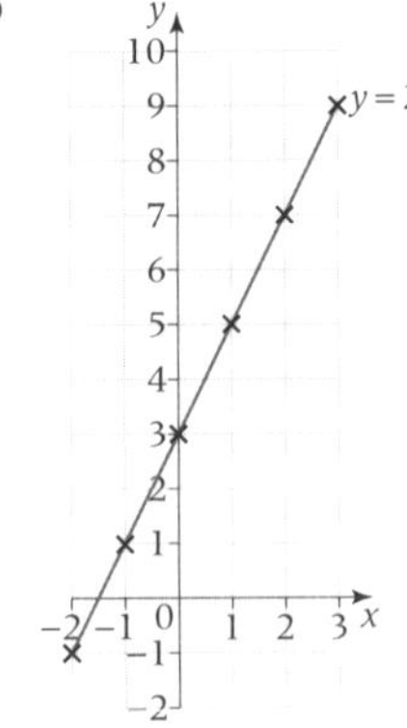

 ci $y = 0.4$ cii $x = 1.2$

N8 Before you start ...

1 Gill gets £400, Paul gets £100
2 33.3% decrease
3 a Batch B, it has a greater proportion of black paint
 b 9 litres black, 21 litres white

N8.1

1 a 17 : 13 b 11 : 19 c 14 : 15
2 11A: 7 : 23, 11B: 13 : 17, 11C: 16 : 13
3 a 1 : 1 b 2 : 5 c 2 : 3 d 3 : 4
 e 1 : 2 f 1 : 1 g 1 : 3 h 4 : 3
4 a 3 : 2 b 4 : 1 c 1 : 2 d 1 : 4
 e 2 : 3 f 3 : 2 g 2 : 3 h 3 : 2
5 a 24, 16 b 27, 18 c 54, 36
 d 60, 40 e 150, 100 f 1200, 800
 g 4500, 3000 h 5550, 3700
6 a 16 hours, 8 hours b 120°, 60°
 c 30 minutes, 15 minutes d €240, €120
 e 164 cm, 82 cm f 80 kg, 40 kg
 g 54 km, 27 km h 36 miles, 18 miles
7 a £180, £180 b £120, £240
 c £240, £120 d £135, £225
8 a 5 : 3 b Karla: £13.75, Wayne: £8.25
9 a £200 b Annie: £120, Ben: £180

N8.2

1 a 1 : 2 b 1 : 2 : 2 c 3 : 1 : 2
 d 5 : 10 : 4 e 2 : 5 : 3 f 3 : 7 : 2
2 a £80, £160 b £96, £144
 c £150, £90 d £40, £80, £120
 e £150, £30, £60 f £140, £40, £60
3 a 50 m, 200 m b 100 m, 150 m
 c 71.43 m, 178.57 m d 214.29 m, 35.71 m
 e 115.38 m, 134.62 m f 160.71 m, 89.29 m
4 Peter: £2700, Bob: £1500, Yasmin: £3300
5 Ann: £377.78, Charles: £283.33, Edward: £188.89
6

Quantity	Ratio	Share 1	Share 2	Share 3
200 km	2:3:5	40 km	60 km	100 km
38 kg	1:2:3	6.33 kg	12.67 kg	19 kg
450 cm	2:3:8	69.2 cm	104 cm	277 cm
£720	4:5:10	£151.58	£189.47	£378.95
95 litres	1:6:7	6.79 litres	40.71 litres	47.5 litres

7 Mango: 444 ml, pineapple: 333 ml, passion fruit: 222 ml
8 a £119.05, £116.67, £214.29
 b £125, £166.67, £208.33
9 Robert: £12 660, Kathleen: £16 880

N8.3

1 a 2 : 3 b $\frac{2}{5}$
2 a 1 : 2 b $\frac{1}{3}$
3 a 1 : 3 : 2 bi $\frac{1}{6}$ bii $\frac{1}{2}$ biii $\frac{1}{3}$
4 a $\frac{1}{2}$ b $\frac{1}{3}$ c $\frac{2}{5}$ d $\frac{3}{8}$ e $\frac{4}{7}$ f $\frac{5}{7}$
5 a $\frac{1}{2}$ b $\frac{2}{3}$ c $\frac{3}{5}$ d $\frac{5}{8}$ e $\frac{3}{7}$ f $\frac{2}{7}$
6 2 : 3
7 a £600 b 3 : 7
8 £770
9 $\frac{1}{2}$
10 £5333.33, $\frac{8}{21}$

N8.4

1 10%
2 a 4 : 7 b 36.4%
3 Nickeline: 80%, US nickel coinage: 75%, Metal bronze: 93%
4 60%
5 44.4%
6 a 80% b 3 : 1
7 a £1140, £2660 b 30%
8 a \$324.47, \$175.53 b 35.1%
9 a 333 g, 104 g, 63 g (to 0 dp)
 b 66.7%, 20.8%, 12.5%

N8.5

1 a £125 b £560 c 46 m d 54 cm
2 a £45 b 20 cm c 360 cm d 102 mm
3 a 1.2 b 0.85 c 1.06 d 0.95
 e 0.94 f 0.83
4 a £59.52 b 51.3 kg c 10.088 seconds
 d 198 m e €260.40 f \$300.90
5 a £50 b £30.94
6 £560
7 a £321.63 b £1749.60

N8 Exam review

1 **a** Shop B **b** Shop A

2 **ai** £1250 **aii** $\frac{1}{25}$

b 12 : 5

S7 Before you start ...

1

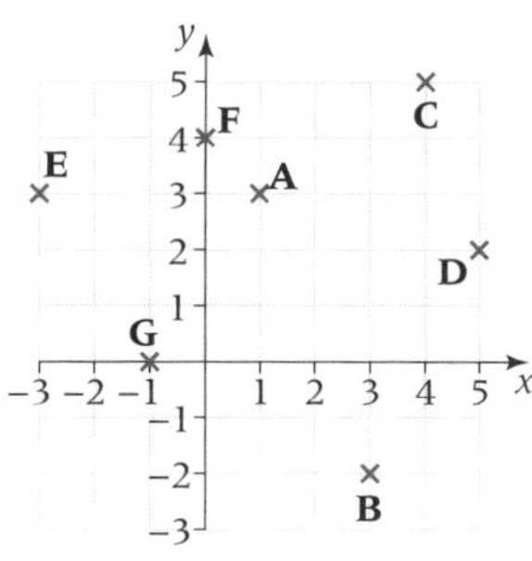

2 **a** 1 : 3 **b** 1 : 3 **c** 2 : 1 **d** 5 : 2

3 8 : 3

4 **a** $x = 14$ **b** $x = 15$ **c** $x = 10$ **d** $x = 18$

S7.1

1

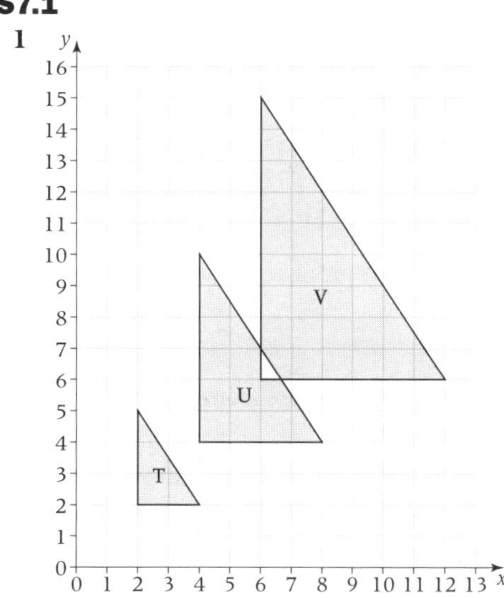

2

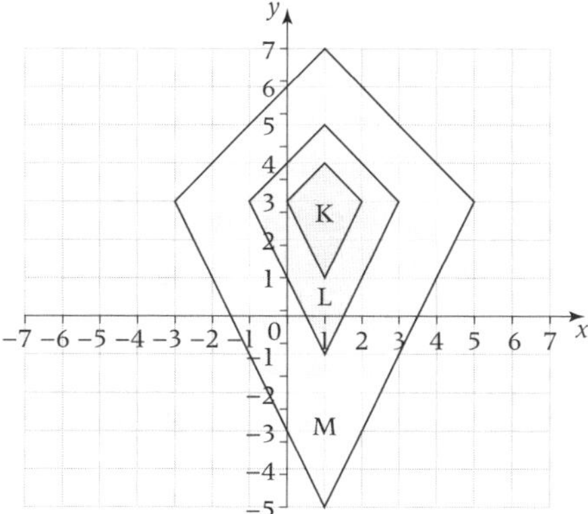

3

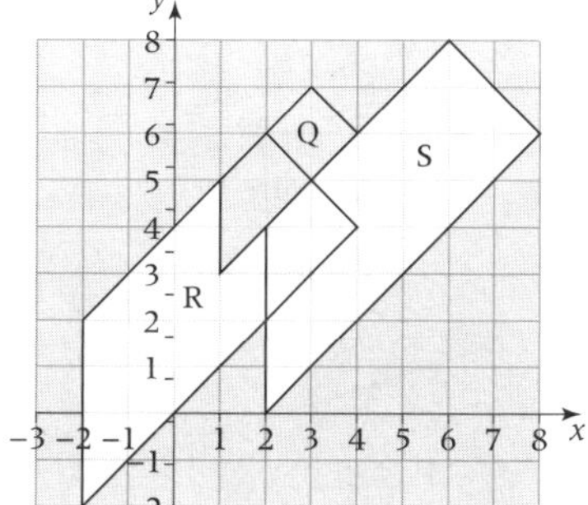

d Twice as large

4

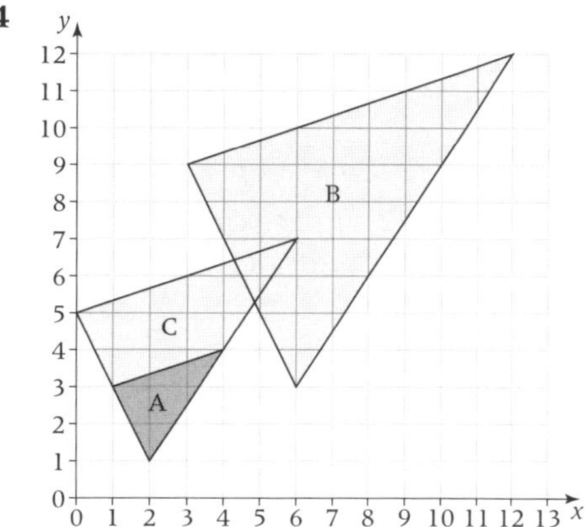

d Three times as large

S7.2

1

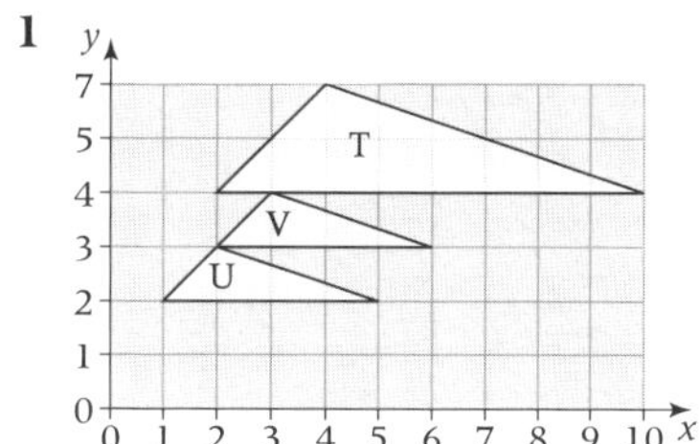

2

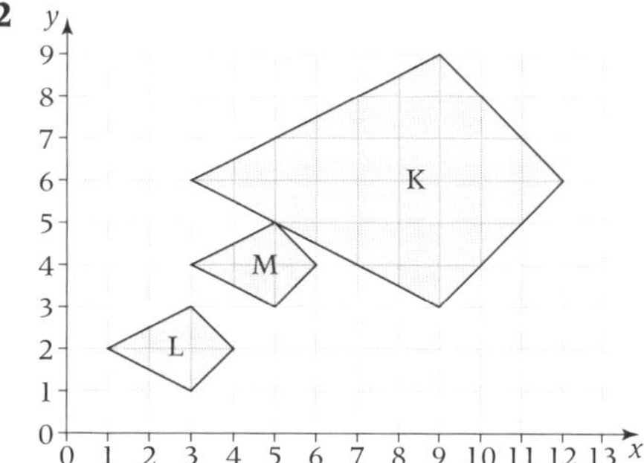

3 **4**

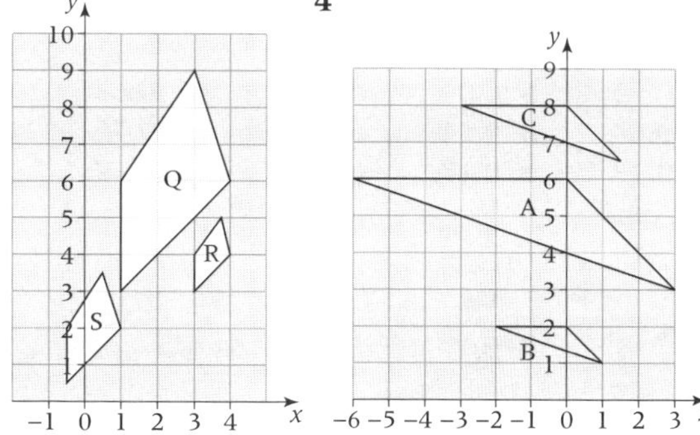

S7.3

1 **a** Enlargement with centre (0, 0) and scale factor 3

b Enlargement with centre (0, 0) and scale factor $\frac{1}{3}$

2 **ai** Enlargement with centre (−3, 0) and scale factor 2

aii Enlargement with centre (−3, 0) and scale factor $\frac{1}{2}$

b 16 units

3 **a** Enlargement with centre (0, 0) and scale factor 2

b Enlargement with centre (0, 0) and scale factor $\frac{1}{2}$

4 **ai** Enlargement with centre (0, 0) and scale factor $\frac{1}{2}$

aii Enlargement with centre (0, 0) and scale factor 2

aiii Enlargement with centre (0, 0) and scale factor 4

b 4 times

S7.4

1 $a = 6$ cm, $b = 4.5$ cm
2 $c = 6$ cm, $d = 2.5$ cm
3 $e = 5$ cm, $f = 2.4$ cm
4 15 cm
5 **a** 3 : 2 **b** $5\frac{1}{3}$ cm **c** 22 cm, $14\frac{2}{3}$ cm
d 3 : 2, same as ratio of lengths of sides

S7.5

1 **a** 4.5 cm **b** 9 cm
2 **a** 2.67 mm **b** 4.5 mm
3 **a** 6 cm **b** 5 cm **c** 15 cm
4 **a** MN = 6.3 cm, JN = 6 cm **b** 15.5 cm
5 **a** 19.35 cm **b** 32.1 cm

S7 Exam review

1 **a**

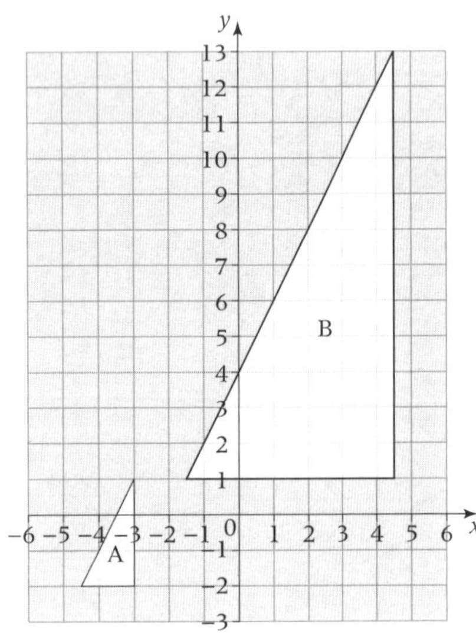

b Enlargement scale factor $\frac{1}{4}$, centre (−5.5, −3)
2 **a** AD = 7.5 cm **b** BC = 3 cm

D6 Before you start ...

1 **a** 0.55 **b** 0.04 **c** 0.72 **d** 0.625
2 **a** 0.6 **b** 0.34 **c** 0.9 **d** 0.75
3 **a** 0.18 **b** 0.17 **c** 0.192 **d** 0.12
4 **a** $\frac{1}{6}$ **b** $\frac{4}{5}$ **c** $\frac{2}{9}$ **d** $\frac{3}{4}$
5 **a** $\frac{13}{15}$ **b** $\frac{11}{12}$
6 **a** $\frac{5}{9}$ **b** $\frac{8}{45}$

D6.1

1 **a** $\frac{4}{10}$ **b** $\frac{6}{10}$ **c** $\frac{7}{10}$ **d** $\frac{4}{10}$ **e** $\frac{8}{10}$ **f** $\frac{9}{10}$
2 241
3 102
4 **a** $\frac{13}{28}$ **b** $\frac{15}{28}$ **c** $\frac{11}{14}$ **d** $\frac{3}{14}$
5 **ai** $\frac{9}{16}$ **aii** $\frac{3}{8}$ **aiii** $\frac{13}{16}$ **aiv** $\frac{9}{16}$ **av** $\frac{5}{16}$ **avi** $\frac{1}{8}$
b the black triangle
6 **ai** $\frac{1}{4}$ **aii** $\frac{3}{4}$ **aiii** 1 **aiv** 0 **b** 64

D6.2

1 **a**

Coin/Dice	1	2	3	4	5	6
Head	H1	H2	H3	H4	H5	H6
Tail	T1	T2	T3	T4	T5	T6

bi $\frac{1}{12}$ **bii** $\frac{1}{12}$ **biii** $\frac{1}{12}$

2 **a**

Red/blue	1	2	3	4	5	6
1	1, 1	1, 2	1, 3	1, 4	1, 5	1, 6
2	2, 1	2, 2	2, 3	2, 4	2, 5	2, 6
3	3, 1	3, 2	3, 3	3, 4	3, 5	3, 6
4	4, 1	4, 2	4, 3	4, 4	4, 5	4, 6
5	5, 1	5, 2	5, 3	5, 4	5, 5	5, 6
6	6, 1	6, 2	6, 3	6, 4	6, 5	6, 6

bi $\frac{1}{36}$ **bii** $\frac{1}{36}$ **biii** $\frac{1}{12}$ **biv** $\frac{1}{18}$

3 $\frac{1}{6}$
4 **a** $\frac{1}{5}$ **b** $\frac{1}{20}$ **c** $\frac{3}{20}$ **d** $\frac{1}{10}$ **e** $\frac{1}{20}$
5 **a** $\frac{1}{100}$ **b** $\frac{1}{50}$ **c** 0 **d** $\frac{1}{25}$ **e** $\frac{3}{50}$
6 **a**

10p/2p	Head	Tail
Head	HH	HT
Tail	TH	TT

b $\frac{1}{4}$

7 $\frac{1}{4}$

D6.3

1

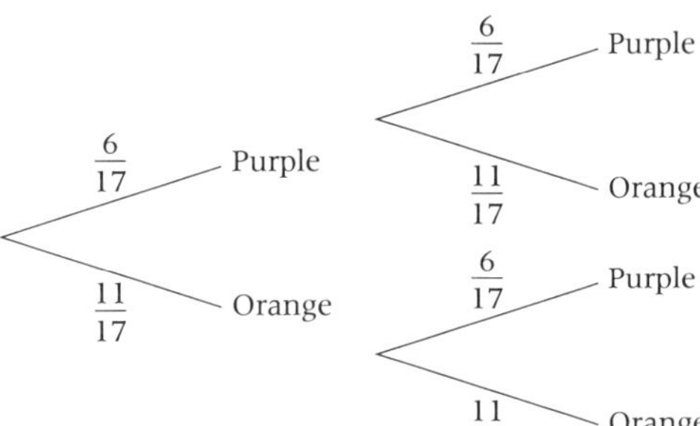

2

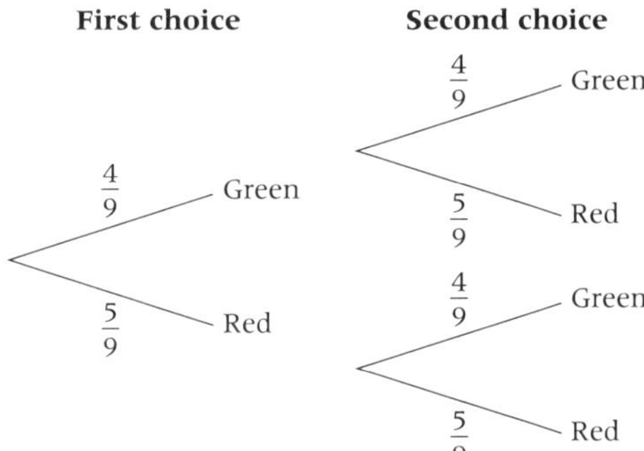

3

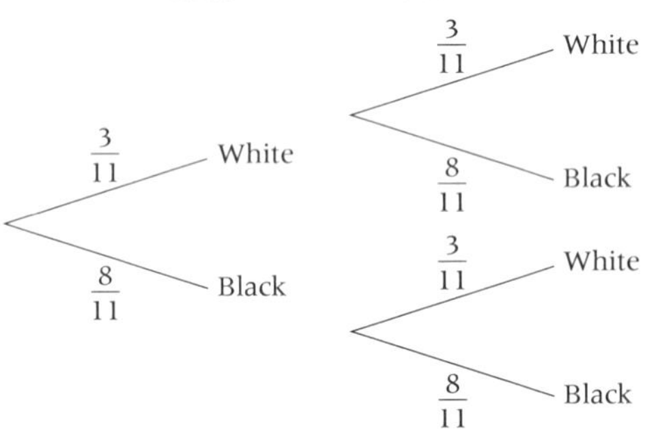

4

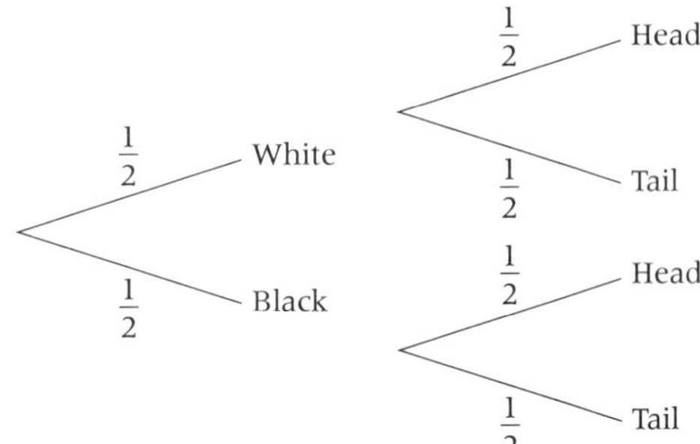

5

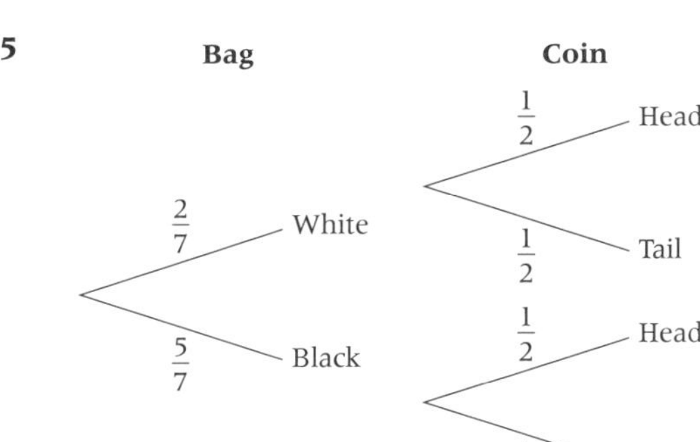

6

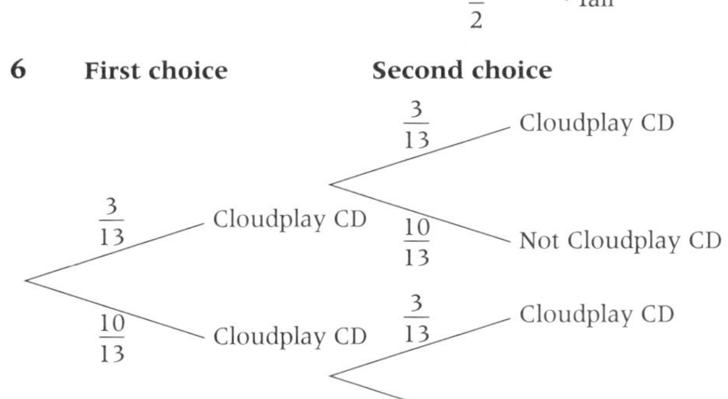

D6.4

1 **a** $\frac{9}{49}$ **b** $\frac{16}{49}$ **c** $\frac{12}{49}$

2 **a** $\frac{16}{81}$ **b** $\frac{25}{81}$

3 **a** $\frac{9}{121}$ **b** $\frac{64}{121}$

4 **a** $\frac{1}{4}$ **b** $\frac{1}{4}$ **c** $\frac{1}{4}$

5 **a**

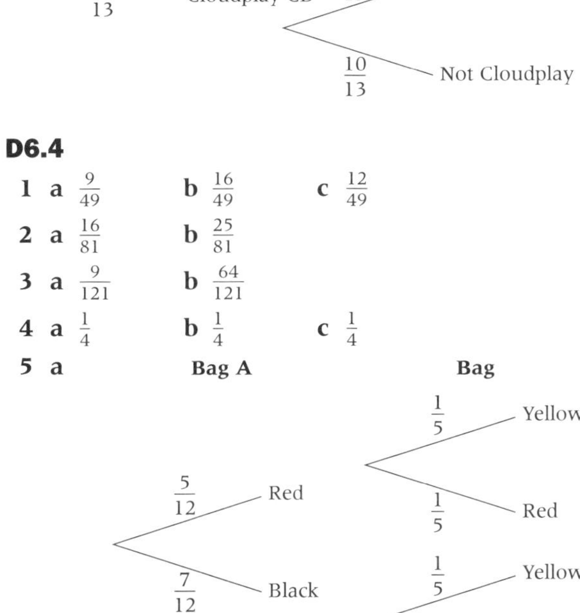

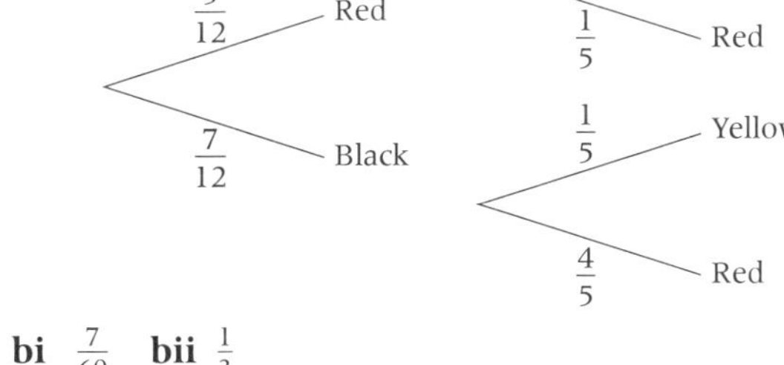

bi $\frac{7}{60}$ **bii** $\frac{1}{3}$

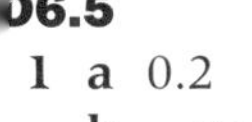

D6.5

1 **a** 0.2

b

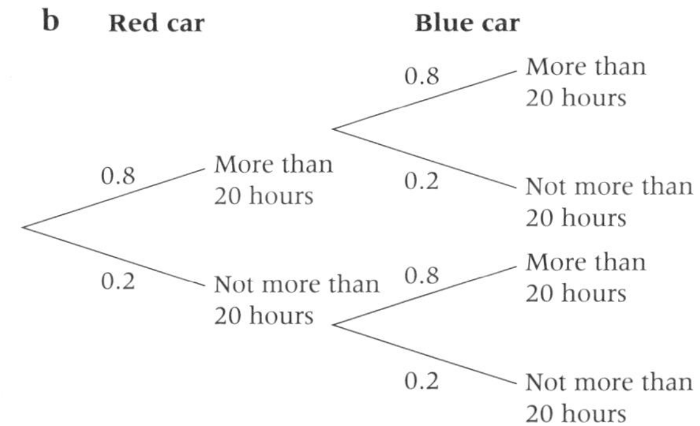

ci 0.64 **cii** 0.32 **ciii** 0.96

2 **a**

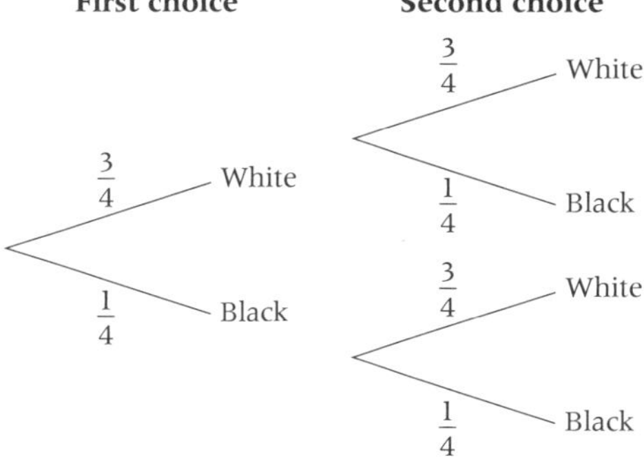

bi $\frac{1}{16}$ **bii** $\frac{3}{8}$ **biii** $\frac{7}{16}$

3 **a** $\frac{19}{20}$

b

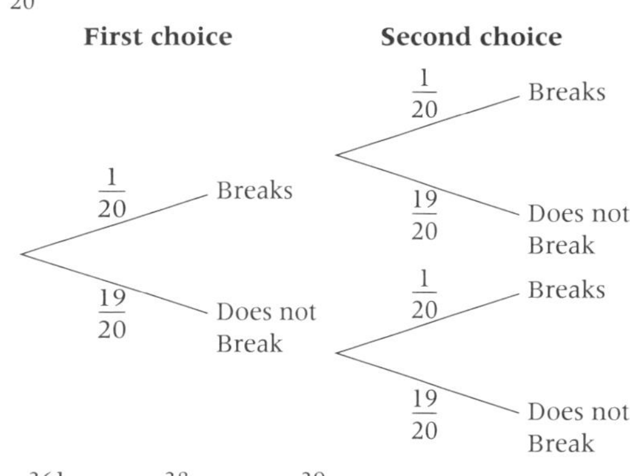

ci $\frac{361}{400}$ **cii** $\frac{38}{400}$ **ciii** $\frac{39}{400}$

4 **a**

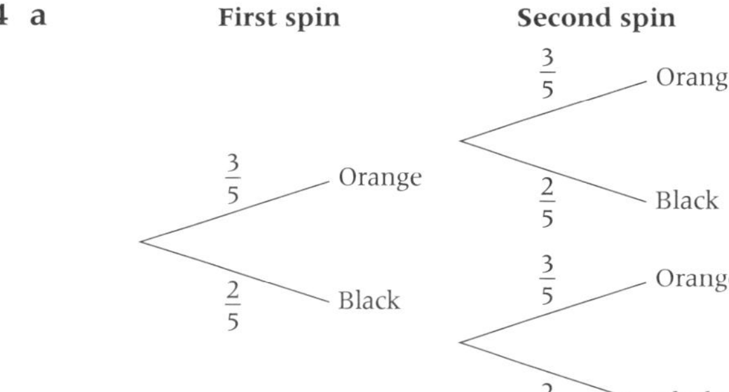

bi $\frac{4}{25}$ **bii** $\frac{16}{25}$ **biii** $\frac{12}{25}$

5 **a** 0.9

b

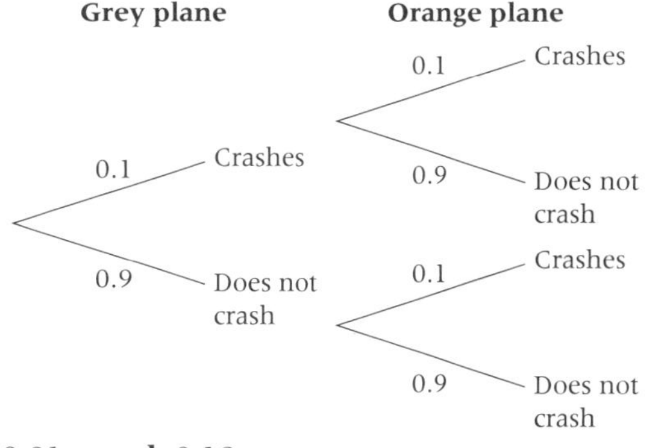

c 0.01 **d** 0.18

D6 Exam review

1 **ai** $\frac{1}{6}$ **aii** $\frac{5}{6}$ **aiii** $\frac{5}{12}$

b 0, there are no purple balls in the bag

2 **a** From Julie's experiment P(6) = $\frac{1}{3}$.

For a fair dice P(6) = $\frac{1}{6}$.

The probability of a 6 inJulie's experiment is twice that expected from a fair dice.

This suggests that the dice is not fair.

b

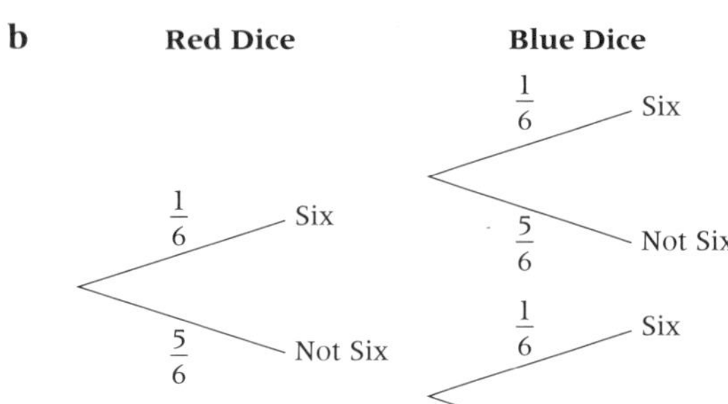

A8 Before you start …

1 **a** 20 mph **b** 80 mph **c** 90 mph **d** 24 mph

2

Equation	Gradient	y-axis intercept	Direction
$y = 3x + 4$	3	(0, 4)	up
$y = 10 - 4x$	−4	(0, 10)	down
$2y = 8x + 10$	4	(0, 5)	up
$2y - 4x = 15$	2	(0, 7.5)	up
$y = 7$	0	(0, 7)	horizontal

3 **a** $\frac{3}{2}$ **b** −1 **c** $-\frac{1}{2}$ **d** $\frac{1}{3}$ **e** $\frac{3}{5}$

4 **a** 5 **b** $\frac{5}{4}$ **c** 11

A8.1

1 **a** 60 km **b** 1 hours 30 minutes
c 60 km/h **d** 11:30 am and 12 noon
e 30 km/h

2 Claire and Christina: Claire's vertical line means she travelled a distance in no time, and Christina's line sloping backwards means she has travelled backwards in time.

3 **a**

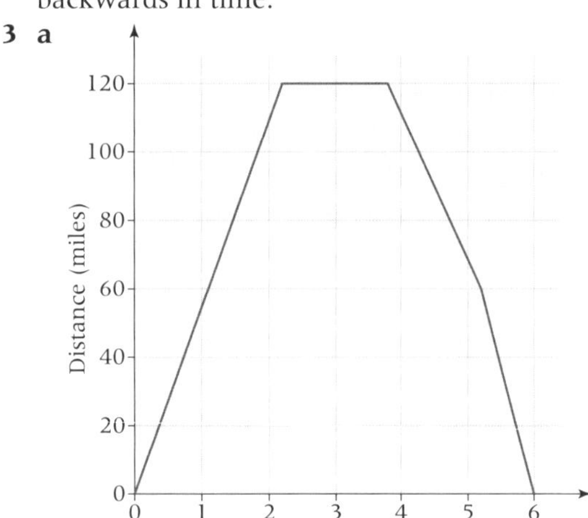

b

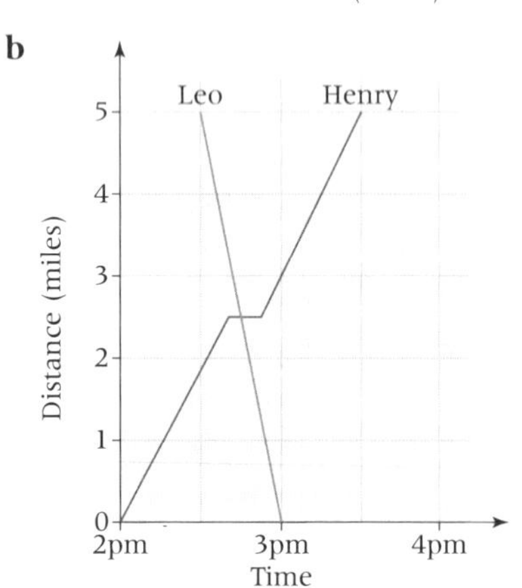

4

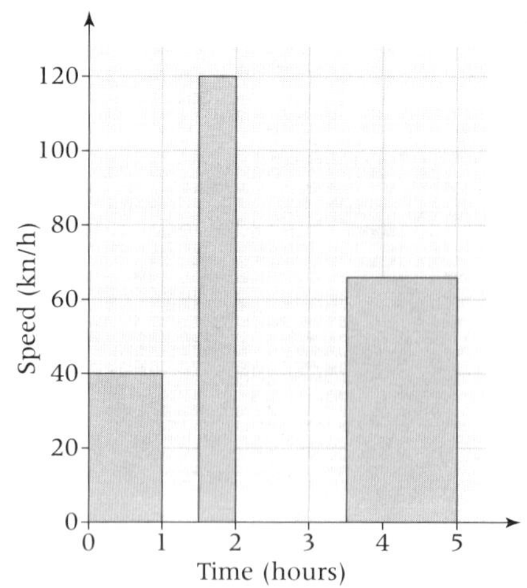

A8.2

1 A3, B1, C4, D2

2 Growth spurt at around 5, steadies down until next growth spurt at about 13–17 years (puberty) then steadies down again.

3

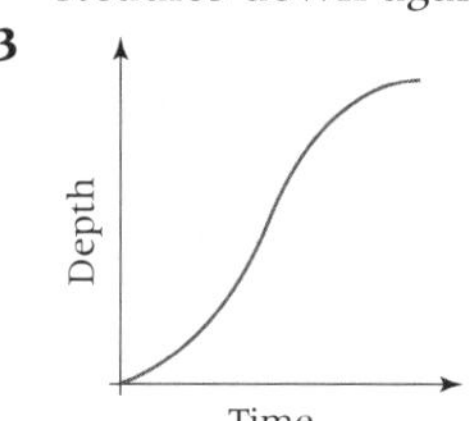

4 **a**

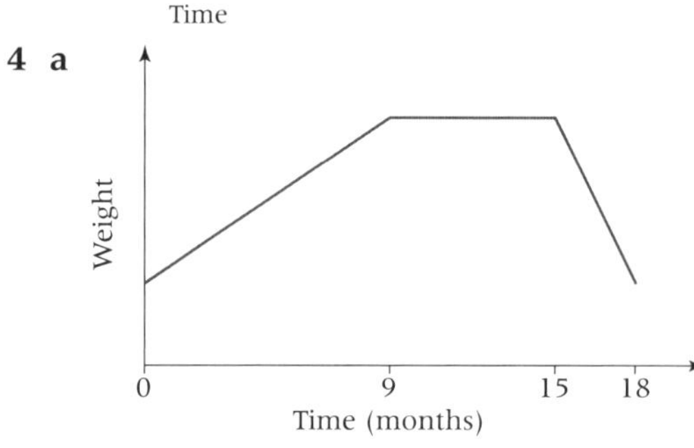

b

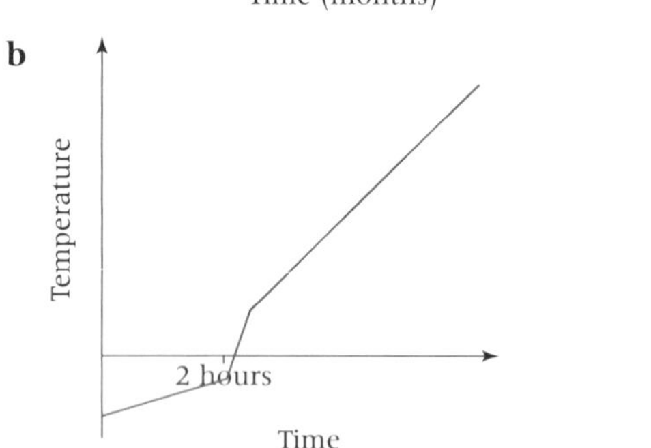

5 **a** **b**

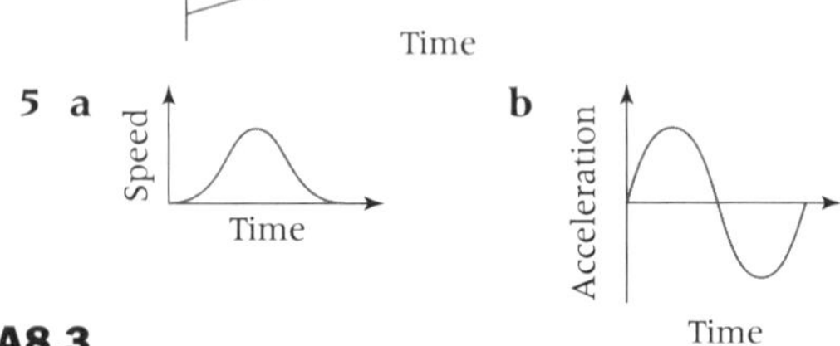

A8.3

1 **ai** 32 km **aii** 37.5 miles **b** Charlie **c** $y = 1.6x$

2 **a**

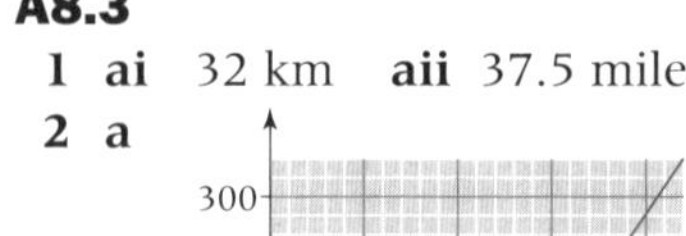

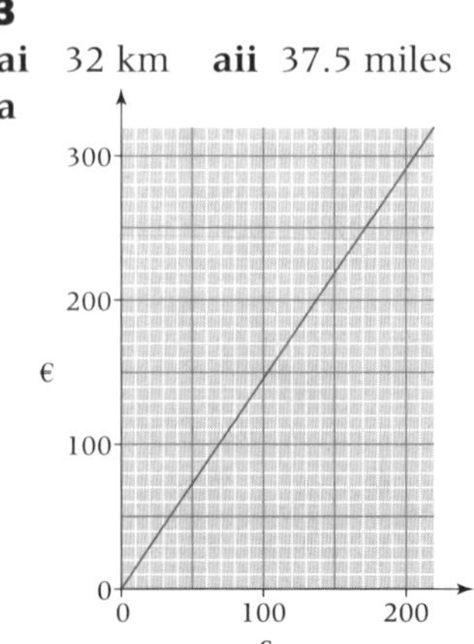

bi Approximately €138
bii Approximately £59

3 a

Number of people	1	2	3	4	5	6	7	8	9	10	11	12	13	14	15
Cost (£)	18	21	24	27	30	33	36	39	42	45	48	51	54	57	60

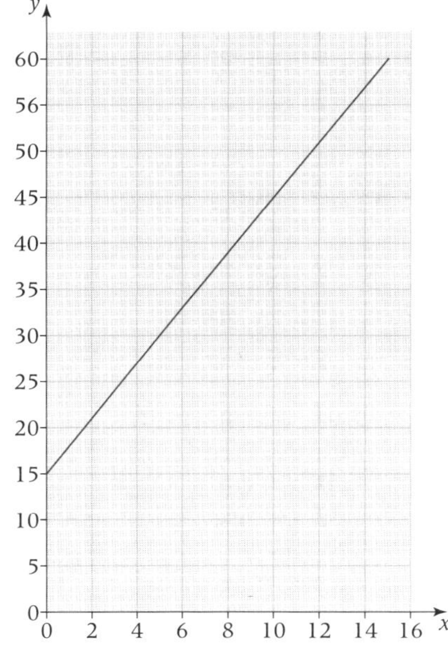

b 7

c This would mean that a fraction of a person stayed in the tent.

d $c = 15 + 3p$, where c = cost per night and p = number of people

e £96

4 Power Up! is cheaper for less than 10 units; they cost the same for 10 units; Sparks Are Us! is cheaper for more than 10 units.

A8.4

1 A: $y = 4x + 3$, B: $y = 2x + 1$, C: $y = \frac{1}{2}x + 3$, D: $y = \frac{1}{3}x + 1$

2 a $y = 6x + 50$; m is 6, meaning that for every year you age, you grow by 6 cm; c is 50, meaning that when you are born you are 50 cm tall, this is reasonable.

b $y = \frac{1}{2}x + 20$; m is $\frac{1}{2}$, meaning that for every two years a person needs one more driving lesson; it is not sensible to interpret c as no one takes a driving lesson at birth.

3 a Pocket money is not given to very young children or adults, so only makes sense between $x = 4$ and $x = 21$

b Does make sense to interpret the equation for any value of x, as temperature can take any value.

A8.5

1 $(-\frac{1}{4}, -4\frac{1}{8})$

2 a

x	−3	−2	−1	0	1	2	3
x^2	9	4	1	0	1	4	9
y	7	3	1	1	3	7	13

b

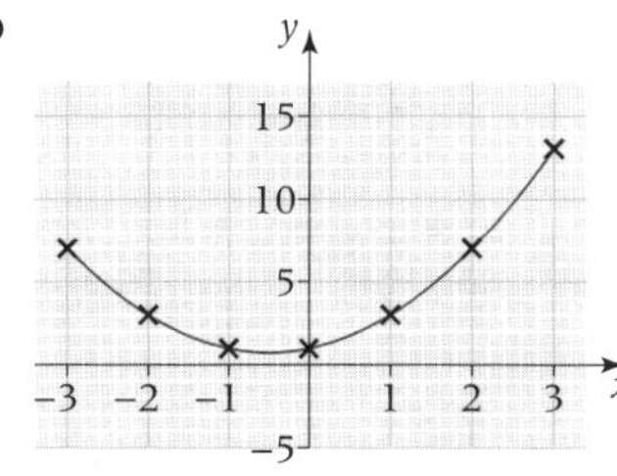

c Minimum is 0.75 at $x = -0.5$

3 a

Time (x)	0	1	2	3	4	5
$20x$	0	20	40	60	80	100
$4x^2$	0	4	16	36	64	100
Height (y)	0	16	24	24	16	0

b

ci 25 m, 2.5 seconds
cii 0.7 seconds, 4.3 seconds
ciii 3.2 seconds

A8 Exam review

1 a 12 m **b** 50 m

2 a Speed = 40 km/h

b

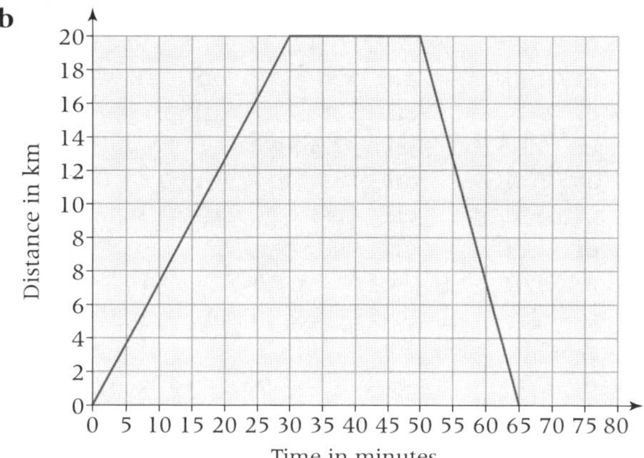

S8 Before you start ...

1 a $x = 6y$ **b** $x = 5y$ **c** $x = 10y$
d $x = \frac{2}{y}$ **e** $x = \frac{5}{y}$ **f** $x = \frac{8}{y}$

2 a $a = 8.1$ units **b** $b = 9.8$ units
c $c = 8.5$ units **d** $d = 13.7$ units

S8.1

1 a 3.63 cm **b** 11.3 cm **c** 10.8 cm **d** 3.02 cm
e 6.38 cm **f** 3.79 cm **g** 30.5 cm **h** 14.2 cm
i 74.5 cm **j** 6.06 cm **k** 7.00 cm **l** 12.3 cm

2 Right-angled isosceles triangles; They are the same as the other shorter side in each triangle.

S8.2

1 a 6.89 cm **b** 8.60 cm **c** 7.88 cm **d** 4.62 cm
e 4.77 cm **f** 39.7 cm **g** 10.6 cm **h** 9.56 cm
i 34.4 cm **j** 13.5 cm **k** 11.8 cm **l** 6.15 cm
m 14.2 cm **n** 7.88 cm **o** 5.12 cm

S8.3

1 a 26.4° **b** 25.7° **c** 56.2° **d** 67.4° **e** 17.6° **f** 23.6°
g 69.2° **h** 55.1° **i** 19.8° **j** 35.7° **k** 27.7° **l** 40.1°
m 73.0° **n** 41.0° **o** 45° **p** 30° **q** 60°

S8.4

1 a $a = 8$ cm, $b = 14.4$ cm **b** $c = 9.57$ cm, $d = 16.0$ cm
c $e = 13.2$ cm, $f = 10.8$ cm, $g = 13.1$ cm

2 a 28.9° **b** 35.5° **c** 61.2°

3 a 15.2 cm **b** 19.5°

4 a 8.5 cm **b** 58.4°

5 a 15.2 cm **b** 6.9 cm

S8.5

1 49.4 cm^2 **2** 63.8 cm^2 **3** 31.5 cm^2 **4** 91.2°
5 10.3 cm **6** 14.7 cm **7** 38.9° **8** 7.5 km
9 a 130° or 050° **b** 20.8 km or 5.4 km **10** 6.6 m
11 Yes, both give roughly the same height for phone mast.

S8 Exam review

1 bi 63.7 paces **bii** 279.3°

2 a DG = 11.7 m (to 3 sf) **b** 36.9° (to 1 dp)

Homework book answers

N1 HW1

1 a $\frac{7}{8}$ b $\frac{11}{35}$ c $2\frac{37}{56}$ d $1\frac{233}{240}$ e $\frac{1}{3}$
f $4\frac{1}{2}$ g $6\frac{2}{3}$ h $5\frac{1}{3}$

2 a $x=4$ b $x=\frac{2}{3}$ c $x=4$ d $x=-2\frac{1}{2}$
e $x=10$ f $x=1\frac{3}{4}$

3 5.83 cm

4 a 0.45 b 0.85 c 8

N1 HW2

1 a 0.004, 0.04, 0.14, 1.4, 4
b 0.329, 3.29, 3.92, 9.32, 32.9

2 ai 490 aii 500 bi 210 bii 200
ci 90 cii 100 di 1050 dii 1000
ei 17 320 eii 17 300

3 ai 138.3 aii 138.35 bi 0.6 bii 0.63
ci 4.5 cii 4.53 di 0.1 dii 0.10
ei 64.0 eii 64.00

4 a 6270 b 94.5 c 267.9 d 3.472
e 0.43 f 0.3529 g 200.4 h 30.45

5 a −11 b 20 c 7 d −2
e −20 f −5 g 96 h 67
i 6 j 31

N1 HW3

1 a −6 b 63 c 4 d −4
e −60 f −27 g 5 h −7
i −20 j 24

2 a −36 b −42 c −9.73
d −5.46 e −74 f −0.0525

3 a 2 b 1 c −2
d −5 e −3 f 7

4 a $60=2\times2\times3\times5$
b $210=2\times3\times5\times7$
c $378=2\times3\times3\times3\times7$
d $504=2\times2\times2\times3\times7$
e $2156=2\times2\times7\times7\times11$

5 $330=2\times3\times5\times11$, so the possible dimensions are
$1\times2\times165$, $1\times3\times110$, $1\times5\times66$, $1\times11\times30$
$2\times3\times55$, $2\times5\times33$, $2\times11\times15$
$3\times5\times22$, $3\times11\times10$
$5\times11\times6$

N1 HW4

1 ai 12 aii 180 bi 15 bii 360
ci 18 cii 1512

2 60; following hint, 1 cm (5, 12) = 60. The instruments play together at every common multiple of 5 and 12; hence the next time they play together after the start is at the first common multiple of 5 and 12 i.e. the lowest common multiple.

3 50

4 $a=3$, $b=5$

5 a and c

S1 HW1

1 a 3.6 b 12.4 c 5.3 d 1.0
e 101.0 f 1032.9

2 a 63 b 0.492 c 3.6 d 0.00853
e 0.086 f 0.1002

3 a 3 b 60 c −0.03 d 37
e 24 f −70 g −17 h −15

4 a $36=2\times2\times3\times3$ b 18 c 180

5 a Prism b Prism c Not prism

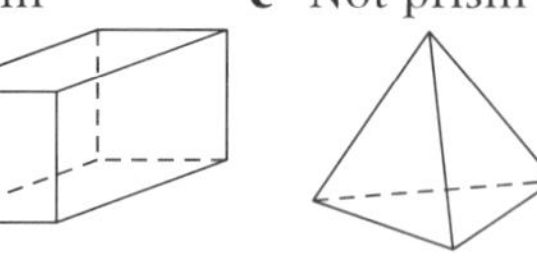

d Not prism e Not prism

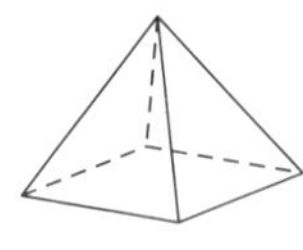

S1 HW2

1 ai 36 cm^2 aii 26 cm
bi 25.83 cm^2 bii 20.8 cm
ci 30 cm^2 cii 30 cm
di 54 mm^2 dii 36 mm

2 a 96 cm^2 b 108 cm^2

3 a 63 mm^2 b 15.75 cm^2

4 52.5 cm^2

S1 HW3

1 a 25.1 cm (answers given to 1 decimal place throughout exercise)
b 11.9 mm c 78.5 m

2 a 50.3 cm^2 b 11.3 mm^2 c 490.9 m^2

3 Area = 170.3 m^2 Perimeter = 55.1 m

4 7.7 m^2

S1 HW4

1 a 12 432 mm^2 b 384 cm^2 c 565.5 m^2

2 30.9 cm^2 (1 d.p.)

3 6 m

4 35.1 mm^2 (1 d.p.)

A1 HW1

1 a 4 b 7 c 30 d −6 e −30
f −6 g −12 h −5 i $-\frac{1}{2}$

2 ai $60=2\times2\times3\times5$
$108=2\times2\times3\times3\times3$
aii 12
bi 326.6 bii 0.3266

3 4350 cm^3

4

Height h (cm)	Frequency
$110<h\leqslant115$	4
$115<h\leqslant120$	6
$120<h\leqslant125$	11
$125<h\leqslant130$	7
$130<h\leqslant135$	2

A1 HW2

1 a 13 b 2 c 15 d 1 e 18 f $\frac{1}{3}$

2 a $2x$ b y^3 c $13x$ d $5a+3b$
e $4m-3n$ f $4x^2+5x$ g $11pq$ h $8pq$
i 9 j $\frac{5x}{y}$

3 a $3x+12$ b y^2-3y c $-5p-10q+5r$
d $4m^2-4mn$ e $2a+5b+26$ f $-10x+6y$
g $9h+2$ h $p-2$

4 ai $6x+8$ aii $15x-5$
b $15x-5=40\longrightarrow15x-45=0$

A1 HW3

1 a $x^2+7x+12$ b y^2+3y-4
c p^2+6p+9 d x^2-25

e $h^2 - 3h - 10$ **f** $2x^2 + 13x + 15$
g $6t^2 + 17t + 5$ **h** $9m^2 + 24m + 16$
i $6p^2 - 7pq - 3q^2$ **j** $16x - 24xy + 9y^2$

2 $77 = (2x - 3)(3x - 4)$
$= 6x^2 - 17x + 12 \longrightarrow 6x^2 - 17x - 65 = 0$

3 a $5(x + 2)$ **b** $3(x - 3)$
c $2(6x - t)$ **d** $2a(2b - 1)$
e $3(2xy + 3x - 4y^2)$ **f** $h(8h^2 - 1)$

4 a $(a + b)(1 + 3a + 3b)$ **b** $(p + qr)(p + qr - 6)$
c $(p + r)(q + x)$ **d** $(y + w)(x - 2)$

5 a 6 **b** 11 **c** 47.8

A1 HW4

1 a $(x + 3)(x + 4)$ **b** $(x + 8)(x - 2)$
c $(x + 5)(x - 9)$ **d** $(x - 7)(x - 4)$
e $(x - 4)^2$ **f** $(x - 7)(x + 7)$
g $(x - 12)(x + 10)$ **h** $(x - 14)(x + 14)$

2 ai $8p - 3q$ **aii** $15ab$
b $20 - 12x$ **c** $10 - 7y$

3 a $24a^5b^3$ **b** $(x + y)(x - 2)$

4 a $P = 2(x + 5) + 2(x - 3) = 4x + 4$
b $A = (x + 5)(x - 3) = x^2 + 2x - 15$

5 a $P^2 + 2pq + q^2$ **b** 16

N2 HW1

1 a 270 **b** 3.4 **c** 7.89
d 50 **e** 9.84 **f** 1.013
g 42 050 **h** 0.0031

2 a $12xy$ **b** $8p^2$ **c** $3a^3$
d $8ab + bc$ **e** $5m + 3n$ **f** $3x^3 - x^2$
g $\frac{2g}{k}$ **h** $8y$

3 154 cm

4 a 4 **b** 5 **c** No rotational symmetry
d 2

N2 HW2

1 a 5000 **b** 70 **c** 63 **d** 24.3
e 4300 **f** 15.28 **g** 0.876 **h** 2.0

2 a 1.4 **b** 25.3 **c** 0.03 **d** 0.70
e 107.7 **f** 5250 **g** 40 000 **h** 96 000

3 a 54 000 **b** 40 **c** 37 **d** 5
e 30 **f** $\frac{1}{39}$

4 a 7.5 **b** 13.2 **c** 3.5 **d** 1.06
e 11.4 **f** 10 **g** 0.41 **h** 2.72

N2 HW3

1 a 8.84 **b** 3.982 **c** 84.852 **d** 1.92
e 5.45 **f** 17.721 **g** 18.892 **h** 35.761

2 a 128
bi 12.8 **bii** 1.28 **biii** 1.28
biv 1280 **bv** 0.000128

3 a 5324.8 **b** 5.3248 **c** 83.2 **d** 0.53248
e 0.64 **f** 8.32

4 a 0.12 **b** 4 **c** 30 **d** 500
e 8.7 **f** 60 **g** 140 **h** 0.396

5 a $60 \times 8 = 480$; $0.6 \times 8 = 4.8$; $48 \div 0.8 = 60$ etc.
b $7 \times 1.2 = 8.4$; $0.84 \div 0.07 = 12$; $0.7 \times 0.12 = 0.084$ etc.
c $0.3 \times 4.5 = 1.35$; $13.5 \div 45 = 0.3$; $135 \div 0.3 = 450$ etc.

N2 HW4

1 Felicity is correct; Andrew has calculated $28 - 24 = 4$, $56 - 7 = 49$ and mistakenly concluded this implies $28.56 - 24.7 = 4.49$.

2 153

3 24.824

4 a 0.759 **b** 16.5

A2 HW1

1 a 5052 **b** 833 **c** 8188
d 253 **e** 17 444 **f** 2451
g 305 **h** 86 142

2 a $5x + 20$ **b** $p^2 - 8p$
c $3a - 3b + 6$ **d** $2h - 24$
e $x^2 + 12x + 27$ **f** $y^2 - 4y + 4$
g $2a^2 - 7a - 15$ **h** $2p^2 - 2q^2$

3 44.0 cm² (1 d.p.)

4 ai 7.36 (2 d.p.) **aii** 8 **aiii** 2 **aiv** 12
b The mode in this case does not accurately reflect the spread of the data; it does not take into account all values in the data set.

A2 HW2

1 a $x = 7$ **b** $x = 2$ **c** $x = 4$ or -4
d $x = 12$ **e** $x = 11$ **f** $x = 6$
g $x = 8$ **h** $x = 7$ or $x = -13$

2 a $x = 5$ **b** $y = 7$ **c** $p = 15$
d $t = 9$ or $t = -9$ **e** $c = 1$ **f** $x = 8$
g $y = 4$ **h** $x = 25$

3 a $y = 10$ **b** $x = -2$

4 $c = -5$

A2 HW3

1 a $x = 5$ **b** $p = 3$ **c** $y = 2$ **d** $a = 4$
e $x = 1.5$ **f** $y = -1$ **g** $t = 3$ **h** $x = \frac{1}{2}$

2 a $x = 18$ **b** $a = 35$ **c** $x = 4$ **d** $y = \frac{1}{4}$
e $x = \frac{3}{8}$ **f** $e = 1$ **g** $x = 7$ **h** $p = 5$

3 a $\frac{x}{5} + 3 = 9 \longrightarrow x = 30$
b $4x + 3 = 2x + 9 \longrightarrow x = 3$
c $7x - 4 = 14 - 2x \longrightarrow x = 2$
d $\frac{15}{x + 1} = 3 \longrightarrow x = 4$

4 $x = 3$

A2 HW4

1 a $x < 4.5$ (0, 4.5)
b $x \leqslant 6$ (0, 6)
c $x \geqslant 3$ (0, 3)
d $x < -4$ (−4, 0)
e $x < 10$ (0, 10)
f $x \leqslant 3$ (0, 3)
g $x \geqslant 30$ (0, 30)
h $x < -6$ (−6, 0)

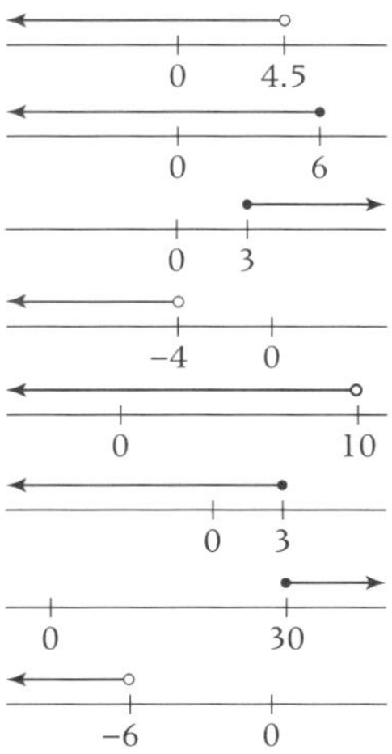

2 a $3x + 2y$ **b** $9a + 8$

3 a $p = -2$ **b** $q = -\frac{1}{3}$

4 a $y = -2$ **b** $c = -4$

5 a −1, 0, 1, 2, 3 **b** $x \geqslant \frac{1}{2}$

D1 HW1

1 a 0.586, 0.856, 5.68, 5.86, 6.85, 58.6, 85.6, 685
b 0.014, 0.12, 0.124, 0.142, 0.41, 1.42, 2.1, 4.12

2 a 2 **b** −11 **c** 12 **d** −1
e 12 **f** −1 **g** −16 **h** 16

3 a 15 cm **b** 18.3 cm (1 d.p.)

4 4 cm

D1 HW2

1 a Which type of music do you prefer?
Dance Rock Pop Classical Other
b How many CD's do you buy per month?
0 – 5 6 – 10 11 – 15 16 – 20
More than 20
c What is your opinion of Radio 1?
Very Good Good Satisfactory Poor
Very poor
d How many music concerts do you attend per year?
0 – 5 6 – 10 11 – 15 16 – 20
More than 20
e In which area of the country do you live?
North East North West West Midlands
East Midlands South East SouthWest

2 ai Trying to influence opinion; leading question
aii Too complicated; difficult to follow or understand
aiii Too few options; other possible colours not included
b Not enough diversity in the sample; all females and all Jaya's friends; she may obtain similar opinions from similar people

3 a Excludes people who used other travel agents/booked online/booked by telephone
b Many people might be unable to shop at 11 am on a Tuesday, for example if they are at work
c School teachers and pupils will be at school

D1 HW3

1

	For	Against
Boys		
Girls		

2

	Rock	Pop	Dance	Classical	Other
Year 7					
Year 8					
Year 9					
Year 10					
Year 11					

3 a 380
bi 47.4% (1 d.p.) bii 20.3% (1 d.p.)
c 33.3% (1 d.p.)

4 ai 12 aii 10 aiii 10 aiv 45
av 4
b Birthday/Christmas. Increased the mean and the range. No effect on mode, median or interquartile range.

D1 HW4

1 a No options to cover negative reactions; options do not accommodate those who disliked the text
bi Peoples views of what is meant by 'a little' or 'a lot' will differ; there is no middle option like 'an average amount'
bii 'How much time do you normally spend on homework?'
0 – 1 hours 1 – 2 hours 2 – 3 hours
More than 3 hours

2 74%

3 $1\frac{2}{3}$ hours i.e. 1 hour 40 minutes

4 $\frac{18x + 14y}{32}$

N3 HW1

1 a 32 b 130 c 0.9 d 393
e 0.027 f 8500 g 6.68 h 60 000

2 a $y = -4$ b $x = 1$

3 1964 cm^2 (4 s.f.)

4 a 75 b 52% c 55.6% (1 d.p.)

N3 HW2

1 ai 10 aii 2 aiii 3 aiv 72
b $\frac{15}{36}$

2 a $\frac{1}{4}, \frac{3}{10}, \frac{2}{5}, \frac{1}{2}, \frac{11}{20}$ b $\frac{4}{45}, \frac{1}{5}, \frac{5}{9}, \frac{2}{3}, \frac{11}{15}$
c $\frac{1}{6}, \frac{1}{3}, \frac{5}{12}, \frac{4}{9}, \frac{3}{4}$ d $\frac{2}{15}, \frac{2}{3}, \frac{7}{10}, \frac{4}{5}, \frac{5}{6}$

3 a $\frac{5}{7}$ b $\frac{1}{3}$ c $\frac{5}{6}$ d $\frac{1}{8}$
e $\frac{17}{20}$ f $\frac{11}{12}$ g $4\frac{7}{12}$ h $2\frac{5}{18}$

4 $\frac{19}{40}$

N3 HW3

1 a True
bi $9 \times \frac{1}{4} = 2\frac{1}{4}$ bii $15 \times \frac{1}{7} = 2\frac{1}{7}$
bii $12 \times \frac{1}{5} = 2\frac{2}{5}$ biv $24 \times \frac{1}{9} = 2\frac{2}{3}$

2 a $\frac{3}{32}$ b $\frac{1}{3}$ c $\frac{15}{16}$ d $2\frac{1}{2}$
e $5\frac{3}{5}$ f 15 g 8 h $1\frac{13}{35}$

3 $\frac{3}{5} = 0.6$ $0.3 = \frac{3}{10}$ $\frac{3}{8} = 0.375$ $\frac{3}{4} = 0.75$
0.35 is the odd card out

4 ai 0.25 aii 25% bi 0.8 bii 80%
ci 0.45 cii 45% di 0.32 dii 32%

5 a 0.375 b $0.\dot{4}$
c 0.3125 d $0.58\dot{3}$

N3 HW4

1 Julie, since $\frac{6}{8} > \frac{2}{3}$

2 ai 0.55 aii $\frac{11}{20}$
b $\frac{5}{9}$

3 a 100
b 1, $\frac{47}{50}$, 0.6, 50%, $\frac{9}{20}$, 25%, $\frac{3}{25}$, 3%

4 $4\frac{7}{12}$ inches

S2 HW1

1

Fraction	Decimal	Percentage
$\frac{3}{4}$	0.75	75%
$\frac{2}{5}$	0.4	40%
$\frac{7}{20}$	0.35	35%
$\frac{1}{20}$	0.05	5%
$\frac{5}{8}$	0.625	62.5%

2 a $x > 2$ b $x \leqslant 6$ c $n \geqslant -5$ d $x \leqslant -6$
e $x \geqslant 12$ f $x < 14$ g $3\frac{1}{3} < x < 3\frac{1}{2}$ h $-5 \leqslant x \leqslant 5$

3 i 153.9 m^2 (1 d.p.) ii 44.0 m (1 d.p.)

4 a Similar opinions may be obtained from similar people.
b Use the school register and choose every nth person (say, for $n = 8$).
c

	Yes	No
Boys		
Girls		

S2 HW2

1 a $a = c = 70°$ $b = 110°$ b $x = 39°$
c $m = n = 142°$ d $f = g = 133°$

2 a $a = 55°$ $b = 119°$ b $x = 55°$ $y = 55°$

3 $6 \times 120° = 720°$

S2 HW3

1 a $a = 64°$ b $b = c = 54°$
c $d = 78°$ $e = 51°$ d $f = 35°$ $g = 55°$

2 a $a = 97°$ $b = 66°$ b $c = 75°$ $d = 60°$
c $e = 90°$ $f = 52°$ d $g = 22.5°$

S2 HW4

1 a $a = 55°$ b $b = 38°$ $c = 52°$

2 a $AOC = 104°$
b Angle at centre is twice angle subtended at circumference

3 a 150° b 9 sides; Nonagon

A3 HW1

1 ai $\frac{7}{9}$ aii $\frac{49}{99}$ aiii $\frac{73}{99}$ aiv $\frac{345}{999} = \frac{115}{333}$
b $0.\dot{4}$ c $0.\dot{8}\dot{3}$ d $0.\dot{1}42\,85\dot{7}$

2 a $x = 35$ b $x = 4$ c $x = 2$ d $x = 36$

3 a 45 cm^2 b 89.1 m^2 (1 d.p.)

4 77.6% (3 s.f.)

A3 HW2

1 a 24, 28 b 23, 27 c 70, 64
d 64, 81 e 64, 128 f 31.25, 15.625
g 13, 21 h 720, 5040

2 a $99\,999^2 = 9\,999\,800\,001$
$999\,999^2 = 999\,998\,000\,001$
b $99\,999\,999^2 = 9999999800000001$

3 a $5^2 = 1 + 3 + 5 + 7 + 9$, $6^2 = 1 + 3 + 5 + 7 + 9 + 11$
$7^2 = 1 + 3 + 5 + 7 + 9 + 11 + 13$
b $k = 99$

4 a 1, 5, 9, 13, 17 b 47, 44, 41, 38, 35
c 6, 9, 14, 21, 30 d 8, 15, 24, 25, 48
e $\frac{1}{2}, \frac{2}{3}, \frac{3}{4}, \frac{4}{5}, \frac{5}{6}$ f $1, \frac{1}{4}, \frac{1}{9}, \frac{1}{16}, \frac{1}{25}$

A3 HW3

1 a $T_n = 2n + 5$ b $T_n = 3n$ c $T_n = 5n - 2$
d $T_n = 4n - 12$ e $T_n = 18 - n$ f $T_n = 23 - 3n$
g $T_n = 1 + \frac{n}{2}$ h $T_n = 5 - \frac{n}{4}$

2 a $10 - \frac{n}{2}$ b $n^3 + 1$ c $6n$
d n^2 e $5n - 1$ f $3n^3$

3 a $S = 4t + 2$ b $S = 2t + 4$

A3 HW4

1 a $T_n = 2n^2$ b $T_n = n^2 + 3$
c $T_n = n^2 + 5n$ d $T_n = 3n^2 + n - 1$

2 a $d = 4n$ b 32

3 $T_n = 4n + 1$

4 $T_n = 8n - 3$

D2 HW1

1 a 0.625 b $0.1\dot{6}$ c 0.0625
d $0.\dot{8}$ e $0.41\dot{6}$

2 Sequence 1 $\longrightarrow 5 - n$
Sequence 2 $\longrightarrow 3n + 1$
Sequence 3 $\longrightarrow (n + 1)^2$
Sequence 4 $\longrightarrow 4n$
Sequence 5 $\longrightarrow 4n^2$

3 a $a = 63°$ $b = 63°$ $c = 54°$ $d = 126°$
b $e = 71°$ $f = 64°$ $g = 45°$

4 14.5

D2 HW2

1 a

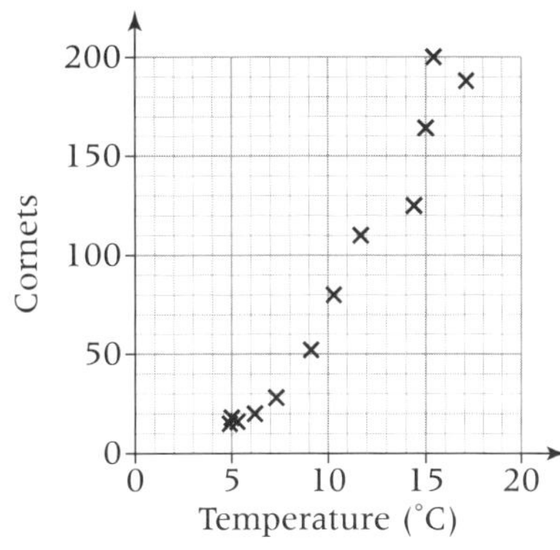

b Positive correlation – the higher the temperature, the more cornets sold.

2 a

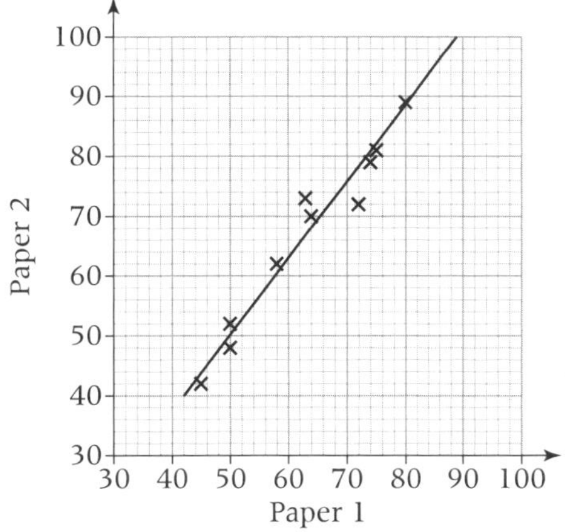

b Positive correlation – the higher the mark on P1, the higher the mark on P2.
c 54

3 a

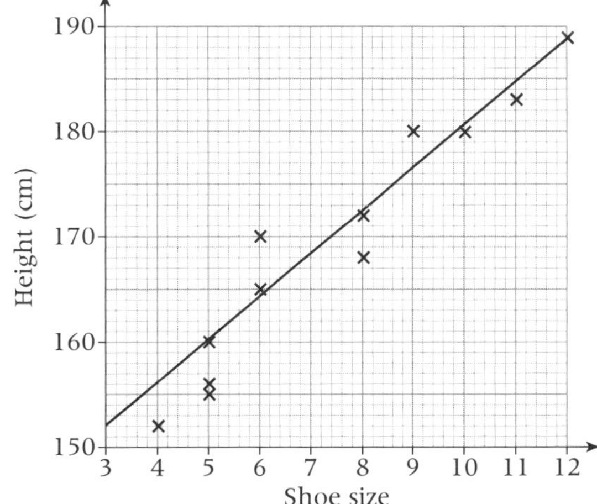

b Positive correlation – the taller the person, the larger the shoe size
c 168 cm

D2 HW3

1 a Key: 7 | 7 = 77

2	1
3	
4	8 9
5	0 1 4
6	2 4 5 5 8 9
7	0 2 5 7 7 7
8	4 5 7 9
9	1 3 5

b Negatively skewed

c Mode = 77 Median = 70

2 b Key: 15 | 4 = 154

15	0 2 4 4
15	5 6 7 8 8 8 8 9
16	0 0 1 2 2 3 3 4 4
16	5 6 7 8 9
17	0 1 4
17	1

c Range = 21 Interquartile range = 8

3 a Key: 5 | 6 = 5.6

1980–1989		**1990–1999**
8	3	
6 3 3 3	4	0
9 6	5	0
0	6	1 9 9
4 0	7	1 4 5 8 9

b Data for 1980–1989 is positively skewed
Data for 1990–1999 is negatively skewed
Mode for 1980–1989 is 4.3
Mode for 1990–1999 is 6.9
Temperatures are higher in 1990–1999
Range for 1980–1989 is 3.6
Range for 1990–1999 is 3.9

D2 HW4

1 i a **ii** c **iii** b **iv** d

2 a

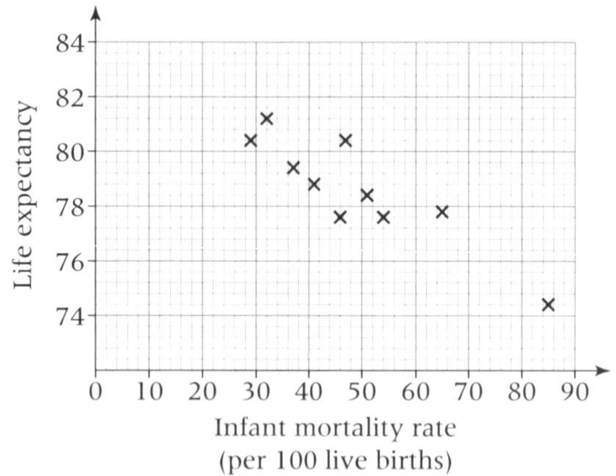

b As the infant mortality rate decreases, the life expectancy increases

3 a Key: 1 | 2 = 12 minutes

0	
0	8 9
1	0 1 2 4
1	5 8 9
2	0 1 2 3 3 4
2	5 6 7 8
3	2
3	

b Median = 20.5 Range = 24
Distribution is fairly symmetric; mode is 23 minutes

A4 HW1

1 a $\frac{3}{4}$ **b** $\frac{1}{2}$ **c** $\frac{7}{12}$ **d** $\frac{14}{15}$
e $\frac{17}{36}$ **f** $\frac{53}{56}$ **g** $\frac{2}{15}$ **h** $1\frac{1}{6}$

2 a $x = 5$ **b** $q = -2$ **c** $a = 1$ **d** $y = -6$

3 $\frac{30}{x+2} = 5 \longrightarrow x = 4$

4 $9 \times 140° = 1260°$

5 Median mark = 63

3	1 5 8
4	0 2 3 8
5	1 2 3 4 5 6
6	2 2 4 5 6
7	0 0 1 2 4 7 9
8	0 2 4 7 9

Key: 3 | 1 = 31

A4 HW2

1 a

x	−2	0	2
y	−1	3	7

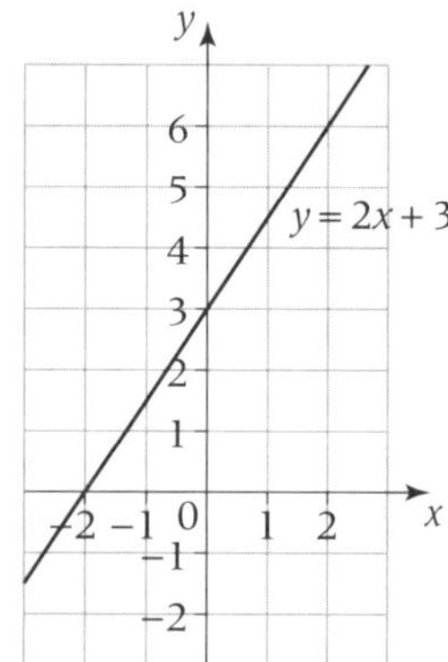

b

x	−3	0	3
y	7	4	1

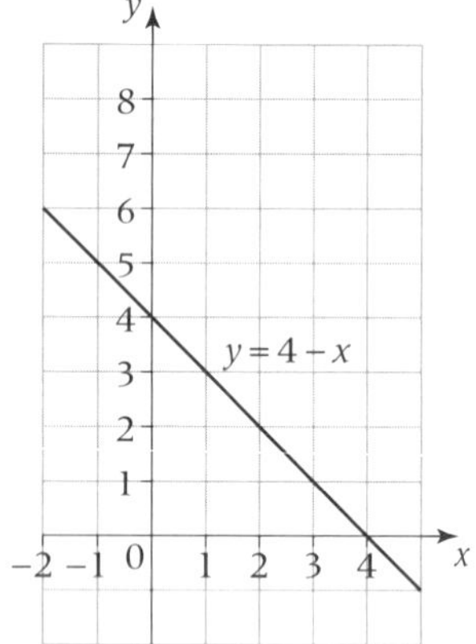

c

x	1	2	3
y	$-1\frac{1}{2}$	−1	$-\frac{1}{2}$

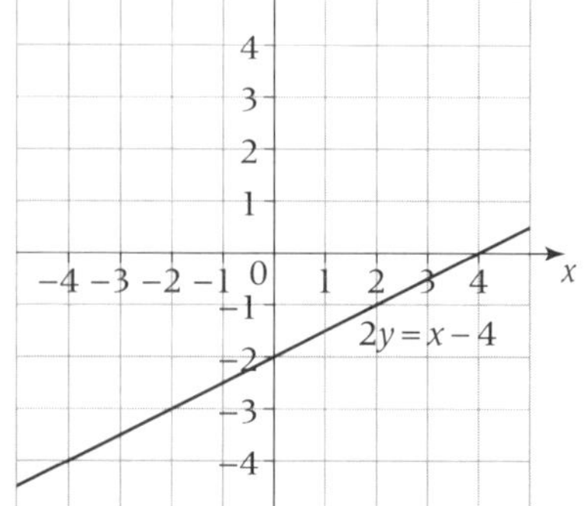

d

x	-1	0	1
y	3	3	3

2

Line	Yes or No
$y=x+1$	Yes
$y=3x-5$	Yes
$2y=x-1$	No
$3y=2x-2$	No
$x+y=7$	Yes

3 **a**

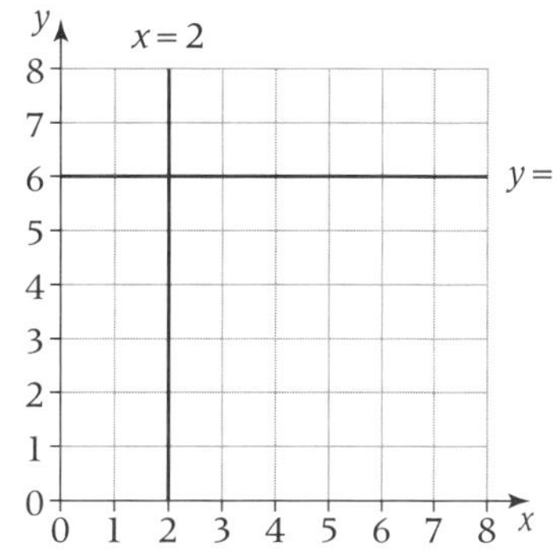

Intersect at (2, 6)

bi (4, 3) **bii** (5, −3) **biii** $(\frac{3}{4}, \frac{5}{8})$

A4 HW3

1 **a**

x	-2	0	2
y	-5	-1	3

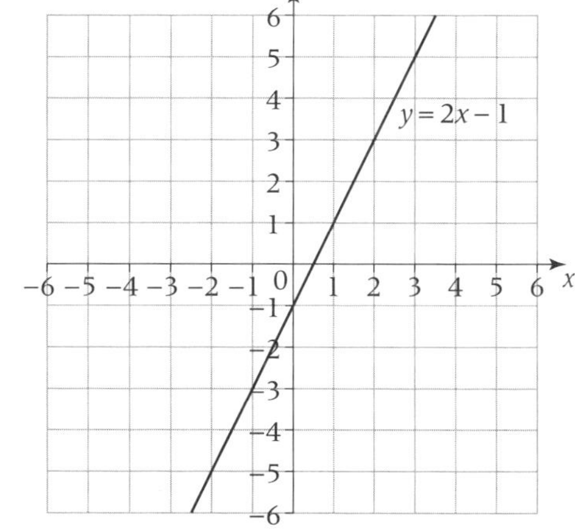

bi (0, −1) **bii** 2

c Gradient of line $y=mx+c$ is m, y-intercept is c.

2 **ai** (0, 4) **aii** 3 **bi** (0, −1) **bii** 5

ci (0, 6) **cii** −3 **di** (0, 2) **dii** −1

ei (0, 4) **eii** $-\frac{1}{2}$ **fi** $(0, -\frac{3}{2})$ **fii** 2

3 **a** $y=2x+c$ **b** $y=-3x+c$ **c** $y=-\frac{x}{2}+c$

4 $y=1-x$

A4 HW4

1 **a** $y=3x+2$ **b** $y=4x-5$ **c** $y=7-4x$

2

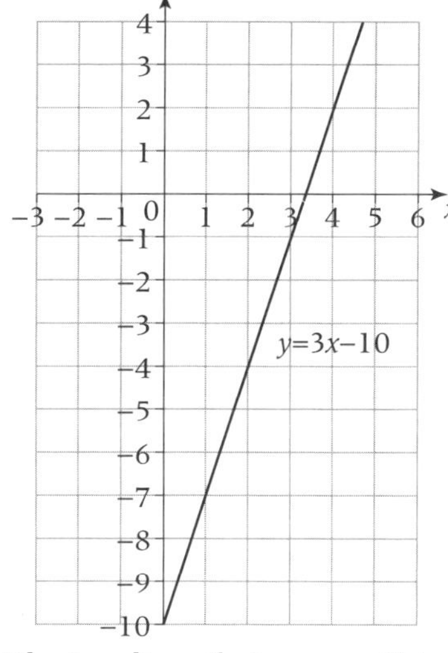

3 The two lines that are parallel are $3y=x-5$ and $3y-x=5$

4 **a** $x=4$ **b** $y=\frac{x}{4}+c$

5 **a** $(0, \frac{5}{3})$ **b** $(-\frac{5}{7}, 0)$

D3 HW1

1 **a** 11.48 **b** 8.78 **c** 4.116 **d** 4.09

2 **ai** (0, 3) **aii** 1 **bi** (0, 5) **bii** 4

ci (0, 8) **cii** −1 **di** (0, 2) **dii** 3

ei $(0, \frac{3}{2})$ **eii** $-\frac{1}{2}$ **fi** $(0, -\frac{5}{3})$ **fii** 3

3 **a** $1152\ \text{m}^2$ **b** $1187.5\ \text{cm}^2$ (1 d.p.)

4 **a**

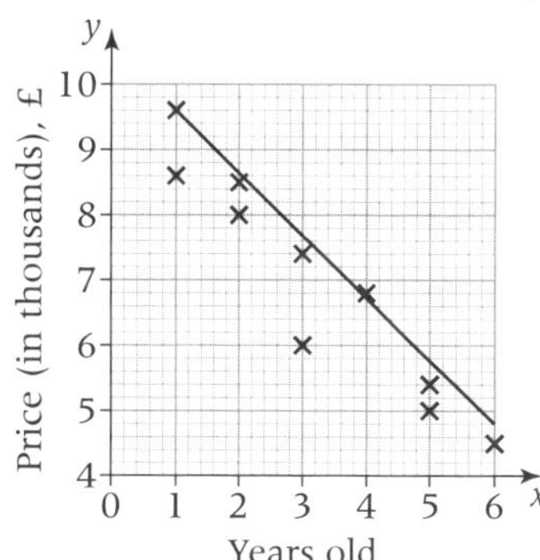

Negative correlation – as age of car increases, price decreases

b £7100

c Other factors such as condition, mileage, service history etc have not been included, hence not all that reliable.

Line of best fit drawn by sight – may not be entirely accurate.

D3 HW2

1 **a** $\frac{1}{3}$ **b** $\frac{1}{6}$ **c** $\frac{1}{2}$ **d** 0

2 $x=0.32$

3 **a** $\frac{2}{5}$ **b** $\frac{13}{15}$ **c** 0

d $\frac{7}{15}$ **e** $\frac{4}{5}$ **f** 1

4 **ai** $\frac{1}{20}$ **aii** $\frac{1}{5}$ **aiii** $\frac{1}{4}$ **aiv** $\frac{1}{2}$

av $\frac{7}{20}$ **avi** $\frac{1}{2}$ **avii** $\frac{1}{5}$ **aviii** $\frac{11}{20}$

b Mutually Exclusive

D3 HW3

1 a

	Bach	Chopin	Debussy	Total
Male	10	6	2	18
Female	4	8	10	22
Total	14	14	12	40

bi $\frac{11}{20}$ **bii** $\frac{3}{10}$ **biii** $\frac{3}{20}$

ci 49 **cii** 63

2 **ai** $\frac{1}{3}$ **aii** $\frac{4}{75}$ **aiii** $\frac{1}{3}$

bi 48 **bii** 84

3 225

D3 HW4

1 Expected number of 3s is $\frac{120}{6} = 20$. Freya scores twice this number of 3s, so the die is likely to be biased.

2 **a** Each level is twice as difficult as the last one (since Georgia has half the probability of completing it).

b

Level 1	Level 2	Level 3	Level 4
$\frac{8}{15}$	$\frac{4}{15}$	$\frac{2}{15}$	$\frac{1}{15}$

3 **a** $\frac{1}{8}$ **b** $\frac{3}{86}$ **c** $\frac{7}{8}$

4 a

	Number	Algebra	Shape	Data	Total
Male	3	8	9	6	26
Female	7	10	3	4	24
Total	10	18	12	10	50

bi $\frac{9}{25}$ **bii** $\frac{3}{25}$

c 120

N4 HW1

1 **a** 3753 **b** 375 300 **c** 0.695

d 3.753 **e** 6.95

2 $c = 3$

3 **a** $x = 106°$ since the angle at the centre is twice the angle subtended at the circumference (212°)

$y = 74°$ since the angle at the centre is twice the angle subtended at the circumference (148°)

b $z = 32°$ since the angle in a semicircle is a right angle

4 a

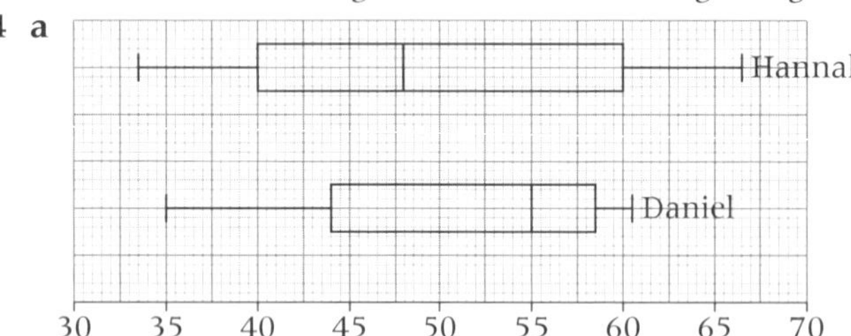

b Hannah has a symmetrical distribution. Daniel's is negatively skewed. Daniel's distribution has a smaller range i.e. he is more consistent.

N4 HW2

1 **a** $\frac{5}{6}$ **b** $\frac{13}{15}$

2 **a** $\frac{7}{20}$ **bi** 64 **bii** 56

3 £112

4 Standard bottle: 0.168 pence per ml

Large bottle: 0.1592 pence per ml

So the large bottle is best value for money.

N4 HW3

1

Country	Exchange rate	Amount of local currency
Australia	1 GBP = 2.34 Australian dollars	AU$269.10
Bolivia	1 GBP = 14.2 Bolivian bolivianos	1633 Bolivianos
France	1 GBP = 1.47 Euros	€169.05
Iceland	1 GBP = 111.43 Icelandic krona	12814.45 krona
Maldives	1 GBP = 22.62 Maldives ruiyan	2601 ruiyan
Thailand	1 GBP = 216.78 Thai Baht	5379.70 Baht

2 **a** $1300 **b** Profit

3 36 mph

4 9639 kg

N4 HW4

1 Yes, a saving of £20.

2 5.70 kg (3 s.f.)

3 Notebook World = £686.4; Asteroid = £680; Balti's = £681.5

Best offer is Asteroid.

4 Jake

S3 HW1

1 **a** $\frac{1}{20}$, 0.12, 23%, 0.45, $\frac{4}{5}$, $\frac{41}{50}$, 1

b 2%, $\frac{17}{100}$, 0.24, 0.3, $\frac{1}{2}$, 65%, $\frac{39}{50}$, 90%

2 **ai** 17, 15, 13, 11, 9 **aii** $n = 15$

b $u_n = 4n - 1$

3 **a** 45° **b** 135°

4 **a** Add in option boxes

b In your opinion, what is the maximum number of hours a day that a child should watch television?

Less than 1 hour 1–2 hours More than 2 hours

S3 HW2

1 **a** Triangle with vertices at (0, 2), (0, 4), (−1, 2) labeled B

b Triangle with vertices at (2, 0), (3, 0), (2, −2) labeled C

2 **a** Rectangle with vertices at (3, −1), (3, −2), (6, −2), (6, −1) labeled P

b Rectangle with vertices at (1, 3), (2, 3), (2, 6), (1, 6) labeled Q

3 **a** L-shape with vertices at (2, −1), (5, −1), (5, −2), (3, −2), (3, −3), (2, −3) labeled B

b L-shape with vertices at (0, −1), (0, −4), (−1, −4), (−1, −2), (−2, −2), (−2, −1) labeled C

c Rotation of 180° about (1, 0)

S3 HW3

1 **a** $\begin{pmatrix}3\\0\end{pmatrix}$ **b** $\begin{pmatrix}0\\-4\end{pmatrix}$ **c** $\begin{pmatrix}-7\\-1\end{pmatrix}$ **d** $\begin{pmatrix}-4\\3\end{pmatrix}$ **e** $\begin{pmatrix}1\\2\end{pmatrix}$ **f** $\begin{pmatrix}3\\1\end{pmatrix}$

2 **a** Parallelogram with vertices at (−4, 2), (−3, 3), (0, 3), (−1, 2) labeled Q

b Parallelogram with vertices at (3, 3), (4, 4), (7, 4) (6, 3) labeled R

3 **a** Reflection in the line $y = 0$

b Translation through $\begin{pmatrix}3\\1\end{pmatrix}$

c Rotation of 180°, centre $(-\frac{1}{2}, 0)$

d Reflection in the line $x = -\frac{1}{2}$

e Rotation or 90° anticlockwise about (2, 0)

S3 HW4

1 a Rectangle with vertices at (4, 1), (5, 1), (4, 3), (5, 3) labeled B
b $\begin{pmatrix}-2\\3\end{pmatrix}$
c $T^{-1} = -T$
2 a Triangle with vertices at (1, 2), (3, 2), (1, 3) labeled B
b Triangle with vertices at (2, 1), (2, 3), (3, 1) labeled C
c Rotation, about (0, 0), of 90° anti-clockwise
3 a L-shape with vertices at (1, 1), (1, 3), (2, 3), (2, 2), (3, 2), (3, 1) labeled M
b L-shape with vertices at (3, 1), (3, 2), (4, 2), (4, 3), (5, 3), (5, 1) labeled N
c Translation with column vector $\begin{pmatrix}0\\6\end{pmatrix}$

N5 HW1

1 a 3.3 kg b 19.8 kg c 33 kg d 79.2 kg
2 a $y = 2x + 1$ b $y = 3x - 6$ c $y = 3 - x$
3 a Flag shape with vertices at (1, 2), (4, 2), (4, 1), (3, 1), (3, 2) labeled G
b Flag shape with vertices at (1, 4), (4, 4), (4, 5), (3, 5), (3, 4) labeled H
c Translation with column vector $\begin{pmatrix}0\\6\end{pmatrix}$
4 a Cannot both occur at the same time; 2 is not an odd number
bi $\frac{1}{2}$ bii $\frac{1}{6}$ biii $\frac{2}{3}$ biv $\frac{1}{3}$

N5 HW2

1 a 2^3 b 4^3 c 3^4 d 10^5
e 6^4 f $(\frac{1}{2})^2$
2 a 2^8 b 4^9 c 8^7 d x^9
e 5^5 f $10 = 10^1$ g $1 = a^0$ h 9^{-3}
i 3^6 j y^9 k 6^7 l 7^7
3 a 1 b 256 c 5 d 2005
4 a 8^3 b 4^4 c 3^3 d x^6
e $7 = 7^1$ f $a = a^1$

N5 HW3

1 a 3^8 b 5^6 c 9^2 d $8^{\frac{4}{3}}$
2 Pairs are 9^2 and 81, $\frac{1}{81}$ and 9^{-2}, 3 and $9^{\frac{1}{2}}$, 9^{-1} and $\frac{1}{9}$. So odd one out is 9.
3 a 1 b 5 c $\frac{1}{3}$ d 3 e $\frac{1}{8}$
f $\frac{1}{81}$ g $\frac{1}{6}$ h $\frac{1}{2}$ i 4 j 8
4 ai 5×10^2 aii 2.14×10^3 aiii 1.2×10^5
aiv 8.953×10^2
bi 4200 bii 502 000 biii 7 000 000
biv 512.5
5 a 6×10^{12} b 2×10^7 c 1.5×10^{11} d 8×10^2

N5 HW4

1 a 3.3×10^5 b 1.98×10^{30}
2 a 9 000 000 b 4.5×10^5 c 20 people/km^2
3 ai a^7 aii x^{-4} aiii y^2
bi 7 bii $\frac{1}{3}$
4 aii $12a^5b^5$ bi 81 bii 7 biii 6

A5 HW1

1 ai 54 aii 144 aiii 96 aiv 198
bi 68% bii 95% biii 56.5% biv $46\frac{2}{3}$%
2 ai $u_n = 2n + 7$
2 aii $u_n = 20 - 3n$ aiii $u_n = \frac{n}{4} + 3$
b $u_n = n^2 - 3n$
3 a Flag with vertices at (1, −1), (4, −1), (3, −2), (3, −1) labeled B
b Flag with vertices at (−1, −1), (−4, −1), (−3, −2), (−3, −1) labeled C
c Reflection in the line $y = -x$
4

Starter	Main
Soup	Lasagne
Soup	Rack of lamb
Soup	Roast chicken
Soup	Beef en croute
Paté	Lasagne
Paté	Rack of lamb
Paté	Roast chicken
Paté	Beef en croute
Brioche	Lasagne
Brioche	Rack of lamb
Brioche	Roast chicken
Brioche	Beef en croute

A5 HW2

1 Identity, equation, identity, formula, equation, formula
2 a 48 cm^2 b 8 cm
3 a 80 km/h b 80 km
4 a $P = 45 + 30h$ b $C = 4 + 12t$

A5 HW3

1 a $x = a + b$ b $x = \frac{q}{p}$ c $x = \frac{A - r}{y}$
d $x = \frac{v^2 + w^2}{t}$ e $x = \frac{p}{q + r}$ f $x = (\frac{b}{a})^{\frac{1}{2}}$
g $x = \frac{d + h}{c}$ h $x = (\frac{qr}{p})^{\frac{1}{3}}$ i $x = \frac{n^2}{p - q}$
2 a $A = 5 + 0.5c$ b £8.50 c $c = 2(A - 5)$
d 10 chores
3 a $y = a - b$ b $y = \frac{p - r}{q}$ c $y = \frac{v^2 - x^2}{w}$
d $y = \frac{k}{t}$ e $y = \frac{b - d}{h}$ f $y = \frac{q^2}{p^2}$
4 iv, ii, i, iii

A5 HW4

1 a $(2n + 1)(2m) = 2(m(2n + 1))$ which is even
b $(2n)^2 = 4n^2$ which is divisible by 4
2 a 17.8776
b He has done 3.15 + 7.1, then multiplied by 4.21
3 a Pat; first square 4, then multiply by 3
b 27
4 $x = \frac{k}{q - p}$
5 $y = \frac{w - kt}{x - k}$

S4 HW1

1 a 0.12 b 0.03 c 0.84 d 3.1
e 15 f 0.525 g 36.5 h 230
2 a $y = 3x + 4$ b $y = 3 - 2x$
3 a $a = b = 77°$ b $x = 60°$ $y = 120°$

4 a

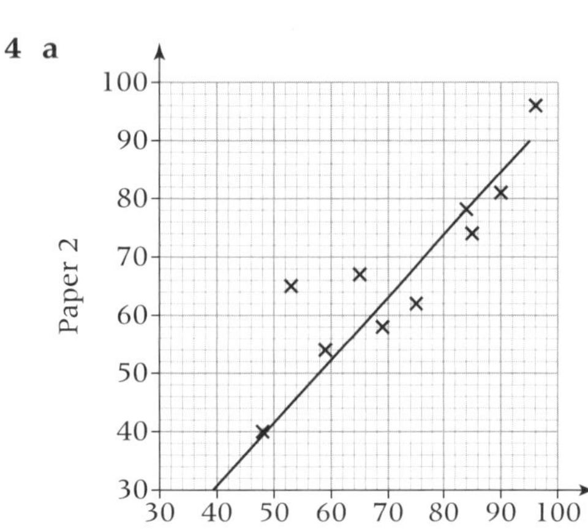

b 66

S4 HW2

1 a Congruent – SSS b Not congruent – only similar
c Congruent – RHS d Congruent – ASA
e Not congruent – SSA
2 a Vertical opposite angles b Alternate angles
c SAS
3 a 38° b 52°

S4 HW3

1 b

a	b	c	Check
3	4	5	9 + 16 = 25
5	12	13	25 + 144 = 169
7	24	25	49 + 576 = 625
9	40	41	81 + 1600 = 1681
11	60	61	121 + 3600 = 3721

c Next two Pythagorean triples are 13, 84, 85 and 15, 112, 113
2 a 10.8 cm (3 s.f.) b 11.5 mm (3 s.f.)
c 22.3 m (3 s.f.)
3 a 5.66 cm (3 s.f.) b 8.28 cm (3 s.f.)

S4 HW4

1 90 cm^2
2 b AB = 5 units BC = 4.47 units (3 s.f.)
AC = 7.28 units (3 s.f.)
3 $96^2 + 28^2 = 100^2$
4 Hypotenuse is 17.2 cm (3 s.f.). No

N6 HW1

1 a 1 b 500 c 256 d 7
e 6 f $\frac{1}{5}$ g $\frac{1}{2}$ h 8
2 a 7, 9, 11, 13, 15 b 11, 10, 9, 8, 7
c 6, 11, 18, 27, 38 d 2, 9, 28, 65, 126
e $\frac{1}{2}, \frac{1}{6}, \frac{1}{12}, \frac{1}{20}, \frac{1}{30}$
3 a 15.8 cm (3 s.f.) b 10 mm
4 a

0	8 9
1	2 5 7
2	1 5 6 7 8
3	0 1 2 3 6 9
4	0 1 4 5 5

Key: 0 | 9 = 9 seconds

b Negatively skewed
ci 30 cii 19

N6 HW2

1 a 77 b 9 c 144 d 39
e 96 f 41 g 10 h 8
2 a $12 \times (3 + 4) = 84$ b $24 \div 6 \times 2 = 8$
c $5 \times (4 + 2) \times 7 = 210$ d $40 \div (4 + 2^2) = 5$
e $5 \times 4^2 \div 8 = 10$ f $15^2 - 10 \times 5 = 175$
g $\frac{8^2 \div (2^3 \times 4)}{2} = 1$
h $(150 - (7^2 - 4 \times 5))^{1/2} = 11$
3 a 25π cm^2 b 288π m^3 c $\frac{6}{\pi}$ mm
4 a 3 b 8 c 4 d 10
5 a $2\sqrt{3}$ b $3\sqrt{2}$ c $3\sqrt{5}$ d $3\sqrt{21}$

N6 HW3

1

	The result is larger, smaller or the same size?
×1.4	larger
÷0.2	larger
×0.8	smaller
÷1.75	smaller
×1	same

2 a True b True
3 a 0.6 b 0.3 c 40 d 300
e 0.06 f 5 g 0.0132 h 150
4 a 64.2 b 14.19 c 91.86
d 8.52 e 2.669 f 7.816
5 a 206 b 3.25 c 6.24
d 56 e 124.16 f 32.8

N6 HW4

1 a 3.46 b 4.48 c 6.72 d 3.8752
2 a 5.75 b 5.75 c 46 d 12.5
3 $2\sqrt{5}$
4 0.9
5 8.91 $g/cm^3 \leqslant$ density $\leqslant 9.06$ g/cm^3 (3 s.f.)

A6 HW1

1 a 28 b 10 c 15 d 55
e 48 f 37 g 2 h 13
2 ai (2, 5) aii (3, −4) aiii $(\frac{2}{3}, -1)$
b $y = 3 - 2x$

x	−2	0	2
y	7	3	−1

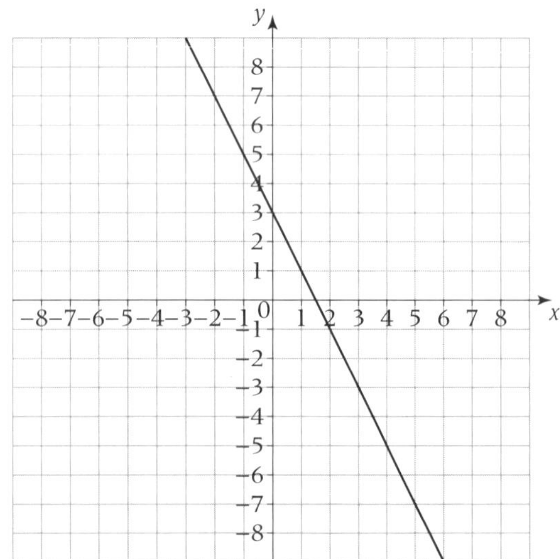

3 a 4.12 units (3 s.f.) b 6.71 units (3 s.f.)
c 5.39 units (3 s.f.)
4 a $\frac{3}{10}$ b 240

A6 HW2

1 a $x = 2$ b $y = 4$ c $p = -2$ d $q = -1$
 e $a = 3$ f $x = 5$ g $x = -1$ h $y = -\frac{1}{2}$

2 Square has side 6 units. Rectangle has length 9 units and width 3 units.

3 a $x = -4$ or -2 b $x = -8$ or -3
 c $x = 1$ or -6 d $x = 2$ or -8
 e $x = 4$ or 5 f $x = 7$ or 1
 g $x = 0$ or 4 h $x = 0$ or -6

4 Solutions are 7, 12 and −7, −12

A6 HW3

1 $x(x + 5) = 33.44$
 Length = 8.8 cm, width = 3.8 cm

2 5.33 (2 d.p.)

3 a $x = 1, y = 3$ b $x = 2, y = -1$
 c $p = 4, q = -2$ d $a = 8, b = 0$

4 Latte is £1.50, carrot cake is £0.95

A6 HW4

1 a $x = 3, y = 2$ b $x = 2, y = -1$
 c $a = 5, b = -2$

2 a $a = 5$ b $b = 7$ c $c = -3\frac{1}{3}$

3 ai $5x^2 + 8x - 4$ aii $4x^2 - 20x + 25$
 b $x = 3$ or -8

4 b 3.8 (1 d.p.)

D4 HW1

1 a $8 + 7 \times 3 = 29$ b $12 \div (5 - 1) = 3$
 c $3 \times (9 - 2) \times 4 = 84$ d $12^2 \times (5 - 3) = 288$
 e $(24 \div 4 - 2)^{\frac{1}{2}} = 2$

2 a $x = -3$ or -7 b $x = 5$ or -9
 c $x = 9$ or -4 d $y = 0$ or 5
 e $t = 0$ or -4

3 a Congruent AAS
 b Not congruent, only similar

4 a

	Red	Blue	Yellow	Total
Girls	5	4	5	14
Boys	6	8	2	16
Total	11	12	7	30

 bi $\frac{2}{5}$ bii $\frac{1}{5}$
 c 49

D4 HW2

1 a 7
 bi 7 bii 4

2 ai 2.43 aii 2.5 aiii 3
 b 5

3 a 24.9 b 1 – 10 c 21 – 30

D4 HW3

1 a

Height	Frequency
$0 < h \leqslant 2$	2
$2 < h \leqslant 4$	7
$4 < h \leqslant 6$	15
$6 < h \leqslant 8$	6
Total	30

 b

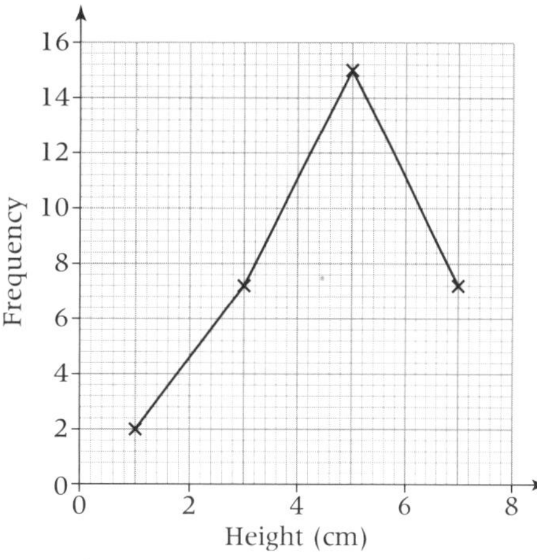

2 a

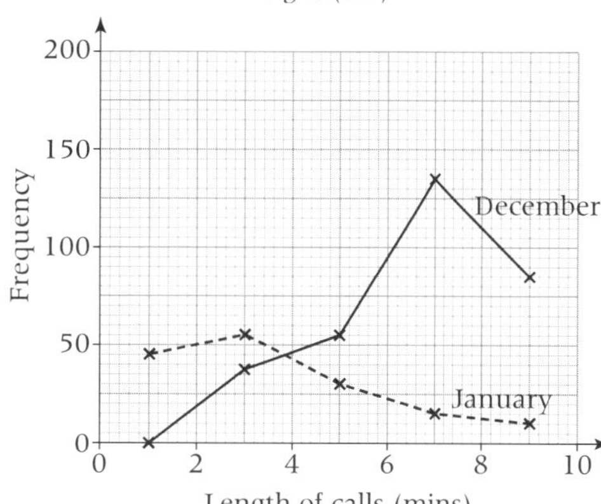

 bi December modal class is $6 < m \leqslant 8$
 January modal class is $2 < m \leqslant 4$
 bii December range is 8
 January range is 10
 c Calls longer in December – due to Christmas mail orders?

3

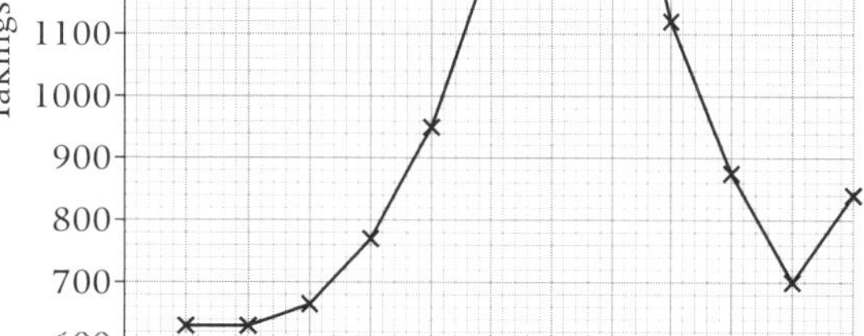

D4 HW4

1 4.09

2 a

 b 199, 189, 206, 275

3 a 270 b $200 < c \leqslant 300$ c No

S5 HW1

1 a $2\sqrt{3}$ b 5 c $5\sqrt{2}$
d $2\sqrt{5}$ e 9 f $27 + 10\sqrt{2}$

2 a $x = b - a$ b $x = \frac{z}{y}$
c $x = \frac{r^2 - q}{p}$ d $x = (\frac{q}{y - p})^{\frac{1}{2}}$
e $x = t - w$ f $x = abc$
g $x = \frac{a}{p - q}$ h $x = h - \frac{m}{r}$

3 a 200 cm^2 b 14.1 cm (3 s.f.)

4 a 2.28 b 1

S5 HW2

1 a b 292°

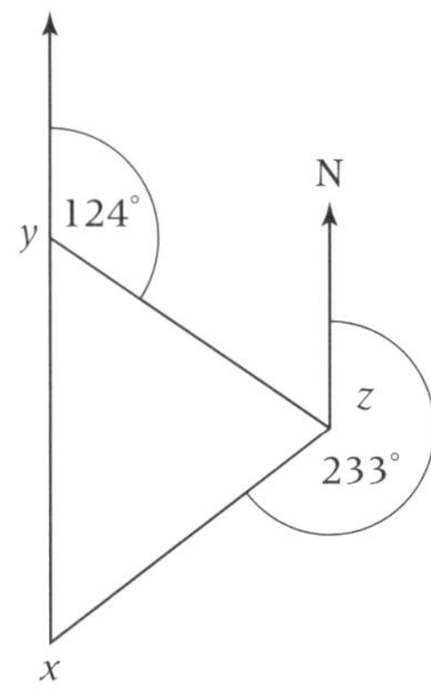

2 b 5.0 cm (1 d.p.)

3 a b 56°, 71°, 53° c 053°

4 b Right-angled c $6^2 + 8^2 = 10^2$

5 False: The sides 5 cm and 6 cm sum to 11 cm, which is the length of the third side, hence will only join as a straight line.

S5 HW3

1 c 45°

3 e Hexagon

S5 HW4

1 a perpendicular bisector of line AB
b circle, radius 3 cm

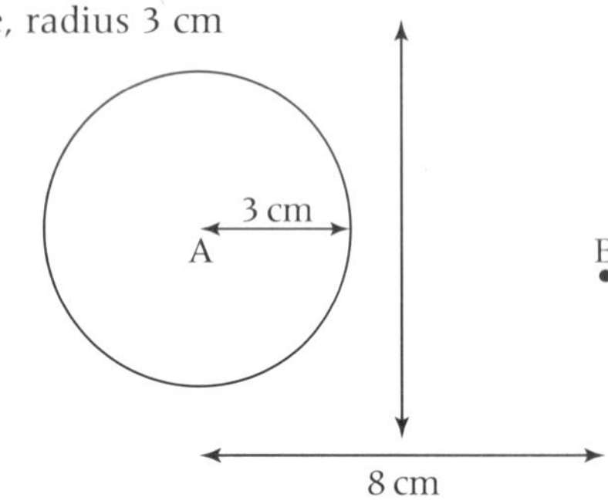

2 a, b

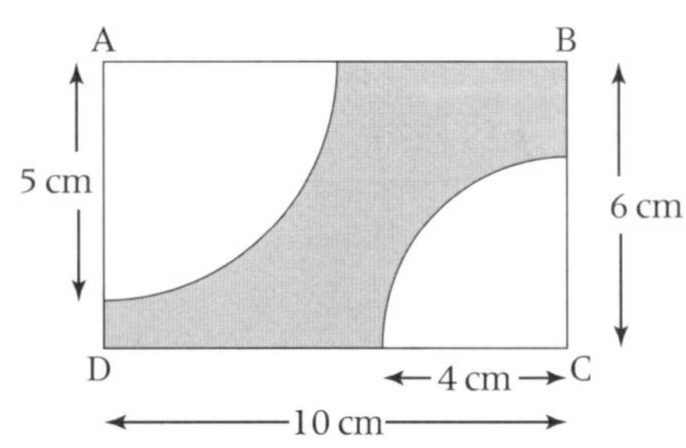

3 b

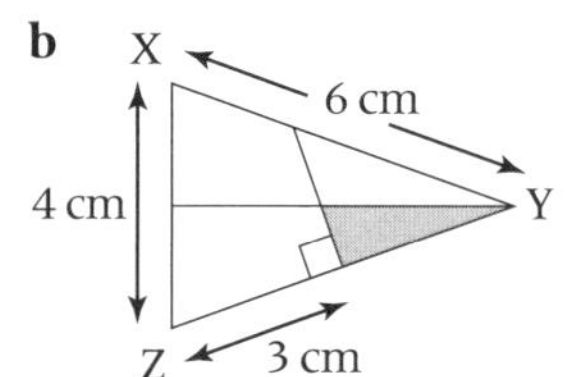

N7 HW1

1 a 5^6 b 3^3 c a^{11} d $8^0 = 1$
e b^{-5} f $4^1 = 4$ g 7^{12} h 2^{11}

2 $u_n = 5n + 7$

3 a b 290° c 110°, 30°, 40°

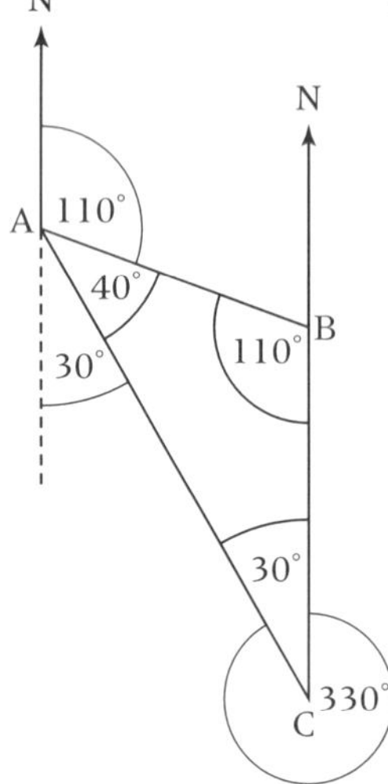

4 a

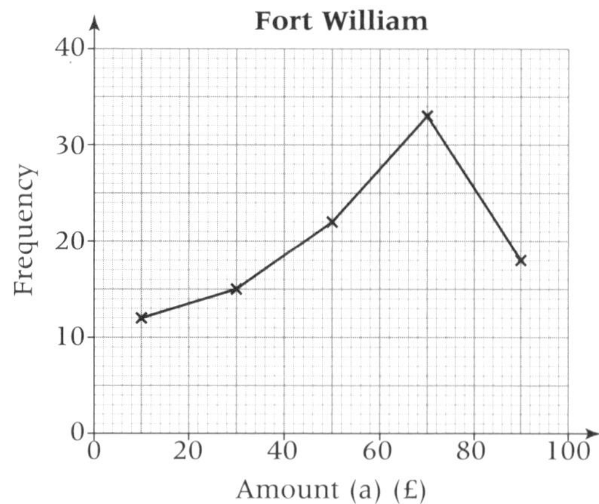

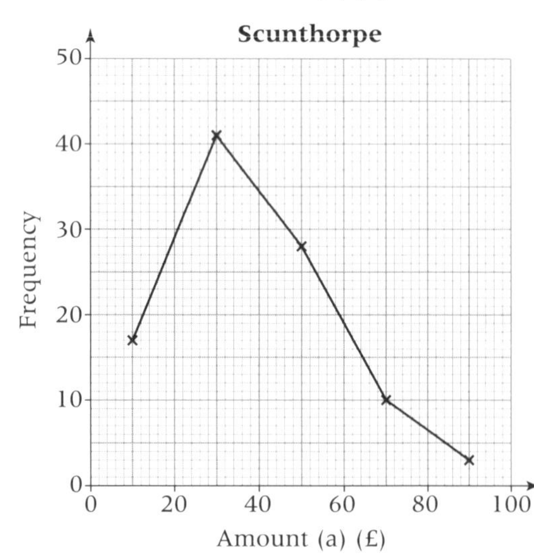

b Modal class for Fort William is $60 \leqslant a < 80$ and for Scunthorpe is $20 \leqslant a < 40$, where a is the amount spent. Modal amount spent in Fort William is higher, perhaps due to Ben Nevis being close by etc.

N7 HW2

1 **a** 35 m **b** 36 g **c** 39 cm **d** 22 litres
e 60 kg **f** 73.5 mm **g** $9\frac{1}{3}$ km **h** $16\frac{4}{5}$ cl

2 **a** 60p **b** £1.15

3 **a** £7 **b** $93 **c** €75 **d** £78
e £143 **f** $1022 **g** €224 **h** £2262

4 True

N7 HW3

1

Percentage change	Multiply by ...
Increase of 5%	1.05
Decrease of 12%	0.88
Increase of 75%	1.75
Decrease of 55 %	0.45
Increase of 150%	2.5

2 **a** £43.92 **b** 18.6m **c** $19.32
d 72.8 kg **e** 235.52 litres **f** 618.8 km

3 **a** £170 **b** £136 **c** 0.68 **d** 0.32 or 32%

4 Option 2

N7 HW4

1 **a** 660 **b** 14%

2 £795

3 Lynda

4 $\frac{13}{24}$

D5 HW1

1 $3\sqrt{3}$

2 **a** $p = 3$, $q = 2$ **b** $a = 2$, $b = -1$
c $x = 5$, $y = 1$

3 **b** Isosceles **c** 4.58 cm (3 s.f.)

4 **a**

Mass, *m*, kg	Cumulative frequency
$2 < m \leqslant 2.5$	1
$2.5 < m \leqslant 3$	4
$3 < m \leqslant 3.5$	7
$3.5 < m \leqslant 4$	10
$4 < m \leqslant 4.5$	3

b

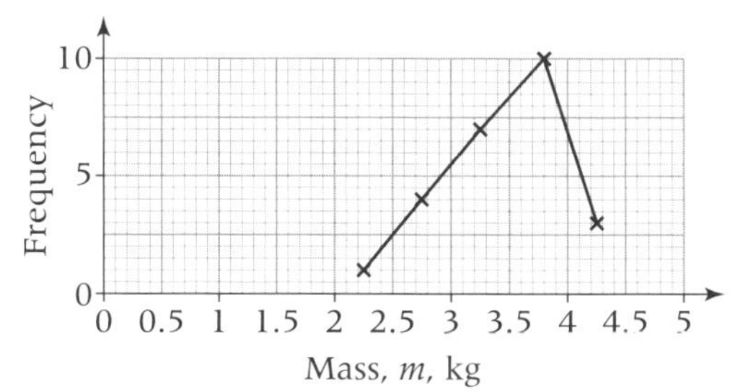

D5 HW2

1 **a**

S	Cumulative frequency
⩽750	12
⩽1000	34
⩽1250	70
⩽1500	88
⩽1750	96
⩽2000	100

b

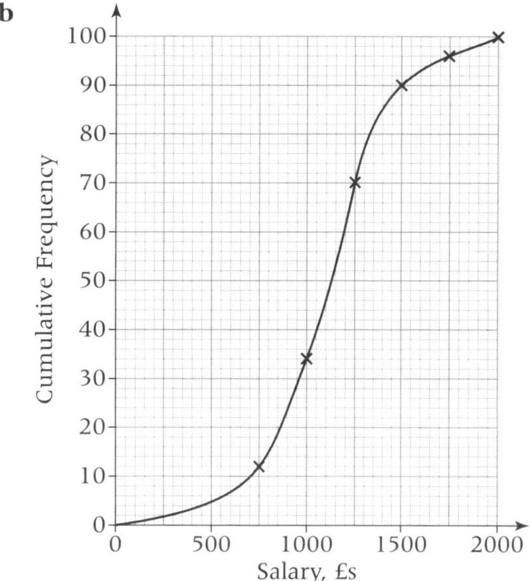

c $1000 < S \leqslant 1250$

2 **a**

Number of minutes, *m*	Cumulative frequency
⩽15	5
⩽30	14
⩽45	26
⩽60	34
⩽75	37
⩽90	42

b

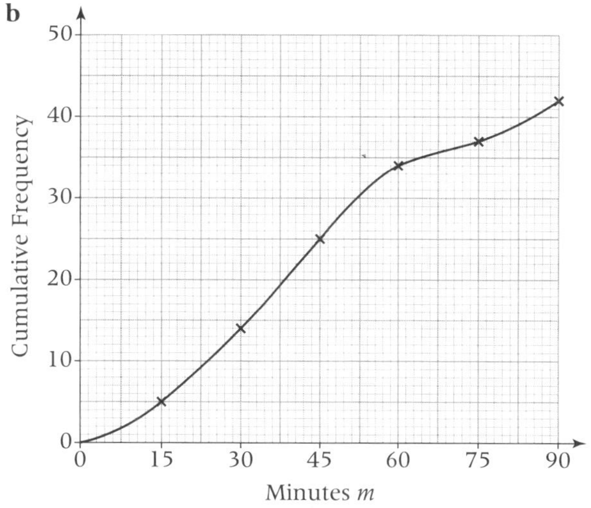

ci 38 **cii** 28

3 **a**

Cumulative Frequency

Test results, *t*%

Test results, *t*%	Cumulative frequency
⩽50	9
⩽60	27
⩽70	55
⩽80	99
⩽90	114
⩽100	120

bi 71 **bii** 5

3 **c** 104

D5 HW3

1 Median yield for Legend = 25 tomatoes
Median yield for Beefsteak = 43 tomatoes, so median yield for beefsteak is larger.
LQ Legend = 18 UQ Legend = 33 IQR Legend = 15
LQ Beefsteak = 35 UQ Beefsteak = 48
IQR Beefsteak = 13
So similar spread. Beefsteak tomato plants generally yield more tomatoes per plant.

2 **a** See above. **b** See above.

c

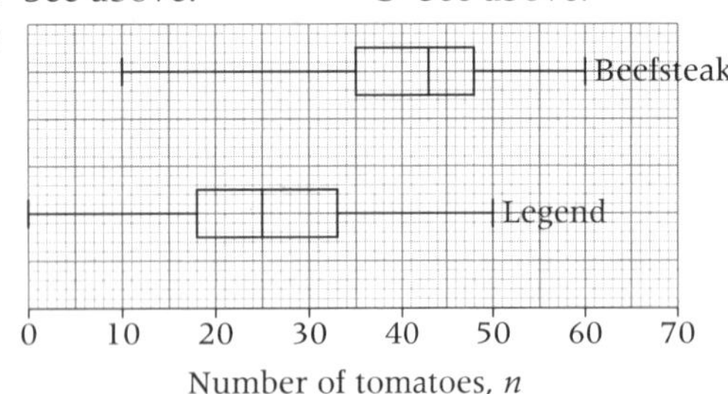

3 **a**

Legend

Mass, m (g)	Frequency	Cumulative Frequency
$200 < m \leqslant 210$	15	15
$210 < m \leqslant 220$	22	37
$220 < m \leqslant 230$	40	77
$230 < m \leqslant 240$	14	91
$240 < m \leqslant 250$	9	100

Beefsteak

Mass, m (g)	Frequency	Cumulative Frequency
$200 < m \leqslant 210$	8	8
$210 < m \leqslant 220$	12	20
$220 < m \leqslant 230$	19	39
$230 < m \leqslant 240$	36	75
$240 < m \leqslant 250$	25	100

b

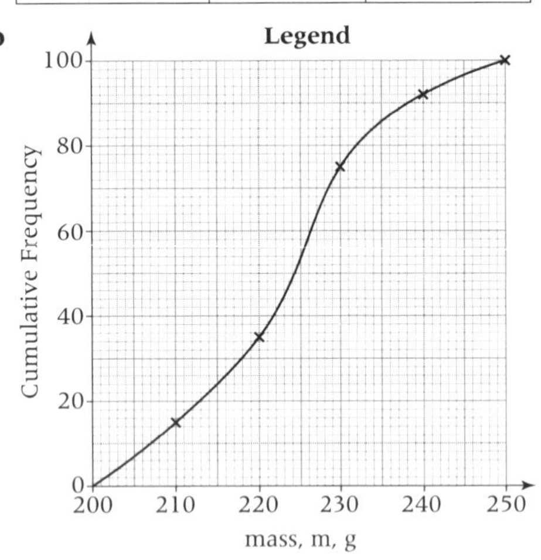

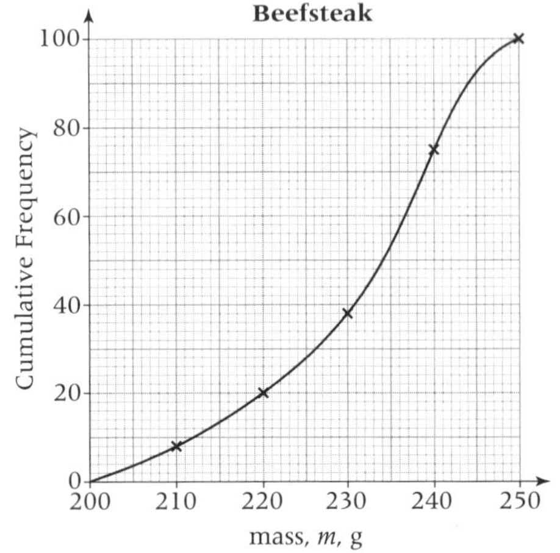

ci Legend median = 222 g
Beefsteak median = 233 g

cii LQ Legend = 216 g UQ Legend = 228 g
LQ Beefsteak = 223 g UQ Beefsteak = 240 g

d

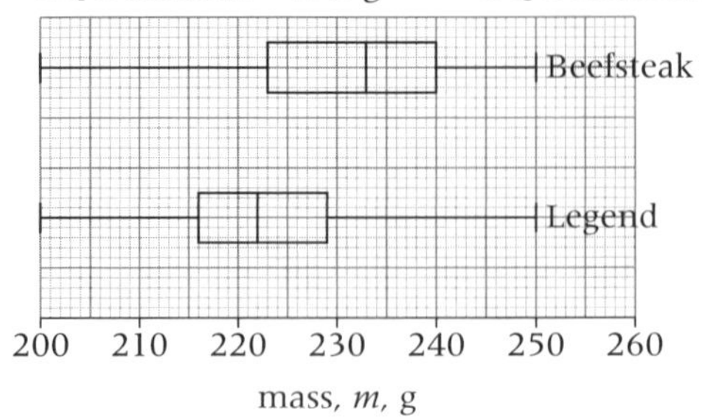

D5 HW4

1 **a** 64 **b** 41% **c** 20

2

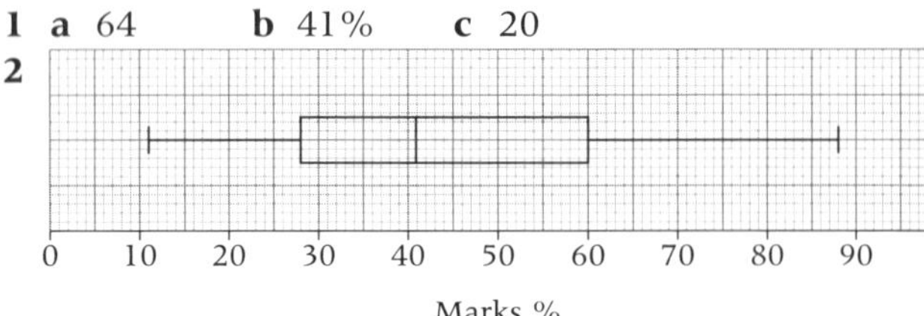

3 **a** 30 secs

b

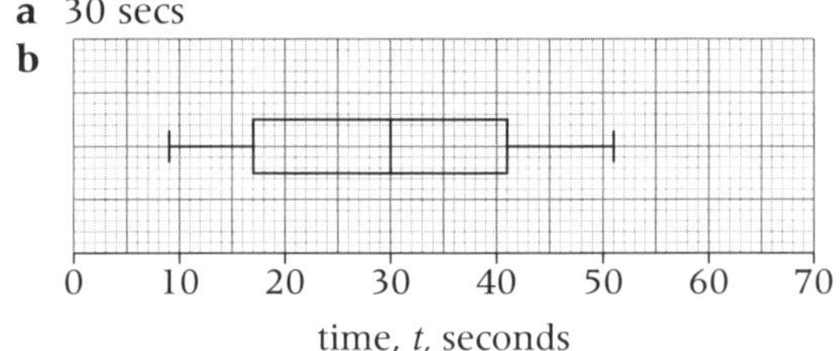

c Boys have lower median and wider distribution.

S6 HW1

1 **a** £160 800 **b** 130%

2 **a** $x = -3$ or -4 **b** $x = -2$ or -7
c $x = 1$ or -5 **d** $x = 4$ or 10
e $x = 8$ or -3 **f** $x = 0$ or 9

3

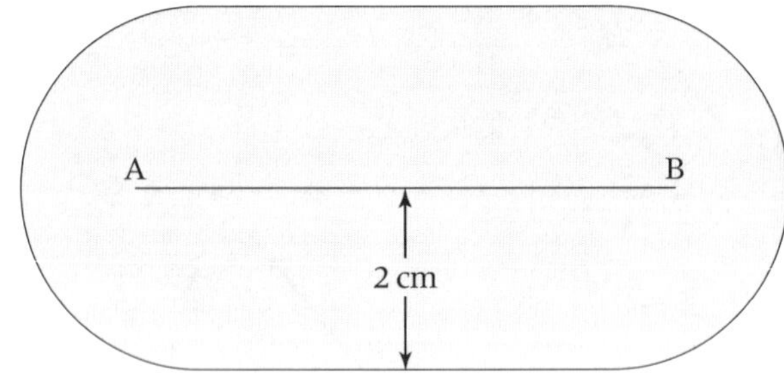

4 **a**

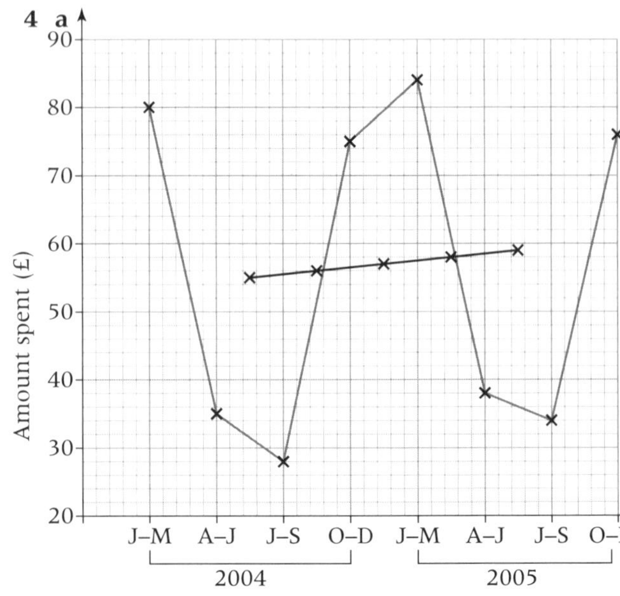

b 54.5, 55.5, 56.25, 57.75, 58
c Increasing

S6 HW2

1 **a**

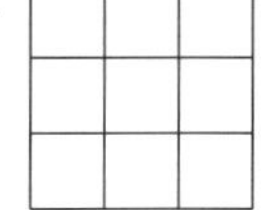

b **c**

2

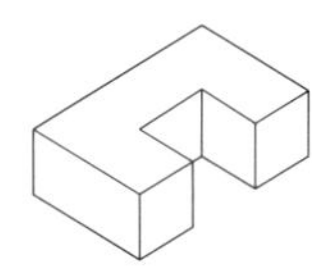

3 390 cm³
4 **a** 1131 cm³ **b** 124.32 m³

S6 HW3

1 **a** 343 cm³ **b** 64 cm³
2 247 cm³ (3 s.f.)
3 12
4 **a** 35 m **b** 18 400 cm **c** 10 mm
d 5000 cm **e** 150 000 cm² **f** 84 cm²
g 2500 mm³ **h** 0.8 m³
5 **a** Volume **b** Area **c** None
d Volume **e** Length **f** None
g Area **h** Volume

S6 HW4

1 Dimension theory says its an area.
2 125 cm³
3 95 000 cm³
4 10.8 g (1 d.p.)
5 $23\frac{1}{3}$ km

A7 HW1

1 373 g cod, 250 g wild Alaskan salmon, 625 g mashed potatoes, 3 eggs, 200 g breadcrumbs
2 **a** $x = 5, y = 1$ **b** $x = 2, y = -1$ **c** $x = 4, y = 2$
4 **a**

Mid-point	Frequency × Mid-point
7.5	22.5
12.5	100
17.5	210
22.5	112.5
27.5	55

Mean = $16\frac{2}{3}$

bi $15 \leqslant t < 20$ **bii** $15 \leqslant t < 20$

A7 HW2

1 **a**

x	−3	−2	−1	0	1	2	3
x^3	9	4	1	0	1	4	9
$2x^2$	−6	−4	−2	0	2	4	6
y	3	0	−1	0	3	8	15

b

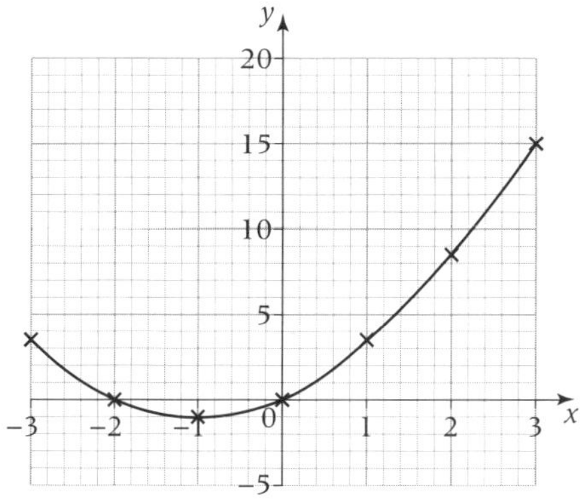

c (−1, −1)

2 $y = 2x^2 + 1$, $y = x^2 + 4x - 3$
3 **a**

x	−3	−2	−1	0	1	2
x^3	−27	−8	−1	0	1	8
$2x^2$	18	8	2	0	2	8
y	9	0	1	0	3	16

b

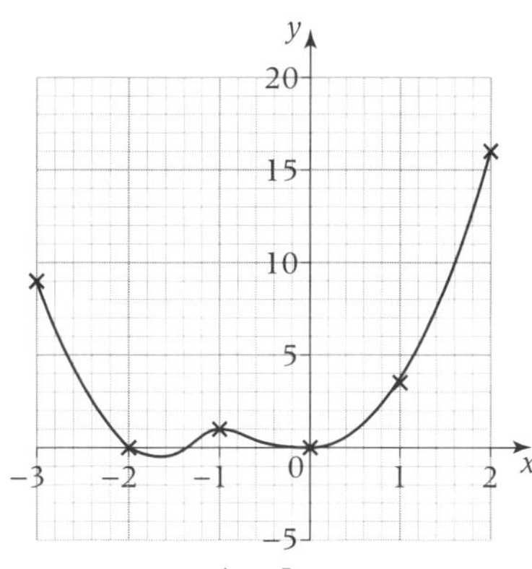

c (0, 0) and $(-\frac{4}{3}, 1\frac{5}{27})$

A7 HW3

1 **a**

x	−1	0	1
y	−4	−1	2

b

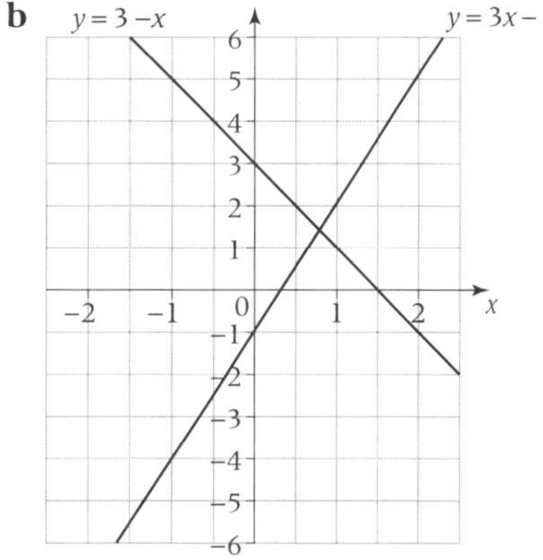

d $x = 1, y = 2$
2 $x = 1, y = -1$
3 Both have a gradient of 4, so the lines are parallel, and so never meet.
4 **a**

x	−5	−4	−3	−2	−1	0	1	2
x^2	25	16	9	4	1	0	1	4
$3x$	−15	−12	−9	−6	−3	0	3	6
-4	−4	−4	−4	−4	−4	−4	−4	−4
y	6	0	−4	−6	−6	−4	0	6

bi $x = -4$ or 1
bii $x = -1$ or -2
biii $x = 1.54$ (2 d.p.) or -4.54 (2 d.p.)
biv $x = 1$ or -5
bv $x = -1$ (double root

A7 HW4

1 $(\frac{1}{2}, -2\frac{1}{4})$
2 (0, 0), (3, 0), (−2, 0)
3 **a** $y = 5$ **b** $y = 0$ **c** $y = 3 - x$
d $y = -2$ **e** $y = 2x + 11$ **f** $y = 4 - 2x$
4 4 and 1

N8 HW1

1 Lower bound is 11.25 cm^2, upper bound is 19.16 cm^2

2 a

x	−2	−1	0	1	2	3	4
x^2	4	1	0	1	4	9	16
$-2x$	4	2	0	−2	−4	−6	−8
3	3	3	3	3	3	3	3
y	5	0	−3	−4	−3	0	5

bi $x = 3$ or -1
bii $x = 0$ or 2
biii $x = 3.24$ (2 d.p.) or -1.24 (2 d.p.)
biv $x = 4$ or -1

3

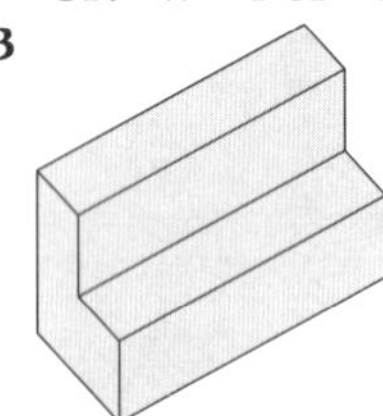

4 a $6 < t \leqslant 9$ **b** 8 hours

N8 HW2

1 a 2 : 1 **b** 10 : 1 **c** 1 : 10 000 **d** 3 : 2
e 3 : 5 : 2 **f** 1 : 20 **g** 5 : 3 **h** 25 : 56
i 1 : 3 **j** 8 : 3 : 5

2 a 1 : 4 **b** 12 : 1 **c** 5 : 2 **d** 14 : 12

3 a $1 : 1\frac{2}{3}$ **b** $1 : \frac{4}{15}$ **c** 1 : 500 000
d 1 : 400 000

4 a £320 and £160 **b** £80 and £400
c £300 and £180 **d** £280 and £ 200

5 a 3 : 2
b Kevin wins £900 and Kathleen wins £600

N8 HW3

1 a $\frac{4}{5}$ **b** $\frac{3}{8}$ **c** $\frac{x}{3}$ **d** $\frac{7}{y}$

2 a $x = 3$ **b** $y = 1.5$ **c** $a = 7.2$

3 3 km

4 a Archie gets £225, Susie gets £200 and Josie gets £125.
b Archie gets £220, Susie gets £198 and Josie gets £132.
c They will eventually each receive the same amount. (Except they would not live that long!)

N8 HW4

1 12

2 a Lois has £500 and Alice has £125
b Lois = £320, Sam = £160, Alice = £80

3 £5624.32

4 £55

5 59.7 million

S7 HW1

1 a LB = 11.5 m UB = 12.5 m
b LB = 14.75 s UB = 14.85 s
c LB = 1150 g UB = 1250 g
d LB = 24.45 kg UB = 24.55 kg
e LB = 3.045 l UB = 3.055 l
f LB = 4.385 m UB = 4.395 m

2 a

x	−2	−1	0	1	2	3	4
x^2	4	1	0	1	4	9	16
$-x$	2	1	0	−1	−2	−3	−4
−4	−4	−4	−4	−4	−4	−4	−4
y	2	−2	−4	−4	−2	2	8

b

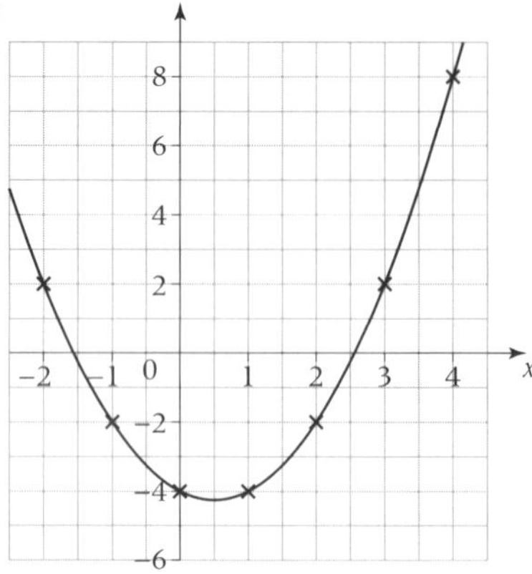

c $(\frac{1}{2}, -4\frac{1}{4})$

3 1800 cm^3

4 0.35

S7 HW2

1 a Rectangle with vertices (3, 3), (6, 3), (6, 9), (3, 9) labeled B
bi 6 units **bii** 18 units
c 3
di 2 units^2 **dii** 18 units^2
e $9 = 3^2$

2 a Triangle with vertices (1, 2), (5, 0), (1, −6) labeled Y
bi 4 units^2 **bii** 16 units^2
c $4 = 2^2$
d Triangle with vertices (−1, 2), (−1, 0), (0, 1.5) labeled Z
ei 4 units^2 **eii** 1 unit^2
f $\frac{1}{4} = (\frac{1}{2})^2$

S7 HW3

1 a Enlargement, scale factor 3, centre (2, 0)
b Enlargement, scale factor $\frac{1}{3}$, centre (2, 0)

2 a Enlargement, scale factor $\frac{2}{3}$, centre (3, −6)
b Enlargement, scale factor $1\frac{1}{2}$, centre (3, −6)

3 a $x = 6$ cm $y = 18$ cm
b Small trapezium has an area of 40 cm^2. Large trapezium has an area of 129.6 cm^2.

4 117.6 cm^2

S7 HW4

1 a 12.5 cm **b** 10 cm

2 $\frac{140}{102} = 1.3725$; $\frac{226}{170} = 1.3294$; Sides not in same ratio, so not similar.

3 Kite with vertices (2, 4), (4, 6), (6, 4), (4, −2)

4 Enlargement, scale factor $\frac{1}{2}$, centre (0, −1)

D6 HW1

1 ai 1 : 1000 **aii** 1 : 12 **aiii** 5 : 2 : 3 **aiv** 3 : 8
bi $a = 2$ **bii** $b = 1\frac{2}{3}$ **biii** $x = 4\frac{3}{8}$

2 a $x = 1$ **b** $x = 6$ **c** $p = -1$
d $y = 1.5$ **e** $a = 3$ **f** $t = -3\frac{1}{3}$

3 $x = 5$ cm Area = 62.5 cm^2

4 a

Number of mins, m	Cumulative frequency
≤1	1
≤2	7
≤3	17
≤4	24
≤5	30

b

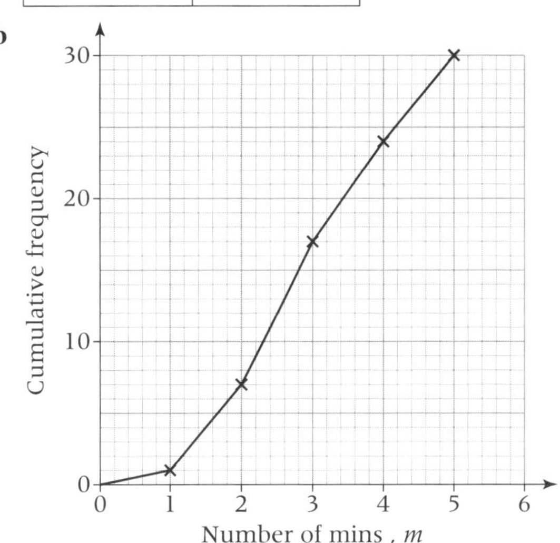

ci 2.7 mins ≈ 2 mins 40 secs
cii 1.7 mins ≈ 1 min 40 secs

6 HW2

1 a $\frac{5}{8}$ **b** 25

2 a $\frac{1}{6}$ **b** $\frac{7}{12}$ **c** $\frac{2}{3}$ **d** $\frac{1}{3}$

3 a

		Dice					
		1	2	3	4	5	6
Spinner	1	2	3	4	5	6	7
	2	3	4	5	6	7	8
	3	4	5	6	7	8	9
	4	5	6	7	8	9	10

bi $\frac{1}{24}$ **bii** $\frac{1}{6}$ **biii** $\frac{5}{6}$ **biv** $\frac{1}{8}$

4 a $\frac{1}{4}$ **b** $\frac{1}{10}$

c Independent – one does not affect the other.

6 HW3

1 a

$\frac{7}{10}$ diamond — $\frac{7}{10}$ diamond, $\frac{3}{10}$ ruby
$\frac{3}{10}$ ruby — $\frac{7}{10}$ diamond, $\frac{3}{10}$ ruby

bi 0.49 **bii** 0.09 **biii** 0.21 **biv** 0.42

2 a

$\frac{6}{10} = \frac{3}{5}$ magenta — $\frac{5}{9}$ magenta, $\frac{4}{9}$ turquoise
$\frac{4}{10} = \frac{2}{5}$ turquoise — $\frac{6}{9} = \frac{1}{3}$ magenta, $\frac{3}{9} = \frac{1}{3}$ turquoise

bi $\frac{1}{3}$ **bii** $\frac{2}{15}$ **biii** $\frac{4}{15}$ **biv** $\frac{8}{15}$

6 HW4

1 a HHH, THH, HTH, HHT, TTH, THT, HTT, TTT
bi $\frac{1}{3}$ **bii** $\frac{3}{8}$

2 a No – would expect to obtain 250 heads, so unlikely coin is fair.

b

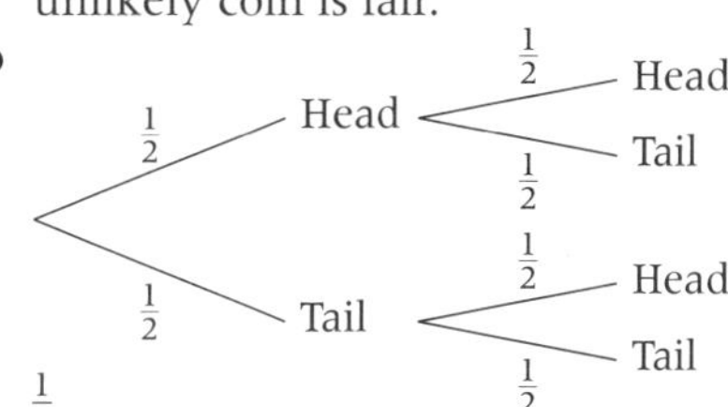

c $\frac{1}{2}$

3 a

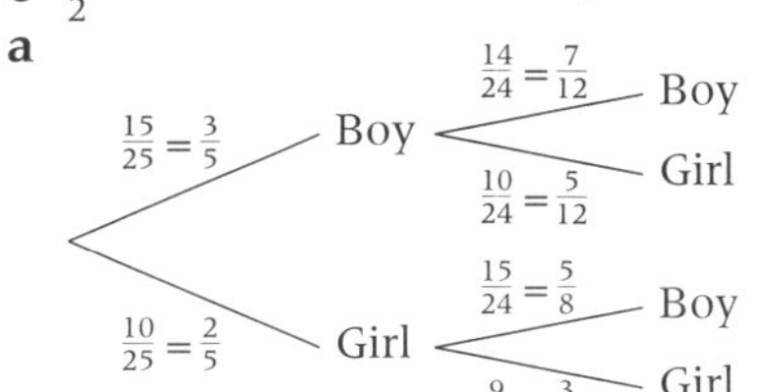

b $\frac{7}{20}$

A8 HW1

1 a £100 and £150
bi $y = \frac{3}{2}x$ **bii** 7.5 **biii** 8

2 4.20 (2 d.p.)

3 a 125 cm^3 **b** 512 cm^3

4 a

	1	2	3	4	5	6
H						
T						

bi $\frac{1}{12}$ **bii** $\frac{1}{4}$

A8 HW2

1 a

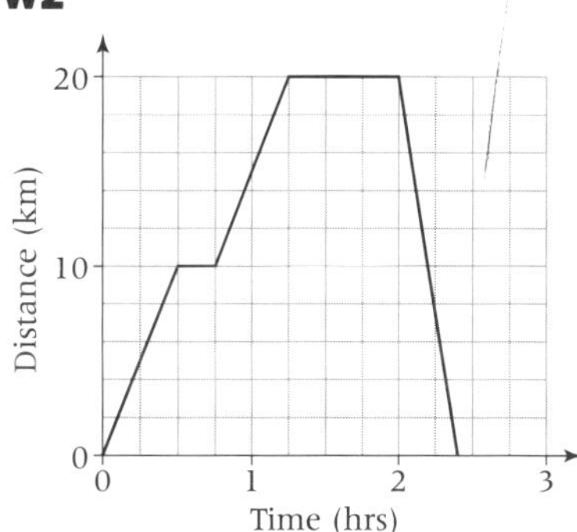

b 10 km
ci 48 km/h **cii** 30 mph

2 a

Joe
Finlay
Distance (km)
Time
1 pm
2 pm
3 pm
4 pm

b 24 km/h **c** 1.48 pm

3

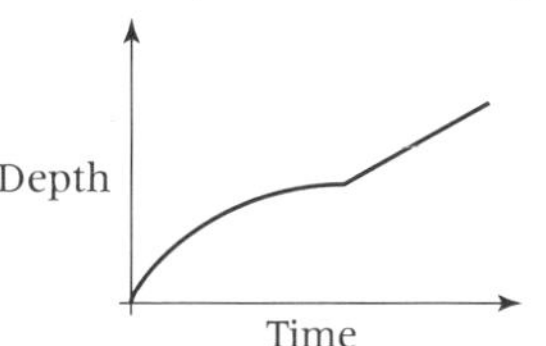

4

A8 HW3

1 **ai** \$120 **aii** £40
 b Julien
 c $y = 2.4x$, where y is Australian dollars and x is British Pounds

2 **a** £90

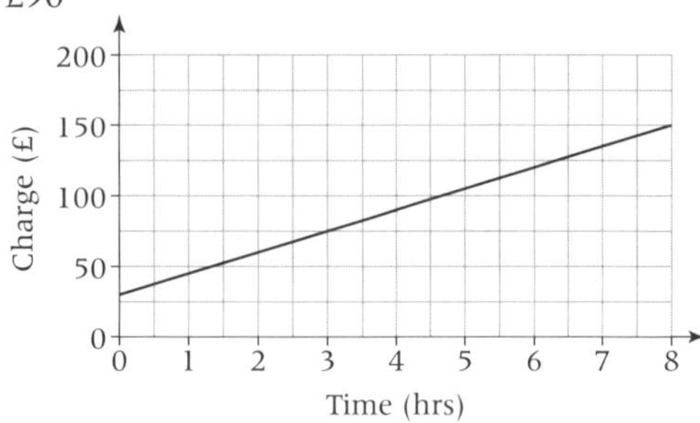

 b 5.5 hours
 c $C = 15h + 30$, where C is cost, and h is hours on the job

3 $A = 0.2d + 1$, where A is amount reimbursed and d is distance travelled
Staff are paid 20p per km plus £1.
Limitation is that staff can claim £1 for travelling less than an kilometer, or not traveling anywhere at all!

A8 HW4

1 **a** iii **b** i **c** ii

2 **a** Called at service station/took a break
 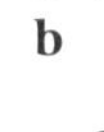
 b
 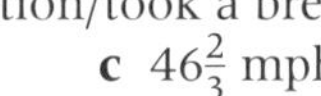
 c $46\frac{2}{3}$ mph

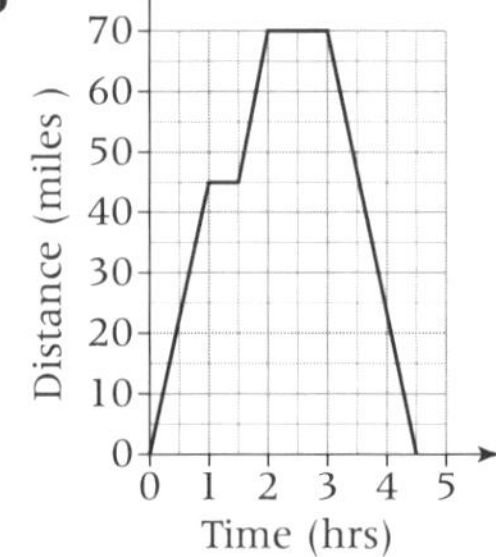

3 **a** $A = x(5 - x)$

 b

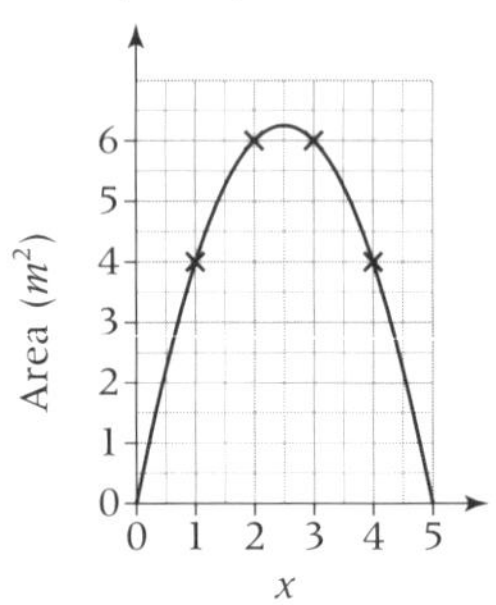

 ci $6.25\ \text{m}^2$ **cii** 4 m by 1 m

S8 HW1

1 **ai** £96 **aii** £36.40
 bi £50 **bii** £39.95 **biii** £89.99

2 **a**

x	0	1	2
y	5	3	1

 b, c

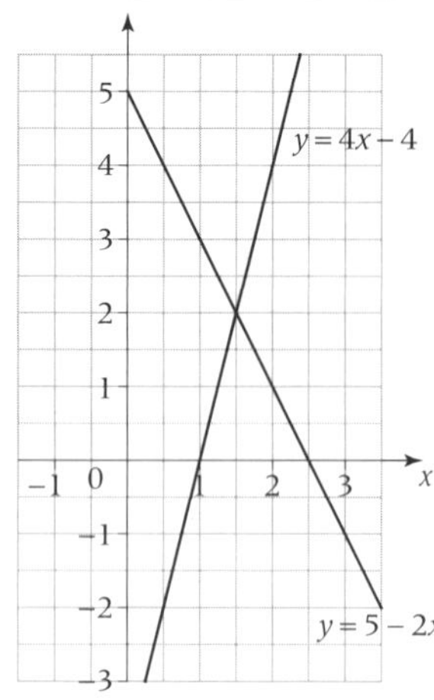

 d $x = 1.5$, $y = 2$

3 The expressions that represent area are $\frac{1}{4}\pi rh$ and $a(r + 3c)$.

4 **a** $4\,000\,000\ \text{m}^2$ **b** $0.25\ \text{m}^2$
 c $0.08\ \text{m}^2$ **d** $55\,000\ \text{m}^2$

S8 HW2

1 **a** 11.1 mm (3 s.f.) **b** 4.33 mm (3 s.f.)
 c 30.0 cm (3 s.f.) **d** 5.44 m (3 s.f.)
 e 5.34 m (3.s.f.) **f** 4.10 cm (3 s.f.)

2 10.1 m (3.s.f.)

3 70 km (nearest km)

S8 HW3

1 **a** 30.1° **b** 74.5° **c** 36.4°
 d 61.4° **e** 32.0° **f** 70.7°

2 326° **3** 20.1° (3 s.f.)

S8 HW4

1 $94.3\ \text{cm}^2$ (3 s.f.) **2** 19.9 cm (1 d.p.)

3 $\theta = 56.5°$ (3 s.f.)

Index